THE CONDITIONS OF HUMAN GROWTH

THE
CONDITIONS
OF
HUMAN
GROWTH

JANE PEARCE, M. D.
and
SAUL NEWTON

THE CITADEL PRESS   SECAUCUS, N.J.

Third paperbound printing, 1980

Published by Citadel Press
A division of Lyle Stuart, Inc.
120 Enterprise Ave., Secaucus, N.J. 07094
In Canada: George J. McLeod Limited
73 Bathurst St., Toronto, Ont.
Manufactured in the United States of America
ISBN 0-8065-0177-4

CONTENTS

FOREWORD

We live in a dangerous world. The technical inventions of the twentieth century have hurled us into the necessity, in order to survive, of inventing hitherto unconceived social forms. Both the invention and the implementation of such forms depend on a capacity for flexibility, cooperation, and rate of change in human personality never before crucial in the history of civilization. The more we can understand about our potential for individual growth and human interaction, and about the personal illusions and resignation that impede it, the greater the chance that people can meet the challenge for their own survival.

Out of this urgency, the authors have undertaken to clarify and schematize the basic theses in man's individual struggle for maturity and creativity, in order to throw light on the process and possibility of personality growth after adulthood.

This book is intended as a contemporary statement of the interpersonal theory of personality. Harry Stack Sullivan, who first clearly formulated interpersonal theory, died in 1949. His work at that time incorporated the two basic propositions of Freud: the importance of unconscious motivation and the active nature of repression. In addition, it incorporated many of the specific theoretical speculations of both Ferenczi and Adler which were current in the literature during the years that Sullivan was formulating his hypotheses.

Sullivan took as his point of departure a study of the causes of schizophrenia in hospitalized patients. He later expanded this to an application of his theories in private practice; that is, in the practice of analysis with ambulatory schizophrenics and so-called neurotics. During the Second World War he became interested in the problems of Selective Service and was confronted with the extent to which his theories of psychosis and neurosis applied to the population at large. His writings after this, until the time of his death, were concerned with the extent and definition of personality damage in the culture.

The present authors were trained particularly in the assumptions and implications of Harry Stack Sullivan's system of thought in the few years immediately before his death. The concepts presented here incorporate the major assumptions of his systematic investigation of the theory of personality. They also attempt and undertake to elaborate some of the gaps in this system and to find more fundamental formulations for certain of the inconsistencies in Sullivan's thought.

In addition to drawing in a major way on Sullivan's theories, the authors were influenced directly by the particular approach to therapy of Dr. Frieda Fromm-Reichman and by certain theoretical formulations of Dr. Erich Fromm. Fromm's emphasis on cultural repression and depression is more explicit than Sullivan's. Fromm-Reichman's consistent relating to the patient's drive to mental health and the therapist's necessity for organizing this drive into a useful dynamic was more explicit than either of her collaborators.

In addition, the authors have drawn directly from other theorists in the field. A teleological strand in the conceptions of neurotic behavior presented here, the goal-directed aspects of the central paranoia in creating particular effects, is strongly influenced by direct contact with Adlerian theory and therapy. The concept of the integral personality has certain important resemblances to Jung's concept of the creative faculty of the unconscious.

Inevitably influenced by pre-existing theory, the concepts presented are based in the authors' own experience in practice and in teaching the theory of psychoanalysis. We have also attempted to formulate a scientific approach to a theory of therapy in terms of the theory of personality development; that is, in terms of the theory of the normal processes of growth or deterioration during the life history of any particular individual. The role of the therapist can only be understood if he is seen as an agent to catalyze or to impede the forces within the individual towards growth. Thus, therapy can be understood in terms of factors promoting growth. We have attempted an initial formulation of an interpersonal theory of therapy in terms of the theory of normal development of adult personality, and the impact of subsequent experience on this preformed personality.

The text is oriented particularly to practitioners of therapy—physicians, psychiatrists, psychologists, and social workers. It is also hoped that it will make a contribution to theory in the social sciences, and to the personal philosophy of the thoughtful citizen.

The central theses of this particular theoretical development were first elaborated as a course in the principles of therapy, given at the William Alanson White Institute of Psychoanalysis in 1955. It was subsequently presented for review and discussion at a seminar of the Sullivan Institute for Research in Psychoanalysis. After the seminar presentation, the central theoretical hypotheses were re-edited in conjunction with Dr. Emanuel Ghent. He made many contributions to the theoretical development. The most striking of these was his contribution to the development of the concept of systems within the personality organization of the adult, particularly of the productive versus the security apparatus within the self-system. This subsequent clarification of concept was implemented in the research and teaching of the faculty of the Sullivan Institute, and subsequent revisions deriving from the application of theory in practice were incorporated into a new version. The current version was edited by Mr. David Cunniff. We are grateful not only for his extensive clarification of stylistic mannerisms but also for his clarification of points of theory which either were obscure in formulation or obscure in content.

JANE PEARCE
SAUL NEWTON

Part One

THE STRUCTURE OF PERSONALITY

I

THE INTERACTING SYSTEMS

HUMANNESS RESULTS from being raised by people.

Interpersonal psychoanalytic theory proposes that the individual human's personality is determined primarily by the quality of his relationships with other people. This applies both to the development of the infant's character through interaction with the adults around him, and to the importance of other people—real or imaginary, past or present—in affecting the individual's behavior in contemporary situations. Interpersonal theory therefore investigates the ways in which and the extent to which a person's cumulative experience with human beings determines his present behavior, perceptions, and capacities: the individual *is* what his experience has *been.*

In analogy with the concept in mathematics, the person at a given time is the integral of his life experience up to that time.

Twin phenomena form the crux of personality structure: the active nature of repression and the importance of unconscious motivation.

Repression refers to the fact that certain aspects of the personality—certain memories, perceptions, experiences, intentions, sensations—cannot easily come into the center of awareness, even under ordinarily appropriate stimulation. If a stimulus does tend to evoke these aspects of the personality, either the repressive activity is successfully intensified or the personality becomes disorganized temporarily.

The repressive activity limits the content of consciousness by commission rather than by omission. Certain memories of past events, certain conceptions of oneself, certain perceptions about one's environment are eliminated from consciousness. Literally, they become inconceivable. However overwhelming the evidence that supports their truth, any attempt to focus on them clearly meets intense maneuvers of repudiation. If these fail, confusion and disorientation result.

Not all experiences call forth such resistance to awareness. Specific kinds of past or present experience—different for each individual—set off the internal struggle. But no matter how apparently victorious the repressive forces, certain conditions indicate that the undefeated underground, the guerrilla fighter for growth of the personality, is never totally defeated. The content of the repressed experience reappears during hypnosis, dreams, and panic states. The perceptions that have been repressed continue to operate in the personality, but they cannot be maintained in consciousness or forced into focal awareness. Thus experience does not necessarily denote consciousness. An experience is any event, consciously or unconsciously recorded, that affects the individual's later functioning. The term will be used also, in a few specifically indicated contexts, to mean the emergence of a specifically blurred or repressed event into the sharp focus of consciousness. But now let us concentrate on that part of an individual's experience that can be focused.

Certain experiences can readily come into awareness. The individual focuses on them without major reorganization. The total of all such experiences comprises the individual's self-system. By definition, this self-system excludes certain experiences. The experiences that it includes are organized into a logical structure that perpetuates and reinforces the fallacy that the other, omitted, experiences never happened and never will. This logic is oriented to maintaining fallacious convictions about oneself and the world that are incompatible with the repudiated experience, and to avoiding events and perceptions that might challenge these convictions.

If these forbidden experiences are forced into awareness, the individual grows anxious. In this sense the self-system can be described functionally as the structuring of consciousness for the avoidance of anxiety.

The self-system also fulfills a highly important productive function. The productive apparatus of the self-system includes those needs, related functions, and satisfactions which can be experienced and executed in full awareness. It includes the potential for exploration and expansion of these functions. It also includes a variety of forgotten or latent experiences which are not usually in awareness, but can be made conscious, if necessary. Certain needs can be experienced clearly, and plans for their satisfaction can be expanded freely and without anxiety. Such conscious functions are easily available for communication to others.

Thus the self-system has a double character: it is both the agency of various productive aspects of the personality, and the tool with which other major aspects of the personality are repudiated.

The self-system originates in the individual's relationships with his parents during the early years of life. The child operates in accordance with the parents' conception and expectation of him.

When the child does otherwise, his divergent behavior makes his parents—on whom he is dependent for food, clothing, shelter, for life itself—intensely anxious. Since survival in the presence of the intensely anxious parent is inconceivable to the dependent infant, he repudiates those explorations that evoke intense anxiety in the parent.

Once established, the infant's self-system expands or contracts according to the quality and sequence of subsequent growth experiences. The resulting self-system of the adult includes all experiences that have received minimal affirmation, in a usable sequence, from the important people in his life.

But that excludes a wide range of experience. Thus the self-system incorporates a security apparatus, like a moat around a castle, that supports and perpetuates its own limitations. This security apparatus is defined by its orientation to constrict rather than to implement satisfactions. Every element of the self-system can be used to exclude. Every activity of the security apparatus indicates other elements of experience that are locked out of consciousness.

The parents' eyes are the mirror in which the child first sees himself; his perceptions of himself evolve in this field. This self-image —the person's central picture of himself—is the symbolic condensation of the self-system. The self-image is the symbol and the focus of the limits placed on the child's growth by his parents— and thus is the symbol and the focus of low self-esteem. Through it the person is reminded at every turn that he must not explore certain paths of development.

Who the individual *says* he is does not identify his self-image. That identity, however, can be inferred from his behavior, especially from "chance" adjectives, dreams, unwitting actions.

For example: a person acts as if he were repulsive. He avoids contact with people whose esteem he values. Experiences with other people are interpreted by him to mean that they find him ugly. He dreams he is diseased. Ugly sub-human images turn up in his Rorschach. All these reflect how this person—who would probably describe himself in quite different terms—really feels about himself.

Most people cannot acknowledge their own self-images. Many compulsive patterns are designed to repudiate or disguise the self-image. Recognizable or not, however, the self-image is the restrictive mold into which our parents cast us as infants. It represents their conception of how much the child could hope to accomplish in life, and that conception is usually: Not much. In spite of the myth that American mothers are overambitious for their children, most five-year-olds feel inept, inadequate, and unattractive—and these feelings haunt them throughout life. As we grow up and life becomes more challenging, the mold of the self-image grows less adequate. It repeatedly bursts at the seams, but we are continuously compelled to force our way back into it.

The logical fallacies that support the self-system around the core of the individual's self-image are indispensable to the self-system's function of limiting experience and censoring growth. These fallacies are the individual's central paranoia, which forms the core of the security apparatus and is implemented in security maneuvers.

The central theses of the central paranoia are not usually conscious. Full awareness of them is very difficult. They often appear in marginal thoughts, but even there with some sense of unreality. But they are nonetheless an integral part of the self-system.

On the other hand, these characteristic paranoid propositions are repeatedly implemented in fully conscious, characteristic thoughts, feelings, and actions. They may even become an obsessive preoccupation. If the individual is abruptly and convincingly confronted with his central paranoia, he characteristically experiences fear, anger, and despair. Such a confrontation can sometimes ultimately result in significant reorganization of the self-system, toward growth or toward further constriction. But the confrontation does not in itself necessarily disorganize the self-system proper. The fear expressed after a confrontation of the central paranoia is quite different from the anxious disorganization that results when the logical structure of the self-system itself is threatened by a move toward growth. The realization that one has always had a hidden conviction need not prove this conviction untrue.

When the central paranoid assumptions are acutely irritated or attacked, they move from the background of the personality to the forefront of immediate conviction. The individual frantically shores them up with supportive evidence scratched piecemeal from the immediate situation. This acute necessity, and the activities it triggers, is called the paranoid episode.

The term paranoid, then, refers not to the fact of the existence of the central paranoia—which is a universal product of our culture —but to the extent to which that paranoia dominates behavior at a given time. We also use the term more loosely, to refer to the expression of the central paranoia, or to individuals who have committed their lives to the pursuit of their central paranoia.

Three vectors cause the self-system to operate as the point of tension between two opposing trends. The security apparatus progressively constricts consciousness. The integrated functions of the personality, incorporated in the self-system, and operating productively, seek to grow to new frontiers that may be less consistent with the logical construction of the self-system. At the same time, the unsatisfied aspects of the personality—the forbidden genuine needs and the related apparatus for their satisfaction that are not incorporated in the self-system—push intensely for recognition, for a place in the sun, for growth. This, clearly, threatens the consistency of the self-system, and the security apparatus operates to hurl back these barbarians from the gates.

The security apparatus is charged with protection of the self-system from these elements of the personality. This tends to require that the self-system grow more consistent internally as time goes by. For survival, the self-system sloughs off its rough edges and becomes streamlined and, in a way, more able to resist the incursion of the forbidden personality elements. Either the data incompatible with the self-system is eliminated, or such inconsistencies are cordoned off to some deserted Siberia of the individual's emotional geography where they can be overlooked.

This process progressively cuts off the adult from self-awareness. The individual progressively commits himself to avoid the satisfaction of certain needs in return for feelings of status and security. The original prohibitions that surrounded one's life limit one's perceptions more and more. If life becomes dedicated to the maintenance of received limitations, the experience that life can expand productively must be avoided where possible and jettisoned where not.

Exploration is progressively curtailed; and so are the hopes, the feelings of curiosity, the aliveness, the great expectations on which all explorations are made. The infant's original capacity for perception is put into dead storage for the sake of safety. The adult becomes relatively deadened.

When an adult struggles to grow, he is struggling to a large extent to reverse this process and to admit into the structure of his personality the experience he has repudiated in the past.

The active process of dissociation is essentially similar to the process popularly described by the term repression. The difference lies in the delineation of areas dissociated and the reasons for it. Interpersonal theory holds that whether a particular experience is admitted to awareness or dissociated is determined by interpersonal factors.

Dissociation is maintained by a security operation. Security operations are any maneuvers, internal or external, that implement the self-system's security apparatus, and thereby express directly the underlying assumptions of the central paranoia. Whether or not the security operation appears destructive (i.e. hysterical paralysis), its true nature is defined by the purpose it serves (restriction of growth), and not by its surface manifestations. All security operations are productive aspects of the self-system, conscripted to serve restrictive ends.

What is the purpose of the security operation? Not the security that functions toward self-esteem or the esteem of other people, but rather the security that makes one feel safely connected with the parents, the security that protects the self-system from disorganization.

There are two broad kinds of security operations. One is the use of the overall organization of the personality for repressive purposes. This is implemented in the overall patterns of interpersonal relations, and is frequently referred to as character neurosis. The other kind includes those particular operations that come into play to answer a specific threat to the self-system. These operations are either specifically selected interpersonal integrations, or more subjective maneuvers for the inhibition of consciousness.

These two kinds of security operations are interactive and interchangeable. Some types of character neurosis provide a better matrix than others for certain security operations, but everyone has some capacity for every kind of security operation. Security operations include the classic symptoms of neurosis: obsession, hysteria, rationalization, projection. They involve a misuse of potentially productive aspects of the personality when the productive aspects of the activity are repudiated or avoided. This use of productive life processes for unconstructive ends is called alienation.

Patterns of interaction with people are often security operations; for example, a predilection for hostile relationships, or a lifelong depression that eliminates access to pleasant experience, that literally "takes the joy out of life." And the paranoid episode that occurs when the central paranoia is threatened is a basic, pervasively used security operation even though it may only be noted in the acute and spectacular instance.

All such security operations protect the self-system from the dissociated system. The dissociated system is the repository for all those needs, perceptions, memories, and experiences that cannot be admitted to consciousness without reorganizing the self-system. These are the experiences that would have been inconceivable to the mothering figures, experiences which could not either be approved or disapproved, experiences that the parent could not perceive or respond to, precisely because they were unthinkable to the parent. Their exile into the bush country of the mind indicates the extent to which the parent was incapable of seeing in his child the beautiful, intelligent, wise, mature, creative, and perceptive. They embody that growth to which the selectively purblind parent could not respond.

Infant behavior of which the parent disapproved is incorporated into the self-system. These types of behavior were and are conceivable, even if their recall or implementation causes fear. "Unthinkable" behavior, on the other hand, cannot be acted on at all consciously.

Relating to something or someone with intense disapproval is an intense relationship. Such a relationship may become indelibly tattooed on the child's personality. But even anger is a form of respect, however painful; sheer unrelatedness is devoid of respect and specifically destructive, in the sense that the parent lets it be known that in this or that aspect of the child's personality the child does not exist.

Respect is the core of the process of validation. Those aspects of the self-system which have been validated with respect constitute the core of the productive functions of the self-system, condensed into self-respect. Conversely, self-respect instigates exploring of the validated functions in the direction of growth.

But impulses which were dissociated in the past for the sake of survival, in response to the parents' anxiety, remain present and active. They are not developed and matured by use, and hence remain in a rather archaic form. When these impulses break through, the event is associated with all the anxiety and childhood terror which made the dissociation necessary. The dissociated system is a chaotic constellation of amputated but ineradicable needs and experiences. It is composed of experiences which were registered by the individual, but which could not be validated by the people who were his central sustenance.

The integral personality is the dynamic organization of all of the individual's experiences, conscious and unconscious. The organization of the integral personality is oriented to expand experience and to satisfy needs. Since those needs that were most blocked in childhood require the most nourishment, the active focus in the integral personality is especially toward the expansion and satisfaction of these particular needs. The integral personality is selective; it forces growth where growth, because relatively stunted, is most necessary. This is who we really are, but we can be conscious of only part of ourselves. Yet the integral personality provides the overall direction of the life line of the person.

The self-system implements the execution of integrated functions and limits their implications to stay within the confines of the central paranoia. The dissociated system is not dynamically or logically integrated, and hence is not truly a system. The integral personality is a true system, logically and subtly organized, and directed unreservedly toward its own expansion and satisfaction. It is constantly assimilating new data and reorganizing its own logical assumptions to incorporate them. It hungers after the infinities of growth. It operates from earliest childhood, and its fuel is experience. It searches for the perception of direct experience, free from the parataxic perceptual apparatus of the self-system. It cares little for the conventionalities of that ominous audience to whom the self-system so dutifully recites its lessons.

The experience available to the self-system is massively limited and weighted in accommodation to the emotional necessities of the parents. The integral personality integrates the data of all interpersonal experience. Its weighting is in terms of its relationship to the individual's true needs. It utilizes observations that are in focal awareness, in marginal awareness, in latent awareness, and fully dissociated with equal facility.

But the integral personality is subject to fallacies also. It is influenced markedly by experiences that have been dissociated in conditions of extreme anxiety. Since most of these experiences were bottled in the nursery and have never been checked against reality, they carry a heavy freight of infantile distortion and fear.

All of this data implements a creative synthesis of a picture of the world. The integral personality then focuses on gaps and inconsistencies in this world picture. In the service of its need for unhampered growth, the integral personality must become preoccupied with formulating the structuring of the central paranoia which frustrates this need. Many paranoid dreams are artistic formulations by the integral personality of the central paranoia.

The central thread of organization of the integral personality is in the continuity, however sparse, of true experiences of growth and tenderness. It is not taken in by the paranoid overlay with which the self-system derogates such experiences. The integral personality is the driving force behind the unconscious reaching beyond the limits of the self-system to satisfy dissociated needs. It seeks situations which will provide additional data in its search for truth. Through the integral personality, then, direct perception is registered, formulated, and reformulated where necessary. This is the heart of the individual's capacity to create. Universally expressed through dreams, the integral personality also finds expression through individual waking creative efforts.

Such conscious formulations are limited by the restrictions on consciousness. Since association with people is the central limit on consciousness, the fulfillment of the creative impulse depends upon commitment to interpersonal growth beyond the self-system. Creativeness is not only a subjective experience. As much as it involves perception of the truth, it involves sharing that perception. If the wider audience does not matter, the secret audience will be mother and the creative act will be limited by the self-system molded in her image.

Personality refers to the process of interaction among all these systems within the individual which incorporate all the individual's mental phenomena, and to the patterns of interpersonal behavior which determine intrapsychic phenomena.

II

ROLE OF CONSCIOUSNESS

BOTH CONSCIOUS and unconscious mental processes derive from the individual's interpersonal relationships. The interpersonal conditions of original experience determine that experience's later accessibility to awareness. To understand how these experiences are allocated either to consciousness or unconsciousness, and hence either included in both the self-system and integral personality or restricted to the integral personality, we must investigate the role of consciousness in the dynamics of character structure.

Consciousness is a momentary phenomenon. The clearly formulated thought is in the context of a particular moment of some audience, past or present, real or imaginary. The context may or may not be conscious.

Thus consciousness is an aspect of communication. It is heavily influenced by the immediate audience, but even in the presence of particularly forbidding people one's conscious thoughts can be addressed to absent or imaginary people. One's perception of this audience, either physical or imagined, however, is subject both to the paranoid and parataxic distortions of the speaker, and to the anti-communication maneuvers of the listener. Characteristically, if the audience is imagined the figure chosen will be someone of particular importance to the individual currently, whether in immediate contact or not.

Two other tendencies limit the content of consciousness stringently. Both are functions of the self-system and of the way it was organized in the context of the original life-giving figures.

First, the thought in question must not be inconceivable to the original authorities. If the thought is grossly incompatible with their attitudes toward life, to harbor such a thought—to run the risk of being "caught"—is tantamount to a rupture with the authorities, the source of all life to the child. This necessarily is inconceivable to the child, and continues to hold terror for the adult individual.

The other self-system limitation on consciousness is that the parenting figures must find the audience of a thought conceivable. If the person to whom the thought is addressed is entirely unacknowledged by the parenting figures, the thought is "unreal" and cannot fit smoothly into the self-system.

These three limits are not absolute. They represent powerful trends in the organization of personality. But there are many borderline situations. Even the most restrictive parent responds to certain things in his child that are beyond his own awareness. While he mulls over the situation, the child can develop his own experience to a considerable degree, and by the time the parent has consolidated and qualified his position against the child's experience, the child has grasped his experience clearly.

Too, the same ambiguity of restriction applies to certain trains of thought and to certain potential "outside" allies. A special kind of audience is the imaginary or literary audience. The parent may be indifferent to such "listeners," but he rarely violently opposes or threatens them. Nevertheless, a friendly, living, bad poet is likely to have more impact in encouraging an interest in poetry than an extremely good dead poet, even if the live poet meets with some parental disapproval.

When the individual can conceive of communicating his thought to any other person, the thought is conscious. When he thinks the thought cannot be communicated, then no audience exists for him and the mind goes blank. The thought may be labelled too bizarre to be communicated. Or he may not be able to conceive of anyone who would be interested in his thoughts. The result is the same: exile of the thought to unconsciousness.

A special circumstance exists. Some thoughts can be expressed to the immediate audience, but such expression constitutes an earthquake of the self-system and isolation from the parents. A break in the stream of consciousness again results. It may be brief, but when it is over it might be impossible to reconstruct the train of thought that led to the discontinuity.

Some functions of consciousness operate effortlessly. No conscious effort to focus on perceptions of space and time is usually necessary, for instance. Eating, walking, driving a car could be included in the same category.

These automatic functions are not kept out of consciousness by active repression. If necessary—if some obstruction to the process arises, such as a skid while driving—they can be focused clearly. But they function best when they are performed with a sort of fringe awareness. When concentrated on, they tend to disintegrate into their components. Watch someone, for example, try to teach a child the proper way to hold a fork while eating, or the simple operation of tying shoelaces. But these operations are part of the self-system and are always potentially available to awareness.

Immediacy of audience has a dominant influence on the availability to focal awareness of thoughts and feelings incorporated in the self-system. This refers not only to physical immediacy. When a person is important to the individual, either in the contemporary situation or in the historic development of his personality, this person can be a part of the immediate audience.

Unconscious phenomena also are experiences originally shared with another person and aimed at either the original or a contemporary counterpart. But that experience, or that person, is excluded from consciousness by the security apparatus. Therefore the experience is excluded from the self-system, and hence from frank communication.

Since the integral personality incorporates the content of both the self-system and the dissociated system, it is denied direct access to consciousness. It functions through the self-system, but in strained, disguised ways. Its clearest and most direct expression is through the magical language of dreams.

The illusion of irrationality that dreams create for their symbolic language is precisely an illusion. A clear, organized, logical conclusion of the integral personality is remembered in vague, inverted, symbolic, or disconnected terms in order to pass through and communicate with the more limited, perceptual screen of the self-system.

Any new experience demands assimilation by the integral personality, with whatever re-evaluation and reorganization is necessary to make way for such assimilation. This integration process takes place normally during sleep. It also operates continually during waking life, parallel to the more noticeable processes of conscious relation to the wakeful situation. If the reorganization needed for integration is urgent or wide in scope, the individual's level of consciousness is markedly affected. The person will grow preoccupied. He may retreat to obsessive occupations that demand little "thought": working crossword puzzles, working in the garden. These functions allow the integration to proceed unimpeded by much need to focus on wakeful activity.

Thus consciousness must be discussed in two dimensions. The first dimension is a concept's accessibility to awareness. Here the concept can either be in central focus, for immediate use; or in marginal focus where the thought is not totally acceptable to the self-system or the audience; or unfocused but available, such as in an "automatic" function; or forbidden. The self-system has access to the first three of these; the integral personality, to all.

The other dimension of consciousness is the relative dominance of the individual's self-system over his integral personality at a given time. This relation creates a sliding scale aspect of consciousness, with clear focal awareness at one end, and frank dream life on the other. In between fall obsessive preoccupation, rumination, and increasingly trancelike or dreamlike states. At no point, however, is the integral personality completely unhampered by the self-system. To remember a dream, for example, that dream must be comprehensible in some way to the self-system. Formulations and conclusions of the integral personality that affect the life line, but cannot be formulated, even in disguised ways through the self-system, cause sleep disturbances suggestive of active, unrememberable dreams.

III

HOW A NEW FUNCTION IS ACQUIRED

BEFORE WE CAN continue our discussion of how experiences are allocated either to the self-system or to the integral personality, we must discuss the process of acquiring a new function, the process of growth.

A function is a dynamism or part of a dynamism for satisfying a need. The push to complete the integration of such a function persists until it is consolidated firmly, so that the related need can find satisfaction.

Growth is defined as the integration of function. Growth includes the organization of a perception, however broad or minute, and the development of the capacity for any human function. It includes those functions conventionally called learning—from learning to eat to learning calculus. Most centrally, this conception of growth involves the expansion of the capacity for intimacy, relatedness, and communication with other human beings. Growth cannot be separated from the physiological structure that is involved. In short, growth involves the expansion of all experience— physiological, physical, emotional, interpersonal, subjective, objective. It implies mastery of a project.

The process of growth in the integration of functions may be covert, in which case the integration occurs in the integral personality. If it is at least potentially overt, it is integrated into the self-system as well.

Now the self-system can exercise any of its component functions, either for productive uses or for security uses. The sheer exercise of function, even if extensive, is not to be confused with growth, since growth always refers specifically to the integration of a new function, or to the expansion of a function.

This holds true whether a function is used for productivity or for security purposes. However, an important indirect relationship exists between growth and extensive use of a function: devotion to productive uses stimulates the elaboration and integration of new functions, which we call growth; and the use of functions for security operations opposes growth and facilitates deterioration.

The limits of infant potential are unknown. Whatever they are, all known adults have failed so severely to fulfill their infant potential that the differences among their limitations are narrow compared to their overall limitation. (Potentialities, conventionally applied only to talents, implies here the total gamut of human functioning.)

Our hypothesis here is that the infant, under optimum psychological and physiological conditions, could develop the capacity to perform any function. A detailed study of the dynamics and interpersonal influences by which a new function is acquired can throw light on this hypothesis. We shall analyze the process by considering its constituent processes, its natural sequence, and its necessary conditions.

There are four relevant constituent processes to the learning process: 1) an alternation of participation and direct experience with rumination and integration; 2) the differentiation of specific functions from broader ones, from the diffuse to the particular; 3) defects in the anlage, the diffuse groping, the early moves toward a certain function, become more pronounced as the function develops; 4) all functions have their natural sequences of development, and attempts to shortcut these sequences lead to distortion in the function's organization and defects in its final performance.

Two operations alternate during a learning process. The first is the acquisition of data, experience and practice. This requires a high level of awareness, a responsive medium and an environment that is at least moderately permissive. The second is the process of thorough integration into the personality. Insight does not contradict trial and error. They are both necessary parts of learning. The process of insight, integration, goes on best when the person is little subject to the demands of society; that is, when he is alone, or withdrawn, or ruminating, "chewing his cud."

If the learning conflicts head-on with the significant dignitaries of the self-system, the integration process can proceed only outside of focal awareness, preferably during sleep, sometimes under the influence of alcohol.

Learning involves the progressive differentiation of a function. An analogy with embryology may help us here.

In embryology, the final organ is progressively differentiated from an anlage which is defined as the first accumulation of cells recognizable as the commencement of the final organ. Likewise, in evolution of functions in personality growth, specific capabilities are progressively differentiated from broader functions. The line of development begins as a diffuse movement in a general direction. It spreads out over the course of a lifetime in a fanlike direction, to include more specific and more elaborate functions.

One illustration is the development of the capacity for motion, from the fishlike total body movement of the newborn infant to the infinity of refined and particular motor skills of the adult.

Our analogy with embryology holds true in another respect. Injured or retarded learning in the less differentiated stage becomes progressively more debilitating as the development proceeds. A line of development which meets great misfortune casts a pall on the later acquisition of a wide variety of derived functions. A relatively minor handicap in walking, for example, becomes a serious defect when one attempts to run.

The development of learning must occur in an orderly sequence, in order to be smoothly integrated into the personality. If a person is pressured to acquire a function before he has mastered its naturally preceding functions, whatever they may be—a child, for instance, who tries to write when his general coordination is still immature—the process will not go well. The individual may be able to pull it off as an obsessive operation. However, it will forever afterwards be performed with strain, and will be highly susceptible to disintegration under stress. This natural sequence of learning for any particular function may be quite difficult to identify. For many functions the sequence is inseparable from the process of physical maturation.

One assumption implicit in this concept of orderly sequence of development should be made explicit. Functions acquired early in the process are not outgrown. Rather, new elaborations of previously acquired functions are added on to the earlier ones. We never lose the capacity for the fishlike body movements of the newborn child. The capacity for the less differentiated response remains active. It is available under regressive situations, and it also remains available as a way of responding to any situation in which such a response may be useful or appropriate.

The normal course of acquiring a function might be outlined as follows. First there is a preliminary preparation. This occurs in marginal awareness. The operation includes a review of previous relevant experiences, and a review of possible obstructions to acquiring the new experience.

Learning to drive a car provides an illustration of this. First of all, there is the process of deciding to learn to drive the car. This involves a certain mustering—making a tally sheet of the related functions involved with the driving of an automobile. It also involves a review of possible consequences, both negative and positive: one might never learn to judge distance accurately and so be a hazard to others; or, on the other hand, one might become competent at driving and thus achieve an anxiety-provoking degree of mobility and freedom.

Secondly, there is a phase of intense preoccupation with the function. This includes experimentation, study, and repetition. A person learning to drive an automobile at first is very much like a five-year-old who is learning to bounce a ball. He can stay at it for hours until bouncing is mastered. There may be distractions, but the person learning to drive the car and the child learning to bounce the ball always come back to the project at hand. This intense concentration alternates with periods of reorganization in which prior and current experiences are dovetailed into a new synthesis.

The third phase involves insight into the new synthesis. This is always an outcome of a period of reorganization which goes on out of focal awareness. This insight involves a conscious recognition that performance of the function has been learned, that it is within the capacities of the person, and that its rudiments have been mastered. One feels that he has gotten the hang of it.

This must be, at least for the moment, a conscious process; and therefore it requires an immediate audience. As we have pointed out before, the audience may be present or absent, may be real or imaginary, but it must be there for the experience to be validated. This process is the gist of validation. Whether the experience remains available to the self-system or is subsequently limited to the integral personality depends on the relationship of both the experience and the validating figure to the logic of the self-system.

If the new function is integrated into the self-system, it gradually lapses into latent awareness. By and large the function is performed without too much consciousness. However, it is always potentially available to awareness. A large number of functions may thus remain intact although latent for decades, if they were thoroughly integrated at one time. They may always be called into operation intact in an emergency. If "forgotten," the relearning may be a capsule recapitulation of the original process. The integration of an infinite number of such functions makes up much of the content of the self-system.

In other words, for any function to be effectively integrated into the self-system it must be clearly in awareness at the time of integration. Once learned, however, the function is performed best if it is performed automatically. Should the function be forced, by some emergency or other stimulus, into focal awareness, the individual involved becomes intensely self-conscious and the efficiency with which he performs the function tends to regress to the learning level.

Consciousness necessarily involves more or less direct sharing with a person of significance. It is by this mechanism that the self-system becomes limited to those functions which are conceivable to the significant figures. Since the self-system excludes those functions which are not conceivable to them, if the function being learned does not receive adequate overt validation, it does not get smoothly integrated into the constellation of what the person can do consciously. When the validation is covert and imbedded in anxious repudiation by the parent, elements of learning proceed only in the integral personality. They remain in a kind of suspension there, until an interpersonal climate more propitious for their implementation develops.

Not every elaboration of the function must be validated as it is added to the personality. But the initiation of the process through the point of grasping that it is feasible requires some direct interpersonal validation. After that, the sharing necessary for the further expansion of the function can go on through internal interaction with a fantasy image of a validating person. This validating person is likely to be the individual who originally catalyzed the particular line of development involved.

Validation of the individual's new interest can proceed by a variety of ways. One form of sharing, and hence validation, is referred to as consensual validation.

Consensual validation is the overt acknowledgment of the new experience as a part of the other person's experience. The terms relatedness and responsiveness imply more immediate interaction between the learner and his supporter. Tenderness is the most highly developed response to the other person's true needs, with much of the interaction characteristically proceeding in non-verbal terms.

Non-verbal communication is based on empathy. The child's feeling of hunger, for instance, is registered by his mother through her own capacity to feel hunger. Empathy involves some degree of identification with the other person. If the mother responds by providing food, she thereby not only provides a satisfaction but also provides validation for the child that this particular gastric sensation is hunger and is to be relieved by eating. Tenderness, then, implies a concern for the other person's needs based partly on feeling them in the most literal sense, utilizing a high degree of attention to the available indications of what is going on in the other person.

Since awareness presupposes this audience, any line of development requires a sponsor who is at least partially acceptable to the people who are already important to the individual. Further, the people who already matter in the person's self-system must not be too unsympathetic to the function itself. This is the machinery by which many functions are admitted to the self-system, and by which other functions are rejected.

Just as developments which involve a break with the original figures result in major personality disorganization, a break with the sponsor of a particular function temporarily upsets the organization of this particular line of development. The very process of growth is itself a function, a complex, self-reinforcing dynamism. Like the integrations of other functions, it must receive some minimal validation to become reliably consolidated.

IV

LEARNING IN THE PRESENCE
OF ANXIETY

WHAT OCCURS if the mothering figure is phobic about a particular function? What happens if the empathic experience provoked in her tends to make her anxious, or causes her to engage in security operations for the avoidance of the anxiety? She is then unable to provide confirmation of the validity of the emotion within the infant, with the result that the function cannot be thoroughly integrated by the child.

Inattention to any significant part of a function during the learning period makes defective integration inevitable, because the function was never fully accessible to consciousness. At best, such a function is integrated defectively and operates with a skew.

In addition, the anxiety present in the mother is transmitted empathically to the learner. Anxiety is profoundly disruptive to normal functioning, and especially to the acquisition of new functions. It creates severe difficulty in concentration because it disrupts consciousness.

Thus the presence of anxiety greatly handicaps and complicates the learning process. Inevitably it causes defects in the ultimate performance of the function.

The miserable experience of anxiety also becomes associated with the function itself, so that one tends to avoid trying to learn the function again. This is compounded with an additional paralyzing factor. Since the anxiety evoked in the mothering one threatens disconnection from her, it places in jeopardy the use of functions that have already been integrated, through the mechanism of regression. This destroys, in the moment, the antecedent learning from which the new function must develop.

The important consequence of learning in the presence of anxiety is that certain essential experiences are never thoroughly mastered. Instead, substitutions are made for them. The exact nature of the substitution, the exact quality of the compensatory movement, is of secondary importance, although these phenomena may be more prominent in the adult personality.

If the adult to whom the infant turns for validation is unable to perceive the infant's movement towards learning completely, appropriately, and responsively, the infant will substitute for his own clear perception one of a variety of responses.

One of these substitutions is selective inattention. When the mother is preoccupied, the result is an omission in functioning in the child at the point where the function comes into interaction with her preoccupation. The effect on the child may be depression or apathy. An example of this is the loss of appetite in a child whose mother is unable to enjoy her food. The child eats poorly and takes little pleasure in eating. Nevertheless, the function is often integrated with a minimum of awareness on the basis of validation by peripheral people.

A frequent occurrence is the preoccupied mother who is busy with her own internal reorganizations and diffusely anxious because of them at the time the child demands her attention. This demand, short of a matter of life or death, is experienced as an intrusion. Sometimes it is the child's level of aliveness and the vividness of his perceptions from which the mother withdraws, rather than the specific demand he makes on her. But in any case, the limiting effect on the particular learning is the same. For example, if the mother is intensely upset at about the time the child is learning to walk, the child may grow up to be persistently unsteady on his feet, and easily subject to dizziness.

The absence of an appropriate response in the mother is characteristically the result of anxiety, but this is not the only possibility. Ignorance, lack of experience, or limited exposure may be the factor principally involved. In these instances, the mothering figure can then respond on the basis of her rudimentary capacities in this area. Thereby she gives the function her blessing. This allows the child to find other sources for its further development. Only anxiety, not ignorance, produces a tendency to lasting blocking in the area of experience involved. There must have been an active repudiation.

If the mother's anxiety is obscured partially by a more elaborate security operation, the child eventually repudiates the function by substituting a misperception for it. For example, if the mother cannot experience tenderness in response to an appeal for tenderness, she misperceives the original need in the child as hostile, and the child also experiences his need for tenderness as a hostile move.

This happens every day in any home with very small children. There are innumerable demands for the mother's attention and the mother's tenderness. Since the adults at hand are too deadened to respond to most of these demands, they register them in a distorted way.

As another example, if the child's appetite for food fluctuates, it may be perceived by the mother as an emotional maladjustment. The entire situation may then be reformulated in terms of the prestige values of the parent. In this case, the child either learns to eat to please the parent, or learns not to eat so as to deny his dependence on the parent. In either case, the child's subsequent preoccupation with the eating problem represents a wistful and misguided attempt to get the mother's permission to eat when he is hungry and not to eat when he is not.

When the function which the child is trying to master is one which is dissociated in the observer's mind, the observer fails to produce the proper affirmation and facilitates the setting up of a substitute repressive reaction which becomes the "expression and repression" that we find in later neurotic symptomatology.

In taking over the misperception and elaborating a reaction appropriate to it, the child has acquired a security operation to preserve his connection with mother. This security operation has a dual basis: the child's own original inability to perceive clearly because of inadequate initial validation, plus the imposed misperception of the parent.

Some functions are more or less dispensable. If they are discouraged at their inception, they may remain relatively dormant for many years without serious inconvenience to the individual. For example, someone may have been prevented from learning social dancing. He can often, without too much trouble, subsequently avoid situations calling for this skill.

Other functions, however, clearly represent ineradicable needs. These needs demand an appropriate response to situations that are unavoidable. When this is the case, several parallel developments may occur:

1) A tendency not only to repress the need, but to sacrifice important related functions in order to organize one's life to avoid situations which would elicit the appropriate response. This involves dulling or distortion of the perception of relevant situations. It involves also an increasing inaccessibility of marginal thoughts and a general loss of emotional tone and clarity of response, which we often speak of as depression.

2) An intense preoccupation with those abilities which are prerequisite to the function in question. The child who is delayed in talking, for example, may become very good at dramatization. At the same time he will remain insensitive to those situations for which drama does not suffice. This elaboration of antecedent stages stops short of solving the problem in question. It has some growth aspects insofar as eliminating defects in preliminary learning can facilitate the ultimate mastery of the next stage. This process is responsible for some of the phenomena that are thought of as repetition compulsion. It develops its own patterning and serves both as compensation for and diversion to the frustrated direct line of growth. We have seen that elaborate functions depend for their development on prior components. If these components are developed, a push to develop the more elaborate function exists.

However, if a function is stopped cold, if a function cannot be acquired at all, this is pathognomonic of serious defects in the prerequisite functions. If a person is prepared to go ahead with the learning process, a powerful prohibition at this point may result in defective learning, but it cannot result in total failure. Failure in itself is a cumulative process.

3) Functions which implement important needs but which are strongly forbidden access to focal awareness nevertheless strive for integration. Most of the striving goes on outside of focal awareness. It occurs in the integral personality and it presses constantly to find conscious expression. As soon as the person notices what is happening, he becomes anxious. He tries to repudiate the entire sequence of events. The person repeatedly runs into situations which would evoke those needs. In order for the function to remain dissociated, the perception of the situation must be distorted. The functions involved represent true needs, they represent important gaps in a person's development; there is a constant out-of-awareness movement to acquire the techniques necessary for satisfying the needs. Elaborate rationalizations may be developed for moving into the forbidden field. When another person responds appropriately to the unacknowledged movement, one is confronted with the true nature of the operation, and anxiety is then manifested. Retrospective falsification or amnesia may occur. In the meantime, considerable experience may have been accumulated as to how to integrate the situation. This body of experience has been accumulated under the handicap that the person cannot more than fleetingly notice what he is doing. He must immediately repudiate it. The experience is therefore very minimally subject to rational evaluation. It remains strongly tinged by childhood terror. Nevertheless, one develops some competence in its performance, which at least lessens embarrassment about ineptitude, if the previously dissociated experience finally comes to be accepted in awareness.

The final block in the sequence of learning may be literally that one cannot conceive of doing it. All the other elements of the situation, experience, facility, opportunity, have finally been developed, but one cannot carry out fully either in action or in fantasy that which one cannot think of.

4) A more illusory solution, especially for those functions that are biologically or socially indispensable, is learning by rote memory.

A child, if he is unable to experience when he is hungry, or if he is unable to conceive for what he hungers, learns to control choice and volume by an elaborate set of social conventions. The adult who has never mastered speed and space perception may become an obsessively legalistic auto driver. He may never break a traffic law. However, he is apt to make the wrong decision in a moment of crisis.

Likewise, many marriages are integrated on the basis of meticulous preoccupation with the rules of the game: competition, cooperation, a subtle keeping-score on whether one is being selfish or giving.

The advantage of this maneuver is that at least it makes it possible to get into situations in which there is the opportunity for some covert learning. But one great liability lies in the investment developed in maintaining the illusion that rote learning constitutes genuine integration of the function. It is a strenuous and tiring way of life, and in fact does not achieve much genuine satisfaction of needs. Such learning can sometimes embody much that is genuine, but it is always rather vulnerable in disadvantageous conditions, and is thus a shaky basis for subsequent learning. A child who has learned the multiplication tables under such conditions may never learn how to think mathematically.

Closely related to rote learning is the substitutive process of pretending to have learned, of dramatizing, of acting *as if*. When an inadmissible function presses for expression, a dramatized imitation of the real function may be substituted. This play-acting usually has an exaggerated, brittle ring to it.

This exaggeration may be overlaid by a reaction formation. Though still an act, it now emerges with a playing-down quality. The most skillful people are able to use this dynamism by so sensitively balancing these two varieties that the outcome can closely simulate genuine expression of the function. The same consequences derive from the as-if technique as from the substitutive use of rote memory.

5) The learning process may go on repetitively, but under the disguise of a particular security operation attached to it.

If mother's response to new friends was contempt, one persists in making friends for whom one can manage to feel contempt, either by selection or by derogation of their true value. Some satisfaction is gained, some experience is acquired, but the basic misperception is maintained. The experience is never completed.

Or, under the camouflage of the preoccupation with achievement and prestige, some direct and genuine experience in mastery can slip by unobserved. But the necessity for camouflage both limits and distorts the integration of the function.

To the extent that a person is forced into operations beyond his level of interpersonal development, he must improvise, distort, and reduce these experiences to the dimension of relatedness with which he has the capacity to deal. He can note in the other person only what he has developed the capacity to feel in himself. This capacity can only be expanded by including additional people into the small roster of those who matter.

V

EXTENSION OF SOURCES OF VALIDATION

THE INDIVIDUAL requires validation for every point of progress in interpersonal relationships. Thus the counterpoint of growth is the evolution of the use of validators. The development of interpersonal capacity parallels the details of the development of the exchange of tenderness, from the intense but unspoken empathy between mother and infant to the more distant but explicit validations that occur among adults.

During the first year of life the infant makes tremendous progress in mastering the world in himself. This learning is integrated primarily through interaction with the mothering person. It includes breathing, eating, excreting, touching, seeing, hearing, cuddling, affection, anger, fear, joy, use of the muscles.

After the child leaves the confines of his crib, his primary concern becomes the mastery of the external world—surfaces, depths, colors, sounds, objects, relationships with pets and, finally, with playmates. Throughout this development the central parenting figures constitute the crucial relationships. Validation of experience still flows from them. But while learning is thus initiated in terms of the original significant adults, it requires contact with other people in order to continue and to become fully consolidated.

Thus, if the parenting figures refuse permission to expand one's source of validation, the later replacement becomes extremely difficult; the more difficult, the more the parents are unwilling to have another person be important to the child.

The most frequent and the most forlorn maneuver that is used to cope with this situation is the dissociation of the need for tenderness. Mother has forbidden interpersonal expansion. Thus the individual tries to free himself from the mother by telling himself he needs no such tenderness. If tenderness were not necessary, the mother would cease to be necessary, and one could expand one's circle of friends. This plan invariably miscarries, because if the need for tenderness is dissociated effectively, awareness of needing any people at all is cut off, and finding new validators or sponsors for independent development grows increasingly less possible. Since she cannot be replaced, the mother continues to be the secret, central figure.

The frequent use of this maneuver has helped create our cultural misconception that one outgrows the need for tenderness. This is a contradiction in terms. By definition one continues to grow to the degree that one's need for tenderness expands. When tenderness is cast out, growth atrophies.

The parents must validate to some degree the child's search for different validators. If they did not, any step toward a new person would be equivalent to a step toward doom that disconnection with one's parents means to the child. Since the new validators cannot be people whom the parents actively ignore, they must be allies of the parents, or people—teachers, storekeepers, neighbors—whom the parents tolerate for prestige or convenience.

Response and validation need not repose in the same person. In activities such as cuddling, both functions reside in the same individual. Where objects are involved, say a watch the child wishes to wind, the response is from the watch, not from the individual audience; but validation comes from the audience. For a large part of life, however, the mother can invalidate everything, interpersonal relationships and relationships with objects, by her contempt.

During the childhood and juvenile years (parents and newspapers columnists to the contrary), playmates ordinarily do not present a significant challenge to the parents' values. The child's self-system operates primarily in direct response to the parents. The first important challenge can come with the dawn of preadolescence.

Then, the important development is usually the child's relationship with the person whom he perceives to be like himself: the same sex, roughly the same age, many of the same attitudes. The challenge to the parents could be couched in these terms: If I can like and respect a person who is so like me, then I am a person whom others could like and respect. The child, for the first time in his life, has discovered love.

The needs of another person are now magically his own, and vice versa. The other person's good fortune is enjoyed without envy. The other person's ill luck is painful but not burdensome. The sense of sharing experience, whether of failure or of success, with another person takes precedence. Prestige in the parents' terms, and also the outcome of the experience, in the moment become irrelevant.

If the parents understand the importance of the friendship and respect it, other friendships can flower for the individual, hampered relatively little by the conventions of the culture. But when parents cannot tolerate losing the status of sole cynosure of the child's universe, the child's ability to grow is crippled. Without this experience of chumship, subsequent acquaintanceships remain subjugated, however obscurely, to the parental values.

Intolerance of the friend may be expressed as contempt, unawareness, anger, jealousy, or puzzlement about the intensity of the friendship. Superficial verbal approval of—or even enthusiasm for —the relationship may serve to conceal some strongly opposing power maneuver on the parents' part.

If the chumship is genuinely achieved, the child's unique emotional dependence on the parents is ended. In our society, however, the child remains for a considerable time financially and socially dependent on his parents. Parents often use the fact of this disparity as a chance to blackmail the young person into submission to their will.

Such is the nature of the frequent threat, more or less veiled, of the withdrawal of financial and social support from the child. The same goes for the very frequent demand that the child now take some financial or emotional responsibility for the parents, thus beginning installment payments on the priceless, and unrepayable, gift of life and sustenance of life during infancy.

The effect of such extortion is to handicap the consolidation and persistence of the chumship. But the process continues nonetheless, for, as in other areas, growth begun with internally recognized validation cannot be stilled by the external limits on this development.

However, the development can be cramped and crippled by such parental actions. Friendships are formed, but exist in hiding and are broken off when made public; or, the friendship continues under a variety of disguises—exploitation, dutifulness, martyrdom, self-abasement. These friendships make their tightly rationed contribution to the individual, but, because they must remain essentially subversive to the truth of friendship, they cannot form a reliable base for future expansion. Through them the individual is fighting a rear-guard action to preserve his right to have any friends at all; and while fights can make comrades, they cannot make peaceful friends.

To the extent that his chumship can be consolidated, the person moves on. He extends his circle of affectionate relationships, of people who can serve as new validators, and of relationships that facilitate further growth. The individual progresses ordinarily from intimacy with a member of the same sex to acquaintance with members of the opposite sex, to mastery of sexual technique, to the discovery and cultivation of a person of the opposite sex who can be identified as being like himself, and with whom he perceives the possibility of continued growth. This process—like the process of expanding one's capacity for friendship with all kinds of people, of whatever age, background or interest—is fundamentally conditioned by the emotional developments during the course of chumship.

Such friendships implement the expansion of consciousness and the expanded productivity of the self-system. One immediate impact, for example, is the affirmation of thoughts, inadmissible in isolation, but possible through communication with the chum.

This expansion continues throughout life through the agency of friendship. In the organization of the integral personality, each person perpetually formulates and integrates his own synthesis of his life's experiences. The small child often expresses these original perceptions out of ignorance of what can and cannot be communicated to adults. Age brings more acute perception of what can be communicated, and hence may lead to the addiction to the use of the cliché as a resolution of the conflict. Nevertheless, the perceptive process continues. In adults it finds expression in dreams and in original productions that we call works of art.

Such works of art often occur in verbally inexplicable modes. Therefore, the implied content is not subject to the parents' contempt, even if the contempt of the general public is expected. But they are addressed to the understanding of one hypothetical, significant member of the audience. Expressive facility depends upon one's ability to conceive that one's own perceptions might influence another person. This can only happen when one relinquishes his image of himself as a child in the adult world. He is then free to seek out people with whom exploration of new boundaries of communication is possible.

Integrity is much more related to the degree of dominance of the integral personality than it is to the self-system's logical consistency. Integrity is the capacity to perceive and to act on a perception in spite of the immediate social inconvenience. The more fully developed a person's relationships, the less dependent he is on that immediate environment. Thus integrity involves not independence from people, but a dependable and firmly based capacity for relationships with them.

VI

ORGANIZATION OF PERCEPTION

PERCEPTION is the organization of sensation. Conception is the organization of perception. This organizational hierarchy includes such processes as interpretation, the organization of both perception and conception; and theorizing, the organization of interpretations. These processes relate dynamically, and we define the dynamism *in toto*, for the sake of convenience, as the organization of perception.

Perception is distorted by the anxieties involved in a situation. A security operation, for example, has been defined as the substitution of an inappropriate response for an appropriate response in a particular situation, which substitution is based on a misperception of the situation. In any given situation the response tends to be appropriate to the perception of the situation.

Direct perception refers to the perception of a situation that would take place in the presence of adequate affirmation and in the relative absence of anxiety. It approximates what is meant practically by the perception of reality, as limited by the prior organization of the integral personality. Thus, it is in opposition to the distorted perception which persists in order to maintain connection with the mother. Such distortions are referred to as parataxic distortions.

All the operations which are enlisted in the process of distortion are essentially normal. They include:

1) fluctuations in the level of consciousness, both as to the total available capacity for concentration and as to the selectivity in concentrating on the various aspects of the immediate environment;

2) the organization of perception, by which is meant the identification and classification of sensation and its integration with related functions,

3) the organization of functions described in the preceding chapters.

Any distortion, once organized, is continually refueled by active misperception. However, correction of the perceptual aspects of the organization is insufficient to disrupt the organization proper. Instead, active intervention at the response end is also required.

Patterns of response tend to persist even after the perceptions involved in the response have been considerably clarified. Sometimes this occurs because the particular patterns of perception have become integrated, prior to the clarification, with other patterns of living which are unaffected by the change in perceptions immediately relevant to the response.

An illustration of this: the homosexual who, in middle age, finally becomes more comfortable with women. Frequently, he still finds the organization of his social and professional life integrated with the homosexual way of life.

Another reason for this inertia lies in the embarrassment about exposing the real level of inexperience in performing any function that would be appropriate to the new and more accurate perception. The overdetermined misperception and the persistent pattern of response to it are rooted firmly in the security functions of the self-system.

The self-system tends to tight organization. Correction of a particular long-standing misperception has wide ramifications. Changes in the correlative response system must follow, and hence threaten the overall stability of the security apparatus. Thus the internal defensive necessities of the security apparatus tend to persist and to re-establish the old response, even in the face of awareness of its incompatibility with the newly clarified perception.

Even so, the real usefulness of re-education procedures often lies in the fact that the attempt to train oneself to a more appropriate mode of response clarifies hidden misperceptions that would otherwise have passed unnoticed.

This attack at the response end is an indispensable tool for clarifying the misperception and forcing a re-evaluation by the individual. The process of reorganizing perception inherently involves a detailed investigation of the apparently inappropriate response.

These normal and necessary integrations of perception and response contribute to the differentiation of the capacity of the individual to cope with the complex world. The non-selectivity of the tape recorder in picking up an automobile horn that no one noticed points up the confusion that might be attendant on the lack of control of awareness.

Another example: the overwhelming multitude of perceptions which intrude on the awareness of someone in the acute catatonic state, in which condition the normal machinery for selective inattention and organization of experience has broken down. Such a person may be entirely distracted from conversation with a friend because he is overwhelmed by the noises which the forks make on the plates during dinner. A disintegration of the usual organization has occurred here. It is an experience that one could have when he enters a completely foreign situation. Some people feel similarly distracted when they go into a large group of strangers. They see a great many details. However, nothing stands out. Nothing is organized. The organization of perception is disintegrated. Many details have "no meaning" and therefore come in as a kind of static.

Conventions might be defined as agreed-upon symbols of communication. Some are more and some are less universal to the human race. Some of them are quite specific to the time and place of the subculture or the family group. They involve, inherently, a certain organization of perception into consensually agreed-upon categories. Objects are perceived in terms of their relation to previously integrated experience. The rationale of conventions is that they are representative of the reality which they presume to express. The word table conventionally represents "table."

Conventions also refer, more extensively, to an entire range of interpersonal operations designed to communicate something other than, or in addition to, the manifest content of the interaction. As a small example, tipping a cab driver can be used to express the mood that has been created between two people during the taxi trip.

The self-system of the individual tends to become consistent within itself. Where a simple rationalization originally sufficed to maintain organization in the face of conflicting data, eventually the rationalization becomes so enmeshed in a web of other such processes as to be almost impregnable. It is entirely a matter of degree or of social consensus, at which point this elaboration is referred to as a delusional system.

Within any culture, conventions tend to take on a certain uniformity. Since communication with the group and the self-system of other members of the group is part of human living, inconsistencies in any one person's self-system tend to be eliminated during interaction. In this process of consensual validation, much universal and useful common sense of the culture is condensed into the cliché.

These clichés may become either useful tools for observation or impediments to observation, depending on the original conditions of learning.

The repressive aspects of the self-system also tend to be shared, reinforced, elaborated and rationalized during the process of group validation. Conventions thus often incorporate the culturally patterned phobias or prohibitions against certain experiences. Moreover, the institutionalization of these phobias reinforces and perpetuates them. The family is the primary cultural agency for the transmission of conventions to the child. The conventions appear in systematized form in the child's self-system. Thus, they tend to perpetuate themselves throughout the child's life.

The development of communication demands the organization of perception into culturally accepted categories. Inherently, however, this process involves considerable restriction of the spontaneity of the original perception and conception. The infant must develop both some capacity for organizing his own direct experience and also some comprehension of the conventional modes of communication.

It is the parent's responsibility to pass on both general and specific conventions to the child. To the extent that the parent is relaxed and comfortable about this, he is able to validate the child's perceptions and also to teach the symbols of communication, as agreed-upon representations of that which they are presumed to symbolize.

Conversely, the child's original perceptions are replaced by substitutive processes to the degree that the original perceptions would induce anxiety in the parent. There will be a defect in the process of organization of direct experience if there is insufficient respect for the integrity of the child's learning processes.

In such a situation, the child is compelled to substitute for learning a preoccupation with esteem in the eyes of the adults. The child perverts the organization of mastery, to this end. If he could take the esteem for granted, he would be that much freer to get on with the business of living and growing.

It is the child's misfortune that adults are to such an extent antagonistic not only to the content of a particular experience, but also to the phenomenon of direct experience proper. Too much aliveness and initiative makes them anxious.

In addition to the phobia about aliveness and about relating directly to the material at hand, parents are also to varying degrees phobic about contact, and about the level of intimacy which is necessary in order to share some part of the vividness of the child's experience. The parents' most frequent defense is to use the teaching of the conventions to crush the child's spontaneity.

One of the parent's most insidious misuses of the teaching function is the demand for premature performance of some function which would have been developed on the child's own initiative if that initiative had been left alone a little longer. The process of direct perception is then subverted to the need to please the mother. This substitutive process in the child is profoundly restrictive and leads to the development of entire congeries of obsessional operations.

To sum up: the mothering one has two functions. First, the validation of the child's spontaneous growth experience in such areas as perception, conception, feeling, sensation, motor activity.

Second, the teaching of a structure or organization, the imposition of the social organization of experience.

A structure is required for the fulfillment of creativity. It is the expansion and the development of this very structure that is creativity. The need for structure is not in itself in a negative context. Unstructured spontaneity is of as little value in life as in painting, music, or fine cooking. Spontaneity per se is not creativity, although it is a necessary component of it.

Unless each infant is going to have to recapitulate the experience of the entire race in his lifetime, the social structure as it exists to date must be taught to him. The structure of language, for example, has to be learned in order to give scope to the child's verbal creativity. A structure, like a theory, facilitates and limits growth at the same time. Growth cannot occur without it, but by the very nature of growth the received structure must be outgrown.

The mother has the responsibility for providing her child with a theory of life. Optimally, she has the responsibility of imparting to him that her theory of life is tentative, and that she hopes his growth will carry him a step further, towards a better organization.

There are numerous prerequisites for the capacity in the adult to facilitate growth in the child. Among the more important are: the absence of alienation from one's own direct experience; the ability to perceive accurately what is going on in the child; the capacity for tenderness.

A continuation of the capacity for learning devolves on the child's development of the widest possible base for additional sources of tenderness. This is one of the many reasons why the development of the capacity for intimacy is the central thread of psychotherapy.

VII

MODES OF COMMUNICATION

COMMUNICATION is an interpersonal process based on the capacity for empathy.

Empathy is the basic mode of relatedness, of communication between people from which, for example, gesture, voice, inflection, and speech are all derivatives. Communication between parent and infant is empathic. Empathy is perception and communication by resonance, by identification, by experiencing in ourselves some reflection of the emotional tone that is being experienced by the other person. Words come later, but verbal communication does not supplant empathy. Rather, empathy continues through life as the basic mode of significant communication between adults. It is because of this that we are limited to perceiving in others only what we can allow ourselves to experience in ourselves.

This capacity to directly and consciously experience the other's emotion becomes largely blurred and flattened after early childhood. The elaboration of the derivative modes of communication, particularly of language, contributes to the fading of empathy into the background pattern. The experience in childhood of intensely anxious prohibiting responses in the parent to empathic insight in the child contributes to the progressive flattening of the empathic capacity.

Nonetheless it remains the underlying mode of communication throughout life, though its reflections in awareness appear in more or less disconnected associations, memories, and intuition. It is the basis of audience response at the theater, it emerges openly during catatonic states, and it is marginally available to consciousness in many adult interactions.

Thus empathy—the direct perception of another's emotional state —is the persistent base for the human community and its communications. The person who has lost the capacity for empathy to any extent is to that extent out of touch with himself and with others.

All the others forms of communication—music, painting, drama, speech, literature—represent an elaborate but convenient shorthand by which one expresses and evokes related parts of the universal empathic experience. These symbolic languages serve to validate the universal, but often unspoken, experiences of people, and to organize the welter of universal human perception.

All humans have the common task of perceiving and organizing the same objective reality. Essential human needs, and the attempt to satisfy them, are ubiquitous. Thus every human seeks to communicate with every other, no matter what the difference is, including language, between them. If Crusoe had not found a way to communicate with Friday, Friday would have found a way to communicate with Crusoe, because all humans need to know what is occurring with other people of whom they are aware.

In the process of organizing and reorganizing perceptions about the world, communication also facilitates itself enormously. As confidence increases in the symbolic manipulations which language requires, for example, perceptual and conceptual facility also increase—and as perceptual and conceptual facility increase, the ability to communicate increases. As each new dimension of perception is consolidated, the elements that it embodies become available for combination and integration with previous and parallel perceptual organizations. Thus the processes of communication, validation, and perceptual reorganization form a complicated dynamic characterized by the positive feedback of self-reinforcing expansion.

Earlier, we distinguished between the integration of new functions, i.e. growth, and the extensive, repeated performance of those functions. The function of communication is no exception to this distinguishing rule. Thus there are two kinds of communication: one is in the service of growth, in which something new occurs in the form or in the content of communication; the other is the repeated exercise of the acquired function of communication.

Insofar as the relatedness involved in communication, or the experience communicated, is new, fresh, and alive, communication occurs in the mode of growth. When the communication adds nothing new to the relatedness on which it is based, it lacks the self-expanding vitality of the new exploration, and, however eloquent the communication, offers nothing new to the persons involved in the communication. Growth demands more sharing, more communication with other people, increasingly extensive and increasingly intense, increasingly deeper in feeling. The communicator must become increasingly sensitive to the infinite variety of his own experience, and increasingly aware of the infinite potential of the people with whom these experiences can be shared. Conventional communication, if it is to support growth, must be increasingly developed to match in facility the empathy on which the ability to communicate is based.

Anything fully available to consciousness, and much only partially suited to awareness, is potentially communicable. Since consciousness is predicated upon previous sharing with someone, anything conscious, in an appropriate situation, may be used to broaden the base of experience shared with any other person. Thus it is useful to organize communication—derived experience from infinite sources shared with infinite numbers of people—in several categories of communication modes.

This can be approached chronologically, by studying the successive development of the capacity for experience and communication in children.

By the end of the child's first year of life, the dimension of motion and the perception of the body image have been added to the primary sensory and physiologic perceptions, as major areas of experience.

By age two, the child has integrated manipulation of objects, motion, posture, gesture, shared emotion, and visual perception

into a relatively elaborate mode of communication by dramatization. At about this time, too, a split occurs in the child's use of
sound for communication. The infant exercises both considerable
capacity for tone inflection and the emotion it represents, and
considerable understanding of the adult's use of specific words
to mean specific things. His capacity to formulate important
thoughts in words, however, remains severely limited.

By three, communication has grown much more elaborate. With
the evolution of speech, time sense, and memory, the self-system
also grows more clearly delineated. This implements the division
of experiences into those that can be communicated consciously,
and those that must be communicated without being fully accessible
to consciousness. This, in turn, causes the elaboration of dreams,
partly as a communications device between the individual's conscious and unconscious personality organizations.

Subsequently, all communications, to some extent, have a double
meaning. They represent both validated needs that are incorporated in the self-system, and also those that have had to be dissociated.

The different sensory systems become progressively integrated and
reintegrated, with the formation of more elaborate symbols that
use perceptions of objects, time and space, and occurrences to
refer to primary emotional experiences. Over the next few years
speech becomes the most elaborate medium for explicit formulations of experience. Progressively wider opportunities for all kinds
of interpersonal contacts further expand the individual's range of
communications media, as well as all the communications abilities
theretofore organized.

VIII

ORDERLY SEQUENCE AND FIXATION

THE CAPACITY for relatedness—to the body physiology, to the object world, and to the world of people—develops in an orderly sequence. Any experience which is seriously blighted, and is seriously incapable of consolidation into comfortable conscious functioning, prevents subsequent experiences from being consolidated into the self-system until the prerequisite experience has been integrated.

Orderliness is absolutely necessary to this, as to every kind of growth. Thus the organization of the body's physiology and its representation in the body sensorium is absolutely necessary for any subsequent experience. Organization of motility and space and sensory perception is prerequisite to any further relatedness to the physical world, and familiarity with the physical world is prerequisite to any feeling of belonging among people. If one is uncertain and anxious in relating to the world of things, he must be so preoccupied with catching up on this learning that people cannot be more than objects in the service of this more urgent task.

Similarly, the individual must sense himself as a member of the human race before he can feel that he belongs to any group of it; and if there is no sense of belonging to a group, individual friendships within it cannot be spontaneous and wholehearted.

Prerequisites to productive marriage would include a predeveloped capacity for friendship, confidence in one's sexual capacity, experience in living with another person significantly outside of the family's pattern. Extension of love to people beyond the marriage alliance is predicated both on prior experience with alliances, and prior experience with love. If these conditions are not satisfied, "friendships" will be subverted for the purpose of solving one of the unresolved prerequisite problems, such as physical integrity, sensory experience, parallel play, or group status.

Developments in areas related to previously consolidated functions can occur smoothly, with relatively little anxiety. But as the tasks of life grow more complicated, experiences subsequent to handicapped functions are progressively less well consolidated. Such experiences relating to later stages of development often occur outside of focal awareness. If circumstances force these experiences into focal awareness, the self-system repudiates them.

One of the usual ways in which such learning is tolerated is for the individual to think he is doing something entirely different. For example: the paranoid adolescent who thinks that he is exploiting a benefactor, when actually he is gaining access to a friendlier set of parents; people who have "accidents," when actually they are finding a way to accept a little necessary mothering.

Guilt is another screen for partial learning. The person does what he must, but manages to feel ashamed of it; the acquired function is alien and unwanted. Thus to hold its uneasy ground, the guilt-laden function must be practiced over and over.

In terms of the orderly sequence of development, there must at all times in every individual be a zone of experience, not already consolidated, which would be the individual's expected next area of development. It is just these experiences that should be integrated into the self-system with a minimum of acute anxiety. This zone is the focus of the integral personality in its push to growth. The integral personality is fixated on this zone of developing functions. In the individual in whom growth is proceeding optimally in sequence and timing, the term fixation is not ordinarily used. But the concept fixation applies in a continuum from this ideal situation to the usual, in which a variety of necessary functions are to varying degrees inconceivable to, and thus blocked by, the self-system's authorities.

Thus fixation refers to arrest in development of the capacity for relatedness, due to early deficits in growth and current prohibitions against growth. This is a dynamic process, since the point of fixation—the stage of growth at which development is held up—progresses during a person's lifetime, insofar as that person continues to grow. The orientation of the integral personality to growth, which forces movements at the point of fixation, is what we refer to as the tendency to mental health.

The attempts to learn an unintegrated function is often focused on as pathological. But, in fact, it is not the attempt to learn that is pathological. A person who has not mastered sexual techniques, logically and healthily attempts to learn them. The pathology, rather, lies in the repeated failures to overcome the prohibitions and to master the function. The new function—say, sex—is so forbidden that the person cannot consolidate it. Therefore, his efforts to progress—to find a partner—are crippled. Thus, in thinking of fixation, what matters most are those things that the person has not yet learned to do.

Similarly, the juvenile character who is preoccupied with becoming president of every organization in his ken is not unhealthy because he has developed skill in group dynamics and political folderol. The unhealthiness lies in his full-time preoccupation with political ascendancy, which at least partially indicates that he feels insecure about really belonging; and, more importantly, indicates that his group successes have not become the base for the next stage of development, which is developing warm, personal friendships within the group.

A person is usually not clearly aware of the point of arrest of growth. To the extent that he becomes aware of it, he must be anxious about undertaking the mastery of the avoided area.

The stimulus for continued attempts to grow is multi-determined. One essential factor is that the point of arrested growth relates to ineradicable human needs. The need for tenderness, for group relatedness, for companionship with another, cannot be written off, and the integral personality remains preoccupied with satisfying the crucial need.

Another factor is that omitted growth zones represent persistent handicaps to the person who must cope with the complicated adult world. For example: a construction engineer, no matter how brilliant his technical grasp of his field, could find it difficult to ply his trade if his sense of simple belonging as a person among people were so undeveloped that he experienced every contact, however casual, with another human being, as a personal rejection.

A third source of the push for growth is the fact that growth stimuli are always present in the real world. If a person is tone deaf, he is nevertheless continually reminded by his environment of situations in which the perception of different tonal ranges would be useful. Similarly, a person who has never had a chum is continually needing and attracted to potential chums. The adult's progress in all previously throttled areas of development is being provoked constantly—and growth would proceed constantly in response to these stimuli, if the progress did not involve anxiety.

Thus, integration of antecedent experience is vital to the assimilation of any experience into the self-system. This applies equally to the integral personality. This is difficult to demonstrate, because of our necessarily spotty access to the content of the integral personality. But, clinically, certain people seem to operate from a broader unconscious experiential base than others, and seem to have greater facility than others for perceiving and formulating this reality, if only in their dreams. Thus we infer stages of growth in the integral personality itself.

IX

DETERIORATION AND REGRESSION

ADULT GROWTH limitations lie much more in the self-system than in the less conscious formulations of the integral personality. Thus, for our purposes, we apply "growth" ordinarily to expansions of the limits of the self-system. However, the self-system's progress is triggered largely by the intrusions of the integral personality.

The integral personality tells the truth much better than the self-system. It is more perceptive of states of the world, and it is better organized, than the self-system. When the integral personality pushes for perception of a fact of life, it provokes from the objective world responses which the self-system can perceive.

Then the self-system either repudiates or validates the perception. In either case there is a temporary loss of coherence, because the individual loses contact with the central figures around whom the self-system was organized. This loss of contact produces feelings of isolation, disorientation and panic—anxiety.

The abrupt expansion and disorganization of the self-system through validation of previously inconceivable experience characterizes the catatonic episode. The episode may resolve itself in the direction of either growth or deterioration. The integration of the new experience into the self-system is growth. If the new experience is finally invalidated, repudiated, the events leading to the new insight must also be repudiated, which reinforces the repressive machinery of the self-system and constitutes the process of deterioration.

Thus the bottleneck to growth lies not in the availability of adult experience but in the process of getting that experience incorporated in the validated precincts of the self-system. This process of incorporation is more or less explosive, according to the degree of self-system reorganization necessary for the incorporation.

Integration of new data into the self-system is a continuing process involving communication between the conscious and the unconscious elements of the personality. Dreams are a common vehicle for this communication, and they may serve their purpose even when they are not interpreted consciously.

Ruminative states often serve the same purpose, even though the integrative function may be masked by the style of the ruminative state; for example, the sleepy, half-awake state of the person who finds it hard to get up in the morning, or the dreamy state of the person who hates to go to bed at night, who likes to get everybody else bedded down. There are inbetween states of low focal awareness in which the day-to-day obsessive interaction with the outside world becomes rather distant, and one is somewhat alone with himself and can attend more to the preverbal, prelogical, symbolic thinking of the integral personality.

The barrier between the self-system and the integral personality is also less sharp when, instead of the sudden disorganizing breakthrough, there is a kind of chronic panic state, a very high level of anxiety, with frequent brief cataonic experiences. During these states, which may last from a few moments to years, the day-to-day common-sense, practical relationships with the outside world are carried on with considerable effort, and the communication of consciousness with the unconscious formulations is relatively free. There is relatively easy access to what is going on inside.

With increasing emotional maturity, the barrier between the self-system and the integral personality weakens. One has more access then to his marginal thoughts, and grows increasingly aware of their meaning and the meaning of dreams.

Growth in adults as well as in children depends on tenderness and validation, but adult growth meets with more difficulty than does growth in the child. Adult exploration of new areas of experience often involves considerable unconventionality. The adult's actual growth project coincides rarely with the problems society expects an adult of a certain age and status to concern himself with. More importantly, the self-system, with increasing age, grows into a more solidly entrenched opponent of the integral personality and its efforts to spark growth. Growth in the adult hence evokes anxiety. Anxiety is conspicuously easier to tolerate in the presence of affection than in the presence of isolation, if only because anxiety, in itself, is the emotional equivalent of the fear of isolation from the parents, and therefore from the world.

On the other hand, in the process of living the integral personality accumulates more and more insights and experiences incompatible with the tenets of the self-system. Thus, the character structure is not static. The self-system and the integral personality coexist in an equilibrium of flux, in which the apparently static point of balance is continually being jostled from one point to another by the relative effectiveness of the two opposing forces.

When a person's self-system grows by incorporating an experience, it is likely that the next experience will be incorporated into the self-system somewhat more easily than it would otherwise have been. But if a new experience is repudiated, the next related experience will thereby occur with more anxiety, and will be less likely to be integrated.

The self-system maintains its limits through the distortion of experience. Modes of this distortion include: selective inattention, lowered levels of awareness, selective choice of experience, manipulation of the situation in order to force the outcome predicted by the self-system, logical or emotional reservations about the experience, fallacious interpretation of the primary data.

Thus a person exposed chronically to a particular experience may seem to have accommodated himself to it. Actually the experience may have been integrated into the self-system, with whatever reorganization necessary for its inclusion. Equally possibly, however, the true perception of the experience may have been dissociated to the point where there could be no response to the reality of it. The process of repudiation is subtle and complicated.

A new experience, for example, may be integrated with residual anxiety. Time passes, during which this anxiety is transformed into nagging doubt of the experience's actuality. The anxiety converts itself into a feeling of unreality and improbability, and a fear of having been taken in. Then, the slightest turbulence in the smooth progress of the new function or the new relationship is sufficient to throw the individual into a tailspin of despair. He grows convinced that it would have been better not to try to grow at all. An observer might say that tenderness has made this person anxious, but the person himself would complain that the tenderness he receives is too little, too late, in the wrong form, or possibly phony. This does not end the process. Every function integrated against previous prohibition suffers from repeated episodes of this kind, until the person finally notes the pattern of his repudiation. At this point the pattern of the repudiation may be repudiated.

The paranoid character is the ultimate creation of the progressive repudiation of experience, and the concomitant progressive loss of the capacity for relatedness. The process of deterioration, like the process of growth, is cumulative. Absence of tenderness is the primary condition for deterioration, first for erosion of the interpersonal, then of the intellectual, capacities. Thus a crucial factor in determining the overall direction of the personality, toward growth or toward deterioration, is the intensity of the prohibition which every individual feels against replacing the mother by acquiring additional significant sources of tenderness.

Development in the direction of such acquisitions is aided materially by the availability of accurate, explicit interpretation of historic origins of barriers to growth, and is dependent particularly on sympathetic interpersonal support. Another factor is the validation of the importance of the individual's quest for the full satisfaction that open-ended growth implies.

Life involves the constant process of either growth or progressive restriction. It moves in the direction of maturity, or in the direction of paranoid deterioration.

Deterioration must be differentiated from the process of regression. The sequence of loss of function in deterioration is not related to the sequence of development of function. Regression can be defined as the return to chronologically earlier, more primitive, modes of functioning under the impact of anxiety. The mechanism for this is the disintegration, in the presence of anxiety, of the later acquired, more elaborate integrations, in more or less retrograde sequence. This leaves the more primitive responses to make do as best they can. The more anxious the infancy and childhood, the more fragile the physiologic and psychologic integrations, and the more easily they disintegrate. The primary indices of acute anxiety—the rapid pulse, choking, uncontrolled urination, motor incoordination—involve a loss of reflexes which were acquired within the first year.

Certain psychoanalytic theorists speak of "regression under stress," as if such regression were a product of yearning for parental protection, for a return to childhood's safety. This concept rests on the assumption that this person's childhood was happy. To the contrary, a high clinical correlation exists between profuse regressive symptoms and strikingly miserable childhoods, and relative freedom from regression correlates with less fearful beginnings.

Even so, in an adult in whom the need for tenderness has been dissociated, regressive symptoms may point to a time when tenderness was not so repugnant to him. To the extent that the anxious adult regresses to earlier modes of coping with life, he loses functions acquired later and returns specifically to the person he was at the time the primary integration was established between himself and his mother. In the main, he regresses to the particular miseries and limitations of his early years, rather than to their joys.

Regression may be specific, general, partial or more nearly total. For example, acute anxiety may precipitate specific retrograde loss of a particular organ's function, or panic can spark a more general regression: physiological instability, such as lisping speech, uncoordinated movements, inarticulate expression, a diffuse susceptibility to infection.

Regression can involve any system, and it can most strikingly involve the security operations. The anxious person, stripped of his usual techniques for organizing experience, thereby is stripped of his usual modes of repudiating unwelcome experience. Thus he may be forced to use more primitive security operations for repudiation.

The retrograde sequence of regression also applies to the productive self-system functions. Shattering these allows the release of those productive functions which are not, in parallel operation, regressed by the impact of the anxiety immediately involved. Hence, what comes through is what has not been disintegrated by the intense anxiety. These residual functions may be extremely primitive.

Part Two

THE NORMAL SEQUENCE IN INTERPERSONAL GROWTH

I

THE INTERPERSONAL GROUNDWORK

Up to this point, we have set forth the theoretical framework of the process of growth. Now we shall relate these concepts to the normal course of development, and consider some of the carry-overs into adult functions of the particular structuring of learning at each stage of development.

These developmental stages can be outlined as follows: infancy, childhood, the juvenile era, preadolescence, adolescence, and adulthood.

Infancy is the period between birth and learning to talk, at about two and a half. Childhood is the period between the development of speech and the age when the need for contemporary playmates becomes intense, characteristically from two and a half to four years of age. The juvenile era, which may last from four or five to eight or eighty, is the period during which the individual learns to deal with contemporaries in the group situation, by the political arts of compromise, competition and cooperation. Preadolescence is marked by the experience of chumship, when a contemporary assumes coequal importance with oneself, characteristically between age eight and puberty. Adolescence coincides with the onset of puberty. Adulthood refers not to maturity but to the age at which the parents have outlived their usefulness as the primary source of validation.

Infancy is subdivided into early infancy and late infancy. Early infancy is the period between birth and learning to walk, at which point the child's knowledge of the position and functions of his own body is accelerated greatly. Many functions are initiated during early infancy, and of these the most urgent is the organization of body physiology, coupled with the organization of emotional patterns, sensory perceptions, and motility.

These physiological functions can be divided into three categories, on the basis of differing susceptibility to interpersonal influence.

The first of these categories includes functions manifest to the mothering person: eating, defecating, temperature control, respiration, pulse, hearing. Validation of these processes is non-verbal. Communication between infant and mother takes place through the relevant zones of interaction, relying in the main on direct physical contact and sound, and later including sight. Such validation is necessary for the full integration of a function into the physiological organization of the body, and also for potential awareness of the function in later life, and for the feeling of aliveness of which people are potentially capable as adults.

The second category includes those physiological functions not directly represented in awareness, but linked inseparably to specific emotions. The validation of these functions is consolidated through parental validation of the emotion involved.

The third physiological group includes functions that have no overt interpersonal connection, or even any trace of conscious representation in the adult. Even here, however, interpersonal leverage probably exists. It may be that in particular historical circumstances awareness of some of these functions could be part of the cultural heritage. The infant has much more vivid perceptions of known sensory systems, internal or external, than the relatively anesthetic adult, and this later anesthesia is partly contingent on lack of consistent validation in infancy. There may also be physiologic systems of which the infant could be aware, which stay out of awareness entirely because of lack of response.

Intense, generalized anxiety transmitted empathically from mother to infant could interfere with the organization of physiologic functions, hampering the adequate development of structure for the functions involved. This would handicap functions which were actively being organized while the emotional climate was stormy.

The physiologic organization may be affected by the interpersonal climate in still another way. Regressive disintegration of functions clearly subject to interpersonal influence may result also in faulty development of related functions not directly influenced.

Both yoga and hypnosis indicate that a much greater range of physiology is available to consciousness and subject to interpersonal influence than we had believed possible. In hypnosis, for that matter, the importance of the validator is so apparent that it is often overlooked.

Processes which themselves are entirely unconscious are progressively organized at the cellular level into larger units of organ systems. Any such system is organized in such a way that the differentiation of function and level of functioning, at the cellular level, is dependent on the cell's integration into its immediate larger unit—lobe, or group of cells. Similarly, this unit is under the dominance of its integration into the next level of organization. In this way, all functions are contributory to, and subject to, the overall patterning and level of aliveness of the major organ systems, which serve for the implicit and explicit interactions necessary for survival and growth, and hence have clearly interpersonal aspects.

The person in early infancy must cope repeatedly with a stringently limited number of adults. Some infants, indeed, cope with one adult only—the mother. Communication between infant and adult is empathic, supplemented by voice and touch. To the extent that the mother lacks the capacity to respond empathically to the infant's need, the function is denied validation and is poorly consolidated.

Here we must consider the concept of awareness in terms of the early infant. He certainly has no self-awareness in the adult sense of the term. His sense of time is vague. His memory is unreliable. It is probable that the early infant's overall level of awareness depends on the responsiveness and capacity for tenderness in the mothering person.

Adult allocation of experience to the self-system or to the integral personality is controlled through the experience's availability to awareness. Early infant experiential allocation is much the same, except that then the mother's repudiation or perception of the experience directly determines this assignment to the anlage of the integral system or of the self-system, as they begin to be differentiated.

Adult awareness functions for better communication and hence relatedness. The refinements of conscious adult communication evolve from the non-verbal communication techniques that begin to be organized at birth.

The organization of the self-system and the integral personality begins at birth. The early infant self-system is organized similarly to the adult self-system into which it develops, to permit organization of physiologic functions by minimizing anxiety, albeit by the most primitive of security operations. The relative ineffectiveness of this budding self-system is attested to by the frequency of more or less catatonic crises in early infancy. These occur because the needs involved are ineradicable, and thus the functions to be integrated for satisfying them are crucial, no matter how much anxiety is produced in the mothering one by having to attend to them. The crises are also predicated on the relative clumsiness of the just-developing security operations.

As the formation of consciousness develops, the infant becomes much more talented at distinguishing between absolutely invalidated, "not to be thought of" experience—Not Me experiences—and the self-system experiences that are validated to the extent that they become Good Me or Bad Me experiences.

With the development of walking, early infancy shades into late infancy. With the development of speech and the consolidation of a sense of identity, the infant forges ahead dramatically in the direction of the self-system.

Many functions integrated primarily during infancy normally operate in the adult outside of focal awareness. An adult does not ordinarily become conscious of his normal heartbeat, although some reflection of it can be brought into awareness by concentrating. The adult may not experience hunger pangs from week to week unless food is unavailable. On the other hand, the reaching, without inhibition, for available food when he is hungry puts this function in the domain of the self-system.

During the first two or three years of life, a large variety of physiologic functions are integrated which could have some reflection in awareness. We may speak then, about functions which get integrated into the self-system in the sense that some representation of them can become available in awareness under special conditions. There are other functions which could, in this same sense, be a part of the self-system but which are cut off from it by lack of validation. Both categories of functions are themselves related in the total physiological organization of systems, to functions which are probably never directly available to consciousness. The optimal operation of the latter may, nonetheless, contribute to a general feeling of well-being.

We can make a distinction between the actual level of functioning of an organ system and the optimal level of its functioning. The hypothesis is that if a function receives not even indirect validation from any source, its development remains inadequate, even as represented in the integral personality. A function cannot be performed if the organ for performing it is not available, at least in the rudimentary form. On the other hand, a structure only develops and integrates structurally in the process of functioning. Further, the continuity of functioning is necessary for the maintenance of structure. Any change in the patterns of consumption of energy must of necessity tend toward a change in the fundamental structure. The dependence of the muscular system on use is a gross example. The interaction between the growth of bones and weight-bearing is another. One could describe the structure of an organ as the high-speed snapshot taken at the moment of examination of its actual functional operation.

The mother's validation of the function at the zone of interaction, by appropriate emotional response, is indispensable. Let us take, as an example, the function of knowing when one is hungry. The newborn has hunger contractions, but these do not correlate with the emptiness of the stomach, and are not alleviated by food in the mouth. They are, however, stopped by putting an appreciable amount of food into the stomach. The infant has to learn that these particular sensations indicate hunger and that they are relieved by food. In order for the baby to grasp this correlation, it is necessary for the mother to be sensitive to the signs that indicate when the baby is suffering hunger pangs. She learns to place the bottle in the baby's mouth with an open mind, and to know by his response whether this was appropriate to his stimulus of crying. It is only after this hunger function is learned that it is possible to get the baby comfortably on periodic feedings. Mother's response, then, is an integral part of the learning process.

The responsiveness of the mother is, in turn, conditioned by the degree of expansion of her own self-system. If the mother is not in touch with her own feelings—cannot feel hungry herself, for example, or cannot recognize the emotions involved in eating—she cannot empathically be in touch with her child's. Her perceptions of his hunger are foggy and inaccurate.

Thus, when an infant is separated from a responsive mother, we see that his initial reaction is a general depression, with some loss of interest in the world about him. Later, he regresses to previous levels of integration. Finally, in the prevailing absence of tenderness, a progressive functional deterioration occurs. The responsive mother is the infant's validator, and without her everything tends to fall apart.

But the baby creates in the mother a remarkable degree of anxiety, which militates against the empathic responsiveness that the baby needs. The prevalence of postpartum psychosis is one index of this anxiety. Having a baby, particularly her first, is a girl's clearest and most challenging exploration into the adult world. Clearly a challenge to her mother's image of her as a little girl, the absolute irreversibility of childbirth intensifies the anxiety that the mother feels about her self-assertion. Although this anxiety is often less with subsequent births, it can become sharpest after the young woman has produced as many children as did her mother. Too, the arrival of the baby demands practical and sometimes resentful reorganization of the young mother's life. Often the stress is intensified by her wish to be seen by her mother as an adequate mother.

Too, the empathic experience of feeling her child's needs may evoke in the mother recollections of experiences—hunger, help-lessness, squirming, cuddling, rage, grief, panic in the night—that she has long dissociated. The concentrated empathic interac-tion with the infant challenges the depression and lack of aliveness with which she has stabilized her own life, and brings to mind the painfully unfulfilled desires of her own infancy.

It matters little to the infant what his mother's anxiety is caused by; whether by direct interaction with him, or whether set off by other events in her own life. The infant is much more defenseless against empathic anxiety than the adult. The mother's anxiety results in disorganization of physiological functions already ac-quired: an infant who has learned to eat well develops colic and diarrhea. Also, however urgent, no new learning can occur in the presence of severe anxiety.

If the function being learned by the infant triggers anxiety in the mother, the function ceases its integration just short of the point where it would become overt activity. If the mother is suf-ficiently anxious, that can act as a powerful, positive prohibition against future attempts to learn the same function.

The mother's security operations tend to be built into defective learning. For example, if the child is mildly constipated, and if the mother is herself subject to constipation, if she is anxious about the bowel function, she will be unable to respond satisfac-torily to the child's problem. The child therefore becomes more tense and less able to relax the sphincter. This increases the mother's anxiety, and in turn the child's. Ultimately, the now chronic constipation becomes an accusation against the mother for her failure to have learned to be relaxed about her own bowels.

The concept of adult residuals deserves some elaboration. Any or all of the functions which the young one is undertaking to master in a particular period may be consolidated or may be de-fective in integration to varying degrees. This defect may be the outcome of stressful integration of a preceding related experience, or it may be contingent on a strong prohibition of this immediate step forward. If consolidated, the function operates inconspicuously in the adult. If damaged, its consolidation becomes an adult pre-occupation. Since this damage makes later development in the same line even more difficult, the subsequent blocking intensifies the repetitive nature of the adult preoccupation with mastery of the function. More severe blocking results in the inability to grasp the function in later life, with consequent elaboration of related, supplementary functions to cover the defect.

Functions which were adequately consolidated provide a base for further development in subsequent periods. Under conditions of regression, these later derivatives may disintegrate, and the childhood mode may reappear in full force. With more severe regression, the childhood operations can also disintegrate, throwing the person back onto even earlier ways of relating.

In considering the historical origin of areas of difficulty, the division into ages and eras is a great oversimplification. Each project overlaps the one preceding and the one that follows. None are completely finished between birth and death, even in a relatively untraumatic situation. All are subject to disintegration under sufficiently unfortunate circumstances. Damage is cumulative, and defects in each antecedent project make the subsequent projects even more difficult.

The adult, then, is busy making up experiences which it would have been easier to have had in the early years. One might think that an adult really cannot go back to playing with blocks again; he must have to write this off and start out where he is.

In any adult, the earlier learned, more diffuse functions persist and facilitate use of the functions later elaborated from them. The two-year-old's capacity for mimicry, gesture and communication— with the shrug of a shoulder blade—if it survives in the adult, is an aid to more elaborate forms of communication. The capacity to hear music facilitates the use of language in the adult. Thus functions integrated at each level normally persist in the adult in adult equivalents. Further, adult reintegration of basic functions which were omitted or later repressed is necessary for comfortable exercise of the functions derived from them.

II

INFANCY

A)   FUNCTIONS INTEGRATED IN EARLY INFANCY

It is useful to review the specific, day-to-day organization of physiologic functions in infancy as it might interact with and might be molded by the emotional climate.

Seriously deficient physiological organization during very early infancy may be seriously detrimental, since many of the functions acquired during that period involve vital organs.

Deficiencies organized during early infancy can remain chronic handicaps throughout life; note the fact that certain schizophrenic children have gross irregularities in the rate of development of physiologic function. Or, mildly damaged organizations can persist into early childhood, as in infant eczema or colic, later becoming organized to function adequately, at least in stress-free situations.

Any defect in organization is particularly subject to disintegration under stress, a phenomenon which has sometimes been referred to as organ inferiority. This includes cardiovascular lability, predisposition to allergy, asthma, deficient resistance to pyogenic infection, and susceptibility to gastro-intestinal dysfunction. We separate functions arbitrarily into three divisions: functions acquired in the first week of life, functions acquired in the first month, functions acquired during the rest of the first year. This division reflects a decreasing speed of integration of physiologic function.

The first twenty-four hours represent the most total reorganization of a person's lifetime. The newborn is suddenly transformed from a water animal to a land animal. The skin changes from an internal organ under water to an external organ exposed to air. The body must cope with a drastic drop in external temperature. The infant must suddenly supply his own oxygen requirements.

The newborn is stimulated to breathe both by small amounts of carbon dioxide and by slight hypoxia. The respiratory center is depressed by more severe hypoxia. A few hours after birth, another great reorganization occurs. The infant's cardiovascular system shifts from the placental circulation pattern to a circulation that uses the lungs as the primary source of oxygen.

At birth the heat-regulating center is also very immature. Its function begins to be organized during the first twenty-four hours, and improves progressively over the next several weeks. At birth the temperature of the body falls. After this there is characteristically a response such that the initial temperature drop is compensated and the infant reaches a fairly stable temperature consonant with the temperature at which he is kept. He thus remains somewhat poikilothermic, that is, his body temperature shifts easily with the outside temperature for some weeks. The smaller the baby, the more need for extra help in maintaining body temperature.

Swallowing and some directional movement of the gastrointestinal tract has been taking place intrauterine, and sucking motions are present before anything has been provided to suck on. The presence of swallowing movements at the time of birth has been demonstrated by the fact that, though the inside of a baby is sterile at birth, by the end of the first twenty-four hours there are always bacteria in the intestines which must have been swallowed.

The skin immediately functions as an organ influencing fluid balance, via evaporation. After the first twenty-four hours, the skin continues its adaptation to the air and to the various foreign bodies that are applied as clothing. Initially, the capillary system is extremely labile in respect to dilatation and contraction. The infant develops a progressive capacity to stabilize temperature, and in this development the skin as an organ plays an important part. Diurnal variations in temperature develop parallel to the establishment of sleep and activity patterns.

The respiration continues to stabilize. The systolic blood pressure comes up slowly during childhood until it reaches the average adult level after adolescence. The pulse rate, high at birth, slowly comes down to the adult average of 72 beats per minute. Both functions, blood pressure and pulse rate, are strikingly more labile during early childhood.

The physiologic patterns of the adult in states of panic and extreme regression can parallel various infantile patterns.

Sucking and swallowing are present at birth and well established. After birth, the directional activity of the gastro-intestinal tract progressively becomes somewhat more efficient.

The enzyme system for the digestion of particular chemical entities develops progressively. Certain foods can be digested and metabolized at birth, and thus the enzymes necessary for their utilization do not depend for their development on those foodstuffs which are present after birth. The capacity for the digestion of other foods evolves subsequently. After birth, there is also a shift over in respect to what types of food can be utilized by the body as a whole, and the infant develops and progressively elaborates the capacity to metabolize more complex foodstuffs.

Another function which evolves during the first few weeks of life is the capacity of the body membranes to adequately filter proteins. At birth, the lining of the intestinal tract is a very poor barrier, allowing the more complex proteins to be absorbed into the blood stream undigested. This early inadequate filtering, and the rate of development of an adequate protein barrier, might be a factor in the susceptibility of some infants to food allergies. The filtering and concentrating powers of the kidneys are also immature at birth.

Starting before birth, and continuing up until twenty-six weeks after, there is a gradual shift over to an entirely different chemical type of hemoglobin. The respiration of the newborn is more diaphragmatic than intercostal, and this shifts progressively toward the intercostal over the next few years.

Some muscle tone is present *in utero*. The infant at birth presents himself with a hypertonicity of muscles which, during the first six months, slowly subsides toward the adult level of muscle tone. The quality of physical handling may accelerate or retard this shift.

Small infants are extremely sensitive to the feeling of touch and the quality of voice tone. Within the first few weeks they can learn to identify familiar persons by either of these modalities.

Certain of the functions discussed above are clearly affected in their development by the interaction between infant and mother, by her style of touch, feeding, and cleaning. Other modes of communication between infant and mother include the use of touch to transmit temperature information, and the visual observation of sweating, shivering, and goose flesh. The young infant's sensitivity to vocal tone makes the mother's tone of voice available for communication between them from the beginning.

During the first year, respiration becomes more stable; the chest muscles come into action and develop appropriately. The pulse begins to slow down. The dilatation and contraction of the capillary system becomes more stable. During this year temperature control becomes relatively well stabilized. The one-year-old baby does not easily shift in temperature in accordance with the temperature of the room. Many mechanisms are involved in this, including the efficient utilization of reflexes of shivering and sweating. The capillaries filter more efficiently. The absorption of whole proteins into the blood stream from the intestines decreases. The skin grows progressively resistant to irritation by foreign bodies.

Many physiological functions are appropriately integrated during late, rather than early, infancy. These include bowel and urinary control, feeding and self-feeding. The period of susceptibility to diaper rash may persist until the child is toilet-trained. One of the urgent necessities of infancy is the development of the body's defenses to bacterial and viral invasions. This process is initiated during the first year, but persists as a necessity throughout life with each exposure to a new population of infections.

By the end of late infancy, the homeostasis of the physiology becomes more established, and the organism becomes less susceptible than before to disorganization by minor traumas. Parallel with this, the functions cited above become more efficient.

The gastro-intestinal system is one of the major preoccupations of mother and infant during the first year of life. By two weeks, the directional activity of the gastro-intestinal tract is rather well established. The more or less continuous activity of the neonatal intestine is progressively organized in the direction of periodicity of eating and bowel movements. This goes on in recognizable stages, first to the four or five-hour feeding schedule, then to the dropping out of the night feeding, and finally to a slow organization of the eating patterns into a pattern of three meals a day, plus some supplementary feeding. Colicky pain unrelated to organicity is the susceptibility to easy disorganization of gastro-intestinal patterns by fluctuations in the emotional atmosphere. This susceptibility usually diminishes after the first four to five months.

Optimally, new foods are introduced a little at a time, thereby allowing time and opportunity for the function of digestion, absorption, and metabolism of each new food to be developed gradually. Most contemporary babies, by the end of their first year, can eat almost any food. Biting and chewing are slowly substituted for sucking. Teeth usually begin to appear at or after six months, and reach the normal set of deciduous teeth at about two and a half years.

Sleep at one year is organized more or less on the diurnal pattern, with one or more breaks during the day.

By eighteen months, gross anatomical maturity of the nervous system is almost complete. A considerable development of all the special senses occurs, together with elaborate organization of the perceptual capacities.

By one year the infant has developed a varied body of experiences with respect to different tastes, textures, and sounds, and has developed definite preferences as to the things he will or will not eat or touch. He has grown increasingly exploratory with his sense of touch. His sight has progressed from functional near-blindness to a fair level of visual perception of form, color, space relations, and motion. Coordination of the eyes and the ability to focus them parallels these developments. The baby's capacity to distinguish among sounds has developed tremendously. His ability to differentiate phonemes and words is rudimentary, but his differential response to tones and inflections is acute.

The early muscle tone and bodily responses, diffuse and fishlike, progress by the age of about four months to the infant's task of mastering his own body, an intense project until about twelve months. The baby learns first to arch his back, then to roll over, sit up, crawl, pull himself erect, maneuver with support, and then, at about a year, to try to walk unsupported. Over this same period there has been a great preoccupation with the shape of objects and the mastery of some concept of three-dimensional space. Here is the beginning of the sense of body image, of the notion, for example, that it is his arm that reaches out to get the object that he wants.

The infant, before he develops enough time sense and perceptual experience to identify individuals, identifies the emotional attitude of the mothering ones. He thus reacts to the Good Mother and the Bad Mother—i.e., the loving, responsive mother and the angry, distant mother. Parallel with this, he experiences his own needs as good, i.e., satiable and responded to with affection, and bad, i.e., unsatiable and responded to with withdrawal or attack. It is this integration of Good Mother and Good Me that is the base for survival in infancy and provides a beginning for the productive aspects of the self-system. Likewise, the system of interactions of the Bad Mother and the Bad Me constitute the origin of the central paranoia. The self-system includes both, and must find logical premises consistent with both. The self-image must incorporate both self-concepts, and define the boundaries of the Not Me—i.e., those attitudes which became dissociated because the appropriate response was dissociated in mother.

Emotional responsiveness develops strikingly during the first year. The newborn infant experiences euphoria and its opposite, sensations of safety or of danger. By one, he is capable of a wide range of emotions, including terror, grief, pain, joy, fun, cuddlesomeness, disgust, distaste, and the wish for mastery. These emotions are learned by empathic communication centrally with the mother, broadened importantly by other mothering figures, and progressively focused. By about a year, the baby's typical emotional pattern is recognizable, if not set. This personality bears a discernible resemblance to, and is an understandable derivative of, the overall emotional patterning of the primary mothering one.

This does not mean that emotions in the child are received from the parent in the same way as preferences in clothing. Only those emotions that the mother herself can feel and respond to freely can become part of the baby's self-system as it develops. Those emotions inconceivable to the mothering one become dissociated, because there can be no empathic validating communication about them from mother to child.

Some of the overt physiological expressions of particular emotions, such as tears with sadness, or the increased muscle tone with anger, can be identified in their initial stages even in people in whom the awareness as well as the expression of the emotion is essentially blocked.

B) RESIDUALS OF EARLY INFANCY

The effect that an anxious mother has on the infant depends to some degree on when in the course of the infancy the anxiety impinges.

The effect of acute anxiety experienced during the first twenty-four hours of life is obscure. To our knowledge, it would be difficult to undertake a detailed correlation of the level of panic of those caring for the infant with the relative difficulty with which the baby makes the transition from the uterus to the atmospheric world.

The practice that someone other than the mother care for the infant during the first twenty-four hours may offer the infant some protection. A relaxed mother would undoubtedly be the child's optimum haven during this period. But if the mother is intensely anxious, fluctuations in the infant's vitality could increase her anxiety, with respondent anxiety and progressive functional failure in the child. Such cumulative interaction could have serious emotional consequences for the baby, with a potential for physiological repercussions.

In the first weeks, whether or not it is the responsive but anxious mother that takes over, the direct effect of the one that is doing the nursing is obvious. During this period the development of the sensation of touch, the general muscle tone, and, most crucially, the functions of eating, swallowing, and gastro-intestinal motility are necessary for survival. Again, not a great deal is known about the relation between the level of panic in the mothering one, and the infant's capacity for acquiring these functions smoothly. Statistics easily available on this point would be thoroughly unreliable. If the mother is not frankly psychotic and ready to be hospitalized, her capacity to function at all depends on her powerful denial of panic, and her necessity to put on the best act that she can for her mother, her husband, her pediatrician, and her baby. Nonetheless, an extremely anxious mother may contribute to the aggravation of such conditions as allergy to milk, generalized skin difficulties, or persistent colic.

After the first weeks, fluctuations in mood of the mothering one are being beamed at the integration of less crucial functions, but nonetheless could intensify a persistent handicap. The functions being integrated are less immediately crucial to physiological integrity.

It is a thesis of interpersonal theory that the crux of acquired defects in later functioning lies in the inadequate representation of those functions in focal awareness. Such focal awareness is directly dependent on the initial validation of the function by the mothering person. The mother who does not enjoy food herself will have a child who enjoys it very little. The function tends to be controlled not by the normal sensations but by mother's obsessive, substitutive reactions. The child eats, not when he is hungry, but by the clock. He accepts or rejects his clothing, not by his own sensation of body temperature, but by his mother's prejudices. When the mother reacts to the child's needs as if they were angry demands, the child learns to express any need angrily. He takes over his mother's structuring of the situation.

Normal adaptive movements toward the solution of any situation are delayed, thus, because the primary experiential defect lies at the point of expression of the need for experience. Occasionally a child does not know that he is hungry, and so does not eat for the primary reason that no one has ever confirmed for him that the sensation he feels is hunger, and is to be gratified by eating.

His difficulty in making the normal postural adjustments to muscle fatigue or circulatory stasis lies partly in the fact that he is relatively insensitive to the sensation of fatigue. He therefore allows abnormal posture to continue longer than it should.

The capacity to selectively inattend a state of feeling in the service of necessary concentration of focal awareness on a particular task is a general constructive human capacity. If one is preoccupied with an urgent task he can forget to eat until the crisis passes, and then suddenly he notices that he is hungry. But in psychosomatic pathology, there is an insidious chronic form of not noticing which can go on day in and day out because of a general insensitivity—a continuous cutoff from awareness of some aspect of the body sensory function.

We can hypothecate here a line of interaction of the capacity for consciousness and the body physiology. If, as above, certain normal overt adaptive movements are persistently absent, this maladaptation may result in chronic functional and structural disorders. There is also some reason to think that subjective unawareness of an organ system, a continual relative anesthesia to its normal conscious representation, could interfere with the internal physiology of the system, even if external adaptations are not crucially involved or the omissions not easily identified.

For example, tactual anesthesia clearly results in accidental skin injury because trauma is not avoided. Psychogenic dulling of skin sensation, associated with repudiation of skin contact between people, seems chronically to parallel neurodermatitis of the potential contact area. It is as if certain physiological activities are intensified in a direction which would force attention to the neglected areas.

What we have been describing could be the interpersonal and physiological base for a great many of the phenomena ordinarily considered in the domain of psychosomatic medicine. Other psychosomatic symptoms derive more directly from physiological regression under the impact of anxiety. Psychosomatic symptoms seem to refer back conspicuously to the organization of bodily functions, especially during the first year of life.

The intimate interaction during the first year between mother and infant organizes in the infant patterns of emotional response which are unique to each individual and tend to persist throughout life. They account, in part, for the adult's continued emotional involvement with his mother, or with persons of similar emotional patterning, long after the relationship with her as a contemporary has become meaningless.

The emotional patterns that characterized the infant's interactions with his mother tend to be experienced as aliveness by the adult. Interactions with adults whose emotional patterns include feelings dissociated by the mother during the person's infancy evoke considerable anxiety in the adult. There is a tendency to seek repeatedly for people who display patterns similar to the mother's and then to try to make those people more loving than they are or than the mother was, instead of finding relationships with people who are more loving in the first place.

The father characteristically makes less impact on the child than the mother, although he can importantly broaden the infant's frame of reference. Too, the adult with the lesser degree of responsibility for the child's immediate sustenance can often relate to the child with more spontaneity than the mothering one. Where more than one mothering figure exists for a child, the baby's actual mother remains the central figure, but the infant's self-system becomes more versatile. Those fallacious perceptions that brought the parenting figures together in the first place become the more entrenched fallacies in the structure of the infant's character.

Much adult falling in love is under the impress of the early infant's organization of perception of the mother into two personifications: Good Mother and Bad Mother. When the object of the falling in love is very like the mother in emotional structure, there is a tendency to regress to the perceptual framework of the initial personifications. It becomes almost as if the person currently responded to separates out into two people: a most wonderful one paralleling the infant's personification Good Mother, who was wonderful beyond expression; and Bad Mother, who was more frightening than the Devil. The intensity of emotion involved reaches back to the adult partner's resemblance to the mother as a young postpartum mother in both her constructive and destructive aspects. Often the adult perpetuates the confusion by a conscious repudiation of the real mother's positive qualities, in the later attempt to avoid being trapped by perpetual reintegration with the mother's paranoid aspects.

The emotional interaction between mother and infant has particular impact on later sexual patterns. The original experience of desire-satisfaction is re-enacted, in however limited a way, in adult sexuality.

Personifications of both Good and Bad Mother tend to be projected onto the sexual partner. In our culture, a frequent compromise is the dilution and restriction of the sexual act.

By cutting down, as far as possible, the intensity of the experience, much positive satisfaction is lost. What is avoided is the danger of re-experiencing oneself as defenseless and terrified as was the infant originally when with "Mother when very angry." Many specific sexual fantasies and phobias are designed to cope with this expected threat of either disgusted rejection or particular forms of physical attack that happened or were felt empathically by the infant when the mother was at her worst.

The first year constitutes the beginning of the organization of the structure of personality. The lines are not sharp, but the anlage of the self-system, the dissociated, the central paranoia, and the integral personality are clearly in existence. Subsequent changes in the central assumptions, and hence in the direction of growth, will require progressively more drastic reorganization from this time on.

C)   EVENTS OF LATE INFANCY

During late infancy—the period between learning to walk and learning to talk, from ages one to two and a half—the newly mobile infant creates more selective interpersonal relationships and devotes more of his energy to the mastery of the object world than before.

This development puts new pressure on the parent. Mothers find that their babies have changed from warm dolls to independent persons. With such a development, the mother may feel released from the considerable prohibition against expressing anger at a babe in arms. Also, particularly with the first baby, this is a point at which relationships between mothers and fathers may be clarified for better or for worse. If their relationship is clarified for the worse, the child can become the hateful symbol of bondage in an unwanted relationship, hence a pawn in parental battles.

The infant's human interactions continue to be primarily with his parents. Body contact is still the focus of much of the baby's interactions with his father and mother. The child craves physical contact and evaluates people largely by touch, smell, and sound.

The infant by now can say a few words. He understands more words than he can speak, but his communications grow more vocal because of his greatly expanded sensitivity to the content of various tones and inflections. Capacity for enjoying and producing music may be initiated in this period, and later elaborated into an effective tool for the non-verbal communication of explicit emotions. Thus the scope of the child's interactions with adults is much more complicated than it was during the first year. Although words themselves hold little meaning for the child, he communicates effectively by dramatization—what later in life we call "acting out." Other children and animals stimulate the infant, but relatively little interaction with them occurs. Animals are considered contemporaries, not property. They are living objects with whom the child feels he must work out a *modus vivendi,* on the rudimentary basis of live and let live.

The infant's capacity to locate parts of his body now grows by leaps and bounds, as he begins to elaborate his own body image. Time sense, memory, and self-awareness continue to be vague, however. The organization of sensation continues to be a major preoccupation, as does the identification and classification of objects by shape, motility, and use. Mastery of space and objects— such as doors—parallels these developments. Relationships to objects occur, especially in terms of appreciation and cherishing of them. Choice of these objects can be as unpredictable and shifting as adolescent crushes, and may involve what to the adult are inexplicable loyalties. This behavior forms the basis for the slightly later development of interest in the manipulation of the various physical media.

D) RESIDUALS OF LATE INFANCY

Damage occurring during late infancy may have diffuse or specific residual effects in the adult, but the adult defects will be in those functions which it was the primary task of late infancy to master.

Deficiencies in the organization of the special senses may appear in the adult either as omitted developments or as preoccupations. Thus the adult may be tone deaf; or, if the damage were less severe, the adult can devote his life obsessively to the mastery of musical expression. Either he forever fails to draw in perspective, or he forever pursues artistic or structural disciplines that involve mastery of space. If the integration of the body image is radically damaged, the result can be bad posture, clumsiness, failure to see oneself as attractive or physically adequate, or general muscular unawareness and lack of positional sense. Lesser degrees of damage to the body image may result in the lifelong devotion to the body beautiful that marks the obsessive amateur athlete and the professional model.

In the dramatized communication of late infancy, the communication is limited essentially to the significant adults, and constitutes something of a private language between himself and them. It is shaped to the quality and the specific necessities of these relationships. These postural attitudes and the perceptions they express may persist into adulthood. This persistence has been described in the concept of the character armor, which is a chronic attitude and posture which was appropriate only to the earliest years. This armor may become a mode of defense against further expansion of the personality.

If this basic communication is not supplemented greatly in later eras of development, it remains as the basic mode of contact with other human beings, overlaid with this personalized reference. On the other hand, if even this mode of communication is not solidly integrated, the person remains dependent, for anything that he feels conscious conviction about, almost entirely on physical contact. It is only by literally touching another person that he can tell something of what is going on with the other person.

After one year, the emphasis is on a more active relating to external objects through more intense and complex organizing of the infant's range of direct sensory perception. This becomes the groundwork for the subsequent development of relatedness with contemporaries. The initial active relating in late infancy is to blocks, dolls, cuddle bunnies, and the like. If the mother disrespects these activities, and expresses the disrespect either with open contempt or by intruding herself as the third party of a triangle, the child fails to develop a sense of personal property. This produces as an adult residual defect either the notion that property lacks any significance, or the obsessive and insatiable drive to accumulate property, a drive fed perpetually by the adult's fear that he really has no right to own anything at all.

Leaving home can be a time of intense anxiety for the adult. This anxiety may be expressed in dizziness, poor posture, incoordination, agoraphobia and claustrophobia. These are all suggestive of the hazards of the time of learning to walk in an anxious climate. The original dangers of passing from a warm doll to a moving entity become the perpetual symbols of the mother's anxiety at his individuation from her. The specific symbolism of the fears involved are the incorporated reflection of the mother's attitude toward this shift in the infant.

III

CHILDHOOD AND JUVENILE ERAS

A)   EVENTS OF CHILDHOOD

Childhood, approximately from age two and one half to four, is defined as the period between the development of speech and the development of the conscious need for playmates. By this time the self-system has become well-defined.

The child begins to form a conscious notion of who he is. The underlying self-image is now initially formulated. His experiences become progressively differentiated into those which are available to consciousness and those which must be denied and repudiated. The directions of future growth are now classified into those which seem conceivable to undertake and those which are inconceivable.

At this age, between two and three, the child's concept of himself and the world and his own scheme of values can only constitute an entirely naive reflection of the picture held by those around him. He has had access to no other frame of reference.

The acquisition of speech makes the previously formed self-system more easily discernible to the adults. It also facilitates validation of perception and continuity of perception, and becomes an important mode of organization of consciousness. Speech is an important tool in the inner organization of the self-system and in the implementation of dissociation.

It may be as an aspect of this sharpening division of the personality that the period between two and three is an age of heightened dream and nightmare activity. Experiences accumulated during the day that are unacceptable to the better defined self-system must be integrated into the personality during sleep.

The adult's earliest memories usually relate to this period of childhood. These memories state the child's crudest initial formulation of his identity and his relationships with the world and with significant people.

Modes of interpersonal communication continue to proliferate during childhood. The major mode of communication has shifted from empathy to dramatization, and now begins to shift to verbalization. Dramatization remains active, however, and emotional communications in particular remain essentially non-verbal.

Sense of time expands greatly during childhood. For the first time, the individual develops a sense of continuity in his own personality, and in his relationships with others, usually paralleled by increased continuity of memory. This is prerequisite to the elaboration of the self-image and the self-system, cited above.

The child's circle of significant adults can now widen to include relatives and family friends as proxy parents. Mastery of speech makes overt consensual validations possible. This also forms the base for childish confusion of truth, however, since what is said may not register with what is felt by the child through empathy. Mother may say that she loves the child, for example—but the tone of her voice while she speaks the words of love may sound angry.

Adults still form the nuclei of the individual's important emotional relationships, and thus the child sees himself as a child in an adult-dominated world. In this milieu the primitive attitudes toward authority are formed, and the individual's methods for dealings with the agents of authority are elaborated.

The patterns of relationship to authority figures—compliance, defiance, dependence, spontaneous independence—the calculation of the degree of disconnectedness from them that can be tolerated, or of how much of a calculated risk to take—are the natural concerns of childhood.

The child still spends much time in consolidating his mastery of shapes, his coordination, and his body image. Intense relationships are now to toys, blocks, teddy bears, modeling clay, colors, sounds, and relatively impersonal, but extremely vivid, day-to-day experiences. Now the child's imagination develops intensely, especially in the area of using, understanding, and investigating the world of objects.

Here again, parental validation is essential. To be useful to the child, the validation must attest the reality of the child's spontaneous organization of existence, and not take the form of imposing the parents' patterns on him. Thus the parents' sensitivity to the child, perception of the child, and respect for the child's natural sequence of learning must be called into intense play, in order for parental validation to be useful. If this does not occur, the child's activities may get organized into durable, obsessive, substitute patterns, the function of which is to maintain contact with the mother, signifying both the repression of and the wish to master these functions.

The self-centered adult often cannot tolerate the intensity of the child's relationships to the objects at hand. He may push other objects at the child as substitutes, with which the child will enjoy a less intense—but, for the parent, more comfortable—relationship. Thus originates the common later complaint that, instead of loving the child, the parent substituted a proliferation of objects for the emotion of love. This means in fact that the parents imposed their own objective tastes on the child in an effort to tone down his too fully budding imagination.

The primary interpersonal growth experience during the period of childhood is the integration of the positive function of parallel play with peers. This refers to the positive mutual stimulation and empathy experienced by two children in each other's presence, while each is, at the same time, apparently focused on his separate individual project of physical mastery. Here lies the first primitive emotional move toward feeling an independent member of humanity—toward experiencing one's peers as like oneself.

To the extent that the function of parallel play fails to be integrated during childhood, contemporaries tend to be seen as good or as bad parents. Their indifference is seen, for example, as if it were parental indifference. The child thus misperceives the situation, because he is convinced that the best of all possible worlds would consist of himself and a multitude of good mothers.

It might be somewhat useful for a little cooperation on the part of the child with other children to be enforced—taking turns, sharing, accepting retaliation for injuries inflicted on others. It is a misconception, however, to think that the child will like this. The child cooperates easily with friendly adults, but pushing him too hard to cooperate with contemporaries tends to set up resistance to the natural development of cooperation in the succeeding period of his life.

Although specific relationships with contemporaries are still only peripherally important to the child, emotions of affection, aggression, and intimidation achieve considerable expression. These contemporary relationships lack significant continuity, and the variation in the child's responses to different contemporaries is in a narrow range.

In families where it is encouraged, the relationship to pets, or even to plants if pets are forbidden, may provide a useful transition from cherishing objects to enjoying other children.

The intrusion of parental phobia about people to inhibit the child's first attempts at parallel play is both more frequent and more damaging than the parental dampening of the child's intense relatedness to objects. The parent's envy of the child's capacity for pleasure in the other child tends to restrict the child, not only to associating with offspring of those few adults the parent can still tolerate, but to restricting the patterns of interaction toward matching the parents' depression. The compliant process in the child is glamorized as "well-behaved."

B)   RESIDUALS OF CHILDHOOD

Specifically, the major areas of mastery between two and a half and four include the following: creativity in the object world, relatedness to pets, parallel play with other children, verbal facility about factual data, a concept of time, dramatization for the purpose of communication, the elaboration of patterns of relating to "authorities," and the accelerated organization of the self-system.

Each of these has antecedents in infancy, and each is necessary for further developments in the juvenile era.

The quality of the relationships to objects provides a good example of the way in which blocking in the integration at an earlier level of relatedness makes more tenuous the integration of a subsequent level. If the cherishing of the object (a normal concern of late infancy and a vital link in the development of direct relatedness to the physical) is seriously frustrated by the mother, it will continue to have an anxious quality. This results in the adult in a mixture of or alternation between acquisitiveness and compulsive poverty.

The capacity to cherish objects, developed normally in infancy, is requisite to the effective integration of the childhood function of the mastery of materials such as clay, paint, dough, mud pies. Making, having, cherishing the object thus become fused as completed subjective experience. This is the initial experiential ground for a wide range of adult productive activity.

If the cherishing aspect evokes anxiety, the mastery function becomes distorted in the direction of an obsessive preoccupation with technique alternating with disjunctive repudiation of competence.

Parental prohibitions vary in style and in content. They may be expressed in the demand to relate to the mother instead of to a toy. Or they may stem from an overall indifference to either child or toy. Sometimes the pattern is of sadistic deprivation, or of the teasing offering and taking away. The parent thus confirms his hold on the child through his ability to wilfully withhold pleasure from the child.

The particular form of the parental prohibition can be inferred from its result, the later style in which the individual seeks to deprive himself of pleasure, but the style matters less than the fact of prohibition. The child's conclusion, in any case, is that he must not want. He is rendered deficiently capable of relating to his own desires. This effect is cumulative with the repudiation of desire rooted in the inadequate validation of his need for tenderness during infancy. This cumulative infusion of wanting with anxiety during childhood can have a broad effect on the adult life pattern.

It hampers the widening of particular potentialities into full experience of them, and tends toward a permanent amateur status. A real commitment to a line of development would lead one inevitably to the frank experience of cherishing and mastery. These experiences must be avoided or postponed. This is accompanied by a vague insatiability. The enjoyment of available satisfactions is vitiated by guilt or derogation. It affects how the adult relates to love and productive work, and contributes to the low self-esteem.

Sometimes one's cherishing of his inner experience can proceed freely because it is hidden, but even this process often mires down. It is difficult to keep one's inner experience hidden, because it must be shared in order to be validated. Yet, objectifying the experience by communicating it makes it taboo. Besides, even secret satisfactions carry a burden of guilt, to which is added a queasy awareness of the self-derogation involved in keeping it secret.

Another development during childhood is the capacity for parallel play. That is, now children relate to other children and are stimulated by their presence, but make no serious attempt at continuity with the particular other child. The adult equivalent to this is found in spectator sports, in large parties made up of relative strangers, in the feelings of audience community at the theatre, in the annual New Year's Eve pilgrimage so many thousands of people make to Times Square. In the genuine integration of the function of parallel play, the core experience is the sense of belonging as of right, of citizenship, of feeling oneself and the other to be more human than not; that in the public feeling of being a member of a community of particular interest, the sense of being a permanent member of the broader human community is expressed and strengthened.

Failure to integrate this function during childhood may result in adult anxiety in crowd situations and compulsive avoidance of those situations. It may also result in a compulsive necessity to stay in the crowd situation and to shape all relationships to fit this mold. Persons afflicted with this particular response will tend to avoid one-to-one intimacy. They select activities in which the people concerned can be related more to the activity than to each other. Also, cooperative living arrangements with many contemporaries often indicate the need to make up the integration of the function of parallel play. If the individual integrated social relationships at this childhood level, but was unable to develop beyond it, he will limit his life to relationships with acquaintances. This can be one of the motivations for the "rolling stone," the person who allows himself no permanent residence, or for the expatriate who never accepts social responsibility in the country of his actual residence.

Communication of emotional content with significant adults is by intricate, non-verbal drama in which words appear but are used in the service of the acting-out and not as a medium for formulation of the inner thoughts. This remains as a normal adjunctive mode of communication in the adult. At the same time it also shows up in the adult in intermittent hysterical symptoms. When it carries over in the adult as the major mode of communication, it defines what is referred to as the hysterical character. Hysterical conversions, since they do not refer back to the period of organization of the basic physiology, need not directly involve important disorganizations of the patient's physiology, but do involve the lability which is characteristic of the small child.

The gross universal impediment in verbal communication has its beginnings in the vicissitudes of the process of learning to talk. Communication between any two individuals involves the capacity to think in the presence of the other one, to formulate one's perceptions about the topic at hand and one's perceptions of the interpersonal field in which one is operating, to get these clearly into awareness, and to find a way of verbalizing them which is appropriate to the other's field of reference. It thus involves a parallel capacity to perceive the other's interaction and field of reference and projections about oneself, and the capacity to communicate in such a mode that the communication evokes in the other useful and expanding associations. Such a capacity is only sporadically achieved in the real adult world.

As our culture is organized, no one ever really teaches the child how to talk. At the time when he learned to talk, the field in which he was learning was disorganized and disinterested. He grows up, then, with the notion that on the one hand the parent figure was never very much interested in what he had to say, or, if so, had no comprehension of the integrity of his processes in trying to learn to say it. The child also grows up with the expectation that the other person, even if he understood, would never admit what his honest and immediate response was. He would be impelled to impose on the field of interaction a variety of hypocritical, conventional formulae designed to disintegrate one's internal thought processes.

Such expectations are very destructive to the processes of verbal communication between adults, and they can carry over with remarkable literalness. The adult assumes that the other will not hear what he has to say, that if he does hear it he will not understand it in context, and if he should understand it, he could never be honest enough, or enough in touch with himself, to be able to respond appropriately. Or if he should respond with his own thoughts, they would be so self-centered in their mode of formulation that they would be totally misdirected towards the first one's formulations.

So-called authority situations between adults are suffused with the authority aura of childhood. The child depends on his authorities for validation and protection. Adult authority situations, were they not imbued with residuals from childhood, could be recognized easily as peer situations in which there may be unequal information, contracts for division of responsibility, struggles for power, or, rarely, attempts to achieve tyranny. The parent has a degree of power over the child which is rarely duplicated in adult life. This includes physical and legal power and the dispensing of food, lodging, tenderness, guidance, and advice. The crucial element in the parental authority is rooted in his inevitable capacity to form the perception of the child. By the bondage of empathy, areas of dissociation in the parent are prohibited to the child. No peer ever has such undermining power over another peer.

Thus, if an adult perceives an integration with another adult as an authority situation, this perception is essentially parataxic. Even under tyranny, for example, the adult does not depend on the authority for validation, nor can he be required to seek affection from the tyrant. In power struggles that do not involve tyranny, the adult can choose to walk away from the "oppressor." Between adults, approximate equality of contribution and interdependence is usually more realistic than between parent and child.

The person in the superior's role in an authority situation, under the parataxic influence of his feelings of being the child in the adult world, often perceives his subordinates as if they were his parents. Or he may aggrandize his expectation of the influence he should wield as if he were an adult with a group of children.

The use and misuse of the time sense as a symbol in authority situations may derive from the fact that both time and authority integrations relate to the same period. The time sense is often directly distorted by the parental demand that the child develop a concept of time more rapidly than his real capacity permits. Also, the angry parent can use the child's observed confusion punitively. The conspicuous thing about the residual deficit is the embarrassment about it and the persistence of it. Where such a deficit in the conscious perception of time persists, the time sense is integrated covertly, thus producing the twin phenomena of persistent, accurate errors in regard to time in the everyday functioning and unexpected time accuracy under emergency conditions.

C) Events of the Juvenile Era

When the child has developed some confidence in his mastery of the physical world and some competence in parallel play, he can devote himself to the task of relating more actively and accurately to his contemporaries, and becomes, in interpersonal terms, a juvenile. These contemporary relationships proceed against the background of the acceptance, physically and psychologically, of the dominance of the parents' frame of reference. The juvenile is also confronted by other values at this time, at the age of roughly four to eight, as he comes more into contact with adults outside the circle of his own family and must learn to accommodate to them.

Meeting these divergent people provides an important new avenue through which the juvenile can validate impulses toward growth besides those sponsored by his family. The parental permissions and prohibitions nevertheless remain the dominant factors determining the juvenile's personality with his contemporaries. Mother must approve playmates for the juvenile to be comfortable with them.

Relatively secure juveniles use these relationships to explore roles different from those played at home. The reliable, responsible juvenile can volunteer, for example, to be the villain in dramatized play. The younger brother can try out being the leader.

If the juvenile's perception of contemporaries continues predominantly in the mode of childhood, he tends to see them as if they were members of his immediate family. He then seeks out a group position similar to the one he occupies at home—youngest, oldest, good, bad, leader, lieutenant.

The need to learn group living is so acute during this era that the juvenile remains occupied with it, even against powerful prohibition—in fantasy, if necessary. He tries to learn competition and cooperation and studies how to compromise in such a way that his self-esteem does not suffer. He develops a concept of fairness and unfairness. Hopefully, he develops a sense of participation that does not depend on buying his way with special talents, an experience that may be rare at home.

The fear of ostracism by the group is one of the dominant motives of this period. Only the child who develops a fairly solid sense of being a participant by right can risk the experience of occasionally opposing the group and risking exclusion, at least from that particular group.

In making these social adjustments the juvenile uses the full spectrum of potentialities he has developed theretofore. The emotional content of the juvenile era is focused, thus, on the group relatedness. The manifest preoccupations are usually with mastery of the physical world, initiated earlier, and its extension into a variety of hobbies, sports, and competitive games.

Formal education, usually begun in the juvenile period, offers another opportunity for elaborate development of the individual's mastery of the physical world. Also, going to school is often the first step in separation of the young person from the domination of his family's values. It also provides, assuming some success in group relatedness, an opportunity for the child to investigate how other families live by gaining access to his schoolmates' homes.

The juvenile adjustment is essentially an adaptive one. It is based on the principle of enlightened self-interest. The goal is comfortable participation in a group. Within the limits of the parent's values, the values of the group are essentially accepted as right. They are to be coped with rather than re-evaluated.

There is no serious concern for the other person's needs as such. During this period children make allies, but love, in the mature sense, is not really involved. Value concepts, such as fairness, justice, and equality are developed and acted upon. The child does develop a conscience and a commitment to the welfare of the group as a whole.

The juvenile still functions with his contemporaries by the intuitive understanding of what is going on in the other child to the extent that this emotional capacity has not already been dulled. But as communication with contemporaries—relative strangers who have no great investment in interpreting one's behavior—becomes more urgent, words take on more importance as a reliable medium of complex communication. The development of language now more sharply than before facilitates the explicit formulations of the content of consciousness. It also facilitates the explicit exclusion of unwelcome thoughts, and the capacity to develop the obsessive preoccupations necessary for this exclusion.

During the juvenile era, in addition to learning about group dynamics, there is expansion of mastery of the world of objects. There is often an intense, widening interest in the world of nature, of the tools of mastery of objects, and in the beginning use of scientific concepts.

Also, the individual devotes much energy to the search for alternate models for his life. This period is characterized often by omnivorous reading, with intense, successive personal identification with fictional or genuine other people. The intent of this exploration—to find a way to change the family tradition in the person of the individual—generally is unclear. Its later implementation is often defeated by this fact.

D)   RESIDUALS OF THE JUVENILE ERA

Relatedness with other children flowers in the group relatedness of the juvenile era. The patterns of active group participation make up the crucial normal learning of the juvenile era. Also, comfort with strangers and facility with the process of getting acquainted, initiated in parallel play experience in childhood, are refined and expanded. To the extent that they are mastered they constitute no problem, but can be taken for granted. To the extent that their mastery was defective or tenuous, they become preoccupations throughout life.

From the juvenile period on, residual damage is especially manifest in the patterns of the interpersonal relationships, rather than in specific deficits in mastery.

If the juvenile does not become significantly comfortable with the learning of competition, cooperation, and compromise, the residual effect is in two directions. First, he is thrown back onto a perseveration in the patterns of childhood. Other people are perceived as if they were literally one of the few members of the immediate family, e.g., the angry father, the exploitative mother, or the competitive brother. The person may persistently integrate in the social group in the position he held in the family group with adults and with siblings. Or, as a compromise, he may play the different-from-home role, never with ease and comfort, always with the wistful hope that one day the magic will take, the old self will disappear and the new one become the reality. If this should actually happen, it is experienced as unreal. The most swashbuckling ten-year-old or thirty-year-old knows he'd better get home on time for supper.

Further, personal possessions and personal comfort take precedence over conventional socialization—whether in the mode of acquisition or in the mode of denial. Language is narcissistic without being autistic; there is no adjustment of the verbal formulations to the audience, and almost no perception of alternate viewpoints. The orientation of childhood carried over in the adult becomes an exploitative one in which other people are manipulated as objects.

The other main residual defect dates from the middle of the juvenile era. The adult juvenile remains devoted to the juvenile values of social adaptability, respectability, security, normality, and popularity, but never feels like a true group member. He feels compelled either to win or to lose, since either competition or compromise makes him uncomfortable. His success as a leader is a disguised expression of the feelings of being marginal, of not really being a member of the group. He mistakes prestige and achievement for membership.

These people organize and administer the background pattern of our culture. On the surface the value system is presumed to derive from enlightened self-interest, but more centrally expresses the obsessional exaggeration of these values intermingled with the adult distortion of three-year-old self-centeredness, in which distortion people are objects for exploitation.

Even if the adult juvenile values were not contaminated with hangovers from the naive exploitativeness of childhood, enlightened self-interest, normal and productive in the juvenile years, is also exploitative if it persists in adults as the basic organizing principle. The defect is in the inability to grasp the concept that another individual's needs can take precedence over one's own.

Specific masteries during the juvenile period are less important to the adult than the specific masteries of earlier periods. Facility with the juvenile preoccupations of sports, games, and formal education is relatively easy to acquire later in life. Defective integrations here constitute crippling handicaps mainly in struggles for economic independence from parents or from people with whom a hostile integration has been established.

Development of the capacity for the obsessional dynamisms is rooted solidly in the intellectual developments of the juvenile era. A smoothly integrated obsession is conveniently available after the juvenile era as a growth-stunting security operation. If regression disintegrates this operation, the adult juvenile will, if need be, retreat to the childhood rooted hysterical line of defense against the expansion of the self-system.

Integration with the parenting figures and their scheme of values remains the dominant influence on the juvenile. Compliance or defiance, attachment or flight may express this devotion, but the parental frame of reference remains the child's. Experiences with different adults may modify it, but only slightly.

If this integration remains dominant, the adult's choice of primary relationships tends strongly to remain among people similar to the parents. The adult also often seems to substitute rigid conventionality and the idea of the power of society for his parents. This indicates on inspection that either the social conventions coincide in fact with the emotional content of the parental values, or that the parental values are projected by the individual onto the screen of society. Thus the Nuremberg Trials offered many examples of adults giving as excuse for their actions an alibi remarkably like the juvenile justification of: "It's all right, because my father told me to do it."

The pathology does not lie in the earnest attempt at mastery of group integrations, but in the inability to experience intimate relationships. If, in addition, prestige or power goals are substituted for the attempt to achieve a valid group relatedness, it becomes, inherently, a self-defeating move which takes the individual farther from his goal.

Juvenile adults constitute a sizeable proportion of the adult population. They invest more energy than other people in political maneuvering. In this case, the neurotic substitution of power machinery for social relatedness becomes a social hazard and a factor in limiting the growth of others, as well as a factor in limiting the growth of the individual himself. Juvenile values and a juvenile concept of life, necessarily corrupted by their extension beyond their real meaning and usefulness, are imposed on the social relationships as the price for the individual's avoidance of isolation from a particular group.

IV

FURTHER ERAS OF DEVELOPMENT

A) EVENTS OF PREADOLESCENCE

Preadolescent experience occurs optimally between age eight and the onset of puberty. The transition from the juvenile to the preadolescent is marked by the development of a love relationship, a chumship, with another individual of the same sex. This individual is experienced as very much like oneself and as of tremendous importance to oneself. For the first time, the other person's true needs are experienced as equal in importance with one's own.

It is a relationship of trust and confidence, essentially devoid of the competitive and the manipulative, and independent of the parental frame of reference. It involves the freedom for and the inner push to be exploring outside of their framework. Thus its genuine development requires a high degree of privacy.

One suddenly finds that one is intensely interested in what another person has to say, in how another person experiences the world, is deeply concerned with his respect, has great admiration and respect for him, and a desire to share one's own inner experiences with him. Communication goes on essentially without effort.

There is now a qualitatively different basis for the validation of one's feeling of humanness; of being more like other people than different from them. Now one finds another person who one experiences as valuable and who is experienced at the same time as so like oneself. Thus, by reflection in the friend, one must conclude that he, himself, is also of some value. One actively joins the human race. The affirmation, the process of sharing and being shared with, and, most crucially, the experience of caring and being cared for, result in the expansion of self-awareness into a new dimension. "Whatever is going on with me must be human and therefore to be cherished, as I can see that the same things go on in my friend."

This expansion of self-awareness under conditions of acceptance is what we mean by self-esteem. With the emergence of this new dimension of experience, the juvenile value system is in the moment disintegrated and displaced as the organizing principle.

Juvenile manipulations, earlier ends in themselves, now are transformed into tools for forming a love relationship. This provides the base for further extension of the sources of validation, for the development of deep feeling for an ever-widening circle of friends. Also, chumship makes possible the extension of the sentiment of love to include segments of humanity with whom direct intimacy is unlikely. This is the development of compassion. Compassion, based in the empathic perceptions of infancy, refined in the common goals of the juvenile, is consolidated in the concern for the chum's needs for growth. It can also, now, be extended to a concern for relative strangers in need and the more abstract goals of human happiness.

Compassion motivates social activities in which juvenile techniques are used for a new purpose, in respect to a new set of values. The impulse toward political mastery gives way to those forces concerned with love, spontaneity, satisfaction, and direct perception. This can lead to manifest conflict with a particular group, or to political retreat for the purpose of regrouping one's personal resources.

The search for intense one-to-one relationships sometimes is used for a rationalization of the repudiation of the broader human struggle for social sanity. Such repudiation actually betrays the love which it presumes to implement, for love of another individual, hence oneself, leads necessarily to engagement in love for humans generally.

Now the process of living begins to make sense to the individual. Experiences once taken on by rote in order to please parent, teacher, or group suddenly acquire personal significance. Interest mushrooms in people, philosophy, history, science, literature.

A particular other person has now become more important to one than one's parents as the source of validation and affirmation. This, therefore, represents the first major point of expansion of the frame of reference carried over from the parents. This experience opens the door to an almost limitless expansion of the self-system through a variety of valid, affectionate, and tender relationships with people other than the original significant adults. Love can now be based not on an unequal dependence, but on equality and freedom of choice, as any genuine commitment in the adult world would have to be. The opportunity for exchange of confidences in itself offers a great opportunity for the correction of personal misconceptions and the expansion of one's original private scheme of values.

This, for many an individual in our culture, is the highest point of humanness achieved during the whole of life. What often follows is an insidious deterioration, marked by increasing bitterness and resignation, as the result of the limitations imposed by a culture that cannot afford to be so alive as the preadolescent boy or girl.

B)   RESIDUALS OF PREADOLESCENCE

Preadolescent defective integrations can hardly be called residuals in the adult. Most people, in fact, are eternally preoccupied with finding a chum. The manifest investment made in the process of finding a partner of the opposite sex more often than not disguises this central quest, or else constitutes a way of avoiding the unresolved problem of friendship.

The other activities of this period are the expanding social, intellectual, literary, and artistic interests. These are not in themselves new lines of development, except as they take on new impetus and deeper meaning as a result of the chumship experience. If omitted, they leave, per se, no major gap in social grasp which cannot be filled in later if it becomes appropriate.

A minor later disadvantage in having had no chum is that the chum, when found, inspires in the individual interests that conventionally are inappropriate to the person's age and status. Thus a person of forty who has just discovered his chum can feel foolish to begin to read Shakespeare for the first time, or to be consumed suddenly with an interest in medieval history, as a result of having fallen in love with humanity.

Often the chumship is not subjected to explicit parental disapproval. But the inhibiting difficulty remains that, for the first time in life, the preadolescent relates to someone more intensely than he does with his parents. This clearly is revolutionary and fearful. Besides, love for the chum reawakens all of the preadolescent's dormant fear and paranoia about tenderness.

The cultural prohibitions to the intense relationship that chumship represents are not consistent. The culture expects the individual to demonstrate a high degree of connubial loyalty, whether he hates or loves his spouse. Similar commitment to one's friends is tolerable to the culture, but its omission does not occasion strong social disapproval. Yet our culture incorporates considerable unstated prejudice, more for men than for women, against the idea of strong feelings between persons of the same sex.

The chumship is permitted as a passing phase in the years of one's youth, to be replaced by a retreat to the juvenile modes of relatedness as one becomes a "responsible" citizen. Affection for particular individuals is usually subordinated to the immediate convenient organization of one's social life. A diffuse non-intimate "friendliness" is substituted for the specificity of deep one-to-one relationships with particular people.

If there has been no experience of chumship in a person's life, he cannot conceive of values beyond the juvenile mode. He never discovers that individuals matter.

More usually, there are a series of abortive experiences, never fully validated, the first usually before the teens. This first experience is often under the handicap that the young person, at the time still in the power of the obtuse or unsympathetic adults, was not free to implement the emotional reality of the relationship. Often this relationship had to be disguised and derogated, even to oneself, for protection from an unsympathetic power.

Later attempts may move, in the overall pattern, toward progressive success or toward progressive discouragement about the possibility of such a friendship.

The effort may be overt or may be disguised as a juvenile alliance. The disguises would include business partnerships, neighborhood arrangements, recreational or intellectual collaborations. Much emotional experience can be explored subversively, but at considerable price. Here, experience of conscious unforced commitment to the welfare of the other is the critical value that determines the relationship's success or failure. The guilt-ridden or grossly disguised attempt for chumship can never be fully integrated.

Active derogation of the relationship is such a disguise of the bid for chumship. The most striking example of this is the so-called homosexual panic. Two lonely people of the same sex recognize their long-sought chum in each other and move closer to each other with unfamiliar feelings of intensity. Suddenly one, or both, becomes convinced that he and/or the other is a homosexual. This triggers precipitate flight from the developing intimacy. This panic indicates that the individual has never experienced intimacy with either sex. The panicked person entrenches his repudiation of the need for tenderness by reflecting that his lack of experience with the opposite sex, plus this near-capitulation to his desire for "homosexual" love, means that he is, indeed, homosexual.

If the chumship is sought in the context of overt adolescent homosexuality, social prohibitions cramp the potential for expanding intimacy by providing the two people with the easy excuse of guilt about their relationship. These relationships characteristically also serve to compensate for previous deprivations of physical attention by the parenting figures. Because of convention, they have an uneasy, defiant, sometimes manifestly hostile quality that tends to thwart the other aspects of chumship. Even so, they make a contribution to the experience of loving other people.

Another form of derogation is the tendency to choose as chum a person in whom one's implicit defects are revealed explicitly. The chum's behavior, especially if conventionally unacceptable, is often used not to validate the human similarity but to derogate the relationship. When one feels that his friends are indefensible, he repudiates rather than consolidates his own feelings of humanness, since he himself is the object of his own scorn.

Fixation at the level of chumship places a severe strain on the adult sexual functioning. Consolidation of chumships is prerequisite to genuine heterosexual love. The omission of chumship hinders this and hinders spontaneous compassion or respect for different kinds of human beings.

C) EVENTS OF ADOLESCENCE

From adolescence onward two main threads appear in the further development of one-to-one relatedness. They are always interactive, but it is to some advantage to differentiate them. One is the developing capacity to love. The other is actual experience in the integration of partnerships. These partnerships can incorporate the experience of love only to the extent that the capacity for it has evolved in the respective individuals. The culture dictates that in the main, these partnerships are heterosexual, but the same considerations apply as well to the homosexual partnerships.

The first experience of love is in the chumship. Its core is in the spontaneous experiencing of the other's experience as one's own. The consolidation of the chumship, then, makes possible the extension of love to other people. But whether or not the capacity for love has been developed, the adolescent necessity is to learn to live and to share with a contemporary. If a reasonable facility in living with another person has not been developed and validated, this lack remains an intense preoccupation. It may be expressed in the attempt to repudiate the need, in the frantic turnover without consolidation, or in stabilizing a particular going arrangement by bribery and intimidation.

The contribution of such partnerships is the mutual gain from the pooling of resources, material and psychological. They provide the opportunity for consensual validation and correction of perception for sharing of everyday experience. The partnership diminishes the paralysis induced by loneliness and fear carried over from childhood terrors. If productive, the partnership must be organized to integrate the most urgent interpersonal functions, whether these be the physiological empathy of infancy, the parallel play of childhood, or the necessities of juvenile experience, available here for integration in this situation of a group of two.

The opportunity for empathy and participation in the partner's life experience in itself provides a valuable opportunity for doubling one's own life experience.

These two evolutions of the capacity for love, and of partnership techniques, are parallel developments. Either can be used, in the moment, to avoid attending to the other. Success in either facilitates progress in the other.

The maturation of the powerful dynamism of puberty precipitates the young person into the urgent need for adolescent experience, and thus presents him with two problems. One is that the culture, in varying degrees, is restrictive of the conditions for adolescent growth. The other is that his previous experience can have provided an uneasy base for what life now requires of him.

Besides, a strong social cliché contributes to the adolescent's crisis; that is, the cliché of the adolescent's need for the influence of home. In reality, after puberty the parents are no longer the main emotional base for growth. No matter what the level of fixation, the child becomes an immature adult. What growth is possible within the parents' frame of reference has already been consolidated, and the urgent necessity is now to expand it. The search for new homes, and, even more, the acquiring of confidence in his capacity to establish them, is urgent.

Puberty arrives, ready or not—although it may arrive late in people who are exceptionally badly prepared for it by their previous integrations, physical or psychical. If one is psychologically unprepared, adolescence is complicated by the fact that the individual must at one and the same time develop experience with partners and develop the experiential bases requisite to this. The focus of this striving to grow has to be at the zone of fixation (childhood, juvenile, preadolescent), but various factors press powerfully to keep this focus out of awareness.

More or less, the period from childhood to puberty involves greater and greater prohibition of the sense of touch, so active in infancy. This specifically includes prohibition and dissociation of genital awareness. Some of this has been made up in disguised modes, such as childhood games and sports involving body contact, but there is inevitably a backlog of active inhibition of a function which had been previously operating, if in a more amorphous form.

Also masturbation is a helpful preliminary in learning the individual patterning and timing of stimulus and response of the lust dynamism. Since it involves reintegration of dissociated experience of touch, it must involve anxiety. Thus masturbation allows the person to rediscover his genitals without the additional complications that interaction with another person would involve. If the person fails to integrate this function, if he remains phobic about masturbation, he will not be comfortable with sexual intercourse.

Next, the person must become confident about his capacity to give and receive sexual gratification with relative reliability. This can be a lifetime preoccupation. When sexual adequacy can be taken for granted, the project of defining the kind of person with whom he could spend his life is greatly accelerated. The initial experiments in partnerships are usually with people who are in some way similar to his family. In this process one progressively disentangles his own emotional needs from the family values. A variety of experience with such relationships is important for an accurate choice of a more lasting partner.

The optimal natural sequence of events is divergent from the cultural expectations, and this fact throws some light on the intensity with which the culture is preoccupied with sexual experience, choice of partners, and the stability of marriage.

The perpetuation of this contradiction and these preoccupations is rooted partly in the primacy of the clan in our social structure. Thus we witness a cultural lag, one of the great ones of contemporary industrial society; for industrial society has made the clan's cultural position obsolete.

From primitive societies through early industrialism, the clan— the family—dominated society through its self-reinforcing power over the sexual unions of the young.

Historically, this power could be escaped only through some sort of priesthood, in which heterosexual relationships were either relinquished outright or severely restricted, through the institution-alization of contempt for the sexual experience, and contempt for women.

Such priesthoods, in spite of their institutional freedom from the clan, have functionally reinforced clan values by incorporating them into ecclesiastical law. This shows up also in their attitude of support, tacit or open, for the practical dominance of the clan structure, judgments, and values for all other people than them-selves.

D) RESIDUALS OF ADOLESCENCE

Even more than the unresolved problems of chumship, the de-fective integrations of adolescence prompt most of the adults' searching. One important reason for the extraordinary difficulty in resolving them lies in the fact that the preadolescent chumship relationships are so rarely fully consolidated. The preparation and foundation for adolescence is almost universally inadequate.

For example, the degree of self-esteem which is consolidated during the chumship experience is prerequisite to adequate sexual performance. Full sexual experience inevitably re-evokes infantile needs and moods, in a wide and varying range, connected both with the initial infantile personifications of Good Mother and Bad Mother as well as those dissociated under the impact of severe anxiety in infancy.

There can be an almost hallucinatory recapitulation of the indi-vidual's central interactions with Bad Mother: fear of dependence and fear of loss of identity, anger, helplessness, disgust, the wish to dominate, fear of punishment, all intensified by the breakthrough of some of the anxiety of the dissociated need for tenderness.

Or the early emotions connected with Good Mother can initially predominate, contributing in the moment to the sensation of com-pleteness and esteem, but followed soon by the intense re-exper-iencing of dissociated anxiety, leading to repudiation of the relationship or constriction of the sexual interaction.

If the emotional core of the relationship is juvenile—if based on alliance rather than on affection—there is a deficiency in self-respect and respect for the partner. The relationship tends to repeat the patterning that existed between the person and parent or between the person's parents. The battle of the sexes is often perpetuated in the familial tradition from generation to generation: *"Plus ça change, plus c'est la même chose."*

The adolescent suffers, clearly, from the culture's expectations of him, and from the concomitant prohibitions which the culture places on his behavior. Adolescent masturbation is publicly derogated as infantile or as perverse. Sexual experience is derogated as promiscuity. This limits the consolidation of physical self-awareness and sexual adequacy, and their future developmental potentialities. Trial marriage, in most communities, is tantamount to a declaration of war on the cultural authorities.

The early marriage is the frequent result of these cultural pressures. Based on great inexperience, determined by the attempt to get enough freedom from the parental authority to have room to grow, the early marriage embodies the wistful hope that it is possible to condense all the necessary preliminary living into this one tight relationship. To the extent that the two young people gain a degree of independence and a beginning of adolescent development, they must feel a great push to make up and fill in some of what they skipped. Even if the inexperienced choice turns out to be a valid one, society and the two people concerned now mistakenly take the sincere attempt to make up enough adolescent experience to grow with the relationship as an invalidation of the relationship.

Thus many marriages settle into juvenile alliances in which pre-adolescent and adolescent growth are stifled, or proceed, at best, covertly. The resulting stress on the values of respectability and of the illusion of normality burden still further the resolution of adolescent problems. Also, the perseveration in juvenile values and patterns is furthered and rationalized in the parents by the legitimate juvenile relatedness of the children of the marriage.

The futility in the juvenile marriage lies in the degree to which the mastery of unresolved juvenile problems proceeds without insight, and is used to avoid dealing seriously with the problem of love.

If the first marriage breaks up, the culture grows more permissive to the estranged couple. Chastity is not demanded with the same intensity. Experimentation is now sometimes considered "wise" in view of the earlier "failure." The outlines of one's demands on the other in a marriage, and of one's own virtues and flaws, have been somewhat identified. It is conventional enough in such an interval to seek out and consolidate relationships with one's own sex. This now occurs in the more or less adult role; one has been married and therefore has achieved some increase of independent status with the parents. Also, almost everyone has more confidence in himself sexually if he has been married than if he has not. This confidence removes one of the real, if minor, sources of insecurity about holding one's own. So that here, too, further relationships with either sex have a better chance.

Altogether, if the intervening period is used constructively, the second choice of a relatively permanent partner may be made on the basis of much more genuine self-awareness. Later marriages often work, where the first did not.

E) MATURITY

Maturity, which occurs so rarely in fact, has come in our culture to be equated with a wide variety of more or less unrelated phenomena. Thus "maturity" has referred to the possession of the capacity for happiness; the capacity to perceive and to relate to the universal loneliness of people; the ability to master the conventions of society, which may involve the corollary that one becomes resigned to society's power, and that one's early attempts at thinking are written off as signs of "adolescent revolt." It has referred to the courage to be unconventional, in that the mature person has less dependence than the immature on society's approval; or to a concept of independence in which the state of having no dependency on other people, the terminal state of alienation, is confused with maturation. Somewhat more understandably, maturity has been associated with understanding anxiety, or understanding growth, since people who do so tend to be the source of advice when other people find themselves in trouble.

The concept of maturity offered here is different from those cited above. Maturity, we hold, is proportional to one's capacity to relate to his contemporaries, regardless of age, in the context of the process of extending love to include more and more people, and to find the common denominator with oneself in more apparently divergent people.

This concept of maturity implies that the individual is relatively free from the original restrictive viewpoints of his self-system, since otherwise each new integration with another person not of the same background would perforce be a shattering experience. The mature person conceives of himself as an independent, interdependent entity, free to choose his own associates and implement his own decisions. The mature person accepts himself as an integral factor in the chain of cause and effect, neither overestimating nor underestimating his contributions to results. Thus the mature person must have shed his concept of himself as a child in the adult world, must have given up the particular distortions of himself and the world which he learned in his family.

Since happiness lies in the unhampered exchange of affection, the mature person has more of a chance for it than most people. Since much sharing involves the sharing of pain, the mature person perceives pain and the universal loneliness. Maturity involves access to the maximum range of empathic emotional responses—fear, anger, joy, ecstasy. It also includes a concept of the possibility of influencing others.

Mature independence is the state of positive affirmation that a person is free from past bondage, and that the person is free to act, assert, love, and be productive. This is not the pseudo-mature masquerade of the repudiation of the need for tenderness, which includes lofty detachment, resignation, compulsive productivity or compulsive unconventionality, for fundamental to maturity is the capacity for full commitment to a person or a project. But this integral capacity for commitment demands that one's commitments to particular matters be conditional. Thus the mature person could never commit himself unconditionally to state, church, family, profession, or even to loved ones. If the commitment were unconditional, it would cease to be the mature act of repeated affirmation of the affiliation—it would become blind obedience, the negation of one's inner freedom, which is the antithesis of maturity and integrity. For integrity means freedom to know and to act on one's inner perceptions.

The mature person would probably appear unconventional and even deviant in the eyes of a juvenile society such as our own. This deviation, however, is much different from immature defiance, the defensive operation of making deviation an end in itself.

Wisdom—the grasp and acceptance of the full range of human complexity, including human capacity for both virtue and vice in the Platonic sense—is inextricably linked to maturity. Wisdom involves faith in man's potential and compassion for his frailty. It involves, moreover, capacity to sense the as yet unknown; particularly, to sense whether an action by oneself or by another is in the direction of growth or deterioration. Thus maturity demands high tolerance of uncertainty and the courage to guess wrong. The wise person has the gift of hearing the latent content of communication.

Maturity implies growth, with no foreseen limit. It involves full acceptance of the need for tenderness and integration of the methods by which that need is satisfied. Thus the mature person, through exploration and consolidation of human relationships, expands his sources of satisfaction of the need for intimacy, continually validates new experiences and continually reintegrates experiences omitted in his past. The mature frontier always stays open and opening.

The experience of chumship now is expanded by example and empathy to involve genuine, mature identification with the problems of groups or of individuals who are distant in history or geography. The mastery of modes of communication and technology—the intellectual and the professional contribution—becomes one of the major implements through which the mature individual can relate to and contribute to the human community, however benighted, that he perceives as made up of people who are recognizably more like him than not.

The mature person, then, is one in whom the central process is the relatively unhampered reintegration of past omissions and expansion of frontiers of experience, interpersonal, internal, and experiential, in the context of the ever-expanding consolidation of freely affectionate relationships. Some of the prerequisites for this include self-esteem, perceptiveness, humility, sensitivity, free access to one's own empathy, and concern for the other's anxiety. Its results involve the capacity for independent action and creative formulation of inner perception. It involves grief, but not depression.

Like the infant, the mature person is free to be in love with life.

V

CRITICAL EVENTS OF THE ADULT YEARS

A) Conditions of the young adult

The primary conditions for growth for the adult are crucially different from the conditions of growth of the developmental eras of pre-adulthood. These differences relate to the adult's responsibility to initiate and consolidate experience, and also to the new significance of sharing. Adulthood rarely coincides with maturity.

The young adult, the young person during the period from age thirteen to approximately twenty-one, has outgrown his emotional need for parents as central validators. By now all adults are, in fact, peers, in the interpersonal sense, regardless of differences in knowledge or maturity. After puberty, if the parent clings to his previous role and undertakes to exert authority by exercising the power to limit the frontiers of exploration and to impose the conclusions to be drawn from daily experience, this becomes a positive impediment to growth. Ready or not, it is the time that each person, in clear knowledge or obscurely, takes responsibility for his own development or deterioration.

The young adult now needs much actual information about how he and the world work. Any adult courageous enough to be honest about these matters would be remembered gratefully by the young adult. Other things are also necessary: education, a place of his own to use for sociability, the goods—clothing, car, hi-fi set—that are required locally as focuses for social interaction. This does not indicate that the parents should pauperize themselves to provide these things, but it does mean that the young adult's request for them should be considered as seriously as the request for equipment of any other adult member of the family. If the request must be refused it should be denied with honest regret, not with hypocrisy. The young adult may be drawn to high society or to Bohemia. This is as experimental, and as constructive, as an interest in Victorian novelists.

Young adult validations come primarily from contemporaries, secondarily from older friends who can genuinely relate to the young adult as equals. The parents themselves are particularly unfitted for this role. They will by now have contributed such validation as their level of growth permitted them, and further contribution from them is contingent on their own integration of further growth. The young adult's need for validation must then be in precisely those areas of experience which they failed to integrate in themselves.

But parents have an insidious investment in continuing as authority. Their prestige is directly involved in the young adult's performances, and besides, it is embarrassing to them to note and to have others note the basic deficiencies in their offspring, made manifest by the young adult's attempts to compensate for them.

Our culture gives parents easy rationalization for this neurotic attempt to extend the period of the offspring's dependency, while in some areas censoring the parent if he relates realistically to the young adult with respect for his new need for independence. The parent confuses power with obligation and uses his control of the economic resources, and social pressure, to enforce his concept of constructive direction on the young adult. In our culture, this process usually continues until legal maturity and sometimes for life.

The adolescent's active participation in this conspiracy is conspicuous. This stems primarily from his sudden awareness of unpreparedness for independent action and his resentment toward the parents for their past failures in facilitating growth. The resentment plays out in a drama of accusation—dependent or defiant. It may be expressed in overt overcompliance with parental standards, by destructive rebellion, by semi-suicidal activities, or pseudo-psychotic demonstration.

The other primary difference between growth in the child and growth in the adult lies in the area of sharing. The child grows primarily by being shared with. His urgent need is for validation of his own direct perceptions for the purpose of organizing, on as wide a base as possible, his own personal orientation to life. By puberty, for better or worse, the individual's orientation has been consolidated. After this, the young adult grows and expands his orientation partly by further direct experience and validation, but also now to a high degree by identification, participation, and sharing of the reference points of people perceived, more or less, as peers. The more people who will accept genuine accurate help from him, the better off he is. He grows by distilling their experience into his own. But the sharing must be emphatic to contribute to growth. If the young adult is cut off from the community of feeling with the other, he learns little, however accurate his observation.

Many parents oppose this process of sharing with people other than themselves as much as they oppose the processes of initiative and independent thinking. They resent the supplanting of their validation with that of experience with contemporaries. They derogate the teen-ager's sense of himself as genuinely important to the growth or needs of another person, and experience it as an added responsibility involving them. In this case they either burden him with their martyrdom, or prohibit the relationship, or cast a pall by taking over and offering their help directly to the other person involved.

The young adult's dilemma is that he is usually considerably confused about the real nature of his immediate situation in life, and finds it hard to come by an informed and honest opinion about it from anyone else. He is desperately aware of the inadequacies of his preparation for adult human interaction, but thinks, naively, that the way to compensate for them is to stay at home longer. It is as if he regretted having not paid attention in grade school, and so tries to reapply to first grade to do it over. He now has, in reality, access to new and better study methods at an adult level. He is still trying to get from the same parents, by reproach, things which they were unable to muster when they were younger and more flexible and he was younger and more appealing.

The parents are especially unfitted to be the object of the young adult's expansion of sharing beyond their frame of reference. The adolescent's legitimate reluctance to share responsibility for his parents' growth is bedded in a lifelong battle to keep the lines clear between his personal needs and the often grim, minimum necessities of keeping the parents propped up. His inevitable resentment of their failures at the moment of his first clear awareness of the damaging effect of these on him makes any apparent sentiment of great sympathy for the parent hypocritical. Nonetheless the adolescent is frequently preoccupied with growing up the parent, but this preoccupation is usually based in the hostile interaction of accusation and neurotic dependence.

Sometimes, as his horizons broaden, he sees his parents more in perspective and is embarrassed about their limitations. In either case, the parents' problem of growth, organization, success, and failure are no longer his legitimate concern either for improvement or for reproachful public exposure.

As our society is organized, the responsibility for financial support, though usually from parent to child, is occasionally a legitimate demand of the parent on the maturing offspring. But the attempt of the younger member to reverse the psychological roles and become a benign authority figure for the parent is at least as inappropriate as the reverse.

B) LEAVING HOME

Beginning with adolescence, the new dimension that appears as the major challenge in life is the assumption of the responsibility for one's own growth. One is from this time on the architect for his own patterns of living. He can no longer legitimately see himself as the innocent victim of other people's patterns of living. This is the central characteristic of the remainder of the individual's life.

This general challenge tends to be made concrete by particular, more or less universal, events, characteristic of the respective phases of the adult years. The first of these phases is the development of some facility in the formation of partnerships beyond the family.

The next urgent project is the problem of leaving home.

Much has been said about the psychological dependence of the individual on the eidetic mother, that is, on the image of mother and her values, permissions, prohibitions, etc., that is built into the young person's psychic structure. But the inhibiting effect of a continued close relationship with the real parent is often overlooked. A continuing intense relationship tends strongly to reinforce the original limitations and restrictions.

The central thread of the inhibitory tie lies in the assumption that at all costs, continuity and contact with the mother must be maintained. If this is a psychological necessity, then growth beyond any point which might involve a major break with her is automatically blocked. One does not allow oneself to become someone who might behave in such a way as to precipitate this break.

Whether or not true growth in the offspring would result in an actual break with the parents depends upon the flexibility of the parent in revising his concept of the child. The young adult tends to be somewhat less hopeful about his parents than is usually justifiable. He misinterprets parental anger for parental inability to grasp the reality of his new independence. In actuality, parents often understand their children much better than their children think.

Parents often make the grade after a stormy period of readjustment. They may even come to accept with relief their release from guilt feelings about the child, and their release from responsibility, which allows them to lead their own lives.

If the parent is entirely unable to grasp the fact of the change in his child, and at the same time fails to prevent that change, then any continued formal relationship between them becomes weirdly unrelated. Say that the child and the parent are sitting in the same room. They are chatting, but they are actually addressing themselves to entirely fictional characters. The parent may be speaking to the child that was, but no longer is. The child, on the other hand, may be speaking to his image of the parent as the terrifying mother that was, but no longer need be. The young adult may act as a child to fit the parent's image of him. If he does so he has, no matter how brief the contact, set a severe limit on his concept of himself as an adult.

Usually, the first step toward independence is the actual moving away from home: away to school, to a job in a different city, to a separate apartment, or into an early marriage. This physical separation allows some freedom to explore new patterns of living, without the daily reminder of how these new patterns would be seen by the parents.

The individual's next urgent necessity is financial independence. The dilemma of our highly technical civilization is that some sort of financial or educational backing continues as a reasonable need for a period of five to fifteen years after the time when the emotional dependence of the young adult on the parent has become obsolete.

After about thirteen, what the young adult needs from his parents is friendliness, a reasonable supply of facilities for the organization of his own life, and financial backing. This is no less true if his emotional development by this time is much retarded. The adolescent, including the one who is poorly prepared for these challenges of adolescence, is no longer an oversized child. He is a young, immature adult who must grow on his own terms.

Actual financial independence powerfully accelerates growth beyond the parental horizon. The prospect of complete financial emancipation is essential to the development of genuine psychological independence. The young adult may be presented with the choice between emotionally relevant but professionally premature independence, as opposed to prolonged dependence in payment for ultimate economic security. A young man may quit school at eighteen to make his own money, or he may go through medical school, just to make sure he can make a living when he finishes.

Alternately, the burden of the parent who must be supported as soon as the child is able to work also imposes a psychological hazard. Psychological continuity tends to be maintained partly to sentimentalize the underlying business arrangement, partly to obscure the joint commitment to the mutual resentment involved, and partly to obscure the young adult's limitation of his own growth.

Problems such as these can profoundly confuse the serious, valid choice of profession which had much earlier been clear. Conversely, choices made in terms of the unclarified emergency of the urgent search for independence may not be revised, or even recognized, until decades have passed.

Another frequent way of leaving home is by early marriage, with or without financial independence. A partner chosen to establish this adult status in the eyes of the parents will be chosen either in compliance or in defiance, or in some intricate intermixture of the two, of the parents' values.

Early marriage has advantages and disadvantages. On the positive side, it can have the meaning of underscoring, for the young adult and for the parent, that he has the right and the necessity to make choices for himself. Again, to the extent that it gives him social permission to establish a physical domicile away from that of the parents, it allows more growth room. Further, even where the unconscious determinants have made it necessary to choose a partner in the early marriage who is in the image of the destructive aspects of one or the other of the parents, it allows the young adult some possibility for dealing with the problem of hostile integrations on terrain of his own choice. This terrain may be subjectively less threatening, hence less paralyzing, than the old home ground.

The choice of partner for the early marriage may be someone who in important emotional dimensions is significantly different from the parent. In this case there is even more room for the constructive use of the first marriage.

On the negative side, the choice of an early marriage as the leverage through which one leaves home can itself constitute a major compliance with the dominant parental values, insofar as neither the parent nor the compliant young adult can respect anything less explicit than legal marriage as the minimum symbol for a grudging and uneasy granting of equal status. He needs the power of the state to validate his adulthood.

Far more damaging is the degree to which the early marriage

can constitute a flight from experience, a flight from the freedom to experiment in the art of choice, a flight from growth at the point of arrest. Early marriage, then, can become the vehicle for the obscure perseverance of the individual in a great variety of modes of the parataxic dramas of childhood with both partner and parent. The spouse may be used as the policeman who enforces the parental values. If the parent is envious and restrictive, he will, in this event, for all his verbal protest, experience the independence achieved by his child's early marriage as the loss of only a minor skirmish.

A dramatically bad first marriage may be used to implement a declaration of independence, with a long-range plan to make better use of the independence after the first relationship has fallen apart. Or, it may serve to invalidate even this rudimentary move, because it is perceived as demonstrating bad judgment. It is often surprising to young adults that the same parent who opposed his early marriage, and derogated it consistently during its existence, is apparently appalled by the divorce. This can be because the divorce implies an expectation of doing better next time. On the other hand, some parents are too stingy to allow their offspring even the freedom to be unhappily married.

Among the many frequent rationalizations used by patients and analysts alike to obscure the necessity of the person's leaving home, one stands out as insidious because of its sophistry. The patient who is having difficulties with his parents feels that it would be cowardly or escapist to move away from home. The courageous thing would be, in these terms, "to remain at home to work it out." The meaning of remaining to work it out rests in the commitment to retain the hostile integration with the parents.

An occasional variation on this theme is the person who, having left home, makes frequent visits home with the avowed intention of looking over the family from a fresh point of view. This can cover the intent to retain the original ties with the parents, and to preclude the possibility of getting enough distance between person and family for the person to be really able to see them in perspective.

C) CHILDREN

Having babies is the next major challenge to growth.

The fact of having a baby, and, for the woman, the process of pregnancy, is an experience which in its simplicity and its complications is a vehicle of a series of affirmations. Its practical effect on the growth process depends on the degree of either conscious or unconscious integration of these affirmations, as against the degree of repudiation.

At the most primitive level, having babies offers an opportunity to reintegrate many uneasy feelings about one's body image and one's physiological capacity. Both men and women are reassured to know that their hitherto untested organs function. For the woman, this reassurance may be offset by a queasy feeling of the distortion of her body. Primitive fears of physical inadequacy are reawakened, and the new experience allows for either their resolution or their reinforcement. Often, this gets projected into obsessive forebodings about the physical inadequacy of the infant.

Having a child of one's own is likely to be the most crucial declaration up to this point in the young adult's life of the possibility of his being equal with his parents. The young adult, now a parent, or about to be a parent, is caught between two pressures.

On the one hand, the quality and intensity of the infant's demands is such that the new parents, in order to meet his needs, are forced to experience themselves as independent adults. The pressure from the child is not only in the dimension of his needs but in the persistent confrontation inherent in the child's perception of the young parent as an adult.

The resistance against seeing oneself as adult derives from one's own accumulated deficits of competence in this role. It is often particularly focused against the acceptance of responsibility for oneself, and of the importance of mutual sharing, inherent in adulthood. It also highlights inadequate experience with intimacy at all levels.

These problems may be only remotely related to current attitudes of the young adult's own parents. On the other hand, his own parents' antagonism to accepting his equality can flare up violently.

Exemplifying this, an obstetrician has commented that the post-partum suicidal risk is directly proportioned to the difficulty of keeping a young mother's own mother away from the situation physically.

The affirmation in the potential sense of equality with the young adult's parents has further explosive implications. The latter themselves may actually be very immature, but the young adult's sense of equality with his own parents opens up the possibility, and therefore, incidentally, the inner terror of becoming different from them, of growing beyond them, of growing away from them.

Biologically, having a child is an affirmative experience in that it is productive and creative work. This is most directly true for the pregnant woman. It is often apparent in her sense of being complacently withdrawn into active, productive, fulfilling work.

We have pointed out that a central thread in the development of genuine human relatedness is a developing sense of belonging as a human being in a human world. In this context, having children has the implication of intense affirmation of one's humanness, and of one's direct participation in the biological and historical continuity of humanity. The affirmation in having children draws part of its intensity from the necessity that all human beings have, consciously or unconsciously, to incorporate the inevitable fact of death in the comprehension of life; to incorporate the brute fact of inevitable ultimate deterioration of this particular center of organization of life experience which is oneself.

Non-existence is essentially inconceivable and can only be absorbed by the affirmation of existence in aliveness and in the expanding quality of experience. The essential adult incapacity to conceive of non-existence is rooted partly in the nature of the process in infancy and early childhood, of the development of the sense of self and therefore of the sense of existence. The development of the latter depends on the continuum of validating response by the significant adults, as likewise does the infant's realistic prospect for survival. Any effort, therefore, to conceptualize non-existence evokes intense, inchoate fears.

The feeling of existence also bears some relationship to the individual's experience or expectation of some effect or impact of himself in the outer world. It cannot be validated purely in terms of the intensity of his subjective experience. The fear of death is often intensified by the feeling of not having lived. One lives in trepidation not only of incomplete development of subjective experience, but of disappearance from the world—leaving no trace, thereby invalidating the fact of life.

The child, then, is an affirmation of continuity, of impact and interaction with the physical world. The act of having a child is an affirmation that one has the right to think of his personal influence as a legitimate contribution to the future, and to accept himself as a legitimate product of the past.

Through the infancy and childhood of one's children, the invaluable gift from the infant and child to the parent is the opportunity to learn a great deal of what life is about, to be confronted with the beginning elements of life and growth through the closeness of the experience. The opportunity is, therefore, for the parent to come closer to his own inner experience, to get more directly in touch with the dissociated aspects of himself, to become aware of what he has missed, what he has avoided and repudiated, and what he needs for his own growth.

The crux of the failure to integrate this experience lies in the dissociation of empathy and identification with the newborn infant and the growing young child. Partly under the rubric of assuming adult responsibility, the parent discards as unusable his own childhood memories and those memories specifically re-evoked by the presence of his child. The manifest content of such dissociation may vary from egocentric self-recrimination for real or potential failures, to bland, defiant, or defensive assumptions of the presumed role of the parent. In the presence of massive dissociation, the subjective emotional tone can vary from intense, suicidal depression to apparent elation. In either case, the opportunity for sharing the joys and miseries of infancy is avoided through the obsessional preoccupation with one's own level of achievement in a performance.

The most common implementation of this repudiation is the incorporation of the grandparents' values in the child rearing. The child is made into the kind of baby who would have been "acceptable" to one's own parents. Sometimes it is even more starkly expressed when the child is turned over bodily to his grandparents for rearing. Thus the individual apologizes for having defied them, or reproaches them for their inadequacies.

One sees many compromises in which the young adult attempts simultaneously or alternately both to give the child what it needs and to get from the experience what he needs, while at the same time trying to be a compliant son or daughter to the grandparents.

The young parent may recoil in defeat from the challenges of the experience. The special hazard in such a defeat, as distinct from the hazards of other incomplete experiences, lies in the great difficulty of walking away from it. Deficits in the mothering function are cumulative unless exceptionally adequate provision is made for alternate care of the child, and the child remains for fifteen to twenty years a reproachful monument to the mother's own defects. Full undoing directly by the parent of the damage to the child would involve even more capacity for self-awareness in the parent than was needed when the child was very young. The growing child takes areas of effective relatedness quite for granted. For the purposes of his own growth he thus throws the spotlight on deficiencies, original or residual, of his parents.

Such intense confrontation often intensifies discouragement, anger, obsessive rationalization, and further moralistic justification of and compliance with the original restrictions with which the parents grew up. If the child grows beyond the level of the parent's interpersonal maturity, the parent is additionally handicapped by the fact that adequate validation of this would involve not only some reawakening of his or her early terrors, but also the slow, painful accumulation of new experiences.

During the period of child raising, the parents try to get ahead as best they can with whatever problems they were personally trying to solve before the advent of the child. It may be the accumulation of property, group relatedness, or the search for compatible individual partners. However, growth tends to go on in a relatively structured situation in which there is always the counterpoint of having, to the limit of one's capacity, to pretend to maturity "for the children's sake." If the parent can be aware of his basic unpreparedness, and of his irritation with the child for having been the agent that exposed it, the child can be released from the burden of guilt projected on him by the parent. However, it does not make up for real deficiencies beyond this, even though it can release the parent and the child from each other to a certain degree so that they can go ahead with growth in their respective projects.

Our culture's usual attitude toward unmarried motherhood exemplifies failure to relate to the concrete growth potential that the experience of having a child involves. Almost invariably, both the lay and the professional attitudes label unmarried pregnancy delinquent, neurotic, hostile, or destructive. In the context of cultural prescriptions it is indeed more socially convenient to marry before one has a child. What is seriously overlooked, however, is that often growth and reintegration through having a child is a necessary preliminary to developing enough feeling of self to be able to choose a permanent partner in marriage. There are situations in which unmarried motherhood is the most pertinent channel through which a woman can grow. Where this is so, her pregnancy is an expression of courage, humanness, and the desire to grow.

D)   THE MIDDLE-AGE CRISIS

Several factors demand reorganization in life during middle age. Emotional responsibilities of the parents come to an end as the children leave home, and the ongoing financial responsibility becomes finite and predictable. Women lose the capacity to bear children. One sees oneself with the passage of time as definitely adult, and, moreover, because of the advance in life expectancy, with many more years of life left. Physical and economic independence is usually a fact by middle age. However defective, working styles for coping with the world have been developed. Indeed, the individual's central problem has shifted from survival to enjoyment. But accumulated discouragement may now establish a powerful impediment to the person's experiencing the situation as an opportunity for growth.

When the children leave home, the parent can no longer

easily use them as an alibi for failure to get on with his own growth. The superimposed responsibility of raising the child has vanished. In order to avoid responsibilities to themselves, parents often maneuver actively to retain their children.

The child may have acted covertly for some time as parent to the parent. This relationship may now become overt, as the parent attempts to maintain contact with the child by trying to force the child to pay the priceless "debt" that he "owes" to his mother and father. The parent exaggerates his age and decline. He minimizes his physical and economic resources. Contrary to actual actuary tables—he projects the aura that he has not long to live.

The transition to an expansive middle age is difficult. After twenty years, many people have gotten out of touch with the techniques for learning, for socializing, for meeting new people. There is the pervasive mythology that the need for tenderness, necessary for any expansion of the personality, should have been outgrown by this age. The result of this is that tenderness must then be sought in covert situations. The freedom of individual action is limited by economic and conventional chains. One may be embarrassed to find oneself in competition, and even at some disadvantage, with people who have twenty years less of life experience.

The culture, on the whole, ridicules the idea of growth during the middle years. "No fool like an old fool" admirably expresses the attitude that every middle-aged person expects to meet if he makes an intimate friend, falls in love, learns to play the guitar, goes to college. These are the "dangerous" years for a man. For a woman, especially in terms of sexual experience, those who are interested in growth can at best be "understood." Mostly her experiments are regarded as inappropriate, immature, laughable, irresponsible, or disgusting.

These factors operate in the context of the ever-present dynamic of the fear of breaking with the parent. Middle-age fears of exploring and feelings of self-revulsion are directly continuous with the original infantile emotions. But now it is much harder than before, because of the many intervening years of role-playing as the mature one, to admit to oneself that one's basic psychological task remains that of leaving mother.

The marriage is now subject to intense re-evaluation. Social and financial respectability have been achieved, and the relationship has frequently outlived its usefulness as a juvenile alliance. The children have been born, raised, and have gone away. Now the challenge that looms silently between the middle-aged couple is for them to determine if the marriage should go on for valid reasons, with the prospect that each of the partners will probably live for the next thirty or forty years.

This question is simply not permitted to be raised in many cultures. Some consideration of it is possible in our society. Still, if consideration reveals a grossly inadequate marriage, powerful pressures call for the individuals to resolve their problem unhealthfully. Usually their own despair conspires with the pressure of public opinion to derogate as illusion their potentialities for growth. They repudiate the possibility of making a better marriage with another partner. They are caught up in the pseudo-dilemma of the socially approved alternatives: the isolation of an unloving marriage, with its chronic inner loneliness and deterioration; or the less approved isolation of choosing to live alone.

If the couple separate, many sources foster the notion that the fact of separation invalidates what had been valuable, appropriate, and constructive in the earlier phases of the relationship. This sense of defeat increases discouragement. Obscurely, the middle-aged become afraid of exposing the desire to live, no matter how the fear is rationalized.

The menopause, which may make a woman feel inadequate or sometimes even mutilated, often has other psychological implications. Culturally, the menopause is thought to depreciate a woman's sexual desirability. Too, future marriages must be made only with men who do not want children.

Still other pressures accumulate on the middle-aged woman. The culture progressively limits her chances to marry after thirty, for while the man is allowed by convention to marry a woman his own age or younger, she can respectably marry only a man her own age or older. Too, while the wife has shared the prestige of her husband's professional advancement over the years, she has usually abandoned her own serious professional development after marriage. All of these forces tend to keep the woman more committed than the man to the going relationship, however stultifying it is.

Thus the woman exaggerates her inadequacies, professional or feminine, in order to maintain the marriage. Her husband often goes along with this derogation of her, both to avoid social censure, and as an alibi for his avoidance of the challenge of growth for himself.

Social prejudice restricts still other aspects of middle age. Adolescents are expected to be changeable in their ambitions, but a forty-year-old man cannot change his professional direction without being considered irresponsible, or oddball. A year of travel is considered healthy at twenty, even necessary in certain European cultures; but it is "psychotic" for a middle-aged man to take a year off to experience something in the world in himself. He is expected to apologize, indeed, if he decides to quit working such long hours so that he can have more time to sit quietly and invite his soul. A shift in the kinds of people with whom he associates is considered either as disloyal or insulting.

Middle-aged growth is culturally restricted through another device—the cultural requirement for apparent expertness. No one lives long enough to be anything but a student of such questions as birth, death, sex, love, war, and marriage. But the culture requires the middle-aged to be teachers—seers—about these subjects and many others. Thus the tendency is for the middle-aged to restrict themselves to concentration on activities in which they can feel relatively confident of displaying expert knowledge without being challenged or discovered.

The secret illusion of eternal youth wears thin during the middle years. Time presses, the future has arrived. Promised projects can no longer be postponed—they are noted to have been already accomplished, imminently to be tackled or jettisoned. One can be a promising young man until one loses his hair. Now repressed needs clamor for satisfaction, and substitutes for them taste increasingly of ashes. One does not want to die without having lived.

Resignation could nevertheless hold the fort if one had only five or ten years left to live, but today one is statistically assured of having twenty, thirty, or forty years to live—too long a time to sweat out in the same old foxhole. Modern medicine has bankrupted the excuse that there is not enough time to make the effort involved in serious personal reorganization in middle age worth it. Now there is plenty of time, energy, and internal resources. Increasingly, one must repudiate life actively in order to avoid using these resources for growth.

The accomplishments of medicine in the increased life span create for some an additional complication—one's parents live longer, too. The secret fantasy that one will grow up immediately after mother and father are in the grave becomes less than satisfactory if one realizes that by that time he may himself be more than sixty years old.

The outcome of middle age depends on the ratio between one's rate of growth and one's rate of deterioration. Time and the culture favor the latter, and standing still is impossible. Too, we have much investment in the retention of long-held mistakes. Yet growth in middle age—accelerated growth—is as possible as it is vital.

The decision to grow or to die is the crisis of middle age. Resignation often masquerades as maturity after forty, but resignation is not growth; it is the person's decision to progressively write off his life as a waste of breath. Therefore, many people die during middle age, although they may wander through the world like zombies waiting to be buried decently for thirty more years.

E) OLD AGE

Little psychoanalytic data exists about the psychological problems of old age.

Few analysts attempt to initiate serious treatment of people more than sixty years old. Conversely, very few people in this age group seek treatment. Those who do are unlikely to be oriented seriously toward the goals of growth and major changes in the personality.

Our cultural cliché holds that growth does not take place in aging years. Thus the quality of life in the later years tends to be set, to be very strongly determined by the nature of the outcome of the middle-aged crisis. If this particular item of the cultural strabismus is not accepted, there remains little cogent reason not to assume progressive expansion of the personality, and even potential reversal of antecedent trends, as a real possibility until the very end of life.

This expansion could take place in any area that does not involve major physical exertion. For example, nothing prevents the older person from engaging in hobbies, in group relationships, in friendships, or in productive work. The expectation of life's end can sometimes dissolve a lifetime shyness by exposing it as a meaningless preoccupation. Too, by this age, the handicap of having to cope with one's own parents is almost universally outlived. Some people do not succumb to resignation in middle age and do, in fact, continue to grow virtually until death.

There is a cultural lag in social response to the fact that a very considerable and increasing percentage of the population is in the old-age bracket. This helps to retain the misconception that old age and deterioration are synonymous. The old person is still, to a large extent, unexpected, unwelcome, and unprovided for. Thus, he still tends to succumb to the cultural conception that his physical age inevitably involves emotional deterioration.

Reversal of lifelong trends, in theory, is both possible and profitable right to the point of major physiological deterioration. But most analysts and almost all friends and acquaintances are insincere with people whom they consider to be old. All the factors antagonistic to the possibility of a middle-aged person's taking growth seriously continue into his old age, now in exaggerated form. Where intense relatedness between middle-aged contemporaries may be considered ridiculous, in old age it may be derogated as grotesque or senile.

Where negative choices have been made by the person in middle age, economic or physical dependence tends to become organized into more manifest patterns of hostility. Subjectively, the inner trend is for life to be more and more on a literal survival basis as the archaic fears of death fuse, now, with the imminence of death.

The frequent result is that much of the person's stake in maintaining face tends to fall away. As a corollary, the central paranoia tends to express itself much more frankly than before.

Often then, the accustomed patterns of the security operations, including the psychosomatic ones, are sharply reduced, and, in the relatively unhampered expression of the deteriorated state, the old person can undergo what seems like a malevolent transformation with respect to other people.

A problem that remains to be disentangled is the complicated relationship between the senile psychosis on the one hand, and on the other, the terminal processes of psychological deterioration in the sense of the conventionalized "giving up on life."

F) THE IMMATURE ADULT

The immature adult is the normal product of the process of acculturation in our civilization. At best, in present known cultures, the individual is still struggling centrally with either preadolescent or adolescent problems, if he is not primarily stuck in the rut of antecedent ones.

The immature adult is not, nonetheless, an overgrown child. The difference between adult and child does not lie in the importance of the need for tenderness. Both need tenderness. The difference lies in another dimension. The child is relatively ready and available to growth, but is unable to control the organization of his life in such a way as to get enough tenderness to make it possible. The adult is a free agent. There is as much tenderness available in the world as he is capable of making use of if he would use his full facilities to seek it out. His problem is that he is constrained from looking for it, and, even more, should he accidentally stumble onto a situation in which it is available, he is afraid to use it. He experiences the availability of tenderness in the paranoid mode.

The immature adult has major defects in his emotional, perceptual, and organizational capacities. He is limited to some degree in all aspects of interpersonal development, and he is, more specifically, stuck at some point along the line. A major portion of his personality is committed to seeing himself as he was seen by his parents until about age two and a half, and in moments of crisis he returns more fully and more literally to this perception.

Performance beyond the point of arrest in any line of development is either subversive and covert or obsessive. Antecedent, more fully-integrated functions remain subject to disintegration under the impact of anxiety.

At the same time, the immature adult possesses covertly much greater capacities than he can conceive of himself as having. Due to exposure to alternate experience, he is potentially, of necessity, more competent, more creative, more loving, and more alive than was consistent with the family tradition. However disguised the search, he is constantly and accurately looking for opportunities to make up gaps in experience, to find situations favorable to growth, and to get more satisfaction. His handicap is that he has the assignment of getting there without being aware of where he is going.

From adolescence on one is as big as one's neighbors; one is actually responsible for one's own patterns of living. However, each of us manages to postpone this acknowledgement of adulthood in various aspects of living for many years. The fact of adulthood is entirely independent of the level of relatedness integrated; it has nothing to do with the level of maturity.

In a literal sense, all of the people in our civilization at the present time more than fourteen years of age are immature adults. Each has his own unique and individual patterns of early experience and his particular methods for organizing later experience into specific stencils to define areas and degrees of growth, and of restriction of growth.

Part Three

LOVE AND HATE

I

FEAR OF DISORGANIZATION

A) THE CENTRAL PARANOIA

In addition to the identification of the orderly sequence of development and of the usual developmental eras, personality can also be described in terms of certain individual patterns that begin in infancy and tend to persist throughout the lifetime. The specificity of these patterns is interlocking and unique for each individual; but this pattern can be separated into several interwoven universal trends.

The infant and very young child are totally dependent on the surrounding adults for life, validation, and tenderness. Conversely, the mothering one has the power of life and death, the power to organize the self-system of the child into relative sanity or to disorganize it into non-functioning and, in effect, insanity, and the power to withhold tenderness. For the infant, then, the dangers of death, insanity, and loss of love are appropriate and inevitable concerns.

The infant's feeling of security in these dimensions will depend on the degree of unhampered availability of physical cherishing, of validation, and of love. To the extent that these functions are provided deficiently, the infant acquires a lifelong inner conviction of being in imminent danger of death, or of insanity, or of total isolation from affection. These three functions provide powerful leverage for the control of one person by another. To the extent that the mother operates irrational control, these then become, automatically, in a variety of patterns, the core of her power operations over the child, whether openly malevolent or ostensibly for the child's own good.

To the extent that the child is operating in this context, it becomes crucial for survival and growth for the child to learn to disguise and deny, in varying degrees, his real or subjective fragility and vulnerability. He devotes his life to the mastery of the fear of death, to the avoidance of anxiety and disorganization, and to the repudiation of the need for tenderness. Thus the child must try to become simultaneously invincible, invulnerable, and invisible.

These three trends constitute the central base of the power of the mother in the struggle between mother and child for her control over his growth. Therefore they constitute the primary underpinning of the restrictive aspects of the self-system. To the extent that they continue to operate in the adult, they represent his persistent allegiance to the mother's prohibitions and restrictive values. Therefore these are the central propositions of the central paranoia.

B) FEAR OF DISORGANIZATION

The self-system is an internally consistent logical system which includes certain premises incompatible with reality. When a particular aspect of reality which does not fit this logic obtrudes itself with such force that it cannot, for the moment, be denied or repudiated, the whole structure is temporarily shaken.

Anxiety is defined as the subjective experience of disorientation, disorganization, discontinuity of experience, dizziness, and nausea, which results from this event. The nausea, as with seasickness, is the usual subjective reaction to the impact of a shifting frame of reference. The disorientation results from temporary shakiness of the frame of reference within which one has always operated. The implied reorganization inevitably results in a temporary re-evaluation of the data of current as well as past experience.

The resolution must either be the expansion of the self-system to include the new experience, or a more thorough defensive reorganization to rationalize it away in terms of the old logic, thereby repudiating the experience. In effect, some section of the integral personality has thrust itself forcibly and implacably into the self-system, and the logic of the self-system must be reorganized either to incorporate it, or to repress it more firmly.

The fear of anxiety does not lie in the fear of discomfort, per se. Anxiety involves regression, disintegration of the modes of organizing experience, and therefore disorganization. Severe anxiety involves disintegration of the core organization of the self-system. It re-evokes the infant's experience of chaos and non-existence which occurred with the mother when she was in a chaotic state. This state of confusion is not inherently so unpleasant as to account for the extent to which life is almost universally organized for the avoidance of anxiety, even at the cost of extreme sacrifice of satisfactions. The experience of danger is rooted in the situation in which the original organization was set up.

It is important to keep the distinction clear between anxiety and fear. Our definitions here differ sharply with the usual distinction between anxiety and fear, by which fear is designated as the response to a real danger, and anxiety to an unreal danger.

By anxiety we refer to disorganization of the self-system with consequent disruption of the stream of consciousness. As the self-system is the institutionalization of the connectedness with the mother, so anxiety is the emotion ensuing from a threat to that connectedness. Subjectively, this threat is experienced as disorientation and disorganization, but not inherently as fear.

Fear, on the other hand, is the appropriate response to a situation that is perceived as dangerous. The fear that almost universally accompanies anxiety is the hallucinatory reliving of the expectation of danger from the mothering one, cast in terms of the complete dependence of the small infant on the mother's good will. It is thus a composite of the fear of death, and the fear of loss of tenderness. The emotion is that of a small child waking from a nightmare. The content is characteristically borrowed from the contemporary situation, accurately misinterpreted to rationalize the hallucinated inner feeling.

The persistent association of the two feelings comes partly

from the fact that both are underpinnings of the central paranoia. Any growth experience which challenges the central paranoia to some degree revives and evokes all of its central propositions, including the fear of death, and the fear of disconnectedness from the mother.

They are, in fact, separate vectors. They can vary widely in their proportions. It is possible to feel intense, parataxically rooted fear with very little disruption of awareness. Conversely, it is possible to be very disorganized with relatively little fear. This happens on those fortunate occasions where the fear, once recognized and felt as parataxic, loses its contemporary force and becomes dissipated. The fear is based on the childhood expectation of retaliation from someone on whom one is totally dependent, and in adult life, the situation of such vulnerability is almost never duplicated.

There are many archaic roots to the expectation of the mother's hostility in response to the acute disorganization inherent in the adult growth process. In infancy, an act of self-assertion that threatened the mother was likely to be perceived as defiance, and so to evoke the mother's rage. In the adult, the perceptions that flood into consciousness as a result of the breakdown in the repressive machinery integrated with her are now expected to evoke hostility from the mother just as they did originally.

The state of disorganization, under the impact of mutual empathic anxiety of mother and child, originally increased the infant's dependency and need for tenderness, just at the time when the mother was least able to fulfill it. The infant's helplessness then tended further to drive the mother to desertion or attack. The infant's anxiety, per se, then made the parent angry, anxious, or distant.

The disconnectedness from the mother inherent in the state of disorganization of the self-system is, in the frame of reference of infancy, equivalent to a sentence of slow death. The infant has no choice of mothers, nor has he any concept of alternatives. His only experience is that both the basic means of subsistence and his capacity to use it stem directly from her and if she severs her relatedness to him, they go with her.

If the mother is chronically angry, consciously or unconsciously, an additional fear of temporary disorganization is set up. It is never safe to be totally in the physical and psychological power of a person whose predominant underlying emotion is anger. Even temporary lapses of alertness occurring in private, if beyond voluntary control, are taken to indicate that they could occur in the presence of the enemy. The lion tamer is never absent-minded in the arena.

In the state of temporary disorganization, the adult feels exposed, defenseless against the hostile maneuvers of the eidetic mothering one. If the mother's hostility was felt to be ever present, at best latent most of the time, the current state of disorganization is expected to offer her unrestrained opportunity to exercise it.

Further, if the mother had been massively subject to the malevolent transformation prior to childbirth, the infant's defenselessness may have served as a stimulus to trigger the mother's sadistic or vengeful impulses, or afford her an opportunity to consolidate her strategic position in the perpetual battle to crush the spontaneity of the infant. She might have used his disadvantage either to demote him from his hard-won status as a going concern, or to extract information about his inner thoughts, in order to improve her methods of control, or to identify his most secret and cherished wishes for use as blackmail at some opportune moment.

The self-system was organized in order to establish and consolidate the connectedness with the mother. The infant experienced many amorphous perceptions, and, in a diffuse or explicit way, reacted to them. Some of these perceptions served to bring him into closer interaction with the mother and were thus validated. They became integrated into functions which thereafter were available to him, even in her absence, for purposes of his survival, wellbeing and growth. Others of these reactions were sensed by the mother but represented dissociated areas in her personality. These resulted in sudden and painful disconnectedness from her.

Growth can take place fully only in those areas which the parents have some minimal capacity for sharing. Other directions of growth die of attrition. They are blocked both by the passive absence of response and by the active thwarting of the function.

This active, anxious repudiation is what is often referred to as envy. To the extent that the adult has lost his own capacity for perception and satisfaction and participation in the world, to the extent that things have become clouded over, dull and fuzzy, he experiences pain at perceiving the enjoyment of spontaneity and aliveness.

The expression of this pain can range from slight irritation to the mother's intense nausea at the infant's feeding habits or at his joy in the color and texture as well as in the taste of foods. The prohibitive effect of the envy tends to vary in accordance with the intensity of the unconscious component of the envy. The parent's presumed preoccupation with the infant's education can be used as the excuse for being too busy to share momentarily the infant's delighted discovery of slugs under the woodpile.

If the mother's perception of the realities of the situation is cut off, she tends to substitute alternate hypotheses or interpretations. As we have said earlier, a response, be it an emotion, a thought, or an overt action, is roughly appropriate to a given situation as perceived by the person, adult or infant.

For example, the mother may, in her intense anxiety, become extremely punitive on witnessing her child putting his finger in his anus. The infant perceives the anus as a place whose curious sensations demand further exploration. The mother, on the other hand, perceives the anus as Satan's earthly residence. All of what follows between the two is the result of this fundamental discrepancy in perception.

The most universal operation for the limitation and repression of spontaneity by the adult goes under the guise of acculturation for civilization. The mother's educative efforts are rooted in two opposite sources, in her productive capacities and her needs for security. These opposing functions of her own self-system show themselves in the educative structuring of the child's world, and later, his self-system. Absolutely vital to the child's development is the validation of his own direct perception and its organization, as is the teaching of the means of communication and modes of behavior which might be expected to have predictable results in a given social milieu. Since we live in a grossly irrational society, it is also a part of the parents' responsibility to teach the child the rules of the game. The fallacy lies in the teaching of these latter, not as if they embodied modes of communication in a difficult world, but as if they held some absolute inherent truth, some ultimate reality of value in themselves over and above their social necessity.

On the other hand, the mother's role as educator affords her maximum opportunity to thwart the child's full development. The impulse for this perversion of her legitimate function derives from her own needs for security, and shows itself proximately in the stultifying and respectable overorganization of perception. This can be effected through either the moralistic, the investment in insincere rituals, or the repudiation of operations as either socially undesirable or socially ineffectual. Most pervasive of all is the ostensibly well-intentioned attempt to facilitate and participate in the child's world, carried out with subtle mistiming and misperception so as to impose the adult's prejudices, conventions and overstructuring on the growth process of the child.

The repressive aspects of the mother operate through her inability to perceive the infant's direct perceptions and the subsequent imposition on the infant of her own. The intensity of her insistence on perceptual acquiescence is in direct proportion to her investment in power operations, her need to control the other person.

The level of organization in the self-system of the function of physical orientation is contingent upon the development of its component elements. These include the earliest developed physiological, perceptual, emotional and motor functions, the subsequent experience with possession and manipulation in the object world, and the evolving concepts of time, space, and the body image. As each system is organized it incorporates the relevant prior data. All of these systems contain both the correct and the fallacious perceptions resulting from the interaction with the mother. As each system is organized it acquires a certain stability and self-perpetuating quality and is subsequently more difficult to reorganize by isolated contradictory perceptions. This also acts to stabilize the original perceptions, correct or not. Later defects in the level of organization of the self-system can be reflected in the capacity to formulate a complicated body of data. For example, the capacity to grasp one's own life plan, that is, to make explicit and conscious the formulations of the historical attempts at breakthrough of the integral personality, is facilitated by the availability to the self-system of competent functioning in respect to time and space.

It must be borne in mind that the tightness of the organization of the self-system has several corollaries. It facilitates the conscious ordering and organization of complicated bodies of data about oneself or the external world. This process may be used either constructively or defensively.

The self-system includes a part of the experiential data from all levels of organization. This includes the primary data of perception and response and their progressively more complex organization under the impact of time and space perception, the capacity for conceptualization, and the condensed organization of the self-image. To the degree that these are incorporated into the self-system, they are available to consciousness. Defective organization in any of these areas does not preclude considerable elaboration of the function in the integral personality, however thoroughly it may be cut off from awareness. On the other hand, the person's experience can have been so restricted that some functions are registered in the integral personality only in a rudimentary way. A frequent and striking division occurs in the person whose direct perceptions are available to the self-system, but in whom organization of these functions into the life plan goes on in the integral personality and must be kept out of awareness. Here the apparent preoccupation with superlative mastery of the perceptual elements is the upshot of the repeated attempts to break through the prohibition against conscious organization of perception. Blockage of the function of perceptual organization, of conceptualization and formulation, often results in the flourishing of the antecedent function of direct perception, as, for example, in certain painters and musicians.

The organization of the self-system can be loose and vulnerable in certain of its dimensions, and the self-image foggy. The damage that results in this vulnerability may derive both from the defective development of particular capacities, and from the defect of the organizational function per se. Both can result, where the damage is considerable, in a persistent inner feeling of disorientation.

This inner feeling of disorientation becomes a lifelong preoccupation and a source of perpetual fear. The fear is experienced as the danger of losing control, the feeling that life is always just about to get out of hand. The external patterns may be frankly disorderly or highly obsessive. There is a tendency to overstructure situations, often in terms of perpetual crisis, real or imaginary, in order to simplify their interpretation. There is often a strong reaction to shifts in the level of consciousness. One may be drawn strongly to alcohol as a method for undercutting the obsessiveness or alibiing the disorganization. Or one may cling tenaciously to a high level of focal awareness or of emotional control.

Situations inherently embodying confusion are characteristically experienced as acutely distressing. Large crowds may be felt not as dull but as actively painful. New areas of intellectual pursuit are avoided because the initial phase of unstructured ignorance is not tolerable, even as a transitional state. There may be a fear of being physically alone, because, in the absence of external demands, one is afraid one would perceive the defects in one's structural orientation. Minor errors in perception, misinterpretations of a friend's remark, losing a familiar object or taking the wrong road are reacted to with greatly disproportionate anxiety or anger. They are interpreted as omens that the control is slipping, that one's inner disorganization is about to be exposed. There may even be the paranoid interpretation that the other person, or object, for that matter, was malevolently trying to show one up.

The fear of becoming aware of one's difficulty in organization and orientation becomes equated in the adult with the fear of psychological disintegration, the fear of going crazy. This irrational fear, if stated more accurately, is the fear of being revealed to have been always crazy. The expectation is that once the pretense and persona of sanity are broken, they could never be put back together again. This, then, becomes a base for desperate avoidance of acute anxiety, thereby placing a severe handicap on the process of reorganization and hence of growth.

In the service of this avoidance of anxiety, a most frequent compromise is a preoccupation with organization in the service of defense. This is often implemented by keeping the spotlight focused on the external world. The perceptual and conceptual apparatus can become highly developed. The formulation of inner experience, including the use of dreams, hypnogogues, and associations; or alternately an emphasis on philosophy, can be used as defense. In any case, the process goes on with paranoid construction. The purpose is to size up the world, to be protected against possible attack, to make out.

The element left out is the person himself: his needs, his sensitivities, his longings, and his direction. He is largely cut off from the use of his capacities to organize the inner and outer worlds for the maximum satisfaction of genuine needs. He tends to take the world as given. He may be avid for certain types of experience, or for expanding his frontiers of perception and mastery, but what always remains foggy is the relatedness to his inner needs for tenderness, for being frankly who he is, no matter how inefficient or depressed or vulnerable. Such inner needs are considered irrelevant or threatening to the project of keeping going.

The self-image is the condensation of the structure of the self-system into a simple, though covert, operating hypothesis as to who one is. It is the formulation of the primary organizing principle of the self-system.

The self-image defines more explicitly potential areas of competence and incompetence. Incorporated into it is the limitation on the development of tender interactions. The characteristic response to danger is indirectly implied in the self-image, as reflected in the attitudes of defenselessness, helplessness, paralysis, fear, or impotent rage. It is not easy to identify, but it is reflected in dreams, Rorschach images, and fantasies. The self-image can be inferred from the overall pattern of behavior. A person acts as if he were a "three-year-old with droopy drawers," "sub-human," a "four-year-old in a battle with mother," "a trained poodle." It can be reflected in reaction formation, as in the extreme fastidiousness which both hides and expresses a profound self-disgust, or as in conscious grandiose fantasies and pressured perfectionism, obscuring the inner feeling of inadequacy.

For a given individual there can be a variety of derived images which suggest flexibility in the self-image. This perception is probably fallacious. The apparent flexibility lies in one's capacity to keep the self-image out of focal awareness and to perform temporarily in terms of the external situation, without reference to one's inner evaluation. When the two collide, either the performance must be invalidated or, with accompanying anxiety, the self-image must expand.

It is generally assumed in lay and professional circles that the usual degree of avoidance of the experience of disorganization is more or less natural and appropriate. A part of the basis for this avoidance of anxiety lies in the persistent, reactive denial of the original defective integration of the self-system. A more central root is the implied break with the mother, and its ramifications. A basic break with any central validator places in jeopardy, until other validators can be found, all of the functions originally organized in the context of this person.

Thus for the infant the break with the mother threatens not only future growth and the means of subsistence, but also the performance of the primary physiological integrations which are indispensable to physical or psychological survival. The adult feels parataxically that, for no identifiable medical reason, he is going to die of paralysis and disorganization of the primary functions of breathing, circulation, and temperature control, which are necessary to survive any interval of transition. In this disintegration of physiological functions organized into the self-system in connection with nurturing figures, the fear of death fuses with the fear of disorganization as the central hazard of reintegration of the need for tenderness.

In the moment of disrupted connectedness, whole segments of experience feel as tenuous and tentative as they were at the time of their initial organization into the self-system. The subjective experience of this disconnectedness is the blankness and disorientation which is the gist of the catatonic episode.

The disconnectedness also raises the question of non-existence. One knew that he existed, originally, in interaction with the mother who made life possible. A break with her raises the question of whether she has the power to take his existence with her. This fear is reinforced by feelings of unfamiliarity with familiar objects and of depersonalization in the presence of known individuals. Either oneself or the surrounding world, or both, feel as if they had suddenly become exposed as having been always unreal.

Any important expansion of the self-system must jar the self-image, resulting in feelings of unreality and depersonalization. The self-image was organized to incorporate the concept of Me, Good or Bad, and to relegate dissociated experiences into the category of the Not Me. To the extent that the breakthrough from the integral personality incorporates dissociated material, it discredits the life-long self-image and substitutes the feeling of Not Me.

The paranoid is a person whose self-system is streamlined to fit his central paranoia. In this case, a vast sector of direct experience is either cut off or perceived with gross distortion. Thus the paranoid mother is particularly unable to validate the child's perceptions in these areas. Further, she has an intense investment in imposing her views on the people around her in order to avoid the threat of being confronted with contradictory data.

Her child is especially vulnerable to this operation, since the mother is to such a large extent his world. And, especially to the extent that the mother is paranoid and experiences contemporary relationships as threatening, she must become exceptionally pre-occupied with her child as the major part of her world.

The paranoid mother's pathology necessarily involves a profound cutoff from people, a nurturing of hostility, and a repudiation of affection. Unrelatedness of this degree is considered by the world, and is even marginally perceived by the child, as insanity. This sets up in the child a confusion of incompatible categories. Spontaneity, especially if it involves a move toward people, is classified by the mother as impulsive, wild, destructive, or crazy. Acceptance of her value system, on the other hand, would lead the child ultimately to a degree of isolation, eccentricity, and deterioration equal to hers.

Caught in this dilemma, the child tends to consider himself and his natural impulses as abnormal, and his mother's as normal. Either way he feels doomed. Both his mystical identification with mother and his empirical knowledge of her crucial influence in his life combine to emphasize his expectation that his potential for living is inextricably limited by hers. The relationship between them characteristically centers in a fight for power, in which the battle-field is the validation of perception. Neither can rest until they both have come to some degree of formal agreement. To agree to disagree openly is inconceivable. This is one of the sources for a lifelong commitment to curing mother or her contemporary repre-sentative, in the attempt to win freedom to grow.

The paranoid mother cannot tolerate a difference of opinion. Thus the child does not dare to express one, unless he feels that he can convince her of it. He becomes, as a result, both litigious and secretive. Passing thoughts cannot be shared until enough evidence has been collected for them to stand up in court. At the same time, the rich secret life that the child develops retains an unreal quality because it can be, at best, exposed only in fleeting and fragmented form to real contemporaries.

In manifest or subversive form, the controlling and restrictive parent uses the threat of her capacity to temporarily disorganize her offspring's thought processes, and hence to induce legal insanity as a lever of power, to restrict any reasonable degree of divergence in the child. In this struggle for power, the danger of such induced insanity lies not only in the feeling of horror associated with it, but also in the danger, often legally supported by civil authorities, of the mother's being able ultimately to establish a legal or semi-legal lifelong guardianship over the grown offspring. The parent's preoccupation with the child's insanity tends to increase his vulner-ability to disorganization, as well as his fear of it.

A variety of factors in the paranoid mother combines to give her an unconscious stake in promoting the fear of disorganization in the child. Such disorganization would create a parallel to her own long-standing disorganization, it would express her identification with the child, establish her possession of him, and would facilitate her sadistic use of power. On the other hand, the same parent may simultaneously make the intense demand on the child that his public performance, before friends, neighbors, and relations, be eminently competent and well-organized, in the common-sense frame of reference. The parent unconsciously promotes and exploits the child's disorganization, and demands further that this remain a private conspiracy between them. This new area of control and domination exercised by the parent makes for added difficulties in the life of the child. As a move towards sanity, the child tries to seek out non-hostile persons to whom he can, so to speak, display and make known his level of disorganization and his irrational perception and bewilderment. The validation of the simple fact of such non-hostile sharing of his subjective despair is vital to his getting through the bewilderment and confusion.

To the extent that this move is met with irritable reasonability, his feelings of bewilderment are intensified. The bewilderment, in turn, is often perceived by the other person as sheer negativism. Such interaction is frequent in the therapy situation and is also present in most marriages. The sudden incompetence, in contrast with the usual high level of competence, real and compensatory, becomes used as evidence for the manipulative intent of the temporary lapse. It may actually be the expression of genuine disorganization and bewilderment because the situation is perceived as safe from the mother. On the other hand, it is manipulative to the extent that the other person is parataxically perceived as the mother. Usually both factors are present in varying proportions.

The more insidious danger of the mother's demand for normality lies in the assiduous avoidance of the experience of anxiety. Any temporary lapse is expected to be a permanent count against one's human right to freedom, initiative, and belonging. Where it is carried over into adult relationships, this fear has validity, even minimally, only if one assumes that it is impossible to disconnect a relationship, no matter how vindictive or retributive the other person. Otherwise, if the person uses a lapse in competence, based on the anxious attempt at reorganization, vindictively to invalidate the person's attempt to clarify his own perceptions, to that extent he disqualifies himself as a friend. There are a multitude of variations on the fundamental malevolent maneuver of precipitating, in adulthood, disorganized states based in unfortunate experiences in childhood.

Since the mother was the principle architect for his self-system, she knows, consciously or unconsciously, how to disrupt it. For example, a mother who has devoted years of effort to limiting the possibility of her child's making friends can at any moment throw the child into a panic by pushing him into a play situation with which the child now, by virtue of her restrictive efforts, cannot begin to cope. The child is especially disorganized, to the extent that the move to relate to peers and hence his wish for friends has already been partly dissociated by his mother's previous restrictiveness. Unable to see through the mother's complicated operation, he experiences himself as thoroughly inadequate and even crazy.

When the mother is deeply involved in this maneuver, the subjective feeling of craziness in the child is a frequent way of retaliation on his part by fusing much of the content of the catatonic explosion into a basic defensive operation. The mother can provoke craziness so that it successfully limits the child's growth in important areas. However, once having set off the craziness, she cannot limit it. The child finds modes of retaliation as threatening to her as they are to him. These may include "hysterics," adult temper tantrums in inconvenient situations, frankly psychotic manifestations with or without hospitalization, and illegal or semi-legal operations that would strike at the heart of the parents' investment in respectability.

We have said that the major source of fear of temporary disorganization is rooted in the specific nature of the original power struggle. There are various supplementary factors. Of these, the nightmarish fear that the state of disorganization will become permanent is one of the most persistent, though least realistic. Disorganizing anxiety is inherently a temporary state. Whether the outcome is growth or deterioration, intense panic and the accompanying disorganization cannot be maintained indefinitely.

In all of its manifestations as thus far elaborated, the fear of falling into the mother's power is also unrealistic for the adult. The real fear is that, once broken and the break survived, the illusion of relatedness with the parent or of safety in the defensive fortifications can never be put together again. The fear, then, is of moving beyond the point of no return.

The early established fear of disorganization has a crucial bearing on the vicissitudes of the process of shifting over from the relative dominance of the self-system to the more direct expression of the integral personality. The self-system can be expanded considerably by progressive integration of marginal thoughts and insights from dreams. However, major shifts in its assumptions can only take place with some explosive quality. The fear of disorganization thus becomes, in itself, a primary limitation on growth and a central underpinning of the stability of the restrictions of the self-system.

The crux of the goal of analysis is for the patient to take over openly and competently the organization of his own conditions for growth. In implementing this process, the dominant influence of the environmental setting tends to be overemphasized, essentially in the service of the avoidance of this basic personal responsibility to oneself. Whether in the integration of functions to implement the satisfaction of needs, or more immediately to experience the desire or feeling of satisfaction, the process of growth necessitates being able to act openly. Each person needs to know what he wants, to go after it competently and without unnecessary reservation, and to experience it fully when he gets it. The complexity of the apparently insuperable barriers and the pseudo-insoluble dilemmas thrown up to obscure the simplicity of this process is infinite.

The self-system could theoretically include the majority of the experiences stored in the integral personality. The growth process is latently continuous, but its manifest adult expression is characteristically discontinuous, marked by crises and jumps to new levels of organization. The anxiety, fear, and pain of this accounts for the relative infrequency of uncompromising commitment to growth. The abandonment of this commitment, and the onset of resignation, sharply increases in frequency in the late twenties in our culture. When the commitment to growth is not abandoned, the organization of the self-system can become progressively more congruent with that of the mainspring of the real personal motivation. There is great difficulty in shifting over to the essential dominance of the frame of reference of the integral personality. As with each revolutionary change, it cannot succeed in isolation, and at the same time each forward move evokes the subjective expectation of total ostracism.

The self-system is partly protected from its limitations by the secret broader organization of the integral personality. From a person's response in moments of stress we can get some sense of the degree to which we rely on these resources and their remarkable competence. Everyone has had repetitive experience of making very rapid, sensitive judgments, literally involving life and death, in a space of time too short to allow conscious formulation. Much creative activity, including its expression in art forms, or in the formulation of theory, is quite inconsistent with the self-system of the creator. There is often the concomitant feeling in the creator of wonder and disbelief, as if it had been produced by someone else.

In the human sense there would be no loss and all gain in ultimately dispensing with the restrictions of the self-system, and the burden of double-entry bookkeeping. This ideal state would imply a full shift-over of the frame of reference, so that all data of the personality would be potentially available to awareness. The machinery of the self-system would then be the instrument of the integral personality as its executive agency.

II

THE FEAR OF DEATH

THE MOST PRIMITIVE underpinning of the self-system has been underestimated in psychiatric theory. This is the mastery of the fear of death or mutilation at the hands of the mothering one.

To understand this, we have to refer to the degree of vulnerability of the infant to the mother's good will, from minute to minute, throughout the first few years. Without a minimal amount of good will, the infant cannot even integrate the basic physiology which is necessary for his immediate physical survival. He is extremely reactive to and completely vulnerable to any hostile attitudes in the mother toward him.

The frequency of postpartum psychosis highlights the emotional strain put on the mother by a new infant. A lesser degree of anxiety, and with it some degree of reactive hostility, occurs in the vast majority of mothers after parturition. The intensity of these emotions tends to diminish as a result of a gradual working through of the mother's anxieties, and a progressive lessening of the infant's demands as he grows and becomes less helpless. No contemporary adult has the capacity truly to fulfill the real needs of any infant. When the mother is unable to produce, she withdraws; if she is in reasonably good shape, she dismisses her withdrawal with a variety of security operations. The infant, however, operates on empathy and is not fooled.

The mother who tends toward the catatonic withdraws essentially into a state of anxiety and confusion which is transmitted directly to the infant. The infant then learns that the price of the need for affection is panic in the mother and hence in himself, and that the condition for survival is to survive panic. He must develop a repertoire of techniques to cope with his mother's, hence his, anxiety. These techniques will include functioning in the presence of high levels of anxiety. Later the child often becomes preoccupied with the modes of therapy, because, caught in this situation, he is pressed to develop tactics for heading off mother's minor catatonic episodes. Because of the disruptive nature of anxiety, he must also learn to limit or tolerate his awareness of his own anxiety, in order to survive the repeated crises.

The mother who tends toward the paranoid orientation sets up a slightly different problem. She also withdraws, but her inner feeling, however well-disguised, is a feeling of murderous rage. The helpless infant, sensing the mother's mood, repeatedly experiences the expectation of being destroyed. The infant's fear confronts the mother, counter to her expectation, with her hostile marginal thoughts. These thoughts are likely to be thoroughly unacceptable to a concept of herself necessary to function with her baby or with her contemporaries. Thus the confrontation with the infant's fear intensifies the mother's rage at the same time that she must exclude her thoughts from her awareness. Out of this dilemma on the mother's part arises a preoccupation in the infant with getting over the fear of death, along with an exceptional vulnerability to situations that evoke this fear.

One frequent expression of the threat of destruction occurs in the mother who finds the small animal of a baby essentially revolting. It is dirty, smelly, and squirmy. Here, in order to survive, the child has to accommodate himself to the conviction that he is disgusting. He will necessarily develop modes for constant reassurance, often partly in sexual prowess, or in a high level of personal fastidiousness or in a great emphasis on esthetic perfection. He schools himself to disregard the presumably inevitable rejection and revulsion of the people around him. He may allow and even elaborate certain traits which are often considered disgusting, partly for the purpose of testing whether the current object of his affections can accept him as he feels he is, and partly to provide an alibi for the expected rejection. He may be insensitive to being unwelcome, since he cannot experience being welcome and thus has no basis for comparison.

The feeling of unwantedness is the continuum, and is only relieved by positive and provable affirmation from the other person present. Since he constantly demands reassurance about being wanted, people often recoil from his pestiness. Being alone may be difficult or tolerable, but it is actively intolerable to be in the presence of another person who is essentially indifferent. The self-revulsion may be concealed by obsessive preoccupations with the body beautiful, in a vain overvaluation of physical perfection and excessive distress at minor imperfections. Conversely, it may be directly reflected in the physical appearance by slovenliness, ungainly posture, aberrations in weight, or poor taste and sloppiness in dress. The functions of eating and digestion come to be used symbolically to express or to deny the mother's original disgust. The resulting self-disgust is one of the roots of compulsive under- or overeating, of voluntary or involuntary vomiting, or of easily provoked nausea.

Perhaps most devastating for the child is the mother with a tendency toward the psychopathic. The camouflage here may be remarkably good. Unhampered by the necessity to appear consistent to herself, she is free to play the role of the good mother whenever she finds it convenient. Clinically it may be hard to distinguish between the psychopathic and the hysteric styles of relating. Also, the psychopathic trends may masquerade as either hysterical or obsessive. If obsessive, the mother may appear superficially as merely controlling and restrictive.

The core operation between this type of mother and her infant is torture and torment, however disguised and denied this may be. Her condition for his survival is that the child be miserable. Mother would rather have him dead than happy.

One of the more frequent methods of denying the child any happiness is through an unrelenting restrictive possessiveness. What makes matters worse is that the child misinterprets, or at least rationalizes, the possessiveness as proof of being wanted. He is then caught in the trap of encouraging the imprisonment, so as to be loved the more, and so at last to be happy. He cherishes the hope and illusion that if only he could be happy in the restricted situation, if only he could destroy his own impulses for growth and escape, all would be well between mother and him.

The reality is the reverse. If, in fact, he were happy with her, she would simply find some other way to make him unhappy. And, if she failed at this, she would find it necessary to try to destroy him. Escape is impermissible primarily because it might result in happiness. Escape to prison or to a mental hospital is more tolerable, because she can continue in fantasy to relate to and be preoccupied with her child's misery.

Such a mother's principal preoccupation is often with the child's state of imagined ill-health, or with fearful anticipation that some injury will befall him. An alternate form is frank apathy and neglect of physical care..

The central experience is the mother's subjective secret satisfaction at observing or imagining the suffering of the defenseless partner in the hostile integration. It can be implemented through physical punishment, or by devastating verbal cruelty. The invariable contemptuous derogation may be gross and directed at anything the child cherishes in himself or the world. Or, it may be incredibly subtle, as, for example, by praising talents he does not have, or by exaggerating talents he does have in such a way as to cause him acute embarrassment. The torture and frustration may be carried out by the accurate manipulation of any particular real need. Mother can exploit the child's relatedness to possessions with pseudo-generosity or pseudo-stinginess, carefully organized so that any object becomes unattainable or is removed at exactly the moment it becomes cherished.

An almost universal pattern in the psychopathic mother is seduction and rejection, the utilization of herself as the alternately available and unavailable, but always desirable, object. This operation is ostensibly for her personal reassurance, but is played out to induce a maximum of frustration and jealous rage in the dependent victim.

An essential property of this mode of interaction is the discontinuity of the relatedness. Many of the right things seem to go on, but nothing is built on them. Mother often justifies the breaks in continuity by blaming the child, who then comes to think of himself as essentially hostile and incapable of relating. To herself she justifies her operations as the need to be needed, and may or may not be aware of her demand that the child remain in a desperate situation so that she can provide or withhold her assistance at her convenience.

What she provokes, promotes, and plays on is fear. When the child makes demands which she does not feel the impulse to satisfy, she becomes threatening and the infant feels threatened. It becomes particularly in this case a matter of survival that he be able to conceal from her any sign of being afraid of her. The state of fear comes to be taken for granted as the normal, and the preoccupation is essentially to learn to live with it. He thus builds a picture of himself as doomed to frustrate the other's needs, and may act out this concept for many subsequent years.

In each of these patterns of pathology—the catatonic, the paranoid, and the psychopathic—the parent's power is vested in the threat of death, in her power to control the circumstances for physical survival, to promote or prevent illness and, in case of illness, to promote or prevent recovery. The variant seen in the "disgusted" mother can be the derivative of any one of these syndromes. Its significance lies in the potentially lethal consequences to the infant of the mother's severe disgust.

The terror evoked in the infant by the threat to his survival makes him centrally additionally vulnerable to the mother's power. The residual adult preoccupation with the infant's necessity to master this terror may take a variety of forms. It may be expressed in the conscious preoccupation with the imminence of fatal physical illness, or of fatal accident. It may involve either hypochondriacal obsessions or pathological unawareness of normal danger signs, or a mixture of both. The expectation of danger from without can be expressed in phobic overcaution or in blatant carelessness. He exposes himself to danger in divided doses in the hope that, if he faces it often enough and carefully enough, someday he will get over minding the imminence of loss of life. He seeks out situations in which the danger and the mastery are combined, as by becoming a skillfully reckless driver, a carelessly intense and competent athlete, or a daring soldier. Minor medical ailments are adequately treated, but symptoms which could have lethal consequences may be selectively unattended to the point of danger.

The conscious attitude toward death can range from excessive, fearful preoccupation, through a variety of rational, philosophical attitudes, to a subjective unawareness of fear. The suicidal pull may be manifest in fantasies expressing the necessity to relive over and again the subjective experience of impending death, in the intense effort to accommodate to it. Or the suicidal pull can be expressed in minor overt suicidal gestures, in repeated carelessness, or in the rare, highly planned, secret suicidal maneuver. The intent is the same: to test oneself in the crucial situation and thus "face once and for all and overcome" the fear. The sport of bullfighting comes close to condensing the entirety of this psychology. It is popular in a culture where the fear of mother and women in general is displaced into worshipping mother and conquering women.

The fantasies and the acting-out also contain the helpless pull to compliance with the mother's implicit demand for non-existence.

There is, then, a wide range in the patterning that is set up out of the preoccupation with getting over the fear of death, depending on the relative strength of these two pressures, toward mastery or toward compliance. In response to real danger in later life, there can be a continuum from denial of fear in the face of imminent catastrophe to excessive fear of a very mild threat.

The child is drawn toward dangerous situations, both in the attempt to accommodate to the sensation of fear, and in the need to externalize it and to disguise its subjective origins. His fantasies tend to be fantasies of torture, of intolerable situations in which the attempt is to relive over and over the details, in order to master the terror which would inevitably be involved.

Where, in infancy, the visible experience of terror increased the parental anger, the later experience of fear tends to connote the expectation of increased danger. Thus there remains a chronic preoccupation with keeping fear out of awareness. This is reinforced to the extent that extreme fear tends to be paralyzing, and to therefore make the person less able to defend himself.

The fear of death often shows up as the fear of castration. However universal deeply dissociated fears of castration may be, they nevertheless are only partially derived expressions of fear of annihilation at the hands of the mothering one.

There are many reasons for the ubiquity of the fear of castration. The genitals, being the embodiment of the generative capacity, become a symbol of life and of immortality, and their destruction, conversely, is perceived as tantamount to death or non-existence.

One result of the conflict between the implacable force of the sexual drive and the intense restriction of its expression imposed by society is the perpetual irresolution of fearful fascination with sex. As a consequence, problems fundamentally non-sexual in nature, regardless of whether they are deeply or more superficially repressed, tend to become fused with the ever-present sexual conflict. The sexualization of the fear of death is one illustration of this.

In many ways the mother contributes to the sexualization of the fear of death. Her own fear of intimate human exchange is often expressed, not only by a paralysis of tender interchange, but by a compensatory preoccupation with sex as a means of contact. This in itself is not harmful. It constitutes a substitution of areas in which she can relate for areas, such as generalized tactile sensation, about which she is phobic. The harm derives from her simultaneous repudiation of sexuality. What ensues is a mixture of quasi-erotic seduction of the infant and child, with a gross undercutting of the child's appropriate response to his perception of her seductiveness. If the child should happen to have an erection, mother becomes panicky and threatening. The residual in the child, then, is that having a sexual response exposes one to danger from the mother, either in terms of disconnectedness from her, or, more explicitly, in the fear of destruction at her hands. The danger here is that the mother's enforced insight about her sexual seductiveness, precipitated by the infant's overt response to her seductiveness, will evoke punitive rage toward the infant.

The mother's substitutive expression of tenderness through sex may be somewhat more complicated with her daughter. It can be very similar to what obtained with her son: the substitution of tenderness by a quasi-erotic relationship, followed by repudiation of the appropriate sexualized response in the child. However, more likely, since homosexual incest is more severely taboo than the heterosexual variety, the mother is likely to go one step further and deny altogether the wish for erotic contact with the female child. This has the advantage of not presenting the child with the double-play operation, but the disadvantage of giving no validation what-soever to the natural impulses for erotic expression in the child. A girl child results in whom the sexual function has been forced into severe dissociation. If the mother repudiates entirely her sexualization of the situation, the daughter, while responding to the mother's deeply dissociated sexual impulses, accommodates to the mother's need to keep these in dissociation by displacing her response onto the father. This is often the basis of the female child's dissociated but acted-out seductiveness to the father.

Both the boy and girl have a second opportunity at the integra-tion of the sexual function later in childhood, when the father be-comes more available. What transpires with him during this period may serve either to accentuate the original castration fears, or to provide a new medium for positive re-evaluation of the sexual function.

For example, if the mother's behavior to her daughter was denial of the erotic between them, the dissociation of sex can be partly relieved by a father who is more frankly seductive. It is the inter-actions of this period that have been described in one form or another as the traditional Oedipal constellation. A third opportunity for integration of the sexual function is during preadolescence, in anticipation of adolescence.

The delimiting of the sexuality of the child is one of the primary modes of parental destructiveness. The parent is often incredibly preoccupied with the child's sexual behavior and fantasies in rela-tion to peers of either sex. In the earlier years, there is often a normal flirtation with the parent of the opposite sex, but later this style of flirtation should expand to include the child's contem-poraries. Often such a parent perpetuates his preoccupation with his offspring's sexual life by being restrictive and prohibitory, sometimes expressed as overt jealousy, though occasionally it emerges as vicarious participation in the child's sexual fantasies and behavior with others.

Whatever the style of restriction, the impact on the child is the same. It serves to retain the particular parent as the fulcrum of the sexual situation, thereby cutting off the possibility of exploration either with the opposite parent or with contemporaries. It almost never results in total restriction. But it does instill the fear of death at the prospect of transgressing this command. This fear later shows up in guilt feelings and castration fears.

The castration fears can be expressed either by compensatory overactivity or by avoidance of the sexual function, but always include a preoccupation with potency. The most direct form of self-castration is impotence or frigidity. To a degree, both the bullfighter and Don Juan are attempting to compensate for the fear of castration. Both are also at the same time attempting to master the fear of death.

In all of these varied situations in which the child is attempting to cope with the fear of destruction, he experiences himself, accurately for his age, as the disadvantaged member of the integration. This inevitably generates rage in him. This rage cannot be fully and freely expressed at the time because of the danger that this would entail. The chronic, displaced, later preoccupation with such fear and rage serves the purpose, in the adult, to inappropriately disguise, express, justify and control these persistent parataxic emotions.

Adult rage serves as a barrier to more constructive experience. The realization of this liability in living may intensify the attempt to keep it out of awareness, and, if so, tends to perpetuate it. Pressure to repress the rage also constitutes a liability in the hostile integration with the mother, and her later facsimiles. In this subjective dilemma any maneuver that might provoke intense rage commits the recipient to putting most of his immediate energy into self-control. He is not free to respond to attack because of his personal reservoir of inappropriate anger.

Fear of destruction and fear of disorganization, along with the absence of common sense validation, are often combined in the same threatening situation. The subjective, inexpressible rage is then taken by the offspring to prove his own insanity. Thus his life work becomes the avoidance of the murder he thinks he would commit. He fears to experience the full scope of his anger, and fears becoming disoriented during the moment of anger.

There are other disjunctive resultants in adult life patterns. Any continuity of relatedness tends to be experienced as being in the power of the other person. Yet there is great difficulty in being alone.

The alternate major mode of response to the fear of destruction by the hostile mother is the decision to avoid danger at all costs. To the extent that aliveness constituted the danger, the way to avoid the danger was to be as dead as possible. If the danger was that the impulse of the infant to satisfy primal needs was focally exploited by the power drive of the person responsible for satisfying these needs, the result is much the same. Needs become flattened in pervasive depression.

What evolves is the philosophy of limited investment; the pattern developed is the pattern of compulsive mediocrity, creative and emotional. The person cherishes the secret fantasy of wanting intense love or high achievement, but avoids any situation, medium, expression, or confrontation that would implement it. We speak of these people as "timid." What we really mean is intimidated, that is, afraid. Their lives are organized around fear and the avoidance of danger. They have decided that, for them, ecstasy is too danger-ous. Their life must be reduced to the avoidance of intense pain. But even the concern with pain and its avoidance has become a cover for the quiet despair oriented to literal suicide. With this hidden, inner focus, they convince themselves that for every good there is some bad, and so, rather than some good and some bad, some joy and some pain, it is better to live life as a medium gray.

The person who tries, in action, to deny the fear of death is more easily recognized by the outsider as suicidal. He takes undue risks: he rides motorcycles, goes off high diving boards, volunteers for war. But the person involved in limiting his investment in life in order to limit his vulnerability, and therefore to remain alive or to avoid being in someone's power, is equally preoccupied with the dimension of survival rather than with the extension of experience. He is usually less overtly suicidal. He may even give himself and others the impression of being, without conflict, entirely committed to life.

The covert suicidal preoccupation is occasionally honored only in the breach; the thought has literally never been allowed to come to mind. Often, on closer inquiry, it can be discovered that the suicidal escape has always been taken for granted; that, though the individual is not preoccupied either with dangerous acts or with suicidal fantasies, it has never occurred seriously to him that he could not at any time commit suicide, if life got just a little bit duller, a little bit more meaningless.

The proof of this may come only at the time when the person has developed a more powerful investment in life. When he realizes this, he can find himself unexpectedly in a rage, because he feels that the back door has been slammed. He realizes now that he no longer wants to commit suicide, that he is therefore stuck with being around for a good while, barring accidents, for better or worse.

Herein is revealed the similarity of the central paranoia in the person whose main operation has been the avoidance of danger through the limited investment in life, with that in the person whose preoccupation has been the active defiance of danger. When the chronic repudiation of aliveness in either of these attitudes has been seriously challenged, the person suddenly becomes acutely and intensely aware of his subjective expectation of being destroyed. He experiences his creativeness, his enthusiasm, his intense wish for succor not only as painful, but in the literal and immediate sense as imminently fatal. He is overcome by a paralysis which would be appropriate to the feelings of someone stepping in front of a firing squad.

These people usually seem to lead moderate, tempered, "well-adjusted" lives. Often they are considered by their friends to be unusually sensible, even spontaneous. There can be common sense in the give and take with friends and acquaintances. These people want things just enough, never "too much." The avoidance of any powerful emotion or any overwhelming investment or experience is automatic. It may be that the only subjective discomfort is a pervasive underlying inertia.

If their activity were not sparked by the situation, their expectation is that nothing would go on. This expectation is often accompanied by a generalized irritability and pervasive dissatisfaction with minor discomforts. Since no strong satisfactions can be felt, such satisfaction is not available to make the inconveniences inconsequential. They never allow themselves to feel that life is not worth living as it is, or that any particular change is a matter of life and death. The experience of desire with sufficient intensity to make the risk worthwhile has been cut off. Any experience which might provoke such intense desire is blandly derogated or carelessly sidestepped at its inception.

The parental prohibitions have been smoothly incorporated. One polices oneself into accepting the parental dictates without serious question.

III

DISSOCIATION OF THE NEED FOR TENDERNESS

THE THIRD CENTRAL underpinning of the self-system is the dissociation of the need for tenderness. Tenderness is response to the other's true needs for growth, with a degree of accuracy appropriate to the urgency of the need at the particular moment of development. Satisfaction of the need for tenderness is crucial for survival in infancy. Tenderness, changing in content with successive eras, is required in a continuum throughout life for the development of relatedness to others and of the capacity for interpersonal satisfaction. This must include the ability to evoke tenderness, i.e., the spontaneous relatedness of the other to one's crucial needs.

Tenderness is necessary for survival and growth. The need for it defines the child's dependence on a particular adult. If the infant has survived, it can be assumed that he got some minimum of tenderness. The tenderness needed for growth and expansion beyond this level is, however, never adequately provided, with the result that no infant is spared the experiences of pain, confusion, terror, desolation, and intense loneliness.

These feelings can be intensified by the mother's vacillation in response to the infant's insistent demands for tenderness. The mother's cutoff of tender response can vary in its expression from obtuseness and withdrawal to acute anxiety or to the angry repudiation expressed in paranoid attitudes of varying intensity aimed at the infant's needs. Out of such experience the infant comes to sense his own needs as futile, bad, or dangerous. The pressure on the infant, then, is to write off the need for tenderness, to become alienated from it, to dissociate it.

This constitutes the primordial experience of rejection—the infant's intense need meeting no response from the sole source of supply.

Objectively all subsequent experiences of rejection differ, in that never again is there only a single source of supply. In the adult, the feeling that the rejecting person is the only source of tenderness can only refer to this original dependent situation. The miserable feeling in the adult derives its painful intensity from the hallucinated re-experiencing of the original rejection by mother.

If the adult is preoccupied with being rejected, this serves the purpose, among others, of reinforcing the renunciation of tenderness by exaggerating the difficulty of attaining it. The chronic preoccupation with rejection owes another part of its strength to the self-revulsion built into the self-system by the disgusted mother. Here the characteristic adult response is that if any one of the people felt to be important conspicuously withholds tenderness, is indifferent or unavailable, this one person becomes the only one to whom one can respond until the presumed insult is erased. Even if the interaction is private between them and has no audience, it is as if this one rejection has proved and exposed the lifelong feelings of self-revulsion, and as if these can be suppressed again only if this specific rejection is reversed. If spontaneous mutual attraction is lacking, this demand only intensifies the other person's withdrawal. This pattern of interaction furnishes the basis for much of what is experienced as unrequited love.

The fear and expectation of death, mutilation, or torture, the fear of disorganization and its inherent vulnerability to attack, and the repudiation of the need for tenderness combine to produce actual isolation, however camouflaged. Each constitutes a barrier to the consolidation of satisfying love relationships. Just as powerfully, though in a somewhat more indirect manner, they evoke in combination an intense fear of being alone, which is often mistaken for genuine loneliness. The fear of being alone feels like the longing for people, or for a few particular people. But the overdetermined push to get together with them does not spring from the expectation of spontaneous mutual enjoyment and sharing.

They are needed, rather, as military allies, as bodyguards, for self-reassurance, and for the postponement of awareness of the inner handicap in giving and receiving tenderness. The relationships formed, then, are in fact integrated in a pattern designed primarily to repress anxiety, anger, disgust, fear of death or torture, and particularly to prove once again that love cannot last between adults.

Such interaction with people has nothing to do with solving the problem of the longing for relatedness. On the contrary, it serves to perpetuate the unrelatedness just as effectively as any other form of obsessional pseudo-productivity.

Basically, this fear of aloneness is the expression of the degree of terror that would attend the recognition of the futile dead-end quality of the neurotic spiral, and the necessity of facing the inner emptiness and meaninglessness of life as the person has organized it.

One might expect that, if a mother's love is inadequate, the child would be then prone to look elsewhere for love. This is good common sense. But the opposite is true. The more loving the mother, the freer the offspring to search for tenderness elsewhere. The more inadequate the mother's love, the more tied the child to her, and the more energetically does he fight off what he wants most—love. He avoids or alienates those who proffer deep friendship. Failing this, he becomes increasingly paranoid in their presence.

The paradox, however, is only apparent. The primordial feelings of self-esteem and the capacity for trust are direct reflections of the genuineness of the mother's love. In the absence of her love, these crucial functions are defectively organized. The ensuing defects tend to be compounded by difficulties which they produce in successive developmental stages.

They interfere, for example, with the development of self-assurance, self-sufficiency and the capacity for the exchange of tenderness. These qualities are prerequisite to the development of new relationships. They are greatly reinforced by the satisfactions inherent in such relationships if such occur. In the absence of mother's love, the child, not having these prerequisites for forming other relationships, has little alternative but to remain with mother. The particular style of his bondage is of lesser significance.

Quite apart from the progressive defects accruing from the non-integration of these basic functions, there is also the active presence of the sequelae of the fear and anxiety evoked in the child by the mother when unloving.

From these develop the investment in the infinite variety of tactics that characterize the later hostile integration. The substituted emotions of cruelty, revenge, domination, submission, prestige, compete with and hinder the satisfactions to be had from the development of new and fuller human relationships. The inadequate care and love of one's garden leads not only to stunted flowers, but also to flourishing weeds, which in turn contribute to the further squeezing out of the flowers. Such apparent secondary satisfactions, then, are both the expression of bondage to mother and the machinery that prevents its undoing.

The reason for the apparent paradox of a person's avoiding what he wants most, tenderness, can be stated in yet another way. The entire structure of the self-system is built around the early established conviction of the relative unavailability of tenderness. Later in life, then, the compulsion is to assiduously avoid tenderness, lest there occur severe disorganizing anxiety and the revival of primitive fears of death.

The need for tenderness cannot be eradicated. Its pressure is continuous, no matter how obscure the expression, and its repression requires constant vigilance.

As the child gets older, he begins to perceive that the only possible sources of tenderness, validation, and support adequate for his own growth are essentially away from the mother. His difficulty is his dependence on the mother, which persists because he was unable to get from her the minimum which he needed in order to develop adequate alternate sources of tenderness. In his shortsighted attempt to overcome his dependence on mother, the child often tries to devaluate his tie to her. It seems to him that the only way to grow up is to repress the intimate and tender aspects of the mother-child relationship, and to dissociate the need for them.

This is an understandable but self-defeating move. If the need is dissociated, the person must reach out for tenderness outside of awareness. Any response to this reaching out which offers satisfaction brings the move, hence the need, into awareness. It also brings more directly into awareness long-standing feelings of loneliness.

The experience of tender response to the dissociated reaching out for tenderness is likely to set off psychosomatic expressions of anxiety since it especially reawakens desires dissociated in the earliest years. Any later validation of the need for tenderness, expressed openly in reciprocal exchange, must necessarily re-evoke the anger at the original mother for her deprivation and exploitation. It is also associated with the dizziness, seasickness, uneasiness, and feelings of unreality which characteristically accompany a major shift in the frame of reference. These feelings are classically associated with falling intensely in love.

The child naively intensifies the denial of his need for tenderness in the hope of getting enough independence from the mother to be able to make up the need under more favorable circumstances. Life becomes, on the one hand, devoted more and more intensely to the attempt to construct these circumstances, and, on the other hand, to invalidate them when they have been constructed—because at this moment they reawaken and challenge the archaic decision that it is dangerous to ever need anyone again as much as one originally needed mother.

The original inadequacy of the mother's tenderness, deriving from the limitations of her personality and implemented in her drive for power and self-aggrandizement, evokes in the child an unremitting, though often unconscious, preoccupation with her. He becomes continuously adaptive to her wishes and demands. His adaptation is not necessarily pleasant and charming; it often incorporates considerable negativistic and teasing behavior which, while it may seem to establish a limited degree of freedom, also intensifies the hostile interaction.

Most important, the hostile behavior, with its aura of pseudo-independence, serves to obscure effectively the fact that its very existence is rooted in compliance. This complicated pattern of compliant hostile accommodation to mother's inadequacy has as an indirect goal the exposure of her inadequacy. The by-product, of necessity, is the child's constant expectation of retaliation. He is rarely disappointed in this; mother becomes the martyr. The child comes to feel that if only he could amputate his need for tenderness, he could win his freedom; if only he could overcome the need for love, he would cease to be dependent on mother, the mutual hostility would be by-passed, he would be free to get on with living. It comes as a great shock when one day the person realizes, if he ever does, that a life of secular renunciation, far from being an expression of freedom from mother, is in fact the quintessence of compliance to her restrictive demands.

Much of what is labelled as "guilt feelings" and "the need for punishment" is, in fact, an obsessional embellishment on the syndrome just described. The self-punishing behavior is the acting-out, in modern dress, of compliance to the parental directive of restrictiveness. It represents the efforts to reinstate the *status quo*. The sensation of guilt is often the cover for the disorganizing anxiety, resulting from seeking tenderness beyond the limits of the self-system.

The real root of this anxiety is the breakthrough of dissociated Not Me impulses from the integral personality. Under this impact, much of the related material existing in the self-system and deriving from the Bad Me is brought into focal awareness. The Bad Me, it will be recalled, has never been dissociated; it was incorporated into the self-system, although not generally into focal awareness. Nevertheless, its access to awareness is much greater than that of the dissociated Not Me.

What often happens is that, under the threat of a breakthrough of dissociated impulses, say the wish for tenderness, consciousness becomes flooded with material from the Bad Me. The person feels guilty. These feelings of guilt are always obsessional, insofar as they substitute for the real issue. This is true no matter how plausible they are in the current situation, and no matter how the individual creates circumstantial evidence to support his claims to guilt.

Destructiveness to others usually accompanies, in greater or less degree, the self-punitive behavior. However, the feeling of guilt associated with some awareness of this destructiveness is, in its obsessional quality, thoroughly insincere. It is more often emotional magic used to entrench and perpetuate the destructive behavior toward oneself and others.

That is, the guilt feelings also provide a cover for the expression of rage. If felt painfully enough, this suffering becomes the atonement and justification for the hostile acting-out of infantile rage. To the degree that such obsessional activity—the self-destructiveness, the guilt feeling, the bitterness and retaliation—occupies the field of consciousness, the defensive activity of the self-system successfully maintains the *status quo*. Lost and forgotten in the midst of this noisy drama is the burgeoning impulse to break down the long-standing renunciation of tenderness and intimacy. The immediate trigger of guilt is the fear of retaliation. The real content, however, is fear of becoming human.

A closely related obsessional pattern is to disguise even more effectively the renunciation of love, by adding self-righteousness to self-punishment. Hostility towards others is in no way diminished, although it becomes more subtle and less provable in court. Here, too, the primary purpose of suffering is to obscure awareness of real needs, and to lessen the danger that forbidden reaching for tenderness would evoke response. The person feels righteous about having had to give up what he would most want— in the name of duty, God, reasonable practicality, creativity, or some other deity—and points to the suffering this self-deprivation has caused him.

Having proved his innocence, he hopes to enlist his friends' sympathy and dissolve their disapproval at seeing him embark on this self-destructive course. What remains thoroughly obscure is that the deity in whose name this is really undertaken is the parent. In fact, this particular style of virtuous renunciation of the need for tenderness is usually borrowed directly from the parent; the mother's martyrdom was the original weapon used against the child to enforce the amputation of his need for tenderness.

Another frequent obsessional constellation serving the same purpose is the substitutive fear of "falling into the power" of any contemporary who is dimly conceived of as a source of tenderness. Anyone who, by his warmth and attractiveness, awakens longing for love is looked upon as dangerous and threatening. Any shortcoming in the new person, such as irritability or vulnerability to being provoked into becoming controlling, is exaggerated. The real mother, with whom the parallel intimidating relationship had been integrated, is now remembered as unthreatening and manageable. The fear of retaliation from mother for the sin of replacing her is projected onto the new relationship, using whatever data is at hand. The more intense the potential for tenderness, the more intense the fear of punishment, and, so long as its source is obscure, the more intense the projection of this fear onto the new adult partner.

Adults display protean patterns of the renunciation of love. There is the person who persistently attempts to milk a tiger. There is the person who persistently attempts to get love only at a great distance, by some sort of radar system, with the understanding that love could never become explicit. There is the person who perpetually falls toward love, but never in it; there is the person who is attracted only to unavailable partners.

A favorite hysterical pattern of repudiating the need for tenderness is the dramatization of what appears to be the exact opposite. The person becomes very demanding of time, attention, service, and may be quite theatrical in emphasizing his needs for affection and care. What he demands is peripheral to what he really needs. By his very insistence, he sets up a smoke screen whereby it becomes even less likely that the chosen partner will develop the insight or the inclination to push through the person's real resistance to accepting what he ostensibly demands. These, and many other maneuvers, however camouflaged, in fact result in some exchange of tenderness, with the limitation that it must come to an end as soon as it becomes explicit.

The more dangerous sequel, and the more frequent, is resignation. There develops a mild clouding of perception, so that the lines are always fuzzy, the distinction between love and hate is never clear. All people are presumed to have some good in them. Any two decent people should be able to make a life together. Nothing is ever quite right anyway, and it must be adolescent, childish, or infantile, to look for the intense and satisfying experience. Relationships which at their initiation were a move toward growth and valid enough in the life course of the respective individuals, become dull and formalistic when continued, having outlived their usefulness. In their stability, they then provide progressively the excuse for the limitation of growth, rather than the base for its expansion.

Thus these relationships slip over insidiously into the frequent, quiet, subtle, hostile integration. Each secretly cherishes his reproaches, and, given the situation, the complaints are invariably valid on both sides. Each one, by the fact of occupying space which should either be filled by someone else or remain unoccupied, is in fact providing the source of perpetual frustration that makes the other's life feel meaningless and empty. Hate sets in.

In the course of therapy, one of the frequent forms of repudiation of the need for tenderness is that the therapist's serious concern for the patient's needs tends to be interpreted as malevolent. It is experienced by the patient as derogatory, as either insulting or invasive, or as implying some unadmitted or undocumented inadequacy in the patient.

This latter distortion derives partly from the fact that the patient himself is trying to compensate for and disguise his inner lacks in those areas in which he most needs help and feels it to be particularly unavailing. He would formulate it that he has suffered such a deficit that the amount of help he would actually need is more than could ever be forthcoming. This formulation is inherently fallacious.

It would be more accurate to say that he expects that if he now fully experienced the intensity of his need after all these years of denial, the pain would be more than he could bear. This also is fallacious. In order to be able to experience it at all, he must be in a situation in which the experience could be shared and in which, therefore, the pain would be tolerable. Those emotions that are truly intolerable in terms of the available support remain firmly out of awareness.

The dissociation of the need for tenderness at the infantile level and in the infantile mode has one specific and pervasive effect in adult life. It places a profound limitation on the fullness of adult sexual functioning.

Sex between adults is a non-verbal form of intimacy and communication. Superficially, it is based on interaction by the sense of touch. The stimulation by touch of specific or non-specific erogenous zones is a highly refined and specialized extension of the communication by interactive physiology that goes on, ideally, between infant and mother. Stimulus and response lie in the basic physiologic functions—respiration, heart rate, muscle tone, skin texture, and the genital response. The interplay of these provides an important medium to express both primitively and symbolically the mutual appeal for and satisfaction of the need for tenderness.

Without this broad base, technical facility becomes empty and ritualistic. The obsessive preoccupation during the sexual act, including the concern with sexual adequacy, is essentially a substitute used to evade or to attenuate the natural physiological interplay. With relatively controlled people, alcohol sometimes facilitates a somewhat freer physiological response, partly by decreasing the capacity for obsessive preoccupations. Concurrently, it acts to depress the sensitivity of all physiological responses. These two effects tend to act in opposing directions. However, the dulling of the intensity of satisfaction may allow for some degree of fulfillment while avoiding disorganizing and disjunctive anxiety.

The full sexual experience must include the mutual build-up of the need for tenderness and the need for satisfying the other's need for tenderness. The core of this experience is the mutual reliving in full intensity, if only for a moment, the infantile needs of each and their satisfaction. If life has been organized to avoid the re-experience of infantile satisfactions, then the whole sexual act must be organized to cut off intense and intimate sharing, the profound flowing union of two people.

The self-denial then becomes implemented by deflecting the sexual function to subserving power, prestige, self-reassurance or self-abnegating submission. The form this cutoff takes can vary widely. The person can suppress his own needs for gratification and relate exclusively to satisfying the partner; or he may callously disregard the other's needs. In either case, his own needs remain dissociated. His disregard of the other may be motivated in lust for conquest and power, in sadistic deprivation. He may use the

partner simply as a vehicle for the implementation of his mastur-
batory fantasy. The partner may not exist as a real person. The
person who is severely cut off from perceiving and responding to
the partner's needs is often perceived to be selfish and only out for
his own satisfaction. This pathology, in fact, represents an exceptionally severe cutoff from his own feelings and their fulfillment.
Selfishness is his style of depriving himself of being loved.

Often the capacity to perform sexually depends on keeping the
need for tenderness out of awareness. Sharing and real union is
then experienced as a threat to the sexual function organized in
the power mode, and may cause temporary impotence or frigidity.
During the course of successful therapy, when the defensive power
operations lose their hold, there may be an intermediate period of
sexual disfunction until the sexual function can be reintegrated
with the need for intimacy and tenderness.

The sexual experience, then, is the subtle and flexible medium
in which an endless variety of limiting character patterns and interactions are expressed and symbolized. The specific patterning of
the deflected sexual experience has received much attention in
psychiatric literature, but is of less importance. The symbolic meanings of the substitutions are of some value in indicating the nature
of the situation leading to the original repudiation. One fear is that
the other person would have no interest in satisfying one's needs.
In this instance, we could expect techniques for shortening rather
than for elaborating the sexual experience, including premature
ejaculation. There would be a phobic restriction of the areas of
physical interaction, or the requirement of special circumstances.
Sex can feel queasy unless it occurs in total darkness, in broad daylight, or at five o'clock Sunday morning when fully dressed, or
when clothed only in a black silk negligee.

The fear may be that the appeal for tenderness would be met
with active cruelty. Where the original threat was cruelty, the
obsessive substitution tends to be in the sadomasochistic mode. This
may be overt in the sexual interplay or in the nature of the personal
interaction, or covert in the fantasy life, conscious or unconscious.
Whether the overt role is sadistic or masochistic, the covert identi-

fication is always with the victim. The deeper meaning is that if the individual could accommodate to being tortured it would be possible for him to fully enjoy sex.

In infancy, mothering is always more important than the avoidance of pain. If the only mothering that the child was able to get was in the mode of cruelty—if every time mother dried him after the bath it was done roughly—the child does his best to integrate and identify the two, using the power of sex to absorb the experience of pain.

The most blatant example of this is the chronically angry mother whose only direct physical and emotional relatedness to the child is when she hits him. The child ultimately learns to provoke this over and over as the only mode of contact which is available. Where this was the early experience, the adult almost universally plays it out, sometimes in overt physical violence, more likely in the covert nature of the hostile integration. Its sexual expression can fall into either category.

One of the barriers to getting over this particular constellation is that, since the preoccupation with cruelty is misunderstood in our culture, it sets off a tremendous amount of guilt. Preoccupation with guilt feelings is more respectable and acceptable than a preoccupation with the imagery of cruelty. The guilt, then, obsessive and insincere, serves the function of a cover for the continued cruelty fantasies. If the mistaken meaning attributed to such fantasies can be cleared up, the defensive guilt feelings are likewise attenuated.

Cruelty cannot be played out in open violence, even briefly, without endangering the actual loss of the partner. Even in fantasy, guilt about the fantasy is usually such that the fantasy itself is repeatedly amputated and not carried to the point where its intent and sources could be consciously identified.

Perversion or inadequacy of tenderness in the mothering one leaves many imprints on her offspring. Among them is a variable degree of sensory and perceptual dulling. This quasi-anesthesia is challenged throughout the lifetime by a world of peripheral stimuli. The natural situation for reintegration of these functions is in direct, physiological, intimate interaction with another person in whom one has some confidence. The residual anesthesias in themselves serve to attenuate markedly full sexual satisfaction. To the extent that the sex act tends to break through them, it may be experienced as disorienting, overwhelming, overstimulating, and dangerous. In retrospect, it is often derogated as disgusting or foolish.

The wider the discrepancy between the level of awareness in daily living and full self-awareness, the more explosive full sexual satisfaction can be. Here, the painfulness of reintegration of experience can be particularly confusing, because of the convention that the sexual experience is supposed to be essentially pleasant, and because of the assumption that the experiencing of distress implies some disjunctive impulse in relation to the partner.

One of the most frequent bargains sought with the destructive parent is to try to purchase technical sanity and physical survival by writing off the search for love. To the extent that this undertaking succeeds, it supports persistent depression, a general flatness of response to all experience. A high level of detached competence in various areas may continue for a considerable time. However, the avoidance of love has a limiting and ultimately a deteriorating effect on such competence.

This occurs because the full and direct commitment necessary to maintain growing competence would always threaten to revive the search for tenderness. Thus, to avoid the latter, the commitment must be attenuated. The very effort to be intensely committed to competence, although defensively designed to minimize the need for tenderness, actually results in reviving it.

This compromise in the sacrifice of love for security usually takes place unconsciously. The conscious formulation is often a false dichotomy betwen the wish for productivity and the need for love. The person then becomes involved in finding some compromise between the spurious alternatives he has created. Sometimes his decision is that love cannot be looked for or even thought of until productivity in other areas is thoroughly integrated.

Conversely, sometimes his conscious thought may be that any concern with competence must be sacrificed until the problem of finding and consolidating a loving relationship is solved. What is overlooked is that the one cannot be fully integrated without the other; that, in fact, productivity, if it is real, and love, if it is real, are one and the same thing.

IV

THE PERSONIFICATIONS OF INFANCY

THE VERY YOUNG infant personifies the good, happy and loving mother and the bad and angry mother before he identifies his mother as an individual. He thus develops the base perceptions for three values: Good, Bad and Chaotic. His need for tenderness was consistent, her response was inconsistent. Under subtly different circumstances, his need evoked tenderness, rage, or chaos in the mother.

The Good personified all experiences of tenderness, competence and satisfaction into one continuum—the Good Mother and the Good Me.

The Bad personified all negative, frustrating, and dangerous experiences into the continuum of the Bad Mother and the Bad Me. The Bad Me experiences are expressed directly in the fear of death.

These Good and Bad personifications persist internally beyond the formation of the self-image, of the sense of identity, and of the identification of the various aspects of mother in one person. They continue to be represented in the adult by the opposing directions of the security functions and the productive functions of the self-system.

The infant's experience of chaos is a response to the presence of the mother when she is herself in a chaotic state because of intense anxiety. It involves an active feeling of disconnectedness. The mother who is absent when needed, on the other hand, is experienced as if she were present but withholding—and therefore as angry and bad.

Those moves that evoked chaos in the mother were the most intensely threatening and were progressively avoided, no matter how urgent the specific needs in question. The primitive personality was organized to avoid them. This third set of experiences, not perceived by the mother, become infused with anxiety, are represented only in the dissociated system, and are experienced as Not Me when they are later forced into consciousness.

The good and bad systems remain in consciousness. In due time, the infant discovers that both sets of experiences derive from the same person, Mother.

As the mother begins to be perceived as a more or less logical entity, the offspring organizes his self-system to maintain connectedness with hers. The original perceptions of both Good and Bad Mother get foggier, but the dynamisms integrated in these perceptual modes continue to function actively.

With the emergence of the self-image and the developing grasp of the integrity of the external reality being mastered, the logic of the self-system undertakes to reconcile and make a logical integral of the experiences perceived in these two divergent modes —the good and satisfying and the bad and frustrating. In the attempt to reconcile these two primitive modes of experience, much verbal and legalistic folderol is brought into play. The self-image must incorporate both aspects.

The logical organization of what will be in clear awareness about this major interactive field, i.e. offspring and mother, takes place primarily in late infancy and childhood. It is strongly influenced by the mother's more cherished fictions about herself. Subsequent experiences, both during childhood and during adulthood, fill this framework with massive data. These later experiences also expand, restrict, and reorganize the major systems to some extent.

The experience of intense desire-satisfaction gets progressively alienated. Memories of experiences with the Good Mother are progressively suppressed.

Repudiation of the need for tenderness begun in the early experiences with the Bad Mother is now reinforced by partial dissociation of the early experiences with the Good Mother, in the service of consistency.

The fear of death subverts much of the interaction between infant and mother into a fight for power. In this fight the exploitation of both the fear of disintegration and the repudiation of the need for tenderness are natural weapons. Thus all three facets of the central paranoia become fused into a major system of defense.

A tolerance for both tenderness and disorganization is indispensable for important moves toward growth and reorganization. Their repudiation in the service of survival becomes the major obstruction to growth.

The central paranoia has as its primary focus the maintenance of dissociation of the Not Me experiences. The security apparatus is the agent of the central paranoia. One function of the security apparatus is to obscure, as well as to implement, the central paranoia. This is partly in the service of preserving the logical consistency of the self-system, i.e., of apparently reconciling the productive with the security dynamisms.

This obscuring is more importantly in the direct service of resisting change. Security operations work to reintegrate the hostile aspects of the conditions of early infancy, thereby constantly reinforcing and refueling the fallacies of the central paranoia. This reinforcement of distorted perceptions and convictions is only effective if the active manipulation of the environment to produce the effect is obscure.

If both the manipulative perpetuation of the central paranoia and the implicit assumptions of the central paranoia are clarified and shared, their fallacious logic is exposed. They become thereby less effective in maintaining dissociation, and therefore in heading off periods of temporary disorganization. For this reason, the machinery of the central paranoia must remain obscure in order to perpetuate itself and serve its function.

The integral personality incorporates data from all these modes, Not Me, Bad Me, Good Me, self-system and self-image. In its struggle for growth, it must relate to defects in the productive apparatus of the self-system, and express needs which this dynamism is not equipped to satisfy.

In this process, it is in direct opposition to the central paranoia, and thus must formulate and, hopefully, discredit the latter.

The integral personality is the agency of growth and is on the side of the therapist. It perceives and formulates the productive functions, paranoid operations, and self-system compromises. It seeks situations in which growth could occur. Its immediate goal is to disorganize partly the rationalizations of the self-system so that they may be reorganized on a more productive and satisfying basis.

To the extent that growth occurs, the productive apparatus is expanded and reinforced by progressive reintegration and subsequent elaboration of previously dissociated dynamisms. To the extent that it fails, the ongoing productive functions are expended in the service of the central paranoia.

All three systems, productive, paranoid, and dissociated, continue to operate throughout life. The self-system imposes a false illusion of logic, excluding the dissociated, and compromising and compartmentalizing the productive and the paranoid. The integral personality organizes a different, broader, and more creative synthesis.

Each was originally integrated in terms of accessibility to consciousness. New experience, likewise, is incorporated or excluded in terms of its immediate availability to consciousness. Conversely, each system has characteristic representations in the stream of consciousness.

Focal awareness can incorporate the present manifestations of all aspects of the self-system, the self-image, the productive apparatus, and the central paranoia. Those experiences of early infancy from which these systems are derived do not appear in consciousness as such. Neither do self-system functions initially integrated and subsequently supplanted as the result of later processes of growth. Such rarely used facets of the self-system, nonetheless, flood consciousness as disconnected moods, attitudes, sensations, and misperceptions. These distortions of experience, displaced from currently out-of-focus functions of the self-system once fully in awareness, actively influence perception of current reality. They thereby influence intensely current patterns of relatedness.

When they come into focal awareness they are experienced with a sense of positive conviction.

Fragments of truly dissociated experiences also float into marginal awareness, usually associated with eerie feelings or with the emotion of horror. The dissociated needs these fragments represent seem entirely unreal, and moves to satisfy them cannot be implemented consciously.

More sensitive, organized perceptions, formulations, and directions derived from the integral personality can also be glimpsed fleetingly in marginal awareness. They are experienced as unreal, and without continuity. Plans to implement this organization must be elaborately disguised in consciousness in order to be implemented.

This organization of perception and formulation in the stream of consciousness is directly implemented in the organization of patterns of interpersonal relationships.

Overall patterns of repetitive interpersonal integration are evolved. These patterns are designed to implement the productive functions of the self-system so far integrated. They also reinforce the paranoid assumptions by manipulating others into attitudes characteristic of the mother when angry during the person's early infancy. They also incorporate machinery for retrograde discrediting of experiences incompatible with the self-system, whether accidental or reached for by the integral personality. These patterns of relating avoid experiences which would have made mother "chaotic." Conscious perception and formulation, however brief, of the resultant relationships become a part of the content of consciousness.

The internal patterning of forces tends to be maintained by the contemporary patterning of interpersonal relations. The hostile integrations are necessary for continuing the central paranoia, and the loving integrations expand the productive apparatus and promote reintegration of discarded interpersonal experience.

Change in the patterning of interpersonal relations, either to disentagle from or to re-explore amputated early experiences, is an inherent part of the process of integrating forgotten memories and forcing marginal perceptions into focal awareness. This provides more data for integration into the continuum of consciousness and thus broadens the base for further exploratory experience.

The actual patterns of current living and their active reorganization and review are thus an inherent part of the analytic process. This process of alternation of new experience with analytic integration is the normal process of acquiring a new function after adulthood.

In therapy, then, the patient learns to identify the various forces represented in any incident, and so to organize his own experiences in the direction of growth.

V

NATURE OF PRIMARY INTEGRATIONS

AT ANY GIVEN moment a person is only who he is in his contemporary world.

At first glance this may seem to contradict the thesis presented up to this point, that a person is the sum total of his past experience. Both are true. Past and present are inseparable functions of each other. The present owes its existence to the past and is shaped by it. The past has real meaning, relevance, and expression only in terms of the present. Who a person is, at any given moment, then, can be defined by what he does and how he does it.

More accurately, his very definition lies in the integrated totality of his actual patterns of interaction with contemporary people. This formulation is theoretical, and its practicality is limited only through the lack of refinement in procedures which could determine with subtle precision the nature and quality of the total current experience, conscious and unconscious. As a result we have constantly to turn to the past in order to attempt to know the present.

The formulation applies equally to the vast area of experience that seemingly has nothing to do with people. The interpersonal thesis, to recapitulate briefly, is that all human activity has meaning and, ultimately, existence in terms of its subserving the relatedness to others, however unconscious, obscure, disturbed, or amputated this relatedness may be. The contemporary audience for an activity undertaken in private is usually one of the current central relationships in a person's life. It may consist of people known to exist but not identified, whom one hopes ultimately to contact through this medium of expression. To the degree that the audience is imaginary, it is almost invariably colored by eidetic residuals. To extrapolate the thesis, all pursuits and interests would tend to cease and life would soon come to have no meaning if a person suddenly found himself to be the only living person on the face of the earth.

In summary, a person's nature and boundaries can be defined by the evolution of his pre-existent character structure, his currently most significant relationships, and immutable external reality.

Little need be said about the circumambient reality as a limiting factor other than that a large proportion of what appears to fall in this category is really a derivative of the other two.

The character structure is a highly complicated organization of the total body of past experience, which centrally includes the now eidetic interaction with the childhood figures. This has already been discussed.

What concerns us here is the second parameter, the currently pre-eminent relationships which we will refer to as the primary integrations. Everything in the character structure has palpable reality only in the nature of the current interactions. The ancients put this more succinctly by saying: "A man is to be known by the way he loves."

At least until preadolescence, the primary integrations are with the parents. In adulthood, hopefully, they are with the few contemporary individuals with whom a continuity of relatedness is of vital importance. The primary integrations are always the relationships with the few particular people who really matter. These are the people who make life meaningful, and to whom a vast amount of one's current thinking and action is addressed.

These integrations may be essentially restrictive and confining —the hostile integrations—or they may be essentially expansive and growth-facilitating—the loving integrations. In either case, all patterns of interaction, both within and without these primary relations, are limited by the need to maintain this basic continuity. Anything such as a new experience, or a change in one's picture of one's self that would operate to make this continuity untenable, must encounter great inner resistance.

At any time in a person's life, primary integrations always exist. The important person may be geographically removed, but psychically the connection is intact. If there is no one of importance in the current life, the primary integration is almost invariably still with a parent, one of the original primary integrations.

The later central integrations tend either to reinforce or to counteract the patterns established in relation to the parents, both for better and for worse. Every human interaction has some impact on the barrier between the self-system and the integral personality. The central integration is subserved by validation or invalidation; exchange of tenderness or of malevolence; cherishing and respect, or their opposites.

Validation is the process of confirmation by one person of another's experience. If it is unavailable when needed, one becomes anxious, confused, and disoriented. The absence of validation in infancy is a severe impediment to the processes of growth, and results in a persistent fear of disorganization.

Tenderness refers to the process of empathic sharing and facilitation of the other's needs for satisfaction and growth. If it is lacking, one experiences pain, sadness, and loneliness. To the degree that tenderness was unavailable in infancy, the individual repudiates the need for it and finds a way to discourage offers.

Cherishing refers to the overall good will as to the other's physical survival, the "life wish," as opposed to the "death wish." Such interest and care for another person is a necessary force for the maintenance, not only of productivity, but of life itself. The range of such cherishing extends from sheer joy in the existence of the loved one to the barely tangible preference that the other person remain alive.

Minimal cherishing is necessary for physical survival. Its absence in infancy generally leads to death or to persistent vulnerability linked to the ever-present latent fear of death. Life becomes occupied with the mastery or denial of this fear.

Absence of such cherishing in adulthood, the recognition of the partner's indifference to one's survival, revives the infantile fear of death, however disguised.

Respect is the process of active affirmation by word, deed, and attitude that the integral personality of the other is more real and meaningful than his self-system, and that the low self-esteem revealed by his behavior is inappropriate to his potential functioning. If one respects a person, one relates primarily to his integral personality, while at the same time recognizing the security apparatus as the machinery adaptive to unfortunate historical necessities.

Respect, then, means respect for the integral personality with tender disregard of the security apparatus. This means that respect is closely related to the pursuit of truth. It is therefore inseparable from knowledge. A show of respect toward a person in the absence of real knowledge of him can only reveal insincerity. One of the most refined techniques expressing contempt for another is disinterest in knowing him. Among very distant people, this is often virtuously rationalized as respect for a friend's privacy.

Loving relationships are marked by the highly developed exchange of tenderness, cherishing, respect, and validation. In hostile integrations, on the other hand, these qualities are conspicuous by their absence. Their opposites flourish. Impact of any particular relationship in promoting or restricting growth depends on mutual interaction in these dimensions.

A loving relationship can be defined as one which promotes growth. It is a relationship in which the dominant integration is between the two integral personalities involved. The actual overt interaction tends to force progressively into awareness those areas of the integral personality which were not previously admitted to the self-system.

Progress of this interaction challenges the central paranoid assumptions of each partner. Each reaches through, behind, and beyond the barriers of the central paranoia to relate to the other's integral personality. To this end, the loving integration includes active affirmation of the other's growth in all its modalities, the validation of original perceptions—tenderness, mutual satisfactions, cherishing, joy, respect, knowledge.

The validity of the central loving relationship depends on the progressive unfolding of the integral personalities in the process of getting to know each other better. Getting to know oneself, and getting to know the other, are always interactive. They are also thoroughly interwoven with the process of expansion of each in many other areas of living.

After the initial mutual exploration, further expansion of a particular relationship depends on the individual growth of the two. Such growth both enhances the relationship and is augmented by it. A relationship is no longer loving when it stops growing. The durability of the love is related to the degree of congruence of the integral personalities, and to approximate equality in the rate of progress in interpersonal development.

A hostile integration can be defined as a relationship marked by the mutual restriction of growth. The integration is between the two central paranoias. It acts to reinforce and perpetuate the paranoid assumptions. Both the durability of the relationship and the preservation of the central paranoia depend on keeping these assumptions out of focal awareness. If the loneliness, frustration, anger, and fear in relation to the other were fully clarified, it would most likely disrupt the partnership.

Durability in a relationship presupposes a certain coincidence in the structure of the respective personalities. Long-term patterns of smooth interaction can only occur where the basic assumptions about life are shared. The personalities can remain engaged only if equivalent systems of assumptions mesh. Durability, then, occurs either on the basis of meshing integral personalities, as in loving integrations, or on the basis of meshing central paranoias, as in the hostile integration.

Most primary integrations start with some elements of growth and some elements of restriction. As the interaction progresses, one or the other tends to become the dominant pattern. Loving integrations can grow, dissolve, or deteriorate into a hostile integration. Change can also occur in a hostile integration. It can become increasingly entrenched, dissolve, or, on occasion, reveal itself as a loving integration, and so seem to change.

The true nature of a relationship between two people at any given time is a matter of fact, not construction. The choice exists only between clarifying or obscuring its reality; never, as is naively thought, between what it is and what it should be or might be. Real change or reorganization of the quality of a relationship only occurs when one or both partners change in themselves—either in the direction of maturity or of deterioration. Such change is a gradual process. What is usually meant by a change having taken place in a particular relationship is that there has been some significant clarification of its real nature. The manifest content of the relationship has become more appropriate to its latent content and to its real potential for enhancing or inhibiting the growth of the respective partners.

One consistent aspect of interpersonal relatedness is its mutuality. Relatedness is not a state of being, but, by definition, a constant back-and-forth interaction. It cannot be static, and if either member fails to respond, this failure stops the development of the exchange in that particular direction. The other person confides in us to the extent that and with the degree of depth and sensitivity with which we are willing to confide in him. We get as much cooperation as we give, and if we are concerned with keeping score and making sure that we are not giving more than we get, we have made the project less cooperative than it was before, and are then, by definition, getting even less.

The mutuality applies to the area of empathy, and of sharing another person's experiences. To the extent that one can be interested and sensitive and related to the other's anxieties, if only for the moment, and can communicate this concern, one's own emotional development and resources have been enriched. One has increased one's life experience by knowing more about how another person felt.

There is no unrequited love. The exchange of love, or of any of its essential elements—validation, tenderness, cherishing, and respect—is on a reciprocal base. The dynamism of growth by sharing the other's private experiences and external explorations involves both partners equally. There can be no dichotomy between giving and getting. One aspect of giving is in allowing the other person to share in one's own experience, from which the gain is mutual. If one partner grows, to some minimal degree he has become a more hopeful, more empathic and more tender person, and these qualities are, to that degree, more available to the other.

On the other hand, if either of the two cannot use the facilities available in the relationship for growth, the other either stops growing or looks elsewhere.

In the hostile integration, the degree of potential for destructiveness in each is not necessarily equal, and the commitment to perpetuating the integration may be dependent on different amalgams of hostility and discouragement. But there are no loving victims of hostile maneuvers. The innocent victim is never a noble character. The direct interaction of anger, resentment, accusation, martyrdom, and exploitation is played back and forth. If the camouflage were cleared away, either the degree of valid relatedness would be found to be roughly equal, or the relationship would cease to exist. The initially less angry partner is also, and with equal intensity, interacting on the basis of his central paranoia.

To the extent that the hostile integration is consolidated and made durable, the active hatefulness tends to equalize, the active interaction tends to become more hostile, and the respective individuals more paranoid.

VI

HOSTILE INTEGRATIONS

A HOSTILE integration is a relationship the central dynamic of which is the inhibition of growth. At its core the relationship is essentially between the central paranoias. The meshing serves to implement this defensive activity and to reinforce its central organization. The partnership is based on the congruence of each other's childhood fears, and the modes of coping with them. Each puts the other in the place of the angry and controlling authority and responds to him accordingly. Each is seen as the censor restricting those impulses which would have been prohibited by the parents.

The achievement of this kind of congruence involves accurate, if unconscious, choice of partners. Each nurtures his reproaches against the other for the lack of cherishing and tenderness and validation. In his efforts to control the outward expression of anger, he generalizes the reproaches to include all people. He concludes that love does not exist except as illusion, and that he might as well resign himself gracefully. This intensifies the conflict between denying the basic needs and implementing the search to fulfill them. Tenderness is repudiated by proving, through a long series of daily experiences with the people at hand, that satisfaction is impossible.

In addition, each hands over to the partner his own inalienable right to decide what his life is about. Does he want to live or die? Does he value experience more than security? Does he believe that the search for love is inevitable? Such basic personal decisions are made contingent on the minutiae of the other's behavior. The perspective is funnelled down so that the other person's mood and behavior are experienced as decisive for one's own decision and overall life orientation.

Thus a person by his own act puts himself overwhelmingly in the power of the other. He then concludes that the great danger in life is to be in the other's power, and that letting another person matter inherently involves being grossly and inappropriately manipulated.

The abdication of one's autonomy, placing oneself in bondage to another person, constitutes the central leverage of the hostile integration. This then engenders the unabating pursuit of power over the very person in whose power one now feels. This is true for both the manifestly controlling and domineering person and for the apparently submissive one. The difference tends to be mainly in the content of the fantasies that accompany the power struggle.

Each partner invariably sets about manipulating the other with the fantasied goal of firmly establishing his own security. This greatly intensifies the other's feeling of being taken advantage of, generates rage, and is taken to justify counter power operations. The resultant rage reflects the archaic rage with the domineering parent. Just as in childhood the conscious rage served to repress the feelings of paralysis, terror, and helplessness, in the adult hostile integration the rage serves not only as a weapon against the other, but as the way of obscuring one's current feelings of impotence and fear.

The rage is not, as it is sometimes oversimplified, "rage against oneself." The self-revulsion and self-blame often associated with such rage are, as with guilt, obsessional substitutions which serve in this case to obscure the rage with the other, and to invalidate any tentative friendly move from the other.

The nucleus of the adult hostile integration is not the destructive manifest content. It is the insufficiency of cherishing, validation, and tenderness necessary for growth. The destructive power maneuvers are the secondary sequelae of the thwarting of the potential for satisfaction and growth.

Destructiveness is the active thwarting, manipulation, and exploitation of the other's needs. It can take many forms. The integration can incorporate a mutual conspiracy to deny a particular satisfaction, or the active frustration of each by the other can be complementary and pertain to different areas.

In any such relationship, all the trends of the central paranoia must be implemented to some degree because the functions of cherishing, validation, and respect overlap. Insofar as the search for satisfaction has been perverted in each into the struggle for security and power, the substitutive power drive becomes an additional point of vulnerability for attack. Much of the manifest hostility is on this level, and often essentially obscures the more nuclear interaction. Over a long period the particular patterning becomes stereotyped.

One method of control is the adult temper tantrum, using exaggerated gross unanswerable insults to which the only sensible response would be to absent oneself. Here the control operates partly through creating a feeling of paralysis in the other, to the extent that he is torn between the opposite pressures to respond either to the manifest or to the latent meanings of the outburst.

A different pattern of interaction is marked by the "innocent" covert provocation of unmanageable anger in the partner. The provocative one can then either root for its overt expression in an uncontrolled explosion, or demand that the partner refrain from any open expression of his anger. In the latter case, the partner is led into devoting energy to the control of his own rage, thus participating in the other's control of him. Where the provoked rage is overt it tends to match the intensity of the covert attack, but is usually hard to justify precisely because the provocation is disguised. The goal is the control and power that this operation confers. The open rage intensifies one's fear of insanity, and the other is free to respond with open retaliation or smug forgiveness.

Inner conviction of insanity and shame about it is a crucial adhesive in certain hostile integrations. The person who denies his irrational or unconscious motivations by the disguise of sanity may think of himself as only trying to keep his head above water in a shifting world. However, he soon finds himself in possession of a powerful weapon over the more impulsive, less controlled and seemingly more insecure partner, and uses it if pushed.

He can use it especially effectively if his partner grew up with a mother who actively promoted the fear of disorganization. The covert unspoken mood could be spelled out thus: "No one understands and tolerates your insanity but me. If you expose yourself to others, they will find out and you will be destroyed. You can get on with no other person but me. It is your necessity, not mine, for us to be tied together forever. It is the gift of my great love that I will devote my life to you, self-effacingly, for your need, not mine, providing the sanity for both of us, and placing you under eternal debt, which you can neither forget nor repay."

The maneuver places the other in a permanent position of inferiority and subjugation in which the inevitable chronic rage is then used as proof of the insanity, and as justification for further enslavement.

There are many such sweet fantasy scripts pervasively superimposed on the hostile interaction. The story line is designed to disguise the underlying malevolence as altruism and to justify the impossibility of separation. Usually the dramatized reason for continuity is either that "I cannot get along alone, and no one else could stand me," or "You cannot get along alone, and no one else could live with you."

The roles are innumerable. They include the martyred, the helpless, the incompetent, the hysterically impulsive, the ingenuous, the innocent, the alcoholic, the depressed, or the manifestly paranoid.

One important mechanism for control lies in forgiving or denying the partner's disjunctive activities. It is partly the implied obligation that forces the partner to stay. More importantly, he cannot get his dissatisfactions into awareness, since they are either forgiven or not noticed. He cannot get them validated. Hence his acting-out becomes more desperate and violent and, because of the absence of validation, he remains unaware of its meaning.

One example is the denial of the partner's extra-marital affairs. The pressure on the partner then is to make them more obvious.

The impact is a mixture of guilt and gratitude that he is allowed such freedom. The apparent freedom expresses a contemptuous disinterest in who he is, apart from his physical and economic presence. The apparent ignorance about who the other person is or what he is feeling constitutes an intense pressure on the other to play out the role in which he is cast. By this device, each can outlaw any activity or even any clearly formulated thought which would disorganize the integration. The illusion of freedom for the subversive acting-out which then cannot be validated in the relationship intensifies rather than counteracts the bondage.

Another pattern is the quiet but accurate withholding of tenderness and compassion, masked by subjective apathy, depression, fear of rejection, and self-effacement. If the hostile integration is nonetheless the base of emotional support, the absence of minimal non-verbal validation or tenderness in the presence of the other causes the adult to revert to underlying feelings of self-disgust, unattractiveness, and unwantedness. To the extent that the self-esteem is thus decreased, the situation becomes that much more susceptible to more active hostile interaction.

Sometimes there is a demand for reparations from the partner for all previous sins committed by parents and friends, and by partners of the opposite sex. Any friendly move is now too little and too late. Grudges are cherished, and helpful actions are selectively inattended. Rage, explicit or implicit, recurrently is the response to that which is known and predictable about the other. Each is repetitively surprised and appalled that the partner is exactly what he was selected for being.

A paranoid integration which often passes unrecognized is the mutual covert resignation to the inevitability of personal loneliness. The fact that there is no absolute resonance and that life can never be organized so as to avoid all experience of being misunderstood and uncared for, is frequently misused in the service of the internal cutoff of the need for tenderness. This can be intensified by focusing on particular incidents of "rejection" and by cherishing them in memory. Much of the force retained in these incidents is derived from the exacerbation of the lifelong feelings of self-disgust that is evoked by them. Resignation to inevitable loneliness requires constant vigilance. It involves a derogation of the actual experiences of tenderness that occur in daily living. It necessitates a denial of the actual continuity of many solid relationships when the friend is not physically present. Whatever degree of intimacy exists is compared, at a disadvantage, to some ideal euphoric state in which affectionate interaction goes on without effort or overt activity and in an unbroken continuum. In a relationship pervaded by resignation, the manifest interaction need not be overtly hostile. The hostility lies in the mutual reinforcement of this misconceived attitude about life.

When either partner is pulled into an inerasable direct experience of affection, the immediate impact is to disrupt the life work of accommodating oneself to resignation. The subjective experience in this moment is an overwhelming resurgence of intense and unbearable aloneness. This is accompanied by anxiety which may be camouflaged by anger or depression, but its painful nature tends to be misused by both partners as a reinforcement of the original paranoia. If the painfulness can be understood as an inevitable aspect of the transitional state, and hence as temporary, this insight can serve toward the acceptance of affection as a daily experience.

The hostile integration substitutes warfare for tenderness. The preoccupation is not how to maximize tenderness, but rather, however disguised, how to maximize control and power. The warfare between the two partners is paralleled by the formation of a military alliance of the two against the world. The conspiracy is that each will see the other as the other wishes to be seen, that neither will jar the other's self-system appreciably, and that the continuity of the relationship has absolute priority.

The patterns of control gradually become interlocked in a complex web, the frustration is cumulative, and the reproaches grow and expand over the years. The interaction involves affirmation of conventional restrictions to growth, and covert exploitation of each other's central vulnerabilities. Often one impetus for such a choice is the wish to recruit an ally talented in the arts of interpersonal warfare for help in the battle with the parents. Sometimes, also, an unconscious determinant in the choice is the impulse to learn to cope with the anger as the presumed condition of being able to experience any friendliness. The attractive force is the other's anger. The move to get together is not to convert the anger into love but to get into a position of control over its expression.

The subjective experience of the initial hostile impact, as with the beginnings of love, can vary from disgust to ecstasy or from intense longing to boredom. Short of thorough commitment, hostile integrations have a certain tendency to a short course and a high turnover. After the pattern is well integrated and serves its repressive function smoothly, the longer hostile integrations continue the harder they are to get out of.

The two people may get together on the basis of hate at first sight. The subjective experience can be of meeting someone with the same values, with whom one does not have to pretend.

More frequently a hostile integration develops when a constructive but limited relationship has outlived its usefulness and has ceased to grow. The only means by which two people who are no longer growing together can maintain a close interaction is by bringing into play their respective central paranoias. Discouragement becomes their currency.

When the hostile integration evolves from the deterioration of a better one, a commitment to appearances is often an important factor in maintaining the continuity. The subscription to conventional fictions, such as the maintenance of the façade of maturity, may be the point of departure for developing the paranoid interaction. Resentment, reproach, and intimidation flourish, so that the relationship becomes indistinguishable from one chosen initially for its hostile qualities.

Sometimes the deterioration is precipitated by the increasing openness of the constructive aspects of the relationship. The point of commitment to the loving aspects often intensifies the paranoid fears and turns the balance. The force of these fears derives from the old expectation of destruction by the envious parents for the self-assertion involved in manifest commitment to independence, experience, and the search for love. A frequent compromise is that the external forms of commitment are braved through, but the loving qualities of private interaction are reduced to dust. Loyalty to the parent is disguised by overt defiance, but the parents' values

and patterns of interaction are copied secretly. The partner becomes the symbol of the frightening push toward growth in oneself; therefore he must be discredited or rejected. He is thus partly
experienced as dangerous and destructive. The more unlike he is
to the parents, the more he is regarded with suspicion and feared.
The content of the fear may either reflect this true state of affairs
or be an inversion of it. The loved one is then seen as being just
like one of the eidetic parents, or as their ally, and becomes to
this extent an object for rage and displaced retaliation. Thus, mir-
roring both poles of the conflict about growth, the partner is feared
both for his resemblance to and difference from the parents.

The quality of interaction in an entrenched hostile integration
can be divided somewhat artificially into the manifest and the latent
content.

The manifest content is an important integrating force. It involves the implementation of a pattern of life which limits and
takes precedence over spontaneous individual moves toward necessary experience. The dedication to social appearances is reinforced.
Activities and moods are subject to mutual judgmental review
rather than to unprejudiced sharing. If an attitude or activity does
not meet the mutually agreed-on standard of maturity, respectability, creativeness, responsibility, status, or efficiency, it is expected to be restrained rather than understood. The mutual
agreement is that the partnership dispenses approval or disapproval, which is implemented by a degree of self-control and control over the other.

The resultant interaction is a fight for power, each one trying
to get some leverage on the dispensing machine of the conditions
of life. The mutual project subtly becomes a conspiracy to produce
a reasonable facsimile of whatever is the agreed-on script for a
proper relationship.

Activities that would evoke new emotional experience are progressively restricted. There is an energetic development of non-
controversial projects which involve no anxiety, and therefore no
important growth experience for either.

Superficially, the manifest content may vary from murderous
quarrels to Hollywood bliss. Open acting-out of sadomasochistic

patterns derived from the childhood fears of isolation and death is more frequent in the hostile relationship than in the loving. Nonetheless, in the hostile one by far the most frequent and conventional pattern is the quiet, cooperative, social marriage in which the mutual agreement is the restriction of life. Nothing shall be experienced too deeply, neither love, nor friends, nor creative work; nothing takes precedence over the formal structure of the chosen illusion.

The latent content of the hostile integration is the mutual exploitation of each other's inner fears and needs. The conspiracy for mutual agreement and review of spontaneous independent action is in itself an invalidation of either member's individual perceptions. The mutual accumulated anger precludes the unhampered exchange of tenderness. Fear accumulates in each, in proportion to the other's latent anger, and is expressed in the demand that the other control the expression of his anger. Between them, open anger is regarded as "being angry," while the chronic, controlled anger is identified as "being on good terms."

The underlying assumption, essentially correct by now, is that the reality of anger is continuous. Lack of respect is implied in the attempt to control the other's behavior and is manifest in the derogation of the other's specific concerns and needs. An essential indifference to the other's survival as such in no way contradicts an exploitative investment in him or her as a source of income, respectability, social contacts, or as a housekeeper, amanuensis, household pet, or hot water bottle. There is nothing personal here; the convenience is what matters.

An almost invariable feature of the hostile integration is the active provoking of the other's vulnerability in the service of power, revenge, and sadistic gratification. Invalidation is timed accurately to exacerbate the other's secret disorganization and confusion and to render him subject to one's own control. One partner's loneliness is the trigger for the other's rejection; one partner's fear sets off the other's anger. A moment of discouragement about oneself is the stimulus for the other's contempt. Self-destructive impulses, under whatever camouflage, are responded to in such a way as to increase them, thus placing the victim in increased physical and psychological danger.

This mutual cat-and-mouse game is almost never conscious. It is in direct opposition to the conscious picture each has of himself. When the hostile integration is well established and feels inescapable, insight becomes dangerous. Any clear experiencing of one's own anxiety, fear, loneliness, or inadequacy, if perceived by the opponent, puts one temporarily in the power of his ill will. Once demonstrated, the other can hold in restraint his ability to break through the defenses to precipitate and exploit such momentary helplessness, and can reserve it for crucial moments.

The quality of interaction intensifies in each the degree of insecurity, fear, anger, and isolation. The purpose of the suppressing of these qualities is not, as it may erroneously appear, to dissolve them by giving priority to love. They are a burden insofar as they are points of vulnerability in the going battle. But they are desperately needed, inasmuch as they constitute the munitions plant for the battle. The defensive requirement is to have them on tap to provide the energy and wherewithal for the fight, while perpetually postponing clear awareness of them. For both these reasons a constant preoccupation with the other person and his reactions is indispensable. If the premium is on durability of the relationship per se, each develops a commitment to winning the control of the going operation. Any momentary victory for one only strengthens the reservoir of resentment and revenge in the other.

The hostile integration can be extremely durable. The very nature of the interaction is to obscure rather than to clarify its basis, and so to postpone change. It was implemented to dissociate the push to growth. Somewhat more superficially, the attractive force was the opportunity for the control of and the postponement of awareness of the inner feelings of fear, confusion, isolation, inadequacy, anger, and frustration, which had been powerful inner emotional patterns since infancy.

In this sense, the adult hostile integration is the direct extension of the hostile aspects of the integration with the parents. The superficial pressure for maintaining the relationship is to keep these incapacitating, persistent childhood terrors in restraint. The hostile interaction, then, increases the reservoir of such feelings, and any break with it involves a transition of more intense discomfort than was being avoided at the onset. The longer it goes on, the more painful is the transition of getting free of it.

There are many cultures in which divorce from either one's parents or one's spouse is inconceivable. In this case, the question of justifying the hostile integration never comes up. Each member is ostensibly involved in nobly making the best of a bad bargain, and secretly in giving the other one a hard time of it. Each one is then being perpetually forgiven for the sin of being who he is.

When separation or divorce are not ruled out categorically, the justifications for continuing a hostile integration become more elaborate and explicit. The core of the contract is that the formal alliance takes precedence over the true needs of either member. Independence and spontaneity are usually interpreted as directed against the partner, and considered hostile. The formulation is that, "Since you know how it upsets me, you would not do it if you had any concern for me." If a move toward growth does take place, the self-assertion involved in carrying it out will likely be accompanied by considerable anger. The resulting defiance tends to confirm the accusation that the intent of the move really was hostile.

Eventually, neither party can tell how much was productive and how much was protest. Such a contract may limit other relationships to include only objects, or acquaintances, or friends of the same sex, or friends of the opposite sex, excluding sexual relations, or to sexual affairs which exclude friendship.

Conversely, the unwritten contract may virtually require extensive outside activity of any of these types. What is now considered hostile and destructive is any attempt to bring more life into the central relationship with the spouse. The requirement may be either to limit the degree of achievement in the world, or, by demanding success, to limit the extent of leisure and hence the opportunity for intimacy.

Another way of limiting growth is to divide exclusively the legitimate areas of development. One partner is to be confined to the children, the other to the outside world. Still another way is the more frankly exploitative formulation of unilateral growth, that one stays home and provides the cushion for the other's growth, without getting and using the freedom for a similar range of development for himself. The results are the same. Insofar as the particular relationship is the matrix, the growth in any area only goes as far as the slower partner.

One excuse for this double standard is social prestige. It does not look good for one's spouse to be frankly busy with an adolescent project. Another excuse is "insecurity," expressed as the fear of losing the partner to the alternate activity, or as the excessive need for full-time contact and reassurance. Undue insecurity is a euphemism for an insatiable need for security, which pushes for total control over the other person. However formulated, the resistance to and the fear of growth in the partner is always in the service of resistance to one's own growth.

Where the relationship is consciously felt as being a poor one, the rationalization for its continued existence is that there is no alternative. This assumes many forms. There may be the presumed obligation to other parties, either parents, children, employers, or neighbors, to maintain the fiction of inevitable continuity. There is the fear of "destroying" the spouse, and the fear of his revenge. There is the fear that the partner would do better if separated, and this would be an indelible injury to one's pride. Often, the fear is of going crazy or of being the laughingstock, and so on and on.

An occasional rationalization for staying together, in sophisticated circles, is "to get clear on the hostile relationship before one tries to make a better one." This is a thin disguise for the commitment to continuity. The anger and despair mount, the interaction becomes more obscured with time.

A variant of this is the patient who says he needs analysis because he wants a divorce. Often the real intention is that by getting into analysis his hope is to postpone any such move for two or three years.

One popular myth is that if one relinquished one's collection of resentments and recriminations and consolidated what little is valuable in the relationship, this would involve a lifelong entrapment in a situation which provides inadequate satisfactions. Partly this is a rationalization to conceal the acquiescence to the lifelong prohibition on satisfaction, and partly it is in response to social pressures. Divorce, in our culture, practically requires proving publicly the miserable moral character of the other.

Resignation to the *status quo* is frequently formulated as the attempt to "work it out." Restated, it is the attempt to negotiate a better non-separation agreement between the two self-systems involved. No amount of "working out" can create a congruence and aliveness that is not there. If, indeed, there is a genuine degree of congruence and aliveness inherent in the relationship, this can only be illuminated if clarification is undertaken with an open mind as to the outcome. Given the commitment to continuity and restrictiveness, no verbal agreement for mutual freedom can be adequately implemented. The restrictions operate automatically. The bland, friendly, distant, unalive central integration which does not inhibit growth is a wishful illusion.

The wish for rupture of a relationship, or for legal divorce, does not always stem from the pressures for growth in the integral personality. The impulse for separation may derive from the security functions of the self-system. Here the motive is the wish to escape the threat occasioned by the possibility of a new level of aliveness and loving interaction within the relationship. Most frequently, separation of this type occurs early in a relationship, before the consummation of legal marriage.

A question arises as to the possibility of a shift over from an essentially hostile integration to a loving one. On the one side, the explosion of all the fear, anger, loneliness, and despair in a hostile integration into awareness could disrupt the relationship. The alternate possibility is that the explosion could begin to dissolve the mutual commitment to the hostile interaction. In any case, such clarification would be indispensable to discovering the real state of friendliness and to exploring to what extent there is a basis for a loving reorganization of the relationship. The very effort to effect such clarification, to speak the awful truth, is the first act of love.

The rarity of such an outcome is no accident. One reason is that in the hostile integration there is, of necessity, a stake in obscuring the truth. As a result, real self-awareness shrinks, as does interest in honest mutual exploration. In this context, some flash of insight into the true state of affairs is likely to be hurled at the partner as an attack rather than to be the point of departure for genuine exploration of the roots of the unacknowledged problems. The attack then begets counterattack, and so on it spirals. Nonetheless, where there is a considerable degree of covert congruence in the integral personalities, the possibility of a shift over to a loving relationship is ever present.

There is a more important reason to account for the infrequency of such a positive outcome to the hostile integration. In most instances its very origin lies in the perpetuation of a relationship after its validity was exhausted. What was once congruent between the two integral personalities resulted in growth in each. The new reality is now that little congruence remains, so that if the relationship is to be maintained the interaction must increasingly go on between the respective self-systems. If the hostile integration began out of the exhaustion of the wherewithal for love, it could only be wishful thinking to imagine that it could return to love.

A further frequent rationalization for continuing a hostile integration is that, if one relationship is dissolved, the next one will inevitably be a replica of it. This may or may not be true. What is certain, however, is that if the validity of the current relationship has ceased to exist, the chances of the next one's succeeding are greater, to the extent that the person's growth was responsible for the termination of the first.

The termination of an unproductive, hostile relationship is hindered by the tendency of the relationship to limit alternate experience in living. Nonetheless, the constructive basis for such separation lies almost wholly in having had some experience, overtly or covertly, with better relationships. While one is fully immersed in a restrictive relationship it is not possible to understand clearly the nature of the interaction. Alternate experience allows some perspective on the central fallacies and some conception of other possible patterns of integration. Without it, one operates under the assumption that everybody is really like this—his parents, himself and his partner—that there is nothing better to be hoped for in the world, that other people really know it but won't admit it, that it is better to maintain what convenience one has than to embark on a hopeless and perpetual search.

While one is developing new relationships there can be momentary insight into the nature of the primary relationship, but this insight cannot be lasting until it is conceivable that the primary commitment could be completely relinquished. Similarly, the quality of the new and better relationships cannot come clearly into awareness until the hostile one can be given up. No move will be made until there has been some growth in the integral personality, if not in the self-system.

Subjectively, it is like the nightmare of jumping from one roof to another, with the terror of falling in between. There has to be a moment of subjective isolation inherent in the transition to another relationship significantly different in its nature.

Another central dynamic for the maintenance of the hostile integration and the fear of its dissolution lies in the residue of childhood fears of making new acquaintances and exploring new relationships. The thought is that "I might do worse and so I'd better hang on to what I've got."

The real fear is that in the more unstructured situation one might be swept into more aliveness than one can conceive of while still in the hostile integration. To the extent that the integral personality has assimilated positive experience and capacity, the goals of satisfaction are now more attainable than in childhood. But, parallel with this, every loving relationship that has been amputated in the service of the security needs has added to the reservoir of grief that will be re-experienced in any extensive coming to life.

When a hostile integration is interrupted, whether by divorce, temporary separation, illness, or death, the proximate emotion is anger or some defensive derivative of it. The most frequent of these are depression, which is mistaken for grief or longing, and pity, which is mistaken for compassion. The pity is a thin disguise for contempt, or a rationalization to obscure fear.

The fear that is felt is partly rooted in the nature of the growth process, partly in the infantile fears that the anger is unsuccessful in warding off, partly fear of the intensity of the angry feelings themselves. The anger is by no means experienced only by the one who was left. Both partners are usually about equally angry with each other by now, although the outward form this takes can vary enormously.

Convention dictates that the one who was left should be angry rather than grateful. Here the anger that derives from hurt pride and the insult of being abandoned is a defense against the much deeper and more pervasive rage. The anger experienced by the person who took the initiative to leave is often less apparent.

The basis for such rage, upon rupture of a hostile integration, lies in the nature of the relationship. At its deepest level, the hostile integration functions to restrict growth and to maintain the dissociation of aliveness and love. However, due to the layering of the emotions and the machinery of the interaction, the impulses most proximately being both provoked and controlled are the angry and disjunctive ones. The immediate function of the hostile integration is to obscure the hostility, and if the thread of commitment is broken, anger is the first emotion to emerge. Because of this, an individual can sometimes maintain the lifelong illusion that the hostile integration is the direct expression of the move toward people.

The fear and rage initially released by the break are intermingled and may be mistaken for each other. One is likely to experience the extent of the other person's rage much more forcefully after the separation. The increased awareness of rage must evoke even stronger expectation of retaliation.

The accumulated anger is rooted partly in the old unresolved rage of childhood, partly in the genuine insult to each person in which the fact has been overlooked that the insult has been mutual, partly in the deprivations and inhibitions to growth, and partly in the waste of time. Often the most important discovery is that there has been no valid relationship for a very long time.

The rage that is liberated at the moment of discontinuing the hostile integration is a transitory phenomenon. The process of getting it clear, however frightening, is the process of getting over it, both in the immediate context and over a lifetime. In the moment, it is projected onto the surrounding environment, and friends suddenly seem destructive, dangerous, unreliable, or crazy. They seem to incorporate the worst defects of the original mal-nurturing environment. When the rage can be fully experienced, it can be sorted out. The parataxic perceptions from which it derives can be identified and re-evaluated, and the rage can be genuinely dissipated.

The alternate choice, the avoidance of this direct experience, has to result in careful preservation of the original childhood rage. The illusion is that if it is buried deeply enough it does not need to distort the patterns of later relatedness. But as any later relationship becomes genuine, warm, and spontaneous and reaches back to draw on the childhood base of the need for tenderness, this subsequent relationship tangles with the carefully hidden reservoir of rage. The only way around this dilemma is to go through it.

It is exceedingly difficult for a person to grasp the obvious fact that the way to end a destructive relationship is to remove oneself from it, to give it up. As an infant, no matter how bad the emotional situation, one could not walk away from mother. She was necessary for survival. The adult break involves a moment of intense fear of nonsurvival as carried over from infancy. If the interaction had served to disguise the absence of relatedness, the feeling of nonexistence consequent to the break is intensified.

The feeling of nonexistence is directly related to the function of validation; the infant initially only knows he exists to the extent of the validation provided. If what was primarily validated in infancy was malevolence, the condition in adulthood for a feeling of existence is a malevolent interpersonal milieu. In this case, the termination of the hostile integration carries additional archaic overtones of self-dissolution.

The fear of nonexistence is also heightened by the sudden disorganization of the external patterns of life. This precipitously throws the individual onto his resourcefulness in creating new content for his life just at the time when the subjective experience is one of confusion and disintegration. The transition involves stepping out into a void, experiencing panic, discovering that the state of panic and the impulse to grow evoke tenderness from unexpected sources, and developing some basis for a beginning reorganization of patterns of relatedness. It is helpful if one tries to marshall some allies in anticipation of the crisis, if only among one's casual acquaintances.

To the extent that the hostile integration grew out of the deterioration of a loving one, the initial loving interaction has moved both persons ahead into new dimensions of exploration. Either or both may be uneasy about returning to the central thread of growth at this new level, with the uncertainty and strangeness of feeling a somewhat different person. This is usually inverted in consciousness into the "fear of being exactly where I left off." The termination is often misperceived as an invalidation, rather than a validation, of the original constructive quality of the experience.

The cultural misconceptions reinforce the neurotic fear. It is taken that if two people were ever genuinely in love, they must still be. Conversely, if they admit that they now are not, it becomes a social admission that they never were.

The commitment to continuity is reinforced by a culture in which faithfulness is extolled, regardless of its human meaning. Changes in the circle of one's friends are thought of as disloyalty or as social climbing. There is pressure to perpetuate already-organized relationships well beyond any constructive use to either. The social opprobrium is especially intense in regard to relationships to one's own parents, to marriages involving children, and to marriages of long standing.

The fear of walking away is often reinforced by one (or both) partner's threat of social blackmail. The fiction is promoted that one person is indispensable to the other. The implicit or open threat may be that the "fragile" one will have a nervous breakdown, will fall apart, will blacken the other's name where it matters most, will retire from all social intercourse, will be punitive and vindictive in demands for money, will do physical violence to the other's friends—whatever might be most effective in the circumstances. The threats may not be explicitly stated, but rather reformulated in some euphemism which is understood by both.

These euphemistic threats can even be cherished as the proof of the intensity of the love between the two.

For one partner of a hostile integration to break it up is usually a favor to the other, no matter how the other may protest. Exploitation is reciprocal; the motive for allowing oneself to be exploited is in itself exploitative. The outwardly possessive partner has invested a great deal of effort in virtuously sitting on his own disjunctive impulses, and feels that he should receive compensation. The hostility flourishes because the degree of implied intimacy is beyond the real capacity for intimacy in this particular relationship. If either has the capacity for fuller intimacy with another partner, he will find one. If he has not, it will be a relief to be free to organize his life to provide that degree of intimacy with friends or acquaintances of which he is capable without pretense.

The maneuvering between the two as to who will take the onus of the decision to separate is dictated by considerations of social censure. Does it look better to leave or be left? The decision in fact is always mutual, even though there is frequently a great deal of subtle manipulation to throw the onus on the other.

Another mode of the hostile integration is the life pattern of isolation. This is illustrated by the person who, no matter what alliances, chumships, mother figures, lovers, or partnerships are available to him, persists in seeing himself as one against the world. Often, with little manifest hostility, he can go through many of the motions and some of the emotions of friendship, but is unable to truly share his inner joys and his inner pain with another human being. This can be acted-out in one prolonged unexciting marriage, in a rapid turnover of apparently intense friendships, or in the literal isolation of the unattached person who would seem to form no adult central relationships. The result is the same. By avoiding the experience of direct and intimate sharing, one reduces the overall intensity of all other experiences. Needs can be perceived, but at a great distance, as if one were looking through the wrong end of a telescope. Friendly moves can often be perceived and responded to, up until the moment when they would have to be experienced as one-to-one, as preliminary to the experience of two people becoming one, if only for an instant.

This limitation can go along for a long time with a high level of objective performance.

Life is seen through a slightly smoky glass. There is a general depression which is so pervasive and unrelieved that the person can fail to notice it. Or he may believe that his state is inevitable and universal or at least that it is inherent and intractable in himself. The rationalizations for maintaining his isolation are both grandiose and self-deprecating. He avoids friends with whom he could share, and he avoids sharing with the few friends he makes with whom it would be possible. Either the other's perceptiveness and capacity for loving is described as not good enough, or the person sees himself as not having the capacity to evoke genuine tenderness from the other. The emphasis is thrown on the unbridgeable differences rather than on the areas of community.

If what he has in common with the other would be validated, the person would come into contact with the totality of his life experiences in a broader dimension than they had been perceived for many years. In the moment this reawakening would be intensely painful, and it is difficult for him to realize that the pain is transitory.

It is hard for the person to come by the notion that the sense of aloneness is a subjective illusion, that the isolation is a self-imposed state, and that it is maintained at the cost of much effort.

VII

THE EXPERIENCE OF LOVE

LOVE IS A function of the interaction between the integral personalities of two people. As we have said, the conscious experience of this interaction can range from some direct awareness of its nature to phobic repudiation. Thus the response to the experience may be avoidance of the other person, or a move to develop the relationship. If the latter, questions arise as to the frequency and duration of getting together, and as to the circumstances. The mode of getting together frequently expresses a compromise between validation and repudiation. A move to get together initiates the process of the mutual and gradual bringing into awareness of a state of affairs which is already in existence, and of implementing it in real life.

Unpredicted barriers to further intimacy may then turn up, or unexpected areas of potential growth may be discovered and explored together. The extent of congruence either is or is not there. The real limits of growth of the interaction are also not matters of decision, but are there to be discovered. A person does not really choose a loving relationship; it happens to him.

The quality of loving interaction between the two integral personalities has certain unique characteristics. The other person's feelings have become a segment of one's own. The two people have known each other in such a significant dimension that neither can ever be utterly unaware of or indifferent to the misery or the happiness of the other. Misery in one person is experienced directly as painful to both, and joy as delight for both.

Moves to alleviate the other's entrapments and to enrich his satisfactions are as automatic and spontaneous as if for oneself. They are not experienced in terms of the exchange of services, as in the juvenile mode, or in reference to being a good or giving person. They are often derogated consciously, and, as with the moves to improve one's own well-being, often take place under a cloud, if not essentially out of awareness. The loving one finds, often to his dismay, that he is in a relationship of resonance which is going on outside of and beyond his conscious control. If his friend is in trouble, his own sleep is disturbed. If his friend has ulcers, he, too, finds food uninteresting. It can be as literal as that. The connectedness happens to them.

Smooth interaction between the two persons of areas of the integral personality makes it possible for marginal experiences to become more available to consciousness in the other's presence. The subjective experience is that "Here is a person with whom I am more alive, more of a person, and in better contact with myself." The pull to extend and expand this dynamism in oneself, in the other, and in the relationship is a powerful one.

A reasonable synchrony in the rate of getting acquainted, an equality in the pressure to get through the initial obstacles, more quickly or less, is important for the relationship to get off the ground.

The individual can react variously to the initial impact. He may deny all of it, depart in panic and undertake to "get over" and firmly dissociate the experience. He may allow himself partial satisfaction, setting up an attenuated and restricted integration. This latter can sometimes slowly evolve into a more open and complete relationship. On the other hand, either or both people may use the frustrations of the artificially-diluted relationship as a lever to deteriorate the initial experience of community and intimacy into a hostile integration.

With extreme luck, both people respond with enthusiasm. This experience may last for a night or a lifetime. Only good can come out of it. The experience of sharing is consolidated. Each has become, to some degree, a permanent part of the other, and the relationship can expand or wither away in accordance with its genuine potential.

If the initial impact is followed by flight or intense repudiation by either partner, the process of burying the experience is painful. It involves acute reliving of the process of repudiating the need for tenderness. The particular repudiation has to be accounted for, often in terms of rejection by the other, or of one's own or the other's inadequacy, or of one's evil fate. Whichever excuse is chosen, the experience has strengthened the paranoia. The pain of this active cutoff of hope has been added to one's private reservoir of despair.

If any subsequent similar relationship does get consolidated, it will release this one from dissociation, and thus involve some reliving of the pain of this separation. The flight from the loving relationship has thus made future satisfaction both less likely and more painful. It also, in fact, prolongs the preoccupation with this particular relationship, by perseverating on the active process of repudiation and on the increased sense of frustration.

If the relationship is consolidated on an attenuated basis and then slowly deteriorates because of the stresses involved, the effect is in the same direction. The proof of the futility of love has been reinforced by being acted out. On the other hand, in both cases, some actual experience in how things could go has been accumulated in the initial phases; this knowledge remains available to the integral personality, if not to the self-system, and strengthens the inner drive to again seek out tender interactions.

The denial of love is often rationalized as the wish not to get involved, or as the fear of long-term commitment to the present partner. The fear of commitment can either represent the uneasiness about integrating a relationship which could be productive and durable, or it can be an accurate prognosis of probable short durability if fully explored. In the latter case, relating with major reservations can only serve to prolong the involvement with a minimum of satisfaction.

Further, a valid but genuinely limited integration, based on the need to expand particular delimited aspects of the personality, tends to deteriorate into a hostile integration if there is the pretense that it is very much more than it is. For example, two people whose real basis for getting together is to learn something about sex often get involved in the notion that they have chosen lifetime partners. Where the relationship is burdened with such superfluous impediments it must be effortful, resentful, and unloving. It can only be maintained by the security apparatus of the self-system.

If the initial attraction is validated and the relationship consolidated, both the relationship and the respective individuals in it continue to grow. When the relationship is no longer growing, it becomes empty, whether this is felt or not.

Some integrations have more inherent durability than others. Some are for the validation of a specific function, emergent in both, and so do not involve evaluating the partner for relevant congruence in any other function. Many one-to-one growth experiences are integrated specifically for getting to know each other, with no concern for either's capacity to serve as a base for continued interpersonal growth beyond this. Some are integrated primarily for the mutual exchange of psychiatric nursing during a growth experience within or external to the relationship, and have no particular relevance beyond the emergency. Most are integrated in terms of the level of interpersonal arrest of both people, and with little regard for the expected rate of progress after the immediate barrier is overcome. Any such factors making for limited durability can be used to derogate the actual potential for expansion in a particular relationship.

It is the natural and constructive sequence that when the fulfilling aspects of a relationship have run their course and no longer contribute to mutual growth, the relationship, as a going concern, comes to an end. There are many loving relationships where the two people meet, enjoy each other thoroughly, and drift apart by mutual consent, without any fanfare. They have both grown; each has become a part of the other. Each is, to some degree, one of the validators of some line of development in the other. The memories are fond, the occasional contacts are friendly, and there is no compulsion to pretend to more than what exists. This happens with acquaintances, chums, teachers, and "proxy parents." It can also happen in the more conventionally recognized primary integrations—with parents, in love affairs, or in marriages.

It may turn out that the central developmental projects of each person are so far apart that the partnership is irrelevant in the long run. Although mutual growth is essential for the continued validity of a loving relationship, it does not guarantee permanence. In fact, it can be the very basis for its dissolution. It is never damaging to the growth process to walk away from a positive experience which has been allowed to run its course.

Validity refers to the latent content of a relationship—the quality and extent of connection existing between the two integral personalities—in contrast to the manifest content, which is a function of the self-systems. Validity of a relationship, then, is clearly related to the dimensions of growth and love. It has nothing to do with the manifest nature of a relationship, say, its storminess, which is a reflection of self-system activity.

Also, validity and durability are quite independent variables. Productive durability is a function of the extent and quality of congruence of the respective integral personalities. On the other hand, the durability of hostile integrations where validity may be virtually nonexistent is of a quite different nature.

Here durability stems from congruence of the security functions of the two self-systems. Specifically, the two security systems may have so meshed that durability is not only ensured; it would require dynamite to cause the two people to part.

Where the relationship is valid, there is no sharp line of distinction between short- and long-term integrations; they lie on a continuum. In the course of the longer ones there is a tendency to shift the emphasis from the initial exploration of each other and the implementing of the immediate projects, to sharing and aiding in the core development of each.

The chance of relatively long-term validity is increased by several factors. These include the amount and quality of previous experience in selecting partners, the common denominator in the level of maturity and the tolerance in each for the specific processes of growth in the other. Experience in choosing partners helps to counteract the hypnotic influence of the parental values, helps to develop adequate confidence in sexual performance and social technique, and makes for some capacity for predicting both how oneself and the other will behave as the relationship progresses.

Any durable, valid integration is contingent on the two members' operating in roughly the same arena of interpersonal values and lines of development. Gross inequalities vitiate the impetus to sympathy and sharing. There is less basis for understanding each other. Each one's lack of comprehension is a major hazard to the other's most cherished projects. No degree of permissiveness can counteract each one's puzzlement about how the other one's activities can be important to him.

Such unequal relationships sometimes occur when the search for parenting is mistakenly substituted for the search for a partner. The immature person tries to find a person more mature than himself in the misguided attempt to recreate the specific directional flow of tenderness from the other to him that was so insufficient to his need in his very early years. A further reason for this attraction to the more mature partner may be the person's need to repudiate anyone like himself, to avoid either the self-awareness or the public exposure of his limitations. This person rests at a certain level, as the best that he can make of life so far. The more mature partner may choose it because, after a prolonged period of anxiety resulting from significant growth, he has regressed to about the same level of functioning. If the association provides some modicum of structuring and security, the more anxious one calms down, and is often then puzzled about how he got there.

Given that the genuine level of maturity of the partners is approximately equal, the higher the level, the better the chance of durability. People of greater maturity have more capacity to provide and to utilize tenderness, validation, cherishing, and respect. They also have less impulse to intrude the exploitative and manipulative values of the childhood and juvenile eras. They have less need to dissociate major areas of perception and experience, hence are more versatile in the variety of experience which each is able to share with the other.

At any level of maturity, a grasp of the nature of growth

processes will facilitate the interaction. This would include: understanding the necessity for both outgoing experience and privacy; some capacity to handle the various expressions of anxiety; some understanding of the roles of hope and despair in one's own and the other's growth; acceptance of the inherent mutuality of growth, that one's own is augmented by one's partner's; and some understanding of the value of mutual confiding, both talking and listening. It would particularly include a degree of freedom from the rigid social conventions of the culture, and some familiarity with the implications of the dynamisms of possessiveness, intimidation, power, and thought control that tend to be carried over from the fight with mother.

All these factors contribute enormously to the durability of a loving relationship. Beyond this, there must be sufficient exchange of tenderness that the necessity to repudiate this vital need is slowly dissolved. Since the expansion of the experience of love inevitably evokes archaic feelings of anger and fear, there must be sufficient capacity for validation without vindictiveness, to make it possible to re-experience the anger and fear and to progressively identify their irrational childhood sources.

To the extent that each one can perceive, respond to, share, and accept the other's most urgent necessities for growth, to this extent the relationship between them can be viable and durable. However covert, our deepest commitments are to those people who are making it possible for us to get ahead with living.

The most active contribution one partner can make to the loving quality of the interaction with the other is his commitment to his own growth. If the individual and mutual growth process does not take precedence over the maintenance of continuity, the relationship is to that extent thwarted.

It is exceedingly probable that the course of love will be stressful and punctuated with crises. Since its full expression was forbidden by the parents and, less overtly, by the culture, love usually takes place under a cloud. The superstructure of camouflage is stultifying and burdensome. Further, the resulting subversive quality mobilizes the central paranoia and intensifies the commitment to the parents.

The challenge of love to the *status quo* invariably reawakens the childhood emotions of fear, confusion and isolation, at times with intense conviction. These emotions may be directly projected onto the partner. They can take the form, for example, of the fear of being controlled, manipulated, rejected, humiliated, or driven crazy. Often these more primitive reactions are disguised by an overlay of anger, or hopelessness and depression. The real roots of these derived emotions are seldom recognized, so that they are instead projected onto something in the current relationship.

This can set off an obsessional spiral of acted-out attack or rejecting withdrawal—sometimes with a bucketful of tears and guilt feelings. If retaliation follows from the partner, it is then taken as the proof of his hatefulness, and so on and on. The net result of this explosive comic opera is the thorough obfuscation of the moves toward intimacy that set it all off. But still more unfortunate is the temporary victory for the central paranoia of each person, insofar as the *status quo* has been reinstated, and distance re-established.

Crises in a developing love relationship can take many forms. To the extent that love stimulates growth in one member, this growth has already been formed in his own inner plan. It is unlikely to be delicately synchronized with the partner's most immediate problem in life, and thus may confront the partner with projects that it would have been more convenient to postpone.

Hitherto dissociated needs which have emerged into awareness can now feel overwhelming. They can even seem to be directed against the integrity of the very relationship which promoted their rebirth. For example, the process of maturing subsumes the extension of the experience of love, sharing, identification, and rapport to people other than the partner.

If experience outside of the partnership validates a new activity, emotion, relationship, or a new line of development, it may be disturbing to the previous integration and be experienced by the partner with anxiety, envy, jealousy, or anger. On the other hand, to the extent that a widening range of genuine and constructive other relationships and activities more securely integrates the function of growth, the expansion of the particular relationship can proceed with less explosiveness, if only because the widened base of relatedness undercuts the paranoid consequences of dependence.

One is likely to think that there is a dilemma inherent in the central integration between two adults, neither of whom has reached the preadolescent level of development. Neither has any great capacity for love, yet the relationship is supposed to provide the basis for growth. Nonetheless, especially if the pretense of maturity can be dispensed with, such partnerships provide both useful experience and a useful base for further growth. There is nothing impossible about getting from the childhood to the juvenile level, or from there to preadolescence, after age twenty-one.

The process is much the same as before puberty, and in some ways the facilities are better. One now has more freedom to choose his contemporaries and his validators, more freedom to compare notes and to develop a perspective on the restrictive family traditions. If he can consolidate, say, the juvenile mode of relatedness at age twenty-five, he is then free to truly explore the new experience of the chumship and so enter preadolescence. This is the way real interpersonal growth takes place, in spite of the handicap that it is usually covert and almost always under the outward camouflage of pseudo-maturity.

The capacity for the consolidation of loving relationships is not solely and directly a function of the level of maturity. The exploitativeness of the childhood era and the manipulativeness of the juvenile adjustment are serious handicaps to loving, but do not entirely rule it out. Their core is in the community of experience, in willingness to use this community of experience to expand one's own horizons by adding the other person's viewpoint and in allowing each other the freedom to use other opportunities to supplement the deficiencies of the individual's limited experience. The impulse to restrictiveness is most fatal to the constructive development of the particular relationship, as well as to the respective overall development of the persons involved.

Integrations focused on childhood and juvenile developmental needs can be very temporary. If the same process were going on in the childhood years or in the juvenile years, we would take it for granted that there would be frequent shifts. If most of it is being done in adulthood, it is not surprising if a partner chosen as a juvenile ally or for parallel play turns out to be someone whose rate and route of growth are not essentially congruent with one's own over the next twenty years.

Role-playing, pretense, secrecy, and dramatization are among the more frequent maneuvers used to limit the full expression of loving interaction.

Often, one pretends to be in love in the hopes of eliciting love from the other. This is always a self-defeating move. It is never possible to make more of a relationship by straining; one can only make less of it and obscure those aspects that are real and could grow.

The search for love is limited by the commitment to secrecy and disguise. One way to keep something secret is to believe that it is something else, and then to dramatize the belief. The particular content of the drama is interchangeable and of quite secondary importance. Its thoroughness in limiting the real interaction is partly because the partner is called upon to play the complementary role in the drama. The repertoire includes the drama of unrequited love, the drama of the perpetual quarrel, the drama of mutual exploitation or of mutual martyrdom.

The theatrical camouflage is, to a certain extent, set up as the condition for the exploration of the relationship. With a partner who won't play the game at all, the gears never mesh. In spite of the security operations, some very useful experience eventually gets assimilated. The superficial dramatics about love, and the feeling that the other person's love is insufficient or insensitive, partly express the inner fear that there isn't any love at all, that what is going on is entirely hypocritical.

If what is real can be acknowledged without prejudice, invidious distinction, or derogation, the camouflage can often be dropped with relief because the worst fears have been shown to be untrue.

The subjective experience of genuine love, its representation in awareness, may range from disgust to ecstasy, and the manifest interaction may be ecstatic, friendly, quarrelsome, or distant. Even sadomasochistic manifest patterns can occur in an essentially good relationship between two people who had remarkably unfortunate experiences with sadistic parents. Love may defensively be experienced as dull, as reasonable, as interesting, as useful filler of no emotional significance, as fascinating but hopeless, as purely sexual, or purely intellectual but regrettably uninteresting physically. Each of these rubrics may represent an inadequate reflection of the true state of affairs, but a fair statement of how the person consciously plans to limit the opportunity for a fuller love relationship.

The integral personality, nonetheless, often makes better use of the distantly structured contact than one had consciously intended. On the other hand, a relationship experienced in one of the above modes could be exactly what one thinks it is. Or it may derive its attractiveness from its similarity to the destructive patterns of relationship with one of the parents. The primary base for this elaborate camouflage lies in the conclusions drawn from the limitations of the primary integration with the mother, with the resultant repudiation of the need for tenderness.

A second significant cause for the camouflage of love is the impact of parental prohibition against replacement of the parent as a source of tenderness. This is a campaign in which fathers often take an equally active part. Fidelity to the camouflage and denial is based on loyalty to the parents, their limitations, and their family traditions. One avoids more rewarding experiences in order to retain the illusion of the validity of the original ones. This can be so literal that, for example, impotence or frigidity after marriage, where there has been no great difficulty before, can occur because the parents "now know we are doing it."

These familial forces are reflected and integrated into supportive cultural attitudes. Being fully involved in the welfare of a contemporary is often looked on either as perverse, ridiculous, or hypocritical.

VIII

PERSONIFICATION OF THE GOOD
MOTHER

THE PERSISTENCE in the adult of the infantile personification of
the Good Mother creates considerable confusion in adult perceptions. This confusion is partly because the original experience
of the Good Mother and the Good Me provided the base both
for the productive functions of the self-system and for the productive organization of the integral personality, before these two
personality systems were sorted into distinct entities.

The early infant's experience of the competent satisfaction of
his needs by the mother was his first experience that needs can be
satisfied. To the extent that his mother was available for such
satisfaction, she facilitated the integration of functions necessary
for survival and validated the conviction that growth, the integration
of new functions, is possible. For the early infant, the dynamism of
intense desire, integration of function, satisfaction of desire, was an
entity, personified in the Good Mother and the Good Me. It was associated with the emotions of ecstasy and elation. The relative flatness
of the mother's response to the intensity of the infant's emotions
initiated the infant's attempt to repudiate the intensity of his need
for tenderness. This repudiation was increased when the infant
discovered that his Good Mother was the same person as his Bad
Mother. Certain aspects of this original unitary experience of tenderness and growth were subsequently incorporated in the produc-

tive functions of the self-system. Other aspects were relegated to the growth-directed organization of the integral personality.

Subsequent events in the evolution of personality tend to disconnect the early experience of the Good Mother from the conscious, productive functions which themselves remain available in the self-system. The infant's orientation to open-ended growth is usually firmly locked in the integral personality. The intensity of the experience of desire usually becomes progressively alienated from the daily exercise of productive function. The basic human physiology, the functions of eating and defecating, of motion and perception, of walking and talking, were all, with whatever defect, positively integrated in relation to the mother. These and many other functions integrated specifically in relation to her persist. Very frequently the affect associated with their performance becomes dissociated.

In reaching out for juvenile and preadolescent experience the child often further intensifies his repudiation of the original experiences with the Good Mother, in the hope of facilitating his freedom to relate to contemporaries. After puberty, the anachronistic structuring of the society often hastens the deterioration of the relatively positive aspects of the integration between the mother and child into an active hostile integration between adults. This adolescent struggle with the actual parents further reinforces the dissociations of memories of helpful and constructive interaction between mother and child. Later sentimentalizing of "Mom" does not undo this dissociation of positive early experiences with the actual mother.

The frequent, continuing, reproachful, hostile integration between the young adult and the real mother also obscures its origin. It implements, in an inverted way, and simultaneously repudiates the young adult's search to recapture the uncontaminated infantile experience of need, integration of function, and satisfaction. The adolescent's battle with mother for the emergency extrication from particular hostile integrations with her serves no productive function if carried on into adult life.

The adult tries, in his attempt to reintegrate this crucial, positive, early experience, to relate to people who are particularly like his mother was during the first six months of his early infancy. For example, this may be expressed in men in a persistent attraction to women who are, or seem to be, as old as the mother was at the time he was born. This can be as much expressed in the attraction of a twenty-eight-year-old man to an eighteen-year-old girl as to a thirty-eight-year-old woman, though the latter instance is more commonly noted as containing the search for mother.

One difficulty in identifying the selectiveness of this integration is that his mother during this period of his life may have been very different from the way she was at any other period in her life. The paranoid and rejecting aspects in her personality evoked by the new baby become incorporated in the baby's central paranoia and his later hostile integrations. They also often persist in the mother's. Her positive capacity to relate to the infant's needs was derived partly from functions in the productive apparatus of her self-system which were solidly integrated and received no subsequent challenge. These functions she passed on to the infant without conflict.

To a certain extent, however, she rose to meet the occasion. The impact of the new baby brought out in her certain aspects of her own integral personality, with secret orientation to growth and to the intensity of desire. These were usually repudiated by her later because they had been brought out into the foreground without adequate preparation and support.

If she subsequently deteriorates, this makes the reconstruction of these expanded aspects of personality even more difficult. The attractive contemporary substitute will seem grossly different from the now thirty-years-older mother.

The adult is attracted to other adults who exemplify the particular intermingling of catatonic and paranoid relatedness characteristic of the mother during early infancy. Because of the subsequent overlay of real mother, sentimentalized mother, and fight with mother, these adult choices are often difficult to trace.

The evoking of the forgotten personification of Good Mother in a current relationship can sometimes be identified by the undifferentiated quality of perception, and by considerable discontinuity with other perceptions of the person. With progressive reintegration of such infantile experience into adult consciousness, the intensity of desire can be experienced without the necessity for distortion of current perception.

The Bad Mother continues to be personified in the hostile aspects of subsequent interpersonal relationships. This personification therefore has a certain continuity and can be reconstructed, in this continuity, in the life history and the current hostile integrations. Identifying the expressions of the personification of the Good Mother is complicated by the fact that many aspects of this personification are firmly imbedded in the integral personality and are not available for a common-sense reconstruction of the interpersonal history. Their projection in particular interpersonal interactions, whether new or repetitious, is obscure.

The concept of the Good Mother belongs partly in the self-system, because of its relationship to the productive functions. This is confirmed by the degree of representation in fantasy life. Its implementation in real life is episodic.

Much of adult fantasy is devoted to the reincarnation of Good Mother—either in reliving past moments of such experience, or in projecting possible ones, or in planning future ones. The fantasies are episodically implemented in glamorous infatuations with "fascinating" people. Such infatuations are characteristically of short duration, either because of the unrealistic demands imposed, or because the purpose of integrating them was to consolidate a daydream into actual experience. It may be that, once a reality, the actual relationship may not be needed.

Fugue states may act out this search. Masturbation fantasies often elaborate it. It is always partially evoked in intense or compulsive sexual patterns or fetishes. The frequent association with sex is because of the natural equation of sex with the situation of infancy—nakedness, cuddling, intense desire—all suggestive. However, Good Mother romances may be intensely intellectual, and sex may be phobically excluded.

A security operation, in addition to its restrictive purposes, also contains a symbolic attempt to formulate and reintegrate needed facets of growth experience which have been lost in previous battles. The more or less conscious search of the adult for the Good Mother derives from the connection between this infantile experience and the productive function of the self-system. The relationships chosen in this search are limited by the unrealistic demands of the infantile personification. The episodes of convinced connection with Good Mother tend to be forgotten and are not usually felt in continuity with other similar episodes. Even so, the attempt to re-experience and reintegrate them motivates and later contaminates much of the drive for intense adult relationships.

The choice is often mutual. That is, each partner often sees in the other person the personification of his own Good Mother. The interaction is initially sympathetic. Later discovered discrepancies confirm for each that the personifications were hallucinated and contribute a bitter overtone to the subsequent disillusionment.

In the adult acting-out, the falling in love is with growth, not with the individual. The growth aspect is the attempt to reintegrate past life experience which is needed for further growth. It does not centrally incorporate concern with, or orientation to, the other's growth needs, though this may be the price for admission to exploring the experience. The person is important because reminiscent of certain patterns of the mother during early infancy.

The overlay of concern for the other's growth may derive from the impulse later in childhood to grow mother up, as a better base for growth experiences with contemporaries. It tends to be hypocritical and abortive.

This search incorporates the germ of its own defeat. For example: in the search for the Good Mother one imposes the demand on any particular partner that the partner never be angry; and that he must be totally tolerant of one's own anger. That is, anger in the partner can always be used to prove that the choice was erroneous, that this particular partner is really the personification of the Bad Mother. It pulls the integration between two contemporary adults back into the frame of reference of attraction and repulsion that was appropriate to the early infant's perceptions. This search for the underground sources of all positive experience can thus be used to invalidate any actual current experience of friendly interest from a contemporary.

Another complication is that the choices—the objects of such intense projection—are determined by the particular patterning of the positive aspects of very early experience. The original mother was likely in a somewhat catatonic state immediately after the birth of the child. Aspects of the current object of infatuation reminiscent of the mother's catatonia are a frequent trigger for the infatuation. It can be easily mistaken and glamorized as commitment to growth. An unrealistic level of maturity is imputed to and expected from the partner. This unrealistic demand speeds up the deterioration of the relationship.

An additional complication is that, if the partner chosen has the particular emotional patterns of the mother when giving, the same person often has, when provoked, the particular patterns of the mother when ungiving. The positive relationship thus often contains the elements of the hostile integration.

It often seems in retrospect, or to the outsider, that the choice was originally decided by such factors.

Part Four

DIAGNOSTIC CATEGORIES IN INTERPERSONAL THEORY

I

THE RATIONALE OF DIAGNOSIS

THE PATTERNING of diagnosis in the realm of personality derives partly from the fact that historically certain aberrant personality patterns, considered psychotic, were cared for in institutions manned by physicians.

If mental patients had been cared for in the beginning by mathematicians instead of by physicians, the science of personality might have been articulated in terms of trends, forces, factors, vectors and the variety of combinations of these, which mathematicians use to understand a complicated body of data. A mathematical problem does not exist in order to be diagnosed, or classified. One thinks of it in terms of the various ways in which universal mathematical truths can be applied to it for purposes of solution, and in order to understand the relationship to cause and effect that led to the concretion represented by the problem.

Diagnosis, on the other hand, implies the existence of categories, the difference among which exceeds the divergencies of the individuals in each category.

Economics, for example, uses operational classifications for its articulation. Thus, for one purpose, economics divides its population of interest into rural and urban; for another purpose, into income groups; for still another, into characteristics of technical skill. Although certain factors are considered more influential than others in determining particular economic results, even these are not taken to form the nuclei of categories which imply the existence of an entity transcending the original operational significance of the classification that defined them.

Over the last hundred years a wide assortment of categories has grown in the psychiatric field, and it is relevant to attempt to relate the more prominent of these classifications to a dynamic concept of personality. Each syndrome is a constellation of symptoms, with particular hypothecated dynamic origins. Each can be understood as a personality trend, present in everyone at least to a minimal degree.

The original personality categories resulted from relatively detached observation of repeated behavior patterns. From this base line, attempts to understand the causes of behavior were made. This attempt was broadened in significance with the insight that the causal dynamisms operated universally. Psychiatry then was considered to be the study of the pathological distortion of universal systems of behavior.

As the concept of determinism came to be applied further to the phenomena of human personality, it could be seen that, in different patternings and degrees, those psychological dynamisms previously considered pathological, were present in all people. It could then be noticed that mental illness involves the normal use of universal dynamisms. Mental illness is the natural, inevitable response to the events of the lifetime, past and present.

If symptoms are the outcome of certain patterns of interpersonal relationships, one might then expect that there would be a correlation between the structure of a given personality, the recurrent patterns of interpersonal relationships, and the particular constellations of symptoms. Once it is seen that symptoms represent universal dynamisms for which we all have the capacity under the appropriate circumstances, the question is how it is possible to set up classifications at all. Why should there be a tendency for a person to specialize in patterns of symptoms? Also, why, even statistically, should certain symptoms appear predominantly in the company of others?

Thus, originally, the categories were descriptions of the manifest behavior of people in psychological difficulties. At first the behavior was thought to be the difficulty. As the manifest behavior then came, rather, to be considered symptomatic of the difficulty, it led logically to classifying patients by their method of dealing with difficulty, i.e., by the "dynamisms of difficulty" which were developed for re-establishing the stability of dissociation. Classification by the dynamisms of difficulty has been reinforced in the practice of psychoanalysis by virtue of the fact that these are the modes of resistance to analysis, and are therefore of particular interest to the analyst.

The individual's preference for certain symptom patterns over other patterns can be accounted for partially by the characteristic neurosis of the subculture to which the patient belongs. Some groups make it reputable to be obsessive, detached, or intellectual, but forbid other defenses against anxiety. Groups exist in which hysterical behavior is called spontaneity, or aliveness, even high wisdom and perception.

A broader base for specialized and apparently categorized responses to life lies in the individual's sequential development during his first few years. Constellations of symptoms can be considered modes of the perception and organization of experience, communication, and interpersonal interaction, characteristic of the individual's stages of development from birth through the juvenile era. Once such a mode, or such a modal constellation, has been consolidated, it remains more or less available for the individual's use, barring special circumstances, for the rest of his life. More or less available, because, to the extent that the original consolidation was defective, it remains either defective or highly vulnerable to disorganization under stress. Regression involves return to more primitive modes, if modes of operation developed later disintegrate under anxiety.

We refer here to consolidations such as: the physiological integrations of early infancy; the mastery of motion and perception in late infancy; the dramatization, the acting-out, characteristically used for communication by the three-year-old; the acquisition of speech between three and five.

The specific mode of interference with the normal course of interpersonal development is considerably less significant in determining later pathology than the fact of the interference. It is similar to the process of learning to swim: once a person is comfortable with being in the water, the fact that he never moves from water wings to independent motion is more important than the details of how this prohibition was effected.

Some recurrent symptomatic patterns refer specifically to early faulty integration of particular functions. Recurrent dizziness may be caused by trauma when the baby learned to walk. Preoccupation with excessive consumption or excessive avoidance of food refers characteristically to difficult infantile eating conditions. Such repetitive symptoms usually constitute the adult attempt to revert to the infant's faulty consolidation of the function, in order to consolidate the function better. But they appear also to carry the symbolic, covert communication, to whatever audience, that the general difficulty with the function is rooted in some catastrophe between birth and the age of four.

There is another possible basis for a specialized range of symptoms. That is, that such specialization is fostered by difficulty in developing more versatile methods for handling anxiety. Overload of the primitive modes of interaction results from the person's characteristic or momentary inability to use more elaborate forms of interaction. This defect is never absolute. No one quits growing at one, two, or three. Yet the earlier the functional disability occurs, the more tenuous the later development. Thus the adult tendency to rely on interactions at a certain modal level occurs partly because of arrested development, partly because of a regression.

There is only one dimension of central pathology in human personality: the degree to which one's development has been limited by the frame of reference in which it has been set.

But the phenomena described by classical diagnostic categories actually relate to two different trends in the human personality. One kind of diagnostic category refers to the balance that exists between self-system and dissociated system at any particular time. The other relates essentially to modes in which dissociation is maintained.

Acute catatonic schizophrenia refers to the moment when the

dissociated system breaks through to awareness. This shatters the carefully self-consistent self-system, which is ordinarily committed to successful defense of the fallacies of the central paranoia.

If the self-system reorganizes itself after the impact of such a breakthrough in such a way that its underlying fallacies are reinforced, we speak of the process as a paranoid integration. This involves both repudiation of the immediate events and retroactive reorganization of memory. This in turn moves toward deterioration, the progressive restriction of the personality. When this process usurps a certain degree of the personality, the person involved is referred to as paranoid.

Paranoia involves, in however disguised form: 1) the malevolent transformation, 2) the derogatory processes, 3) the mechanism of projection, and 4) grandiosity. These are indispensable to the validation of restricted life, and to the invalidation of life experiences incompatible with it.

The acute paranoid episode occurs in the moment of decision as to whether a new insight will be used for growth or for deterioration. The episode involves intensification of the forces of the central paranoia, stimulated by the anxiety engendered by the threatened breakthrough of insight as to the fallaciousness of the central paranoia. This includes considerable catatonic experience, during which earlier experiences are relived and transferred to present objects. One feels, for example, that his mind is being read, that he is being controlled, that he is being followed; these are all daily experiences of late infancy.

A minor impediment to a leap forward often triggers the paranoid episode. Through the lever of this small impediment, the total move toward growth can be invalidated. Sometimes this involves intense repudiation of gains already apparently consolidated.

The paranoid episode is loaded with rage, for past frustrations and for frustrations anticipated if the paranoid maneuver either should or should not succeed. The paranoid episode is characterized by self-destruction acted out as destructiveness toward others —and painful retaliation is expected from them. It is triggered by impulses to free oneself from the dependence on one's parents— and if one succeeds here, one expects painful retaliation from one's parents. This rage, characterized by destructiveness and anticipated or real retaliation, in fact recreates the experience of the small child in the hostile but necessary adult world.

The second symptomatic diagnostic categorization concentrates on the machinery by which the self-system's fallacies are maintained. This security apparatus uses all those modes for the organization of experience that are acquired through the child's interaction with adults during his early years.

Thus many clinical diagnoses, and a wide variety of maneuvers which have not been so extensively described in the literature, represent facilities of the conscious self available to all people for the purposes of organization of perception and conceptualization of data from the outside world. They also serve to restate data from the dissociated system, or conclusions of the integral personality which are not congruent with the self-system, in a form which might be at least partially acceptable to the conscious self.

Here as before the tendency to specialization is based essentially on a defect in normal human versatility. One could focus the question rather as: What is the person's difficulty in being adequately obsessive?, rather than as: Why does the person characteristically choose an hysterical mode of expression?

A third basis for classifying personality, involving both of the above categories, is the examination of the individual's organization of patterns of interpersonal relationships for the perpetuation of these particular personality patterns.

These descriptions have been referred to as character neuroses. The relevant descriptions, then, center on what sort of people the patient chooses to associate with, and on what basis. They refer to the extent to which he is still involved in the mastery of barriers to growth that were prevalent in the original home situation, and to what extent he is getting on with the project. This would include also which project he is trying to get ahead with and whether it is, in fact, the one he had had difficulties with as a child, or whether he has been pressured in trying to achieve the appearance of a level of maturity for which he is essentially unprepared.

Here again the significant dimension is not the accurate description of the unprofitable patterns of relatedness into which the person repetitively falls. The repetition itself is based either in the fact that he is working intensely on the wrong project, often one ahead of his true level of maturity, or in the fact that there is some obscure interference with the consolidation of the operation. He is unable to grasp how things should go in a more elaborate, subtle, and versatile pattern of integration.

II

NATURE AND MANAGEMENT OF THE CATATONIC EPISODE

WE HAVE SAID that schizophrenia is the only mental illness, present in various degrees in every person as a natural response to the events of his lifetime, and that some kind of schizophrenic experience parallels inevitably any abrupt expansion of the self-system. Schizophrenia is therefore inseparable from the process of growth after puberty, and if one hopes to analyze someone without the appearance of schizophrenic phenomena, he must limit himself to analysis of unanalyzable patients. Appearance of catatonic phenomena during analysis may indicate an error in timing, or may indicate that things are proceeding on schedule. But growth and catatonic symptoms go hand in hand.

Everything that gets into the self-system was admitted through the connection with some significant person. The subsequent evolution of a particular line of development is influenced by the connectedness with the original validator. Throughout life the self-system tends to become integrated into a logical system incorporating certain central prohibitions or fallacies. These fallacies tend to be formulated more or less concretely in the self-image. The self-image thus incorporates the deficiencies and inadequacies of the individual, and serves as a sort of policeman to keep him from attempting what he conceives of as either impossible or fatal.

The catatonic episode has been described as the phenomenon

accompanying the breakdown of dissociation. This formulation implies that the precipitating event is that the neurotic defenses have proven to be inadequate. More accurately, it is the breakthrough of the dissociated material against the lifetime resistance of the self-system, as reinforced by the eidetic images of the original significant people. Energy for the breakthrough accumulates from the continued existence of dissociated needs in the integral personality.

The outside world responds to the person's indication of his dissociated needs. The person, through perception of this response, becomes aware of previously unrecognized potentialities within himself. This upsets his low self-esteem. It upsets not only the logical construct of his self-system, but also the individual's relatedness with those people who collaborated in the design of his self-system.

When this relatedness is not merely threatened, but is disrupted; when, even for the moment, it cannot be maintained, one's entire basis for conscious relatedness with other people is shaken.

Then the logical structure of the self-system—awareness—is forced to collapse, because its foundation has been abolished. The person returns temporarily to the childhood situation in which the entire smashed wall of prohibition was first constructed. Momentarily, he is as dependent on the prohibiting mother for life, survival, sanity, and education as he was at the time when he was learning to think.

Subjectively, the experience is of a break with people, a total isolation, a loss of sanity. There is a massive disruption of both the productive and the defensive functions of the self-system.

Sanity is composed of a person's accumulated memory patterns and interactive experience—meaningful experiences with other human beings, objects, animals. The complete loss of the conscious continuum with people is what we mean by the loss of sanity. This can occur on the basis of organic damage with a loss of memory, senility, or panic, which temporarily obliterates connectedness with all past experience, and breaks the current of communication between this person and all other human beings.

In infancy the dissociated needs became dissociated because

they were associated with "mother when chaotic." Thus, when they break through they are initially strongly associated with the inner feeling of chaos. This chaos is cumulative with the intense confusion from disintegration of perceptual organization.

The catatonic experience involves the terror of the unknown. "If this which I have always assumed to be true is not true, then I do not know who I am and I have no basis for operating." Every single individual act has to be re-evaluated. "I do not know who I am in relation to a knife, a fork, a plate of ice cream, a person, a job, a telephone." One's relationship to every familiar object is in terms of who one is in relation to such objects. A tremendous number of things in which the action that seemed to be appropriate had been taken for granted, suddenly have to be re-evaluated in terms of "How does this change affect my relationship to this object?"

Dizziness and nausea are almost direct expressions of a shift in the frame of reference. The shattering of the defensive machinery, which was itself developed in interaction with adults, is another expression of the disruption of previously developed capacities. Its loss leaves the patient stripped of the usual modes of organizing experience into usable categories.

Thus the catatonic symptoms are of two kinds. One involves the loss of higher levels of integration of function for maintaining a homeostatic equilibrium in both physiological and psychological realms. The other is the hallucinatory reliving of previously dissociated experiences with the emphasis on the warning experiences of fear of the parents' anger or the dependence on the parents' attitude.

The small child takes it for granted that the adult can read his mind, and only regretfully acknowledges the necessity for verbal communication as life gets more complicated. The catatonic has the delusion that the other person can read his mind, and feels similarly about the necessity for verbal communication, that it is an unnecessary imposition—why should he have to say what the other obviously knows? The night terrors of early childhood take over, with an intermingling of historic and contemporary data for documentation and example. The dilated pupils are one index of the degree of terror.

The catatonic finds it difficult to concentrate. This lies both in the breakdown of acquired modes of organizing experience, and in the powerful distracting force of the experiences which are being relived. Experiences of self-revulsion are direct re-experience of the parents' emotional response to him.

At the same time, the reintegration of the positive experience which set off the catatonic episode can cause relief or even ecstasy, intermingled with feelings of horror. What is being reintegrated is essentially positive, as, for example, the capacity for mastery, or perhaps the capacity for tenderness. The relief lies partly in the release from the lifetime effort of maintaining the dissociation.

But the patient's powerful conviction that he has always known the absolute truth of the material he is experiencing indicates both the intrusion of the previously repudiated reality into the self-system, and the reassertion of the parental omniscience that represents the "absolute" truth of the self-system's central paranoia.

The catatonic episode may last for a few moments, a day, or for several months. If it lasts for months, the patient may be dangerous to himself or to others, or unable to care for himself. He remains in an essentially preverbal state, but can comprehend communications that are sufficiently simple and direct.

Such an episode results in one of two possible outcomes. Some of the intrusive material may be integrated into awareness, but the rest dissociated. He emerges with more potential for life and relatedness and freer access to his true feelings. The other possibility is that the invasive experience is redissociated, along with the events that led to the episode. This is the paranoid integration. Thus the constructive management of the acute schizophrenic episode is one of the cornerstones of psychotherapy. The patient is a person for whom hope has occurred with reference to some aspect of his life about which he has felt chronic despair. This hope has challenged his lifelong despair. This despair may, in turn, be reinforced if the therapist suffers from similar feelings of despair about himself or from his feelings of hopelessness about the patient he is treating. Such an interaction works out to confirm the worst fears of both patient and therapist.

One common patient-therapist source of despair is the conviction that the schizophrenic patient is basically incapable of sharing intimate experience. Yet the experiences that precipitated the schizophrenic episode always were experiences leading to potential closeness with other people. For example: one patient, a writer who thought he had written a dud, whose book was not only a best seller but a highly regarded contribution to his particular field; or another patient who was hospitalized after routine promotion in a job that seemed anything but routine to her. A classic example is the lonely person who suddenly meets a friend, and explodes into "homosexual panic." The patient may disparage the incident using his mother's original clichés for revulsion or intolerance.

Another interesting example is the heterosexual panic experienced by the patient who had been a practicing homosexual for twenty years. Although he disparaged the incident variously, his therapist discovered that his panic had been triggered by a situation in which his relationships with women could be improved.

The varieties of the experience are endless. The lecturer who feels that he cannot conceivably speak well is warmly applauded, only to return home feeling that he must have made a fool of himself; the girl who surmounts parental prohibitions and enjoys sex feels like a prostitute the next day. The therapist's business is not to be beguiled by such inversion.

Anxiety is inherent in analysis. Even major catatonic episodes may not be avoidable if the person is committed to growth. At the point where a catatonic break seems possible, however, even excellent psychotherapists often assume that they should routinely veer away from the topic at hand and reapproach it more gradually. If there is a constructive approach that is more gradual, this would certainly be worth a try. But time is often urgent. Neither analyst nor patient lives forever and, actually, the worst fate for the patient is not the acute schizophrenic episode. That, although more spectacular, is much to be preferred to the person's insidious deterioration, the progressive restriction of his life, the ultimate expulsion of all interpersonal emotions except those of bitterness and of envy.

It is the therapist's responsibility to try to direct his patient's insights to material that can be integrated into the person's self-system within the week, if not within the hour. One way to do this is to direct the patient's attention toward the interpersonal problems he needs to solve. The therapist himself must evaluate those problems according to their needfulness to the patient, since the patient's definition of his urgent problems is almost always muddled by status ideas.

Now the patient did not get into therapy because he was aware of his difficulty in becoming a member of the group. Had he been clear on this problem, quite possibly he would have been able to solve it for himself. However, the patient may have gotten into therapy because he felt that a central relationship—parents, a marriage—had so eroded his feelings of prestige that he could not associate easily with a group.

The therapist's emphasis should be on fostering growth in those areas of his patient's personality where growth can proceed quickly, but a little at a time. It may be both futile and cause for extreme anxiety, for example, for the analyst to pressure a person who has never solved the problems of the juvenile era, who has never grown comfortable about sex, to become overnight a loving partner in marriage.

This emphasis on growth for which the patient is ready—growth in sequence from the patient's base line—does not exclude the use either of massive confrontation or of comprehensive summary of the operations of the person's self-system. There is a difference between anxiety and the fear of being found out.

Sometimes the person's accumulated anger and despair should be brought into focus gradually. However, the therapist must realize that the presence of these powerful emotions indicates the patient's growing awareness that all his loneliness and unrelatedness were imposed by inept and damaged parents.

Fear of being found out is either the patient's fear of the observer, or his fear of realizing the pervasiveness of his own low self-esteem, both of which are in the self-system. When the patient grows angry at confrontation, his inner reaction is that the analyst's implied suggestion of change is conceivable and possible, regardless of parental dictate, even though he may suspect that his therapist is trying to humiliate him by revealing the bitter truth.

The therapist's knowledge of his own life is directly or indirectly relevant to his ability to avoid the precipitation of unnecessary extensive anxiety in his patient. This includes the therapist's knowledge of his own prejudices and limitations.

For example, when the therapist begins with a patient who exhibits signs of current or potential severe anxiety, the therapist should evaluate whether or not he has the necessary resources to see the patient through a successful analysis. Is there the possibility that analysis may require hospitalization? If so, would this be too embarrassing for the therapist? Or, might therapy be expected to take ten years? Would this cause deadly boredom in the therapist? Would successful therapy involve a period of non-payment, and, if so, would the therapist resent his patient?

Sometimes the patient and the therapist both panic. The therapist who allows himself to be unprepared by his patient's condition often transfers his responsibility to a more experienced therapist, or to the even more "secure" confines of a mental hospital, with no serious evaluation of such a transfer. This therapist needs to discover in himself why he did not expect his patient's panic, and why, when it arrived, it so frightened him—for his own and for his patient's sake. Panic states in the patient ordinarily are bad periods for transfer of therapy, if it is at all possible to continue work with the patient.

Explosive schizophrenic phenomena can be avoided also by the promotion of less explosive means of communication between the integral system and the self-system of the patient. Dreams are a basic mode for such communications. Marginal thoughts and ruminative states are other such modes. A wide range of artistic productivity also serves this purpose.

The patient may need to sit dazed for a few hours every day. Exactly these periods of privacy and low focal awareness may make possible his functioning for the rest of the day. Often it is not wise to promote alternate ways of life which would interfere with the possibility of such privacy.

But what can be done if the patient does grow acutely anxious?

Since anxiety usually projects the patient's sense of disconnection from his mothering figure onto the screen of available audience, a vital emergency maneuver is to shore up the patient's waning self-esteem. After the patient makes an embarrassing admission, it is useful for the analyst to clarify for him that what he has admitted is universal experience, and that every honest person admits it. It is also useful to find out what about the embarrassing experience might be constructive.

Secondly, since anxiety is disjunctive, the patient has moved a long way off. The therapist must figure out how he can move closer to the patient. This does not involve joviality, and interpreting the patient's move as distance is usually futile. The analyst must be open-minded here: physical contact, for instance, might be as fearsome to one patient as it is reassuring to another. Certainly, the therapist must be acutely sensitive to the patient's need and ready to abandon any move on his part that seems to induce distance between him and the patient. The therapist should certainly be willing to dispense with the formalization of the patient-therapist situation if it seems to intrude unnecessary distance between him and his patient.

During catatonic episodes, the therapist can communicate to his patient if he keeps his messages simple. But the patient's ability to communicate to the therapist is impaired seriously. The situation therefore demands of the therapist an acute awareness to "cue" phenomena observed in the patient, so that the therapist can discover what the patient has in mind and what are the patient's immediate, continuing, and ultimate needs.

In the more routine course of therapy, it may even be useful to the patient for the therapist to be, for the moment, slightly preoccupied. The therapist can allow himself time for rumination on the data in order to give an intelligent response, or to leave a space which pulls the patient into formulating the problem. In the presence of acute, severe anxiety, such rumination by the therapist is an indulgence that cannot be afforded. What is appropriate is the kind of alertness that is involved in driving a car at eighty miles an hour on a crowded highway in response to an emergency, using every available sense to anticipate possible change in the situation before it happens.

If the therapist becomes anxious he, of course, also moves away in the opposite direction. If the therapist can be aware of this interaction, he should move with and towards the patient, so that the distance between them can at least stay approximately the same as it was before.

A third procedure, sometimes quite reassuring to the patient, is for the therapist to reformulate, undramatically, some of the experiences being lived through by his catatonic patient. Thus, depersonalization can be discussed as a distance maneuver, hallucination as a literal repetition of some childhood taboo, as a symbolic communication, or as a postulate necessary to explain an even more incomprehensible experience. It reassures some patients to point out that respect shown for them to which they are not accustomed, summons up feelings of unreality—"It couldn't be me."

The symbolic communication in any patient's behavior is secondary in importance to the situation in life that made the behavior necessary. What new growth, insight, or satisfaction did the triggering situation involve for the patient? When the therapist gets at the event or the train of thought that touched off his patient's acute anxiety, he can aim his interpretation at some halfway house between total validation and repudiation. If the therapist is overly timid he may tend to make a repudiative interpretation. The overambitious therapist often makes the error of pressing for total integration, when it is patent from his patient's condition that he has gotten hold of an insight that he cannot handle all at once. If the therapist can find a midpoint between the two absolutes to allow the patient to grow without the shaking consequences of too much integration too fast, the patient can keep some of it.

A fourth principle, then, is to try to limit the insight to something that could be integrated into the self-system with less shaking consequences. This is in line with the principle of attempting to plan therapy so that the particular insight involved would not be more than the patient could integrate in a finite period. If this nonetheless happens, it is sometimes possible to latch on to some significant but not central aspect of the profusion of memories that are flooding the patient, and let the rest slip back into unawareness; to latch on to something that gives the patient some opportunity to grow and still implies that the change is manageable. This can occasionally be accomplished by quickly recapitulating the precipitating interpersonal incident, what was going on when the patient started to get upset, and reinterpreting the patient's interpretation in a less threatening or less revolutionary way.

A fifth maneuver is to focus on the simple, physical realities of life. The patient really knows his address and street number, and how to dress and undress. He may be using the physical world symbolically for some elaborate communication, but he probably knows which room in the ward is his. The simple relationships with the physical world, for people who got far enough through infancy to learn how to talk, are usually not centrally disorganized even in the open schizophrenic break. A derivative of this is that, despite his difficulty in concentrating, the patient still has the responsibility for his own decisions. These decisions may include not only the simple physical life operations, but also many activities concerning the job, or the living arrangements, or the marriage.

All patients, no matter how apparently disorganized, have some secret plan about what might get them into a better state of mental health. The plan may be misguided, but it often turns out, after prolonged and intensive therapy, that it wasn't quite as bad as it seemed. As far as possible, the therapist should cooperate with what seems to be the direction of what the patient is doing and try to understand it. At the same time he must identify the fallacies in it and set up limits for it.

III

THE AMBULATORY SCHIZOPHRENIC

THE TERM *schizophrenic* is applied also to people whose daily life includes more than the average of psychotic experience, in intensity and in duration. These people live with a high anxiety level, which involves acute parataxic misinterpretations, symbolic thinking, night terrors, fugues, strong withdrawals, and preoccupied states. These schizophrenic phenomena emerge in various ways.

The "pseudo-neurotic schizophrenic" organizes noticeable security operations into either a piling up of mixed psychoneurotic symptoms, or a caricature of one of the classical diagnostic syndromes which serves as a thin and bulging container for his chronic state of panic. The therapeutic difficulty with this patient lies in his intense investment in keeping this façade going. He is not concerned with appearing normal. He may be preoccupied with demonstrations of how upset he is. Direct interpretation of his maneuvers is usually futile. Direct reassurance is harmful, since this response tends to validate the patient's delusory façade.

Like the drowning man, this patient is trying to pull his rescuer down with him. The therapist must doggedly elicit, by direct questioning if necessary, the patient's truly relevant interpersonal data, and then, with equal tenacity, stick to the job of reconstructing that data. By such validation of the patient's interpersonal frame of reference, the brilliant behavioral display grows less necessary, and finally can be relegated to the unimportant by a sort of good-humored disregard. Then the screening behavior can profitably be interpreted.

Often long periods of restricted "common-sense" living alternate in the schizophrenic with short periods of intense panic. Here the analytic impediments are the unavailability of marginal data and the patient's investment in maintaining the front intact. The person is convinced that under this front he is like an undrainable bayou of water, sand, and quicksand, and that the therapist can only help him to hide this confusion of unchartable elements. He may deny a sense of the limitations which his attitude implies, but actually he sees no way to broaden them without suffering disintegration.

Here, every scrap of autistic data must be latched on to and utilized to the fullest. The therapist's engineering project is to reduce the pressure in his patient by expanding his life to include more and more of his integral personality. Explosive episodes must be expected. They are inevitable protests against growth, which decrease gradually as the base of the patient's life is broadened, and new experiences consolidated. Thus the course of his therapy is stormy—but, if the patient does not abandon therapy as a dangerous project, and if the therapist avoids growing so cautious that no progress is made, the project is feasible.

The schizophrenic behavior pattern may be less well organized. The patient may simply be operating on the margin, postponing from hour to hour intense surges of experience from the integral personality or the central paranoia, or limiting them to some convenient period of time, as does the businessman who collapses every weekend.

Relevant analytic material is here easily at hand. It is both safe and urgent to deal with it, if it is interpreted prosaically. The therapist must assume considerable responsibility for identifying reality for his patient, and for confirming a frame of reference about it. He must guard against catastrophic misreading of a word or remark, because the patient has ready access to his feelings of despair and self-revulsion, and is looking to his analyst to validate them.

The patient is not disoriented, nor is he particularly committed to denying lifelong behavior patterns. He has undertaken a specific project that will, if completed, involve consolidation of functions already latent in the integral personality: a good marriage, a divorce, a new job, a new house, a new book.

The analyst's job is to keep track of his patient's machinery for growth and repudiation, to connect the patient's paranoid behavior with both its immediate trigger and its historic source, to affirm the new undertaking in detail. He must, in time and place, keep himself fairly available; a lot is going on in his patient.

IV

PARANOID AND PARANOIA

THE PARANOID personality results from the accumulative effects of repeated paranoid integrations. This kind of person is restricted, rigid, and suffers from low self-esteem. His life is insensitive, unmeaningful, spent in derogation of others and the transfer of blame to them.

The lives of some people seem to have been so restricted almost from the beginning, at least from the time of their earliest memories. Although these people show no signs of imminent breakdown, their way of life is such that almost any new experience would upset the apple cart. They have never had a friend. They have never learned even the rudiments of getting along in a group. They cannot even form a father-daughter relationship, or a mother-son relationship with a boy friend or a girl friend. Such a person is extremely vulnerable to a major schizophrenic break. If one occurs, he has limited inner resources for reorganization on a better basis. This major psychotic episode, then, usually results in more manifest deterioration unless it is handled with sensitive therapy.

More frequently, deterioration takes place with less intense anxiety and more frequent explosions. It is the process of repeated acute paranoid episodes, each involving in the moment much catatonic experience, each culminating in a more restricted personality than the one previous to the paranoid episode. Each episode is precipitated by some insight or some incident of getting closer.

Paranoid loss of spontaneity tends to be progressive in direction but may be quick or gradual. Some paranoids remain impressively creative, even warm, between episodes over many years. Others capitulate early to stormy hatefulness. In treating the patient outside a hospital, a major hazard is that the paranoid will leave therapy during such an episode. Since, between episodes, he is legally sane, it is usually impossible to keep him in the hospital long enough for therapeutic effect on the paranoid's accumulated deterioration. Another hazard to therapy of paranoia is that, even in a hospital, the therapist tends to be intimidated by the paranoid's explosive anger, thereby entering into a conspiracy with his patient that central material will be handled gingerly.

Ambulatory therapy between episodes must be undertaken after evaluation of the risk that treatment might precipitate not only flight from therapy, but long hospitalization. This possibility is not in itself a reason to refuse to embark on therapy. The patient knows that he needs therapy, in spite of his great investment in projecting his difficulties onto his environment.

In consultation the paranoid always, implicitly or explicitly, tells a tale of persecution. There are certain obvious pitfalls for the therapist to avoid. One is the intimidation of the therapist by the anger he feels under the paranoid's overt civility. Therapists may mistakenly try to placate this anger by too much compliance and friendliness. Nothing makes the paranoid more uneasy. He is being confronted by the apparent response of love which he knows could not be genuine because he is painfully aware of the intensity of his own feelings of anger. He then becomes preoccupied with what malevolence in the therapist could lead him into putting on such a disguise. The patient also is not aware, and this is repetitive in his life, of the extent to which he radiates his own inner angry feelings. Thus, if the therapist is angry, the patient rarely realizes that the therapist is responding to the patient's latent anger. He experiences himself inwardly as frightened and defenseless, and thinks of his anger as a very adequate protective barricade.

The therapist is also confronted from the start with the dilemma of how to express a mild demur, a non-participation, in the patient's manifest paranoid illusions and distortions, without immediately precipitating a fight. Some of the workable formulae for this problem include: "I can think of alternate interpretations," "I can't be too concerned with these other people's problems with hostility—we can only work on what gets their anger focused on you," or "Your problem is to find out what there is about you that brings out the worst in people."

Another useful maneuver is to isolate in the patient's story the actual facts from which the particular delusional construction was confabulated, from the symbolic communications. To the extent that the valid can be validated in the paranoid's delusional system, his intense investment in selling it to the analyst in its original form can be moderated somewhat.

When paranoid deterioration occurs episodically, there are periods of acute catatonic experience during which, by skillful therapy under good conditions, the analyst can establish good contact with his patient and exert leverage to reverse the deterioration process. But when the deterioration proceeds insidiously, when the manifest catatonic and paranoid experience is minimal, it is difficult to make the patient notice that anything serious is going wrong. Nonetheless, these patients sometimes get into analysis under pressure from relatives, after an unusually effective consultation, or for "didactic" training to become psychoanalysts.

Once in analysis, the difficulty remains much the same; that the amount of anxiety involved in reorienting the life pattern and reintegrating the lost spontaneity will seem to the patient at any given moment to be entirely out of proportion to the degree of discomfort that he is currently experiencing. Even if he is verbally and intellectually committed to such an undertaking, he can only operate to postpone it from week to week, session to session, or year to year. The therapeutic maneuver has to be a series of confrontations; that is, the therapist must, about once every three months, review with the patient who and where he is, who he used to be, who he could be, what the actual current direction seems to be, and what the necessities are for intense participation in the project of change. It occasionally works. The effectiveness is probably contingent on the fact that the patient has become more deeply related to someone than he had thought to be possible.

Another route to ultimate deterioration is what is known clinically as true paranoia.

Here the patient, physiologically, functions adequately. However, all data relevant to the exchange of tenderness are misconstrued through the agency of some central fallacy, which may be that the person is unlovable, that he is Jesus Christ, that people are no damned good, or that people are all in a conspiracy to mislead and misinterpret him.

Maintenance of the fallacy demands considerable manipulation of data. The paranoid feels an affinity for hostile integrations. He avoids more pleasant relationships, and, when they occur, depends strongly on the malevolent transformation, the misconstruction of any friendly advance as hostile. His charm and intelligence—often originally considerable—slowly deteriorate under this imperative.

Early in deterioration, the paranoid considers withdrawals from people temporary respites that let him muster his strength to move again toward people. But as deterioration accelerates, open malevolence increases, and the paranoid's investment in maintaining even the appearance of continuity with people is progressively less. In the end, the paranoid may, like the psychopath, find it necessary not only to repudiate affectionate contact with others, but to try to destroy those who offer it.

V

THE PSYCHOPATH

THE ADULT PSYCHOPATH is characterized by an intense commitment to the facsimile of relatedness and an intense repudiation of the actuality of it. The manifest content of the interpersonal interaction is frankly exploitative. People are used as tools at the psychopath's convenience. The less obvious, but nonetheless clear, content is manifest destructiveness. The psychopath wages an active war against any impulse to tenderness, growth, or creativity in himself or the other. The literal fact of survival of the other person may seem outrageous to him. The other's joy is an insult to his scheme of values. His intent, then, is less to break the contact than to destroy either the physical or the psychological existence of those people with whom he is in interaction.

His interpersonal relationships may last a shorter or a longer time. He is devoted to the superficial appearance of bland affability. In implementing this superficial drama he is able to use aspects of all forms of communication, including the hysteric and the obsessive. This apparent commitment to the forms is counterbalanced by an intense contempt for the substance of the relationship. The psychopath lies for the purpose of being found out and thus of making the other feel that he has been made a fool of. His petty stealing is in the same design. On a more serious level the psychopath is actively sadistic.

His sadism is directed toward the active terrorization of the other, exploiting in the other any real needs or moves toward growth and all the archaic fears of the other's central paranoia. He plays this cat and mouse game with a bland denial of its intent. But it is irresistible to him to move in on the other's disadvantage.

Thus his life is pervaded by violent rage, often blandly denied, but occasionally breaking through into his own awareness in terrifying proportions. He is constantly afraid, wittingly or unwittingly, of either committing a murder or provoking someone to murder him. His rage is stated not in terms of the chronic, reproachful, hostile integration, but in terms of active physical or psychological torture or of actual life and death.

The active repudiation of the phenomenon of relatedness to objects, materials, strangers, or friends perpetuates in the psychopath a continuum of terror of total isolation. A tolerance of being alone is inconceivable to him.

However much it may seem that he has declared war on people, the psychopath's actual concern is with the values of infancy, physiological survival, and the possession of objects. People then become the most important objects and are related to as acquisitions to be collected like beads on a string, or to be rejected because even acquisition was a function essentially denied to him. The psychopath has neither the interest nor the capacity to be creative with physical materials. Neither is he able to comfortably cherish the ownership of particular objects—a pipe, a vase, a jacket. His problem is that literally nothing means anything to him except the integrity of his own physiology. Even with this he is compulsively careless.

In relationship to property he tends to be avaricious and hoarding. Characteristically, however, this avariciousness is punctuated by violent repudiations of it. He builds a career for the purpose of exploding it. He creates a successful business which he then drives into bankruptcy. He may hoard all property he can get his hands on with a meaningless accumulation, never simply experiencing that he has finally validated his right to possess it. At the same time, his inner emotional affinity is with the Bowery bum and he often ends up there.

He lives perpetually caught between the terror of being found out and therefore ostracized from the human race, and the compulsion to self-exposure. The impulse to self-exposure derives partly from the impulse to make the first move toward humanity, to try to find another human being who can accept him even as he is. It also derives directly from the sheer weight of effort of maintaining the façade and from the severely impaired motivation for even physical survival.

In the adult, the psychopath is very difficult to distinguish from the particular form of deterioration described as true paranoia. The psychopath, however, tells a different story of his life. The paranoid can remember in childhood, and often in early adulthood, acts of kindness, pleasure, and productivity, which, as it seems to him, mysteriously deteriorated over the years. The psychopath has no such conscious memories. His earliest memories describe the world in the paranoid mode. He first tells a story of mysterious, consistent persecution of him by the human race. On careful inquiry, this mysterious persecution can be shown to have been repetitively provoked by atrocities perpetuated by him on the human race. He set fire to his mother's house at the age of five, he tortured the cat, he squealed on his best friend, he remembers no simple acts of kindness directed toward him, and early established himself as unfit to be associated with by contemporaries.

At a minimum, during the childhood years, there was a repeated preoccupation with overt physical violence and with both provoking and avoiding violent retaliation for it. He was probably an actual danger to younger siblings.

In etiology the crucial damage lay toward the end of early infancy. These people are usually physically adequate or even superior. Their perceptual and intellectual capacities are at least initially well integrated. Usually they handle their bodies well, which would suggest that the immediate transition from early to late infancy—the learning to walk—was relatively permitted. The cherishing of objects was not.

Whatever went on constructively between mother and infant during early infancy, by the age of two the young psychopath had solidly integrated the hostile integration with mother. Subsequent learning, then, was essentially dominated by the paranoid mode. Facility in dramatization and in talking, in the techniques of living, was acquired, essentially in the service and style of security functions rather than of productive ones.

In the psychopath the consistency of the paranoid tinge of his experience is striking. We would thus have to hypothecate that his mother made no essential recovery from her initial repudiation of the infant's attempt to become a separate person. She must have continued the war on his aliveness with unabating energy during the early formative years.

Insofar as a catastrophic transition from early infancy to late infancy is a fairly frequent experience, the capacity for psychopathic operations is a recurrent difficulty in all of us. The occasional violent repudiation of the whole phenomenon of relatedness to anything except one's most rudimentary physiology is a universal experience. On the other hand, the total commitment to this as a way of life is a fairly rare personality type in our culture.

The psychopath's affinity for drugs and alcohol, though not universal, may have some reference to his attempt to return to the conditions of early infancy, in which fluctuations in the level of consciousness were a prominent feature, in an obscure attempt to start the growth cycle over from the beginning. His preoccupation with the servicing of his own bodily needs derives from a similar source. In the adult the avoidance of grossly illegal activity derives partly from the chronic terror with which the psychopath goes through life. He is, in fact, a coward. It derives also from his real lack of investment in the object world. He rarely cares enough about a person or a possession to risk his physical survival to destroy them. His capacity for perception, organized in early infancy, is constantly sharpened by the necessity to carry on this elaborate style of life without being too openly found out.

The psychopath rarely gets into therapy. It would strike him as an irrelevant project to the course on which the ship is set, merely a handicap, increasing the risk of being found out for no gain of which he can conceive. On occasion, the impetus for therapy is the fear of retaliation for the destructive impulses. Occasionally one gets into analysis with the fantasy of using his skill at manipulating people as some sort of social engineer.

The process of therapy consists of detailed, explicit, intensive, and extensive speculation and confrontation. This must be kept up as a continuum throughout every interview and accented by repeated summaries. If the therapist is not thrown off by the flood of righteous indignation which this elicits, he can get some validation from the patient to the effect that all of the interpersonal experience that he can remember so far in his life is irrelevant. It was all perceived in the paranoid mode. He may have achieved a high level of technical facility, but he is in total ignorance of what life is really about. If this can be established between patient and analyst beyond controversy, without rancor and with a full grasp of the historic origins, the patient can then set out humbly to begin to observe what does go on between other people. He can begin a tentative exploration of what relatedness is about. He can make a start on the assignments of late infancy and early childhood, ownership, manipulation of medium, and parallel play situations and can build from there.

In the course of therapy, the patient, and perhaps the therapist, may be intimidated by the danger of major psychotic episodes, violent acting-out or precipitate withdrawal from therapy. Any of these constitute the lesser danger. There is no pleasure in his life as it is organized, and without drastic reorganization he can look forward only to the progressive loss and strangling of even the technical facilities on which he relies.

VI

THE PATHOLOGY OF NORMALITY

EVERYONE IS an immature adult and is in the active process of either growth or deterioration. Nonetheless, there is a frequent personality syndrome characterized by relative stability at a particular level of immaturity, and by the fact that manifest anxiety is not a conspicuous feature. This syndrome is popularly considered the normal. This category does not include people who place a high value on the social denial of anxiety. These latter may be exceptionally anxious, fighting off a frank psychotic break, or they may be involved in a particular power struggle, in which the frank admission of anxiety would put them at a great disadvantage.

The illusion of normality involves an apparent relative stability of interpersonal patterns, which patterns are not strikingly deviant from the culture, the relative absence of manifest anxiety, and modes of containment of anxiety in which the paranoia is not too socially divergent.

Many people have achieved some minimal level of relatedness, and apparently have no great impulse to increase it. Examples of such adjustment would include some psychopaths who seem to be comfortable with the notion that other people are objects for the provision of physical comfort. They value neither the esteem nor the affection of their contemporaries and they seem to make no serious moves to catch up. Some homosexuals seem to stabilize with a compromise in which they accept some sort of tenderness from the same sex but cannot accept any exchange of tenderness with the opposite sex. The attempt at stabilization can occur at any level of interpersonal fixation. If the manifest performance is consonant with the central conventions of the particular subculture and the level of maturation is within the average range for this group, they are considered normal.

It is not possible to consolidate one phase of development and to repudiate the next, without ever looking forward. This would be against the laws of nature. Whenever a preceding phase of interpersonal development has been consolidated, there is a move in the direction of the next stage, no matter how much anxiety this may produce. There is, then, the problem of accounting for the apparent absence of this push.

The intense preoccupation with the stage of development at which these people are stuck may be rationalized as a repeated attempt to maintain and expand the necessary runway for a flight into the next stage. But the plane never gets off the ground, no matter how it races its engines.

One conspicuous feature is the relative absence of manifest anxiety. They experience almost no anxiety from day to day; the episodes of acute anxiety occur as rarely as once in several years, and it is almost impossible to provoke anxiety by therapeutic maneuvers. One possible origin of this defect may be that what was most dangerous in their childhood was the experience of anxiety. Each of these people grew up with a bully. Someone in the immediate family would close in on any symptom of weakness in the child and make him suffer for it. The avoidance of anxiety became so necessary and the machinery so efficient that they smother the tendency to grow in the service of this. They expect themselves to grow up without ever being made uneasy by it. Their efforts to grow may have a quietly frantic quality and are often carefully misdirected. They try to get somewhere insidiously without having to notice it.

Since the true consolidation of any particular phase of development inevitably leads into the initiation of the next, it is also necessary to postulate that, in fact, the level at which they seem to function is an illusion. The performance is the reflection of some primitive, secret repudiation of the process of life.

One possibility is that these are people who underwent a major paranoid integration relatively early in development and acquired the later techniques of living within this frame of reference, and in the service of the central paranoia. The paranoid integration took place between late infancy and childhood. They subsequently developed some manipulative knowledge of interpersonal techniques. The result is that all of their adult relationships are cast in the form of a mild, chronic, hostile integration which they and their fellows in society consider the only possible relatedness.

Therapy would involve a reliving of the panic of the original paranoid episode. The acute disorganization involved seems to the person who has a minimal degree of comfort to be entirely out of proportion to his discomfort. His self-system has settled into a pattern of little pain, little pleasure, and minimum commitment.

The magnitude of the reorganization is made more fearful if any disorganization was originally dangerous. Both of these factors are often congruent with the overall life pattern of limited goals as the direct expression of the fear of death from the mothering one.

The therapeutic problem is: 1) to reintegrate the experience of anxiety; 2) to clarify the paranoid nature of the interpersonal relationships, however well civilized the hostile integrations may be; and 3) to elicit or expose at any level some direct investment of the patient in satisfaction—oral, sexual, material, perceptual, creative, or interpersonal. If he does not recognize his needs, he will have no reason to risk his security in order to expand the satisfactions. Security and status are not enough. The patient will not undertake a major security risk on the chance of bettering himself in the same dimension.

The prognosis depends on the success of these maneuvers.

These patients do not often come to therapy. If they do they may stay passively, whether or not progress is being made, as long as the therapist will participate. They may stay because of status, loneliness, or the wistful fantasy of inventing some way around their dilemma. If at some point during the process of therapy the patient does not learn that he is working for his own increased satisfaction, nothing useful will come of it.

Some people who are apparently stabilized at a particular level of interpersonal development are relatively susceptible to anxiety. It can be evoked under moderate pressure. These patients are focused on a level of activity in advance of the appropriate one. They may be relatively alive and experience some genuine affection in their relationships. The observer is misled by the apparent level of activity. The patient appears to be working at juvenile activities, presumably trying to progress from the juvenile to the preadolescent, but is, in fact, stuck somewhere in the transition from childhood to the juvenile. The juvenile is already beyond his inner capacities. The apparent search for the unique special friendship may represent an attempt to circumvent adequate coping with the juvenile era. He does not know how to cope constructively with group dynamics, and the searching is partly an expression of the resultant sense of isolation. He is stuck somewhere else, and this never gets noticed.

If the problem is very primitive, this patient may have to go back to an affirmation of the physiological needs by the affirmation of a preoccupation with food as an attempt at self-mothering, or a preoccupation with doctoring for validation of the right to physical survival. Once the problem is clearly in awareness and he rids himself of his pretensions, he may make up lost experiences with remarkable speed.

Another group focus on manifest activity below the relevant point of movement. This is frequently a flight from success and its implications. The patient goes back to doing something at which he is very competent. It serves as obsessive filler and hopefully, as a holding operation against further deterioration. It causes little acute anxiety. He suffers from a general depression. Life was previously more meaningful and he is unclear about why it has lost its zest.

The therapeutic road in this situation runs uphill. The patient's maneuvers to stabilize his deterioration are, in fact, not effective.

Acting obsessively on the integrated function will repeatedly move him toward the dangerous challenge and thus progressively imbue the integrated function with anxiety. Thus the patient looks for relief from difficulty by abandoning position after position, falling back to his previously prepared lines of defense.

VII

DEPRESSION

THE DEPRESSIVE syndrome includes in varying degrees: a flattening of emotion, a feeling of intense discouragement, a preoccupation with suicide, and a subjective feeling of internal inertia.

The cutoff of emotion and perception, internal and external, includes a decrease in the physiological sense of well-being. The depressed person is not in touch with his own internal sensations. His perception of the outside world is flattened—bright colors are seen grayed out—and there is a general flattening of intense emotion.

The immediate focus may be on the repression of some particular emotional response—such as joy, anger, ecstasy, fear, or anxiety. It is not possible to cut off one emotion selectively; they all have to be reduced together. A corollary of this operation is that the patient seeks out more powerful external stimuli, apparently to break through the cutoff. These patients often get themselves into dramatic situations. They may seek out strong colors and highly seasoned food. The mechanism of seeking out powerful external stimuli occasionally effects a breakthrough of intense sensation. The patient may even consider himself particularly sensitive and perceptive, rather than as having lowered the overall intensity of his perceptual experience. He assumes that the intensity at which he usually perceives is universal.

A second element in the depressive syndrome is the degree of subjective discouragement. He perceives a great disparity between his goal and his expectation of achieving it. The subjective feeling about this may vary from apathy to intense longing. He recognizes that happiness exists in the world, but he is convinced that it is not for him. This discouragement may not be at all appropriate to the objective reality, and frequently tends to increase at the moment when the goal becomes more attainable. On the other hand, the discouragement can represent a moment of genuine insight, that something which one has always wanted will probably never happen.

The depressed patient feels intensely inert. He assumes that if he ever stopped going, and no one prodded him, he would never get in motion again. He says that he could sleep for a week. His real expectation is that he could sleep for the rest of his life if no one took the responsibility to wake him up. This feeling of inertia is subjectively indistinguishable from the feeling of genuine chronic fatigue. The inertia also becomes cumulative with actual fatigue. He learns that rest does not restore his energy and so becomes disrespectful of the common rules of exertion and recovery. He abusively organizes his life so that the chronic fatigue becomes actual and uses it as a sort of drug addiction. The fatigue then reinforces and disguises the subjective inertia that leads to it.

A fourth element in the syndrome is the preoccupation with suicide: in explicit fantasy, in the affinity for taking risks, and in a variety of more or less effective suicide attempts.

A common dynamism links together these various aspects of the depressive syndrome. The preoccupation with suicide derives from fear of death at the hands of the mother. Mother, however unconsciously, was in conflict about whether or not she wanted this child to survive, and the child responds by his secret pull to oblige her by relieving the world of his presence. At the same time, he must master this fear of death in order to function, and he devotes much of his life energy to this project of mastery.

Moreover, it was especially the aliveness of the infant, his vivid perception, which was the threat to mother's depression. He bought survival by paying the price of being as little alive as possible. The fear of death was handled by the person's limiting his investment in life. This is implemented in the general cutoff of awareness, perceptual and emotional. Subjective feeling is reduced, both in the service of being less alive if one survives, and in the service of caring less whether or not one survives.

To respond to an overwhelming external stimulus, be it a dramatic movie or a dramatic life situation, is less disobedient to mother's command, and involves less initiative, and less awareness of one's own needs, than to be equally responsive to one's own inner dynamisms. Hence the external stimulus can sometimes break through. Relatively very peripheral stimuli—a sentimental movie—can provoke the person into inconsolable tears. Since the emotional trigger was external, the depressed person feels safely irresponsible, and detached from the emotional outburst.

The conviction that important goals are unattainable also derives partly from the fear of death. The attaining of goals has been centrally forbidden. Further, the depressed person curbs his enjoyment of these attainments because to enjoy is to commit oneself to life. Thus, he seeks to get enough satisfaction to sustain life, but not to fall in love with it and its dangerous satisfactions. The satisfaction must be falsely reconstructed as unattainable even in the moment of attaining it.

This combination of discouragement and intense longing implies that the patient in fact knows that life could be better. He either has had some good experiences which he has dissociated, or he knows someone noticeably capable of having a good time. Thus the depressed person must actively circumnavigate pleasure. He knows what he wants, but someone has intimidated him out of reaching out for it.

A more direct inversion of the inner continuum of fear is the depressed person's feeling of inertia. Any practical action that validates that one's life is a going concern—the signing of a letter, the dialing of a telephone number—suddenly seems to be an overwhelming task. The person feels as paralyzed as he would be if the task of walking through the door involved stepping in front of a firing squad, or stepping out into thin air. Even such simple mechanical tasks seem to involve him dangerously in life.

In proportion to how frightened he is of taking routine action, the depressed patient becomes obsessively preoccupied with winding himself up to perform the act which has now become a burdensome chore. He castigates himself for his laziness, discouragement or perversity; he reviews lists and prods himself to further efforts. This obsessive activity obscures the fact that the prospect of the minor act, representative of life, causes his heart to pound and his hands to sweat. The clear noticing of these indices of fear can have a tremendously releasing effect. For the first time the depression does not seem bottomless and inexhaustible. He can now allow himself to directly experience the fear he has been living with. He notices that the intensity of his fear is entirely disproportionate to any rational implication of the act that triggered it. As different from the inertia, the duration of the fear, if allowed to run its course, is finite. Even in the moment, once the fear has been experienced, it is not quite as paralyzing as the previous inertia.

No one in our culture is immune to depression, and for many, depression is the most conspicuous syndrome. An index to the degree of depression is the degree of discrepancy between a person's operating level of aliveness and his full potential for aliveness. Depression is a preoccupation with renouncing what we know we want. Since it involves a cutoff of strong feeling, it often is evoked for the purpose of control of an imminent emotion. One example is the control of intense rage, when acting on the rage is perceived as dangerous. Likewise, postpartum depressions serve partly to control the anxiety evoked by the aliveness of the baby, with the result that the mother can often perform necessary duties without major breaks in continuity of performance.

In middle age, the end of the manifest activity of raising and bearing children may allow a lifelong inner depression to come to the surface of the personality. If the marriage relationship is played out, this emptiness then triggers the struggle for power, the result of which is rage, which is throttled by depression. These two factors interlock to fuel the middle-age depression, which is manipulative to the extent that it is beamed to reduce the independence of the offspring.

The therapeutic problem with the depressed person lies initially in helping him to notice his general emotional flatness. To a certain extent, depression can be recognized only in retrospect. The depressed inertia must then be identified as a way of containing fear. The experiencing and dissociation of needs, the experiencing and dissociation of satisfactions, the reasons for disparity between needs and satisfactions, the maneuvers toward presumably unattainable goals, all must be charted during analysis. Thus the therapy proceeds by analyzing contemporary data in terms of their historical references, and by encouraging the patient to collect new data and to experience new enjoyment.

We could say, then, in summary that the core element in the depressive syndrome is fear.

It is the intensity of the fear that accounts for the illusion of intensity of the feelings of longing, grief, rejection, or stimulation, which feelings are in themselves moderately flattened by the machinery of the depression. The pathology in depression is centrally located in gross intimidation about genuinely experiencing that which one could feel clearly, and in using this as a basis for expanding the horizons further.

VIII

ALCOHOLISM

THE INFANT'S LEVELS of awareness are determined by the central response to contact between him and his mother, and the infantile emotional alternatives of euphoria and anxiety are communicated between the two people by empathy. In terms of this hypothesis, we speculate that drug addiction—including addiction to alcohol—acts on the basic physiological accompaniment of euphoria, anxiety, and the level of consciousness.

Alcohol is used pervasively for three purposes: suicide, reduction of anxiety, lowering the level of consciousness.

When used for purposes of suicide, either through deterioration of the liver or through alcohol's enhancement of the possibility of "accidents," alcohol serves as a means to master fear of death, or as a means to capitulate to expected destruction. The alcoholic feels as if catastrophe is around the corner. He drinks to reduce this sense of doom. Even if the manner of reduction is recognizably poisonous, it affords the person, for the moment, triumph over his fearful sense of mortality.

Alcohol also seems to exert direct leverage on levels of anxiety. Excessive use of it is frequently one symptom of an imminent psychotic break or an ambulatory schizophrenic episode. Alcohol is also used in abundant quantities by anxious people in order to allow them anything from bravery at large parties to success in bed.

It can thus be useful in many situations. The use of drinking to cut the level of consicousness is suggestive of what takes place in the first months of life. It resembles the stupor of the well-fed infant. It can express a compromise pull toward mother, a return to early infancy without the one-way passport of suicide. It can produce a level of relaxation attained otherwise only in euphoric situations or with sexual satisfaction. This gives many people their only chance to rest their chronically hypertonic musculature, which itself indicates early lack of access to tender petting by the mother.

The capacity of alcohol to lower levels of consciousness may also constitute a mild regression to earlier periods, caused by its capacity to dull the person's obsessive machinery. This is experienced as a loss of control, a decrease in alertness, guardedness, and concern with one's impact on the audience. This is probably the reason for the strong feelings for or against alcohol held by most people. Some people find that alcohol puts them usefully in touch with their own inner organization. Others, to whom maintenance of self-control feels crucial for survival, find that the slightest beginnings of loss of this control are terrifying in themselves and are cumulative, with the anxiety released by lessened control. These people either avoid alcohol entirely or they remain cold sober until they pass out, having permitted themselves no tipsy halfway house between obsession and unconsciousness. Sometimes these latter people stay drunken on the margin of unconsciousness. This lets them find release from obsession but spares them anxiety.

The therapeutic management of alcoholism depends on its use in the psychic economy.

Suicidal alcoholism must be treated as a resultant of the fear of death. The sources of this fear must be investigated completely, but this is difficult, since the only time the patient can tolerate access to the sources of his terror is when he is drunk. Thus the alcoholic is two people to the therapist—the drunken man who communicates; and the sober man who cannot remember what the drunken man heard, or, if reminded, remembers it as alien and Not Me.

The severe alcoholic, partly in recognition of the intent to self-mutilation, is often also severely preoccupied with the damage he already has inflicted on himself. Is his higher intellectual capacity irrevocably lost? Is his liver so damaged that he'll be a chronic invalid? Often these patients—who have very shaky body images—feel that suicide is preferable to physical impairment, and that the die is already cast.

On the other hand, the therapist must realize that, if it entails little physical damage, a considerable dose of alcohol can be the simplest, easiest, and most useful sedative for patients during crucial periods of therapy. The patient can control his consumption of alcohol according to his needs. His friends and family can be helped to understand the use of it.

The limitation on the use of alcohol, like that on the use of sleeping pills as a prop to therapy, lies in the therapist's estimate that the patient's need for the drug will be temporary, that is, that the crisis will last for a night or a year. If the patient leaves therapy in the middle of such a crisis, he retains such gains as he had integrated up till that time. However, continuation of sedation if therapy has been discontinued is a risky procedure. It will eventually take its physiological toll unless there is, at the same time, some alternative method for working through the conflict. It can be a useful, if limited, adjunct in directly undercutting anxiety for the ambulatory patient, in occasional release from hypermotility for the overactive person, and as a mode of investigation and exploration of the direct inner perceptions unhampered by vivid awareness of the immediate audience. If the individual can arrange his use of alcohol constructively, it can sometimes facilitate his clearest perceptions and his most creative formulations.

IX

THE PSYCHOSOMATIC

WE USE THE CONCEPT of psychosomatic involvement in relation to the concept that physical health, like mental health, is an ideal which few of us approximate. It is thus no more appropriate to take the average level of physiological functioning in a culture to be normal, and hence, by implication, healthy, than it is to assume that the average level of fixation in a given culture represents maturity. The psychosomatic is that proportion of discrepancy between ideal and actual physical functioning which is attributable to the individual's interpersonal experience. The point of contact between the individual's physiology and his psychology is in the area of sensory representation of the function in question.

All theories of somatic neurosis must include the impact of the individual's interpersonal relationships on the mechanism by which he dissociates somatic perceptions from awareness. Current psychosomatic theory emphasizes three factors.

One is the active repression of emotion. A repressed emotion heightens its concomitant physiological responses to the point of pathology.

The second examines the specific symbolisms of the somatic symptom: the organ involved and the specific activity of that organ are held to refer to specific interaction with a parent, and to the person's attitude toward that parent, under the particular circumstances that cued the psychosomatic display.

The third approach emphasizes organ inferiority. Certain of the person's organs are particularly susceptible to difficulty under stress. The fact that psychosomatic troubles tend to run in families is taken to support this hypothesis.

Each of these aspects of psychosomatic pathology can be understood in interpersonal terms. Repressed emotions do tend to persist, and the blockage of their physiological function tends to lead to anatomical pathology. The specific conditions of validation in early organic integration lay a basis for the possibility that specific organs could react in specific ways under similar later conditions. Some inequalities of functioning of organ systems are present at birth. The typical emotional conditions in the family at the time of early integration after birth, rather than inheritance, could affect lifelong organic fragility. Areas of dissociation of somatic sensorium in the mothering one may become focal points for later pathology for the infant.

The classical distinction between the psychosomatic and the hysterical—that the hysterical involves no organic pathology other than that of disuse—cannot be relied upon absolutely. The same organ may be used simultaneously or alternately by two emotional systems. The distinction may lie in the fact that the hysterical somatic symptom refers to a slightly later era of physiological development than the psychosomatic.

Adults suffer from various patternings of psychosomatic disorders. Some patients are psychosomatic museums. The most extreme, the psychosomatic suicide, has in the moment the physiological instability of the infant. The physiological crisis is a primary mode of communication, and it is often presented as the initial focus for therapy. The patient is testing whether or not anyone wants him to survive. He is acting out the drama that if no one cares if he lives or dies he does not care either. These people characteristically exhibit naive pride in physical integrity and undue fear or humiliation about minor physical defects.

Clearly, adults with severe psychosomatic pathology have developed competency beyond the level of early infancy. The major area of fixation is quite primitive, however. Such people are frequently preoccupied with asserting their right to experience primary sensory data and to accumulate and own objects. They usually have organized their lives to duplicate the absence of cherishing conspicuous in their early years. They characteristically take thin consolation for their deprivation by bribing others to care about them by offering themselves for exploitation. However, they are embittered by the inadequacy of the bargain.

Intimate cooperation between the therapist and the patient's medical physician is indispensable in this situation, since sudden insight may either free the patient from his symptoms or precipitate an acute and dangerous attack. The patient's point of growth is focused on the validated exploration of perception, the arts, and physical mastery. If the patient can take a brief vacation from his lifelong preoccupation, he may discover to his surprise unexpected talents and enthusiasm in these areas.

The interpersonal relationships of the predominantly psychosomatic patient pervasively repeat the familial pattern. Therapy should introduce some perspective about the extent to which contemporary adults literally replace members of the family in the patient's life. The patient tries hard to disguise his true picture of himself and others, and the reality of his relations with others. The therapist has the double task of sidestepping the patient's machinery for denial of the reality, and undercutting the patient's guilt about the self-absorbed quality of his relatedness, before he can get the truth into the open.

The psychosomatic's most valid relationships with people are characterized by interactions with large groups of people, and by sexual play. Close rapport with a friend is rare and short in duration. His primary integrations are characteristically carefully struck bargains.

Prognosis is fair, assuming adequate medical care, because the patient by his symptoms indicates that life, to him, is not worth living as it is. But much new experience must be accumulated before the patient can believe that someone could want him to live for his own sake.

The person whose somatic awareness and physiological responses are generally and broadly flattened is also psychosomatically damaged. It is a frequent residual from the indifferent, apathetic, frankly rejecting mother. This depressed physiology goes with other indices of depression.

Psychosomatic pathology is frequently classified by organ systems. Some psychosomatic illnesses have as their key point lack of representation in the body sensorium, with resultant relative insensitivity of the afflicted organ. The resulting internal pathology may involve loss of normal adaptive movements, or a proliferative response of the antecedent reflexes, or a retrograde atrophy and low level of integration of the physiological and organ systems.

For example, in the area of muscle tone, a considerable body of data exists in the field of "body analysis" on the relationship between aching muscles and the difficulty in subjectively experiencing the muscles involved. Treatment oriented toward facilitating awareness of specific muscles and muscle groups has often had remarkable effects in alleviating difficulty in "sensitive" areas which, apparently, are actually insensitive areas.

Chronic backache can respond to analytic therapy. A striking example was a man suffering from scoliosis. When he started analysis, he had spent most of the preceding year in a cast. During analysis, it became clear that in fact he was often unaware of the tension in his back until it became acutely painful.

At first, a moment of interpersonal insight was associated with sudden acute awareness of pain in the back. As analysis progressed, this reaction decreased, the general experience of aliveness improved, and the capacity to function was, at the end, in the normal range. The etiology here partly involved his mother's fury about the scoliosis, and his embarrassment about it. It also involved to some extent the mother's sadistic beating of her children. It was not possible to reintegrate fully the man's experience of his back until he could cope with some of his rage against his mother.

The situation was more urgent for a middle-aged woman for whom a disc operation had been recommended with a definite time limit on how long it was safe for her to postpone it. After a year of intensive therapy, partly emphasizing the relative insensitivity to muscle fatigue, the operation was entirely eliminated from her prospectus, and the only thing she was unable to do was to play golf.

Neurodermatites have been described as correlating with a particular type of personality, described as tense, overresponsible, overactive, controlling, domineering, and insomniac. The phenomenon has been attributed to a fatigue reaction based on compulsive overactivity, that is, insensibility to the sensation of fatigue. The depression of the sense of touch because of inadequate cuddling in infancy is an equally likely mechanism of etiology of neurodermatites. When physical tenderness becomes acceptable and accessible, the neurodermatitis tends to clear up quietly and unobtrusively.

In respect to ulcers, the original pathology could lie in the child's integration with his mother—with the emotional climate at feeding time during infancy.

The pathology often includes a temporary or durable dissociation of the normal animal sense of the full or empty stomach. Since the patient does not know, from his own sensations, either when to start eating or when to stop, he tends to eat either by or in defiance of his mother's instructions or his doctor's orders. This eating can provide only the grossest approximation of the amount of food his stomach really needs.

Again, more interest in the sensations of the stomach, more flexibility and permissiveness in searching out the type of food that really appeals—not the type that had prestige value in mother's eyes—often result in considerable improvement. Likewise, in constipation, a relative insensitivity to the normal impulse to defecate at the appropriate time seems to be involved.

One of the striking examples in the importance of validation for conscious representation of sensory data lies in the occasional failure of a parent to validate the experience of pain. The result, in the adult, is the inability to notice physical pain when it occurs in normal intensity. Thus, sinusitis remains undiagnosed because the adult's headaches do not hurt. Or the adult is subject to a variety of repeated cuts and bruises, because the pain involved in being cut or bruised is not sharply in awareness. The adult, so deprived by his parents, is put into danger by his inability to respond to warning signs.

At the same time, since the full experience of illness is unavailable to him, he learns that he cannot tell when he is sick, and hence tends to become a hypochondriac. Certain isolated signs of illness are overinterpreted, because if they were in fact part of a true illness he would experience them no more fully.

The undefined psychosomatic body image—sketchily filled in because of areas of dulled sensation—becomes a screen on which other feelings about oneself can be projected, feelings such as self-revulsion, or the expectation of catastrophe. These may find their formulation and their rationalization in "objective" medical terminology.

Clinical experience indicates that when, in the course of analysis, anger can be experienced directly and express an effective asser-tion, the tendency to suffer migraine headaches begins to disappear.

There seems also to be a relationship between sinusitis and tears. The person who is moved to tears but blocks them develops the physiological swelling appropriate to crying. If chronic, this can block off the sinus passages and create a seat of infection. It sometimes happens that when a person can weep, not occasionally but freely and without inhibition, sinusitis begins to clear up. The capacity to weep depends on many factors, including the capacity to relate to another person with sufficient trust, and the capacity to feel sufficient tenderness to make the sharing of grief feasible.

Thus we emphasize awareness of the emotion. This, in turn, is contingent on some minimal degree of freedom in implementing the emotion. This requires the possibility that one can conceive of escape from the original situation which conditions the emotion. In the situation in which the emotions become repressed, the imple-mentation of such emotion would have been dangerous—and so it still seems to the adult.

The simplest example of this is the emotion of rage. Mother's rage set up appropriate counter-rage in her child. It was unsafe for the child to express it at the time it was set up. Thus the rage has been carried in massive repression throughout life under the dom-ination of the initial impress that the repression was necessary for survival.

For an emotion to get into awareness and to be dealt with, it must be capable of some direct expression. Thus there is the apparent paradox that direct expression of emotions of long standing, while necessary, can emerge with explosive intensity, involving, in the current context, irrational acting-out. An alternative mode of expression is the full subjective experience of the emotion and the sharing of awareness of it with another empathic person. Clarifying the negative emotions underlying a hostile integration depends on one's freedom to disconnect the relationship, if necessary. This freedom, in turn, is based on one's own capacity to experience joy and self-respect.

In the process of therapy, there is a circular operation of degrees of freedom, full depth of emotional experience, and implementation in reality interaction. This occurs with both original integrations and their contemporary representatives. One limitation of this process is the commitment to relative rationality of behavior. Antagonism to the irrational is based partly in the patient's investment in society. He feels it as humiliation to act as if he were one year of age, to admit openly to still experiencing primitive feelings. Additional therapeutic confusion can be created by moralistic response in the therapist to his patient's irrational dictates. An additional complication may be that, according to the structure of society, full expression of the patient's repressed emotions would often be illegal.

The repressed emotion must, to some degree, come into awareness in the context of reliving the original situation, whether in contemporary example or by hallucination. The extent to which this is true varies with the different emotions. The experience also must be shared with another person to get fully in the open.

In therapy, in the process of explosively unrepressing the repressed emotion, isolated breakthroughs do not accomplish change. The emotion must repeatedly break through the barrier surrounding it. This will occur initially in the paranoid mode and must be progressively reintegrated in the context of sharing in order to get consolidated fully into the patient's self-system.

Anger, fear, grief, and despair are the lifelong emotions at the core of the central paranoia. They are not all anti-social, but they are all more or less blocked, partly because they feel irrational in the adult context. All of them would have been dangerous to experience in the childhood situation. Their suppression refers primarily to the mother's power of life and death over the child. When they have been repressed, it is frequent that joy, freedom, and self-esteem have also been made more fragile, or have tended to be obscured or repudiated by the patient.

Regression can also have an important impact on physiology. It may affect the level of physiological integration of the total organism, or of specific organ systems. The specific symptoms may involve a loss of later acquired, better balanced integration, and a return to the early distorted integrations with the mothering one. On the other hand, the symptoms may simply involve the disintegration of the more mature functions, however integrated, and a return to the unstable infant patterning. Choice of the symptom may be on the basis of faulty equilibrium due to generalized anxiety at the time of integration, or may refer symbolically to a particular aspect of interpersonal interaction.

The surgeon's carefully calculated casual air, and its importance to him, is a suggestion that the surgeon thinks that the repression of his own anxiety and the undercutting of his patient's anxiety are two of the most important elements in his operating success. Conversely, one conceivable explanation of unexpected and unexplained surgical shock after a relatively mild operation might be that a very high level of panic existed in the surgeon, or the patient, or both.

The daily indices of anxiety, those things by which we know that another person in a social gathering has been made anxious by some incident, involve disintegration of some function that had been previously well-organized. Some examples are: a stumble, the dropping of food, upsetting a glass, a cough, a choke, a cinder in the eye from the temporary breakdown of the blinking reflex. The physiological symptoms of more severe anxiety, such as rapid breathing, rapid pulse, and flushing, might also be related, at least in part, to a throwback to an earlier physiological level. The diarrhea of acute anxiety has some similarity to the intestinal function of the three-months-old baby and his susceptibility to colic, and it also responds to the same paregoric. Anxiety also might be a clue to the sudden loss of resistance to infectious diseases against which one has apparently previously been relatively immune. Anxiety may also explain an unexpected breakdown in the efficiency of forces designed to maintain the localization of an acute bacterial infection. Too, there seems to be particularly clear clinical correlation between resistance to the common virus infections and a temporary period of acute anxiety, which involves lower body resistance and inadequate body temperature control.

Symptoms of malaise, chilling, sweating, and depression in many illnesses are essentially indistinguishable from the symptoms of a period of withdrawal, regression, and rumination following major psychological insight. In addition, an illness—migraine attack for example—can set up that state of mild intellectual confusion and emotional withdrawal which considerably facilitates integration of some new insight in the personality, with a minimum of the usual necessity to cope with external reality. The fact that withdrawal at such times is socially acceptable facilitates the process.

Now there are two even more speculative applications of this theory. One is that there is a mechanism that accounts for possible prenatal influences in determining the "temperament" of the newborn infant. A very high level of anxiety of the mother throughout the pregnancy, or any considerable period of acute panic, could have a detrimental effect on the development of whatever functions were being integrated most actively during that time. After the baby starts to move, it can be noticed that the general level of anxiety on the part of the pregnant mother may be responded to by the baby, either by marked increase of motility or by noticeable inertia. The pregnant mother also tends to avoid certain situations which she might ordinarily be able to cope with adequately, but which could invoke intense anxiety in her condition. It is also conceivable that the overall emotional tone of the mother—angry, sad, fearful, or contented—may facilitate in the infant the easier integration of appropriate physiological patterns on either a resonant or a reciprocal basis.

We can also speculate as to whether or not these concepts might provide a theoretical mechanism for interpersonal influence on the development of cancer, or on the choice of organ for the seed of a cancerous growth. To review the concepts related to this possibility: some representation in the background of awareness of the body, experienced as aliveness, or contentment, is hypothecated as necessary for the full, healthy functioning of any particular organ. Many psychosomatic symptoms represent partial integration of the function with some distortion, associated with pathology paralleling the blocking of its accessibility to awareness. Further, we suggest the hypothesis that a profound dissociation might result in a minimum of integration into the body organization of any organ, and thus could promote the conditions of clinical atrophy.

Examples of correlations can be cited, at least with respect to the sex organs. Mild neurotic dysfunction can be associated with hypertrophy of the prostate, flooding and dysmenorrhea of the uterus, and oversensitivity of the breasts. Parallel with profound repudiation of the female role, the breasts may atrophy and the external genitalia may become dry and insensitive.

Now the forces maintaining a high level of functional integration of any organ must involve not only the general function of the organ, and its central representation, but also must involve organization at the level of subdivision of lobes or units. These in turn must interact with the overall organization at the local or cellular level. All of these forces are involved to some extent in maintaining a high level of specialization or differentiation of the individual cells, and their integration into the local, the sublocal, the total organ, and the total body system. Since structure is maintained by flux equilibrium, if any of these forces, the constant activity of which is useful in maintaining the structure of an organ, are missing, there is a tendency for the structure to regress.

We add two additional hypotheses: one, that the full integration of a particular cell or small group of cells into the organ is a factor that tends toward the maintenance of specificity of cellular differentiation and thus could be a factor that tends toward the inhibition of cellular regression and hence toward the inhibition of development of cancer. To put it differently, perhaps all or perhaps a very large variety of cells, or types of cells, are potentially cancerous if they were not subjected to certain inhibitory influences that depend on their position in the organ, and on the forces immediately surrounding them. This is somewhat in line with the fact that the move toward cancerous degeneration involves ultimately a move toward de-differentiation. Those cells that are more embryonic are those relatively more susceptible to cancerous development. In these terms, it could be that the cells, de-differentiated because of a lower level of organization, become more susceptible to cancer. Or, the lower level of organization as a whole may fail in its usual function to hold in check the cancerous development of potentially embyronic elements.

Second, an unusual and diffuse stimulus to cellular growth may be a factor in the development of cancer. This would include specific carcinogenic chemicals. There is a high tendency for embryonic remnants that got into the wrong organ in a young child to develop into cancerous growth. Here, we see the general growth stimulus of the infant body, plus the failure of the organ—it being the wrong organ—to apply influences restraining cancerous growth. A parallel situation would be the tendency for placental remnants to develop cancer.

Specific physiological growth stimulation of particular organs, as with the breasts in pregnancy, can exacerbate a known tendency to cancer in these organs. It is even conceivable that psychological reintegration of long dissociated organ systems later in life could provide a similar stimulus to growth beyond the scope of the concurrent level of the organization of the organ involved.

Cancer tends to develop with age, paralleling the frequent weakening of the integrative forces, psychological and physiological, and the general progression of the body towards atrophy. Whether any more specific correlations could be identified between the probability of the particular organ's developing cancer and this organ's accessibility to interpersonal relatedness and therefore to interpersonal repudiation, would depend on a much more extensive investigation of currently available data. Documentation of these hypotheses would lie, first of all, in improving communication among, and organizing individual collaboration among, good internists and good analysts, rather than in the moment, setting up long-range experiments.

The analyst, in order to investigate the interpersonal influence on psychosomatic and organic disorders, must select people who are sufficiently motivated toward interpersonal growth to get into analysis. Some of them will suffer a disease—sinusitis, for instance. With sufficient numbers of such patients providing cases documented in collaboration with a specialist, one might establish or disprove the hypothesis that this conditon would clear up when the patient is able and willing to weep freely in proportion to the true sadness and regret of his life.

The crucial experiments defining the possible interrelationship between interpersonal relationships, early and current, and specific psychosomatic pathology cannot be identified until there exists the intellectual understanding for much more extensive collaboration between medical specialists and interpersonal theorists. Such experiments cannot be set up in delimited areas but must evolve deductively from a broad base of data collected in collaborative preliminary research.

X

THE HYSTERICAL SYNDROME

THE CORE OF the hysterical syndrome is that the communication is in the context of the patient's behavior and not in words. It goes by demonstration, and the point of the script is in the story line. The content of the communications is variable and can include the drama of rejection, of the waif, of the innocent victim, of insecurity, of overresponsibility, of unrequited love. The interpersonal attitude to be communicated is not easily available to direct sharing —it is dramatized. The words used are a part of the script, though a silent movie would do almost as well. The context of the drama is the child in the adult world. The language is motor: by pointing, by shrug, gesture, mimicry, facial expression. Words are considered very unreliable. This is the mode of communication of late infancy. It is elaborated and amplified in childhood, and continues to be used for significant emotional content during this era.

One may use the hysterical syndrome predominantly because of arrested development in the prejuvenile era, because of regression from later levels of organization, or for the purpose of camouflage.

If interpersonal development stopped at the prejuvenile level, the patient of necessity uses the hysterical mode as his best level of communication. Such patients' characteristic interpersonal relationships approximate the parallel play interaction of childhood. To the extent that they are cut off from knowledge of their own inner organization, their orientation is to outside stimulation, but they have not achieved the high level of organization of the obsessive. They are not responsive to the factual details and long-range considerations. Their behavior is considered impulsive; it seems to lack continuity. They have a poor grasp of group dynamics and may be much more comfortable in the one-to-one situation.

Therapy with patients arrested in the childhood era must emphasize the history of the very early years. The focus of growth is the initiation of genuine group relatedness. The therapeutic difficulty lies in learning how to interpret hysterical communication. Without this understanding, the therapist tends to be taken in by the words without accurately relating them to the general context, in which their intent is often symbolic. The hysteric's role carries conviction, partly because at the moment he believes he is telling the whole truth and partly because the patient is using all of his faculties to adapt the role to the particular audience. When the unaware therapist later realizes that he has been taken in, he often feels antagonistic toward the patient.

A necessary technique in therapy with these patients is the art of being pleasantly rude and incisive, with enough good nature to brush aside whole areas of the patient's verbalizations in order to focus on the genuine content of his production. One therapeutic task is to facilitate the patient's learning more about the use of words for communication, partly by clarifying that what he has done so far is not that.

With people who have matured beyond the childhood era, the hysterical syndrome is often the product of regression under severe long-lasting anxiety. With increasing panic, the patient experiences difficulty in concentrating. He is no longer capable of the high level of organization which is necessary for effective obsessional control and he falls back progressively onto the hysterical mode of organization and relatedness. The elaborate, obsessive preoccupations become fragmented into a few verbal or ritualistic formulas which are then secondary to the more general phenomenon of the hysterical acting-out. He clings desperately to what is left of the obsessive machinery in order to restrict the acting-out. The hysterical phenomena take over in full force in the borderline state just short of frank, massive breakthrough of dissociated material, that is, just short of frank catatonia.

The self-image becomes fuzzy and shifting, and tends to regress to the prearticulate. The patient attempts to keep going by being overadaptive to the momentary situation. He cannot depend for orientation on either his own feelings, which are sheer anxiety, or his own perceptions, which are highly colored by the parataxic, and so he attempts to play roles. The choice of roles is strongly affected by the dream life. He is very reactive to the external situation, and suffers a loss of normal self-direction. He cannot get clear on who he is, or what he wants.

For this reason, there is a tendency to take on the role that the other person casts him in. Conversely, the acting of the role becomes a part of reality testing. It is in a way a question, "Am I really who you seem to think I am?" He regresses to the period before the articulate use of language for communication, somewhere between two and three, when the self-image was being formulated.

Frequently, over a period of days or months, the pressure of the anxiety slowly calms down, and the patient returns to a better organized, if somewhat obsessional, level. This return may be essentially to where he was before, but it often involves some real expansion in the personality, some real reintegration, incorporating some aspects of the experience that precipitated the crisis.

This regressive phenomenon can be thought of as a form of acute ambulatory catatonic episode, and its therapeutic management is essentially the same. If the patient is already in this borderline state in the initial contact with the therapist, it is not always possible to reverse the direction of the move toward a full-blown catatonic break. With these patients the hysterical phenomena are the result and the expression of acute anxiety, the loss of continuity is based on regression, and the patient has previously experienced relatively durable periods of better functioning.

It is important to differentiate this syndrome from the patient whose development was arrested in the prejuvenile period. The arrested patient is often not particularly anxious at the time of undertaking therapy, however much he may protest his suffering.

The patient in the acute regressive syndrome can only be severely anxious. In the former case, part of the therapeutic operation is to aggressively search for and elaborate any material that seems to come genuinely from the integral personality. Where the syndrome is due to anxiety, such material must be dealt with, but should be played down and understated by the therapist. Clear, incisive interpretation of the context of the total situation is extremely reassuring for the anxious patients. They have recently lost a frame of reference to which they are well accustomed, and are searching for it frantically. Anything to help them re-establish its outlines or to clarify a new and expanded one will cut down the anxiety and greatly improve the level of communication.

A third group of apparent hysterics have found it necessary, in the power struggle with a repressive authority, to camouflage their real level of competence by "acting crazy." As in the regressive group, there will be a clear history of more effective performance, after which their lives seem to have deteriorated, but here the deterioration is false. The inner organization may be elaborately disguised for purposes of manipulation or self-protection. The level of anxiety is not catastrophic. Those things which are absolutely indispensable to the life plan get done, however much behind a smoke screen. The competence is real, and the incompetence is a part of the act; it is for the audience.

The patient's maneuver cannot be entirely conscious, otherwise he could not play the role so effectively. But he will respond to common-sense communication, if the therapist is not beguiled, and if the therapist can disentangle himself in the patient's mind from the patient's central dangerous figures. The hysterical person's convictions about his role need not be validated, but neither should they be treated with scorn.

XI

THE OBSESSIONAL DYNAMISM

OBSESSIONAL DYNAMISMS involve use of words, thoughts, rituals, and organizations of time and space as substitutes for, rather than as tools of, direct inner perception. Thus the frame of reference of the person's audience and of the exigencies of the situation are substituted for that of the person's own needs and perceptions.

The intellectual capacity for making sense out of complicated data develops during the juvenile era. Previously, verbal communication took place between the child and a few well-known individuals who were interested in and responsive to his private and autistic symbols. The juvenile is now concerned to relate to many relative strangers, contemporaries, who are not concerned with being understanding, and this orientation involves a high capacity for conscious thought and for verbal communication. Whence, the primary concern of the juvenile is with the universals of communication. He has already learned the separate elements of speech, numbers, and primitive social interaction, and during the juvenile period these are reintegrated into universal modes of contact.

The intellectual capacities developed during the juvenile era, and their application to the study of human relations, are normal tools of adult living. The pathology of the obsessional lies in using them to limit emotional experience rather than to expand it. The pervasiveness of the obsessional in our culture is partly a measure of the degree to which our lives are persistently organized in the juvenile mode of interpersonal relating—that is, for enlightened self-interest, with a minimum of compassion. In our culture no one is immune to the obsessional dynamism except those who have never achieved it. This dynamism remains available as a security operation after the experience of a reasonably satisfactory chumship, but its pervasive use is incongruous with the direct interaction of the loving relationship.

The obsessional is by far the most efficient of the security operations and if available it is used by preference for limiting experience. If the obsessional operations are smooth and efficient they can constitute a reasonable facsimile of integrity and spontaneity. If obsessional operations are fragmented by anxiety or amputated by deterioration, the repetitive, disconnected nature of the inefficient fragments becomes conspicuous. These fragments, the conspicious but inefficient obsessional dynamisms, are the source of textbook descriptions of the obsessional dynamism. The use of intellectual modes of communication for adaptation to external pressure is rarely thought of as a deficit in personality organization. This is partly because it is very efficient in inhibiting individual creativity and satisfaction of personal needs and in siphoning energy into maintaining the elaborate superstructure.

The therapeutic problem is to distinguish spontaneity from the obsessional way of life. This is difficult because to a degree—short of ecstasy, full maturity, or immediate creativeness—all acts are a composite of spontaneous impulses implemented by intellectual maneuvers which are equally available for obsessional diversions. To the extent that any action is limited by deference to irrational authority, it is to that degree obsessional.

In a broad sense, much obsessional activity is indispensable, for practical purposes, in modern culture. It is impossible to organize one's life so that there is complete congruence between the implementation of growth and satisfactions and the presence of cultural facilities or the absence of cultural barriers. There must be postponements, compromises, and choices. Even so, much more spontaneity is possible, and it evokes much less social reprisal than any of us expect. In a freer culture it should be that much more feasible. The highly technical nature of civilization does necessitate learning out of the sequence of spontaneous interest and performance, and inappropriate to the immediate creative need. At the same time, this high level of technical development has solved the major problems of physical survival and comfort. This recent technical advance has made it possible to seriously raise the question for the first time of developing creativity and personal happiness on a universal scale.

In analytic theory the above polarity, obsessional-spontaneous, has been formulated variously. It has been said that the obsessional is inversely proportional to the degree of spontaneity. It has been said that the obsessional is directly proportional to the degree of malevolence in any given interpersonal interaction. Obsessionalism has been described as the substitution of convention and rote memory for the direct organization of perception; or as the substitution of the frame of reference of the environment for the frame of reference of the integral personality. Put more cryptically, it has been described as the substitution of the form for the reality.

Since the obsessional dynamism is the most universal mode of resistance to growth, experience, spontaneity, and analytic insight, the major part of psychoanalytic theory and technique was invented and evolved essentially to understand and circumvent the obsessional mode of resistance. Because of this fact, the preceding discussion of therapeutic management of other diagnostic categories elaborates special cases and special modifications of this overall technique, the routines of which are highly adapted for the management of the obsessional.

The specific form of management of the obsessional style of resistance in therapy depends on the situation in which it occurs.

The obsessional defenses may become fragmented and desperate as a result of deterioration. If so, therapy is primarily focused to deal with the underlying paranoia. Such fragmentation can also be a regressive phenomenon, caused by mounting panic heading in the direction of an acute catatonic break. In this case, the immediate task is to undercut the mounting panic so that the intellectual processes can come back together in one piece. With both the deteriorated and the regressed obsessional, the immediate task in therapy is to cut through the verbal static and to learn and reformulate some emotionally significant data.

In another group of obsessional patients, the mental and interpersonal life is entirely consumed by this defensive operation, but the obsessional operation per se is relatively intact. They operate well in a crisis and tend to structure their lives into a series of crises, which keeps the external reference points clear and obviates pressure to formulate inner experience or reach out to satisfy the inner needs. These people have either written off their possibility of creativeness of have written off their expectation of expressing it. They are uneasy in situations of lowered awareness, which involve less facility for remaining highly organized and force less constant contact with the external reality.

If such an obsessive system is put under pressure gradually, the result may be a regression to the hysterical organization. If it is suddenly shattered, the immediate effect is the catatonic experience. The patient fully expects that once shattered, he will never be able to put the pieces together again. In reality these are competent people and the disintegration is essentially temporary. After a relatively short period of time they may reconstitute the previous organization, either with or without the addition of some new insight. With somewhat longer time, they may begin to implement the organization of the integral personality and begin to deal with the world in their own terms. They come into analysis essentially through awareness of the effort involved in maintaining the superstructure and through noticing the gradual, progressive loss of texture in their lives. As with the depressive syndrome, to which this is closely related, this loss cannot be perceived by the individual unless something good has happened to lighten the pervasiveness of his subjective cutoff of experience. This is the leverage that the analyst can use. In spite of the highly efficient organization of the resistance, and in spite of the patient's conscious conviction that it will come to no good, analysis with these patients is a feasible project. Once the patient begins to be aware of what he is missing, the initiative shifts from the analyst to the patient.

Obsessional resistance can also be prominent in patients who use obsessional machinery under pressure of anxiety, but whose life incorporates much spontaneous activity and genuine exchange of affection. They often make constructive use of minor obsessional operations, as, for example, the use of minor distractions, to allow for periods of rumination, for the purpose of integrating new thoughts into the integral personality. The obsessive machinery is in good condition and well-oiled and the analyst's dilemma is in identifying it when it is thrown into play to avoid a particular insight. It will always be plausible.

XII

HOSPITALIZED PATIENTS

PATIENTS IN ACUTE psychotic states get to be hospitalized because of a combination of factors. The precipitating incident is always some insight which cannot be incorporated quickly into the self-system without major and extensive reorganizations in the patient's life situation.

One factor influencing hospitalization is the amount of logical reorganization of the self-system which this insight entails. Hospitalization is less likely to be necessary if sympathetic and supporting friends are available. Another factor may be the manageability of the job situation. Does it allow for periods of temporary vacation, if the patient finds it difficult to concentrate? Availability of therapy may be crucial in heading off or working through an acute episode outside the hospital.

The precipitating insight may be attendant on something good being forced upon the patient in the natural course of living—a baby, an advance in a job, a boy friend. Occasionally psychotic episodes can be precipitated by misdirected therapy. Therapeutic techniques designed only to break down the character armor without improving the quality of the patient's real relationships, sometimes trigger a latent psychosis into the open. The application of an unguided free association technique combined with emphatic stress on interpretations which the particular patient feels are unresolvable, such as latent homosexuality or intense, murderous impulses toward someone with whom he feels trapped, or incest attachments, can also throw the patient into a temporarily disorganizing panic.

Another crucial factor in determining hospitalization is the life situation of the patient at the time. The patient may feel entrapped in some specific interpersonal situation. He may have a destructive wife and several children and no money to pay for alimony, or demanding and clinging parents combined with responsibility for the younger siblings. Under the powerful illusion of indispensability, caught up in intense rage and the fear of retaliation, incarceration in a hospital can seem to offer the only protection from the family's demands and from his own inner conflict between compliance and the pressure for freedom. He may secretly hope that the family might learn to get along without him while he is away. A very frequent pattern is the young adult who is profoundly stifled at home but afraid to leave home.

Another factor in hospitalization is the patient's expectation that he will receive adequate therapy there for his difficulties. It is then both a public appeal for help and an attempt to get help. The particular circumstances contributing to hospitalization, the organization of the hospital facilities, and the psychotic condition of the particular patient all contribute to determining the plan of therapy while the patient is in the hospital.

The course of therapy during hospitalization involves a little of everything that is covered in intensive analysis, but in a simplified and abbreviated form. It might be called the first cycle of analysis. The patient's self-system is fragmented, he is unable to organize, so it is entirely clear here that it is up to the therapist to structure the situation and set up the plan. He will have to establish the character of the relationship with the patient, the content, emphasis, and sequence of therapy, and must help the patient get a grasp of his immediate environment.

There are four major topics that have to be covered. One is a grasp of the simple outlines of the history, including the precipitating factor or factors concerned in the immediate break. The second is the patient's interpersonal relationships on the ward with the other patients and with the staff—that is, with contemporaries and perceived authorities. A third is the patient's capacity to cope with the parents and the immediate relatives. Fourth are the problems to be expected on release, particularly relationships with parents and other relatives and the problem of financial and physical independence.

The initial procedure is to get the history. This should be from the patient, if possible, even if it can only be done by direct questioning. This should be put together with a simple structuring of the most obvious interpretations, including an attempt to relate the character of the integration with the parents to a statement of the patient's problems. It should also refer briefly to the patient's expectations after discharge.

After therapy has been initiated, the major content is the quality of the patient's relationships with hospital personnel, with the therapist, and with other patients, with the physical demands of the situation, and with visiting relatives. Toward the end the emphasis goes over into specific plans for life after discharge, with exploration of the difficulties which can be anticipated.

The reconstruction of the patient's personal history provides a framework which the patient and the therapist can fill in later in therapy. It is not concerned with too much subtlety in exploring the events involved. It is not aimed at evoking any more reliving of the incidents involved than is inherent in collecting the factual data. The patient is already having enough trouble with the reliving of childhood terrors without the therapist promoting more.

The hostile integration with the parents is one of the primary items of content. Here, too, the emphasis is on the factual interaction with a nondramatic statement of the practical effect on the patient. At this stage the therapist does not get any more involved than necessary in the patient's feelings of love or hate toward the parents. He must tread a thin line between condoning atrocities and clarifying the hateful aspects of someone with whom it will turn out clearly the patient is very much identified. Sticking to the actual interaction and the actual sequences of events is one of the safer ways of avoiding this issue. The therapist must take a stand against the parents' brutality, and not be too interested in explaining it as the inevitable result of the parents' unfortunate childhood. At the same time the brutality must be treated as something in the range of human experience, and not outside of it.

The rest of the content of the history includes attempts at relationships with people outside the family and how these were discouraged by the parents, that is, specifically, by what threats and by which attitudes? The other relationships include acquaintanceship and group membership with other children, chumships, boy friends, girl friends and marriage.

An important thread to investigate is often that of the patient's and the parents' relationship to pets. The child, more or less cut off from contemporary relationships, often turned to pets as the source and objects of tenderness. The parents' attitude toward this can be some index of the parents' real orientation to tenderness.

It is important to identify which friends, physicians, or relatives gave the patient some accessory support. In laying out the chronological history of interpersonal development, it is important to locate periods of better adjustment or of relative happiness, to note the circumstances under which they occurred, and to identify turning points of increasing discouragement with life.

Transference and counter-transference flower in work with hospitalized patients. Transference is almost never a topic of primary interest for discussion, and is rarely useful as primary evidence in reconstructing with the patient the character of the relationship with the parents. The therapist will have all he can do to keep the relationship between himself and his patient straight, without throwing in analogies. The emphasis has to be on playing analogies between parents and therapist down.

Counter-transference, whether in anger, irritation, or intense affection, is usually one way of expressing the therapist's discouragement, and the patient tends to react to it as such. In the process of structuring the practical situation, the therapist should keep in mind the real interaction between them. One index of this might be how they would relate if they had met socially. The therapist should avoid a personal commitment beyond this.

The relationship between the patient and the therapist should be simple and practical. It also has to have a certain durability. If the therapist is only at the hospital for a certain period of time, this fact should be clarified and the limits of that time established. The various eventualities of the patient-therapist relationship must be discussed. If the therapist will leave before the patient is discharged, what will happen? If the patient is discharged while the therapist is still working at the hospital, what will happen? Will therapy with this therapist be available after they both leave the hospital?

The hospital patient is realistically more dependent on the therapist himself than is the patient in private practice. Usually he does not pay the therapist himself. At best, the relatives pay the hospital and the hospital arranges with the therapist, thereby making the patient's claim on the therapist's time relatively tenuous. If there is no pay involved, the patient will be concerned with why the therapist should have any commitment to him at all, especially if it is obvious that there are many more patients than there are available therapists. The therapist must also provide some reassurance about his availability in an emergency, although it is not at all necessary for him to put himself on twenty-four-hour call.

The patient feels intensely dependent on the therapist, and is terrified either at overt friendliness or at overt unfriendliness. The slightly friendly move on the part of the therapist can be interpreted by the patient as a demand for sexual payment for his services. If so, it should be noted by the therapist that at least one parent was probably aggressively seductive. It is usually not helpful to tell the patient this. The patient who claims to be upset at the discovery that the therapist has a wife, and to be concerned about whether he would have a place for her in his heart, is in fact reassured when told that the therapist leads a pretty busy life, and would not have the time to invade the patient's life emotionally. The patient who constantly accuses the therapist that the love he offers, or the interest, is too little and too late, is in fact busy repudiating his own need for tenderness and derogating the tenderness that his therapist offers. In other words, transference misinterpretations can be guessed at by a knowledge of the most obvious distortions of the parent-child relationship. The therapist should try to reassure the patient that his distortion is untrue, without throwing the emphasis on analogy to parental attitudes.

Often the transference problem is a simple inversion. The patient often twists this reference point around so that he states that he is afraid of exactly the opposite of what he is really worried about. This is also true of many of the patient's other preoccupations. For example, the patient's expressed preoccupation with fear of failure is actually a preoccupation with the fear of success. This occurs because successful performance would be attacked by one of his parents. Therefore the patient becomes more obsessive and perfectionistic in the effort to be invulnerable to attack and thereby obscures his fear of retaliation for effective performance. Even the fear of insanity can sometimes essentially be the fear of sanity. If mother was genuinely disoriented, if she was recognized by the child as a very poor model for living, the child still did not dare to question her sanity. This was too frightening. By preference the child became obsessively preoccupied with his own insanity. He kept wondering whether his own valid perceptions of the world, since they differed from mother's, might be insane. It is touchy to share this interpretation until the patient is really in a position to recognize the extent of his mother's handicaps.

Overtly or covertly, to a certain extent, the patient will use his therapist as a substitute mother. He will turn to him for validation, for permission to grow up, and for the alleviation of pain. This is dangerous only if the therapist has his own neurotic reasons to cherish this role, or to be made anxious by it. The therapist also, whether he deserves it or not, will be used as a model. The best he can do is to accept this honor with humor and to debunk it gently but consistently. By doing so he can help the patient identify more appropriate and available models from his own environment.

Verbal or mute, the disturbed patient is extremely inarticulate about communicating what is really going on inside him and how he really sees the world. Much of the bizarre character of the psychotic production is clumsy communication and formulation rather than distortion in perception. On the other hand, if the therapist sticks to simple formulations, the patient can understand him much better than might be expected. This is one of the reasons that the direction of formulation must come from the therapist to the patient during the early phases of therapy.

With a totally mute patient there is nothing else to do but collect data from other people, family, staff on the ward, and attempt to structure this to the patient as well as one can. One must always emphasize that the validity of the data must be in doubt. The patient who talks, no matter how psychotically, can usually answer direct questions which can be answered with yes or no, and, by this pedestrian method, the persistent therapist can put together enough data from the patient himself to provide a base for getting communication going. A different difficulty exists with the aggressively obsessive patient. He demands that the therapist pay attention only to his obsessive preoccupations. Here, the therapist has to be willing to be quite aggressive and firm about putting his own project first.

For each interview, when some data has been collected and formulated, it is usually useful to leave. One simple interpretation of good data is frequently more useful than two interpretations, because these give this patient too much to cope with in the interval between interviews. Once the therapist is sure that nothing is going to occur during a particular interview, it is better to leave than to stay and try to disguise his or the patient's irritation or discouragement.

The data of therapy is primarily intended to help the patient sort out and restructure the realities of his situation, past and present. The profusion of data that is coming through into consciousness direct from the dissociated material should not be ignored. If it can be restated in less cosmic and more manageable terms, this can be very helpful. For the purpose of clarifying his own picture of the patient, the therapist must use all data available to him, including dreams, associations, and marginal thoughts, delusional content and acting-out. However, he should keep in mind firmly that this material has a reality for the psychotic patient that it does not have for the ambulatory patient and that the therapist's overt interest in it will tend to validate and reinforce this feeling of reality for the patient.

The patient is unsure of his own perceptions, capacities, and values, and therefore will be involved in a good deal of mind-reading of his therapist. To a certain extent, the simplest thing to do with this situation, is for the therapist to say what he thinks, and then to help the patient to formulate his own thoughts. The patient is intensely interested in learning a few stable rules to operate by and is not too hopeful of ever understanding the mess that he considers his life to be. Because of his intense discomfort, he is terribly anxious to do something about his troubles, but is confused as to what to do. Therefore he tends to respond to interpretations as if they were commands. The therapist can utilize this unobtrusively to encourage the patient to try out some new experience.

However, this fact almost categorically prohibits certain types of interpretations because of the danger of their being acted out, the more so if they happen to have some truth in them. These would include major modes of destructiveness, murderous impulses, suicide, incestuous fantasies, and, for some patients, homosexuality. Unless the patient is exceptionally sophisticated, this also includes any serious discussion of diagnosis or the presence of psychosis. This is because the patient is apt to take the diagnosis as an official invalidation of all of his perceptions, no matter what the truth in them is. If he has to mistrust every one of his perceptions and every one of his judgments, he can only become completely paralyzed.

Persistently the therapist must throw the spotlight on the patient's mode of coping with things, and validating what in his operations is useful. The therapist must make very clear his respect for the patient's efforts, and must help the patient clarify his own values. Perhaps the most legitimate area of guidance is for the therapist to try to sort out what experiences need to be made up by the patient, and then to guide the patient in the direction of making them up in an orderly sequence.

The therapist tries to start with those realities on which the patient is clear, and tries to build the patient's confidence in his own direct perceptions by eliciting and affirming these. Usually, the direct perceptive data is accurate enough but the patient cannot get it in context. This is just as true of the empathic perception of character of another person as it is of physical perception. The patient is often extremely responsive to the overall emotional tone of the other person, but cannot get perspective on it. He misinterprets anger in the other as the intent to murder, he misinterprets tenderness as the commitment to love. The patient's interpretation always incorporates some aspect of truth. Usually this has to be validated before the patient can give up his distortions. The distortions are partly exaggerations that reflect the patient's conviction that the therapist will not acknowledge that aspect of his experience which the patient feels sure is true.

In this interchange the therapist cannot afford to stint on sharing his own opinions. He must put them simply and he should limit himself to opinions which the patient could conceivably integrate at the time. The patient needs the therapist to make clear where he stands. The therapist should commit himself only to what he is fairly sure of and should try to spare the patient the therapist's unnecessary doubts. If it is something that he really cannot evaluate, or if the alternatives are a matter of choice only for the patient, then he can say that he does not know and give the patient a simple analysis of the factors involved. Even so, the patient will probably assume that the therapist can read his mind.

As far as advice is concerned, some advice is needed and should not be withheld if it seems relevant. There are situations that the patient has to cope with that he cannot yet be prepared for. These include many tangles with the relatives on visiting days, and the therapist can be helpful in suggesting how to handle them. Warnings as to risks involved in alternate methods of handling the relatives can be quite essential. Again, the difficulty arises only when the therapist is either overinvested in this or allergic to it.

One theme in the content of therapy is the clarifying of the patient's lack of experience in living, the identification of how he was cut off from experience, in order to find where he can begin to make it up. In this connection, the individual and group relationships on the ward provide important primary data. Very often the cutoff from the external world began somewhere between one and three. Mother could not tolerate the baby relating to any object other than herself, and therefore claimed constant priority even over his toys. This can give the patient a lifelong feeling of being physically amputated, unable to reach out and touch and mold the physical world. Here, the physical program of the hospital, including occupational therapy and athletics, may provide useful primary data for reconstructing the problem, as well as the opportunity for gaining new experience.

A part of the content of therapy will consist in validating the patient's impressions of the therapist's colleagues and the hospital personnel. The therapist usually must find ways of validating what is real in the patient's impressions of the therapist's colleagues, without saying anything quotable. If the patient has information about his therapist which can be used against him, he will be tempted to use it in acting out his anger with the therapist at some later point. If it is a situation in which such validation could make genuine difficulties for the therapist, it is no favor to the patient to place this kind of divisive burden on the therapeutic relationship.

It should be assumed quietly that the patient is an independent person responsible for his own actions, even in his dealing with the hospital staff. It should be assumed, further, that certain activities on the patient's part produce certain predictable patterns of response in the staff. A patient should be considered a responsible person in both the hospital community and the outside world, who is free to investigate his own real impressions of the people around him, and his own impact on them. His method of handling his parents can be somewhat clarified if it can be established that the patient will not live at home again.

Certain other topics almost always have to be taken up. One is the patient's fear of loneliness, which is the result of mother's lifelong prohibition against socialization. The fear of dependence relates back to the original repudiation of tenderness, and the miserable experiences when the child was, in reality, dependent on the mother. Fear of invasion, or of being dominated, relates essentially to the problems of self-assertion. The patient may need some coaching on how to limit the other's invasiveness. Self-assertion on the child's part in a hostile integration with mother was often treated as if it were hostility from child to mother. These two factors will have to be disentangled over and over again, as the patient develops some capacity for self-assertion with his contemporaries and with perceived authorities. He may need coaching in the less aggressive modes of self-assertion. If he expects that self assertion will be considered hostile, he is likely to assert himself in the most defiant and provoking manner possible, since he has no experience in how to be firm and calm at the same time.

The therapist's expectations of what the patient can ultimately accomplish have an influence on the course of therapy. The patient's self-image was formed and can be changed by a reflection of the attitudes and expectations of the significant persons toward the patient. Its persistent inflexibility to favorable experience, the patient's machinery by which more positive opinions are invalidated, is important data. The patient is unaware of how he rejects friendly moves toward him and of what he did to provoke a negative response, and hence is unaware of what is going on in the responder. If the therapist has only a partial grasp of this process this can lead him to a sort of Pollyanna attitude of grimly overlooking the disjunctive aspects of the patient's personality. This does not produce change, but commits patient and therapist to a fiction as to the level of the patient's competence in interpersonal relationships.

Useful therapy requires an accurate awareness of what moves and what aspects of them are positive for this particular patient, and a relative inattention to anything else. Limits on the therapist's expectations become one of the primary factors in the ultimate success of therapy, just as the limited expectations of the parents set the original limitations on the patient's development. Patients can, and often do, go beyond their therapists' expectations. This occurs only when the therapist does not consciously or unconsciously prohibit such development.

The psychotic patient is especially vulnerable to this interaction. Even when the therapist's expectation of the patient is clearly valid, the patient is in danger of feeling pushed. Conversely, easily discouraged, he may be swayed by his therapist away from satisfactions which are attainable and necessary. He is unclear about his own capacities.

One can move toward most projects with the attitude that this is something that the patient can probably do eventually, if he wants to do so intensely enough to invest the effort in learning to get there. Then one can freely discuss whether the patient is ready for it. At the same time, it is useful to warn the patient that he is likely to become anxious, and to explain to him some of the reasons why this is so. Such discussions have to be held about privileges in the hospital, management of parents, discharge, where to live after discharge, jobs, friends, marriage, or babies. A dangerous fallacy, but a common one with hospitalized patients, is the discouraged assumption that the patient's potential for undertaking major projects in living has necessarily been cut off. This is not necessarily true. No less than others, these people can have friends, chumships, can be married, can have children, can do productive work. They need the available impetus to change, and sufficient tolerance from their therapist for the anxiety involved.

The patient may tackle his problem of growth either in a superconventional way or in an extremely unconventional way. One patient may have to get married while she is still a virgin. On the other hand, another patient can only get married if it is possible for her to have an illegitimate baby first. The conventions do not really mean very much to the patient since they have caused him so many difficulties already, but he may be very badly intimidated by their power. Another point of conventionality is that it may be psychologically necessary to help the patient try to meet his family's professional standards, as to job and social status. With a different project it may be equally necessary to help the patient repudiate these standards entirely. It is well to keep an open mind about the person the patient is going to be after he gets over his psychotic episode. Who he was before his episode may or may not give an index to this. It is useful to try to make an estimate, however.

Except for the fact of the intense integration with parents and a severe handicap in some dimensions in getting along with people, we cannot predict what the patient's character structure will turn out to be after his episode is over. He might turn out to be destructive in his interpersonal relationships, or a very loyal friend. A patient who has needed hospitalization may turn out to be a good mother, or a hostile one. We do not have sufficient data to make sure predictions of what patients who get into hospitals can do with their lives with competent psychoanalysis.

If the therapist gets committed either way in his expectations of what the person will become, the patient tends to get stuck with

these standards. If the patient turns out to be a very hostile operator, and the therapist has persistently assumed that he was going
to be a nice boy as soon as he calmed down, the patient is in the
very tight spot of having to pretend to be nice. Conversely, if the
therapist assumes that anyone who is this sick must be pretty hostile
with people, it gives the patient very little opportunity to validate
that he really doesn't have to be that way.

In interviews with the patient's relatives, the therapist cannot be
entirely unresponsive to the anxiety or self-esteem of the relative,
but it is never useful for him to enter into conspiracies with the
relative that exclude the patient. If for no other reason, relatives are
conspicuously unreliable about telling patients what the therapist
has told them. There are some circumstances in which it may be
helpful for the therapist to run interference for the patient with his
parents. It should never be done, however, without getting the
patient's consent first. It is also important to keep clear with the
patient that the therapist has less actual influence over the parents
than the patient does. The therapist may have the advantage of
being less afraid of them, but he has no real leverage. The parent
can usually get rid of a particular therapist, while they cannot get
rid of the patient. They continue to be related to him.

Patients are sensitive to any contact between the therapist and the
parents. On the one side, the patient is afraid that the therapist will
be influenced against him by the parents' hostility. On the other
hand, he is afraid that the therapist and the parents will get into a
fight, and he will be forced to make the impossible choice between
them. When the parents are involved in paying for therapy, the
situation grows even more tense for the patient.

Discharge of the patient is contingent partly on the accessibility
of a reasonably favorable environment outside the hospital. Problems of discharge that must be solved are a job, a room, the relationship to be expected with the patient's parents, the availability
of therapy. The job is important because it is almost impossible
for the patient to clarify his relationship to his relatives until he is
financially independent. If he happens to have independent means,
this may provide the minimum independence. If so, he then must

clarify some purpose in life other than that of earning money. A real need to hold a job, if the patient can meet the challenge, may be a stabilizing and socializing force in the patient's life, and thus some real advantage. On the other hand, the therapist cannot bully a patient into taking a job just because he thinks it would make therapy easier. Almost any patient who can be discharged from a hospital can hold some kind of a job. In fact, it is remarkable how disturbed a person can be and still perform competently in responsible positions.

Patients almost always relapse if they return to the parents' residence on discharge from the hospital, and on this proposition the therapist must take a very definitive stand. Where hospitalization has been an escape from an apparently impossible marriage, a much more careful evaluation of the situation is necessary. If the patient has gotten clear on what he really wants, he has a better chance of solving the problem of marriage without the temporizing recourse to hospitalization. Any patient who can be discharged is able to set up some mode of living away from his family, if this is indicated.

As far as the choice of the patient's lodging is concerned, in addition to the requirement that it be away from the parents' home, it is preferable that it provide for some acquaintances. The hospital has usually provided some opportunity to make up some experience of parallel play, and some aspects of the period of juvenile adjustment. The patient has been living in a dormitory. He has learned something about the rights of others. A gradual transition into group living with greater freedom can be helpful before he has to face the self-assertion involved in setting up his own place. The therapist thus may need to take responsibility for some knowledge of lodging facilities, and some responsibility for evaluating the choices with the patient. Sometimes it is a useful transition to encourage the patient to live at a resident club in the interval between living in the hospital and living in his own apartment. However, temporary return to the parents' home will not suffice for this transition.

Otherwise, discharge arrangements should be as gradual as possible. The patient should ideally have freedom to come and go from the hospital before he is officially discharged. Ideally, he should get a job, work at the job, and sleep at the hospital for a little while before he moves to his own room. The analytic problem with respect to both the job and to physical independence will be in coping with the parents' opposition to acknowledging the patient as an independent entity, past and present. Adequate management of feasible living arrangements at the time of discharge is probably one of the crucial factors in determining whether the patient can stay out of the hospital or not.

If hospitalization has been prolonged, the patient has been cut off from the use and development of many of the techniques of life. This confronts him with a burden added to the prior discrepancies in his life, and to his particular difficulties in solving problems. It can be helpful if the patient can come to understand that everyone suffers from a discrepancy between experience levels and chronological age.

The problem of continuing therapy after discharge is often the problem of love and money. That is, the patient understandably pushes for greatly reduced fees as proof of the therapist's avowed interest in him. If the parent pays for therapy, this perpetuates dependence of the patient on the parent. If the parent does not pay for therapy, though he is in a position to do it, it sometimes results in the patient's reproach against the parents being projected on to the therapist. The patient and the therapist can then get into a subtle wrangle about the numbers of hours, the fees, what is a reasonable expectation, and what should be free. If the parents clearly cannot support the therapy, the wrangle is then openly between patient and therapist about appointments and reduced fees. It is not as simple as the decision not to start with a patient who cannot pay one's fee, for therapy is a going concern here, and the juggling is very hard to avoid. In discussing this, the therapist must be realistic about how much free therapy he can support without resentment. Usually, he tends to overestimate his altruism. Such therapy is a long-term job and may not be feasible in view of other pressing projects in which the therapist is involved. If the therapist can afford it, the only other caution is that he not slip into under-estimating the patient's capacity to earn and to pay after the patient is better, thus feeding into the patient's secret exploitativeness or feelings of inadequacy.

Part Five

THE THEORY OF THERAPY

I

THE THERAPEUTIC RELATIONSHIP

A) As a primary integration

The adult personality is organized into dynamic systems: the self-system with its significant sub-systems, the security functions, the productive functions, the central paranoia; and the integral personality. The interrelationship between these systems and their relative access to consciousness is a major determinant of the stream of consciousness. The stream of consciousness is also both shaped by and shaping of the immediate patterns of interpersonal relationships. It is a reflection of the personality systems, and of their expression in the ongoing interpersonal relationships. The stream of consciousness is an active force in organizing both the personality structure and the interpersonal activity.

This is a constant interactive process. New influences come from the evolution of new patternings of the interpersonal relationships, both before and after adolescence.

The structure of therapy proceeds along two main lines: the study of the conditions and situations necessary for the expansion of the productive functions of the self-system or, to put it another way, the process of incorporating dynamisms from the integral personality; and the study of the modes of resistance to growth as organized in the security functions of the self-system. Both involve detailed study of the patient's stream of consciousness.

A study of the actual interpersonal relations, present and in historical perspective, constitutes most of the factual content of therapy. Patterns of relatedness that are consistently avoided or misperceived must be explored.

When the patient enters therapy, the relationship with the therapist becomes one of his primary integrations. The patient's experience of the relationship with his therapist has a direct impact on thoughts available to consciousness, and on the balance of forces within the patient's personality. The relationship between patient and therapist is also a useful field in which to study, and, experimentally, to change, the patient's characteristic patterns of relating. The explicit data of investigation are focused on: 1) the patient's interpersonal relationships; 2) the patient's stream of consciousness; 3) the motivational personality systems of the patient, historically and in current operation.

The central responsibility of the therapist is actively to promote the organization, from whatever scraps of experience are available to the patient, of a growth dynamism within the patient's patterns of living. This growth dynamism, if effectively integrated, can become self-perpetuating and self-expanding.

The organization of forces is perpetually in motion. It also progresses in an overall direction, and in immediate and intermediate interactions, toward growth or toward deterioration. The pressure for growth lies in the patient, but the efficacy of therapy may determine the outcome of the struggle.

The interpersonal field of the therapeutic situation must be organized to include cherishing, validation, and tenderness: the elements indispensable to growth. Unless they are, in fact, present in the therapeutic situation, it is impossible either to challenge seriously the assumptions of the central paranoia, or to grow in the presence of new experience.

Validation of the patient's inner perceptions must be undertaken in such a way that their scattered and disconnected quality can be filled in and the patient can recapture and integrate his own inner perceptions.

Tenderness in analysis is expressed in many ways. It may be expressed in a genuine response to the other's psychic pain, in such a way that the pain is alleviated. In the analytic situation no less than in the surgical one, the response to pain is often a gentle and regretful awareness of the inevitability of experiencing pain which has long been postponed. The context of analysis should make it possible for the patient to be slightly less distressed, in order to be able to experience how much distress has remained unnoticed for lack of sympathetic attention.

Another mode of expression of tenderness between patient and therapist lies in a spontaneous sadness at the other's distress and a spontaneous pleasure in the other's pleasure. That is, the other's needs become of direct personal concern to oneself. Tenderness is also expressed in the accuracy of each one's perceptions of the other's feelings, patterns, and dilemmas. The self-centered therapist who relates with difficulty to the other's needs has been described frequently in psychoanalytic literature. The self-centered therapist who cannot allow the patient to respond to the therapist's needs, or to truly perceive what the therapist's needs are, limits the patient's growth no less, though more nobly. He sets up a situation which perpetually re-enacts the true tragedy of the patient's life. That is, "Because I got no love I could never learn how to give any."

Such a therapist either will forbid any expression of insight about himself or will ignore it when expressed; or, if it nonetheless is obtruded on his consciousness, will undertake "altruistically" to analyze the reasons for the patient's interest. Is he intimidated, is he compliant, is he afraid, is he obsequious, is he apple-polishing, is he trying to avoid his own analysis, is he trying to convert the relationship into a friendship and thereby subvert the analytic intent, or whatever? It is overlooked that, within the limitation of time and space and the fact that there is much work to be done, and little time in which to do it, the patient, in attempting to respond to the therapist as a person, is asserting his own right to grow in the context of the situation.

Any interaction which is defined in terms that make the flow all in one direction, and thereby eliminates the naturally inherent process of the free exchange in the area in question, is set by definition to limit rather than to promote growth in this area.

The patient, usually, presents initially the persona that he developed in the childhood years in interaction with mother and promotes the particular fictions that she promoted about himself, his family, and his history.

As a working relationship evolves, the therapist can usually sidestep much of this façade by focusing his interest on known, remembered, objective data. The major constructive area of contact between patient and therapist is between the productive functions of the respective self-systems. They both begin with productive use of functions that have already been validated in both. These include what each already knows about communication, about organizing and reorganizing data, about the normal processes of growth.

The nature of the interaction will, at least initially, be within the bounds of the last level of interpersonal development which the patient has achieved. This may be friendship, juvenile alliance, or parallel play. Attempts on the part of either to integrate a level of intimacy beyond the other's capacity results in anxiety, confusion, and play-acting.

The working relationship is integrated on the basis of mutual sharing. The patient shares the facts of his life and perceptions, the analyst shares his expertness in understanding them. Both gain in total life experience and both expand their experience by sharing.

The analytic relationship is integrated for the purpose of the patient's growth. Both patient and therapist are committed to this responsibility. However, evolution of the active, positive relationship between patient and therapist challenges both to grow.

The patient undertakes no commitment to the therapist's growth. But the inherent mutuality of the growth process gives the patient some direct investment in the expansion of the therapist's frame of reference. If the patient is preoccupied with maintaining an entirely unilateral focus, this preoccupation is a serious handicap to the therapeutic process. The primary data consist of the events of the patient's life. The events of the therapist's life are primary data only indirectly, but importantly insofar as the totality of the therapist's life experience shapes his participation in the interaction. Beyond this unilateral choice of data, the sharing of perceptions and formulations about them and of deductions and theory relevant to them, must be mutual. The quality of the analytic relationship is defined by this basic interaction.

Thus the analytic relationship is optimally a primary relationship integrated for mutual growth. The concepts of the loving and hostile integrations are as applicable here as in any other primary relationship. The degree of growth does not have to be exactly equal, nor the central areas exactly identical, but the continuing of the process of growth in both individuals is crucial to a productive interaction. As in other primary relationships, to the extent that is a loving one it involves a relationship between the integral personalities of the two individuals, and is oriented toward counteracting their respective central paranoid misconceptions. To the extent that it is limited in its potentialities for promoting growth, it involves a relationship between the two central paranoid systems of two individuals, and to the extent that this is so the integration will fail in its original purpose.

A hostile integration between patient and analyst can take on some of the interlocking aspects of durability which we have described. However, here the factor of limitation in total exposure is a very important one. The hostile analytic integration need not be as difficult to break up, need not involve as much fear and disorientation in its termination, as if the two people were a more extensive part of each other's lives.

The process of growth in the context of a primary integration, then, involves the process of validating progressively larger areas of the integral personality, mutually, and of increasing in each person the potential for experience in an orderly sequence. This is contingent on the availability of reciprocal responses to marginal moves either in one's own or in the other's integral personality.

Nonetheless, love is not enough. The analyst operates with more sophistication than the patient's life partner. On the other hand, he has less leverage. There is less mutual commitment, less exposure. The analyst's leverage lies in certain other factors.

He is, hopefully, talented in translating obscure modes of communication. His position vis-à-vis basic purposes of the relationship allows him freedom to attack the security operations and the central paranoia, directly and indirectly. He assumes responsibility for the strategic organization of the campaign, in terms of orderly sequence, the preparation for crises, and the handling of crises. Another advantage that the analyst has over the patient's life partner is that he is usually somewhat less accurately chosen by the patient for congruence of the central paranoias than is the life partner. Also, even more clearly and insistently than in other primary relationships, the only valid condition for the continuation of the interaction is the condition of mutual growth.

Thus the therapist's central dynamic is his own growth. The more the therapist learns about the actual, detailed, explicit processes of growth and deterioration under a multitude of different circumstances, the better chance he has of getting a good grasp of, and therefore influence over, these processes in himself. For this project, the opportunity to study them in detail and to try to facilitate them in another person is a golden opportunity. It gives the therapist a good grasp of the field. In the process of dealing with the patient's growth problems, the analyst notices that he himself is less clear on some than on others. Frequently this is an expression of the analyst's own dissociation in those areas. If he misses this, the patient is sometimes kind enough to point it out.

A second crucial force is the internal dynamic of the relationship itself. The exciting process of interaction, of getting progressively more intimate, better and better acquainted with another human being, is a compelling and exciting experience. The interaction of response to the other's response, of each contact serving as a base for a small opening into a new dimension of sharing thoughts, feelings, and perceptions with another human being, the process of discovery of new facets of life in conjunction with progressive sharing of experience, has a powerful pull to draw both participants into the core of the project at hand. For the therapist as well as the patient, here is an opportunity for experience in the anatomy of sharing, and in its value.

A third central motivation pressing the therapist to relate in the situation without reservation is his commitment to the patient's growth. An important blind spot in the analyst can be the reef on which the main push for reorganization in a particular patient founders. If the analyst has come to value this particular patient enough as a person, he may undertake a considerable degree of self-revision. This is less in order that he, personally, will not fail the patient, but rather that the efforts of his friend, the patient, should not founder. The therapist seems to be the one in a position to make a difference in the outcome. Such experience in the meaning of love can broaden the therapist's relationships in many other contexts.

The therapist's necessary capacity for patience, for repetition, for sidestepping the derogation of himself and the undertow of demolition of his work, does not stem from any superior tolerance or great magnanimity on the part of the therapist. It involves a technical understanding of the reason for the operations of obtuseness or demolishing. But its force is that the therapist has come to care whether this particular person becomes more happy or less.

B) COMMUNICATION

Consciousness is an aspect of communication. The immediate conditions of communication have a strong influence on expanding or restricting the content of consciousness.

The sharing of marginal and intimate thoughts takes place in a general context, and the whole quality of the analytic relationship will determine, to a large extent, the patient's freedom in participating in this.

Empathic communication always continues between two people, whether or not its content is conscious. In amputated form, the therapist knows directly what the patient is feeling by noting what he himself is feeling when the patient walks into the room. Does his heart beat faster? Do his hands feel cold? Does he feel nausea? Does he feel reassured? Is he more at peace with himself? Do his muscles relax, or does his left toe begin to twitch? Even more relevant for communication is for both patient and therapist to find their mode for observing and acknowledging such direct empathic interactions without setting off so much anxiety as to disrupt the interpersonal field.

Many more specific communications may be non-verbal. These include the use of vocal rather than verbal communication, that is, the use of the tone of voice—the utilization of tone to communicate to the patient an element of friendly consideration that might be absent in the literal, typed transcription of what is said.

The therapist must be sensitive to physiological shifts in the patient. These have been emphasized as indices of anxiety. It is also necessary to respond directly to changes in physiology, and to the chronic, static, unchanging physiology of the patient as a mode of communication. The physiological responses of the therapist can be a mode of answering these communications.

Some patients need to find, at some time during the hour, a social formula for some brief physical contact with the therapist, both for expression and for perception and confirmation of what has been going on during the hour. Rather more frequently, the actual communication of such central and empathic exchange of feeling comes in perception and exchange of fantasies, memories, associations, dreams, daydreams and anecdotes which occur to one or the other in the context of the shared emotional experience. The validation of such sharing is no substitute for the search for simple, common-sense verbal communication. It is one necessary part of the field. It is the background interaction that makes possible the perception, conception, and expression of thoughts and artistic symbols between two people so that they can learn how to talk to each other. This interaction is the tool of therapy.

In order for growth to take place, experience must be conscious. In order for this experience to be used as a base for further growth, it must remain available to consciousness. Otherwise one keeps repeating it to try to make it stick. Therefore the communication system between patient and analyst must be organized in the direction of explicit formulation and sharing of non-verbal and marginal processes.

At the beginning of therapy, neither person knows how to talk to the other. The process of integration of any analytic relationship recapitulates, for each of the two people involved, the original situation of learning to talk. This involves many of the anxieties which were originally in the situation. Analyst and patient begin as strangers who must get to be more than strangers before anything happens. Thus, the primary condition of therapy, the situation of a more or less accurate verbal exchange between two individuals, already is a process of intense mutual growth and new experience. For both, therefore, it must involve inherently the validation and/or repudiation of such new experiences, and the acting-out and the activation of the original paranoias in each about the possibility of simple, uncensored communication between one human being and another.

In childhood, many soon learn that they are rarely heard, and that their wish to be heard evokes parental anxiety. The adult, therefore, often expects indifference or anger, and therefore often evokes it.

The process of reading back from manifest to latent contexts in therapy consists of a gradual evolution of language between two people with widely divergent experience through the intimate sharing of attendant sentiments and implications. Private contexts must be shared to the degree they can be made to overlap, so that the symbol expressed by one may evoke related experience in the other. Two people cannot agree smoothly on symbols when one of the two cannot be aware of their meaning. Private contexts which cannot be openly shared retain an unreal quality.

One basic condition of communication is mutuality. The patient will formulate accurately to the same degree as the therapist. The therapist does not necessarily need to formulate his own problems, but his frank perceptions and serious syntheses of his impressions of the patient's problem is necessary to any continued communication.

Much of the theory of psychoanalysis has been designed to save the therapist from becoming aware of his own handicaps. Some theories absolve him from talking. Other theories, in throwing the focus overly on the importance of formulating, absolve him from listening. Almost all theories sidestep the inherent mutuality of the process of growth in this crucial area, as an antecedent to the utilization of communication for facilitating growth in other areas. Both patient and therapist must learn to be able to listen, think, and formulate in the other's presence.

Formulations must evoke responsive thought in the listener. They must, thus, incorporate awareness of how the patient tools his thoughts to what he thinks is the therapist's frame of reference, and vice versa.

Much of the therapist's artistic skill lies in his feeling for the possible dimensions of communication. He needs a broad and explicit grasp of its manifold modes—verbal, vocal, dramatic, or physiological; conscious or direct from the integral personality; perceptual or artistic. Communications from the integral personality to the self-system of the patient include marginal thoughts, trains of associations, and fantasies. It may be difficult to persuade the patient to note them and share them. In the context of therapy, dreams always involve, to some degree, communication from the patient's integral personality to the therapist.

C) Sharing, transference, and resistance

The analytic relationship is integrated between contemporary adults with the intent of sharing their insights about the data of the patient's life. Any inhibition about sharing on the part of either is an impediment to the process.

An indispensable mode of growth after puberty is doubling of experience by participating in another's struggles for growth. If sharing the other's experience is perceived as a loss, this perception is entirely in the paranoid mode. For the adult, this sharing is pure gain.

The infant's need was for his experiences to be shared by the parent; that is, for validation. The parent's demand—or the child's push—to develop understanding of the parent's struggles, grew out of the parent's inadequacies and the child's needs to keep his parent functioning. The child's need was not to love the parent, but for there to be the appropriate structure in the home for him to be able to get on with earlier learning. It is the projection of this old and desperate situation onto the current one that makes a person misperceive the privilege of participating in the other's growth processes as an imposition. He still clings to the picture of himself as a child in an adult world, as a child who is limited in his participation with equals and is forbidden access to the natural channels of growing up.

Another source of prohibition in the original context was in the degree to which the child, by his very existence, was a threat to the mother. To this extent, the child was prohibited from even manipulative sharing with the parent. His contribution to the mother's sanity was to be invisible, inaudible, and to make no demands. He could only watch and wish. Any participation would defeat his purpose. The child was thus involved in the compliance of repudiating his own need and of being preoccupied with the parent's fragility. In this situation there was often the additional dilemma for the child that if he acted out the preoccupation with the parent's need, this also evoked increased irritation or disorientation in the parent, who had to keep her needs dissociated. The chronic assumption, later, with contemporaries, is that the impulse to share would be an imposition and a danger to the contemporary as well as to himself. It either would drive the other away or throw him into confusion. The primary contribution, then, is not to exist.

The prohibition inherent in the original experience was reinforced by the mother's contempt when the person tried to share with a contemporary. His adult handicap is not, then, the search for missed mothering, but fidelity to the original mother's command to avoid satisfactions.

Subjectively, it is often put in terms of "investing in" a relationship and the expectation of "getting back" tenderness. This formulation obviates the whole point, which is that the one who "invests" cannot fail to learn and grow, though the recipient may or may not be able to make use of it.

The therapist's responsibility to share the patient's life experiences is manifest. Conversely, if the patient fails to grasp the style of thought, philosophy, and perceptions of the therapist as a person, he effectively sabotages one of the central experiences of therapy.

This incapacity to share creates gaps in the relationship which are filled by resistance and transference maneuvers.

The analytic study of the interaction between patient and therapist provides important data about the patient's styles of relating, his misperceptions, and their historic origin.

Resistance is an active, reciprocal interaction of the security operations of patient and therapist. It implements the hostile aspects of the relationship and evokes, for each, the personifications of the Bad Mother.

Transference phenomena, likewise, involve mutual reinforcement. They may be based similarly in mutual paranoid systems. Transference can also implement the mutually active but disguised search for the personifications of the Good Mother, suppressed after early infancy.

Both personifications introduce major distortions into the analytic field. Their identification provides useful data for understanding the history of these adult motivational systems. Until they are identified, they impede seriously development of a useful, friendly, cooperative working relationship.

In view of the reciprocity of interaction between patient and analyst, concepts of transference and counter-transference are artificial entities removed from the context of interaction between the two people. Both terms have been used to cover various aspects of the total relationship. Positive transference has been used to include both the real and the synthetic attractive forces between the two people. To the extent that the relationship is real, the whole analytic process includes the implementation, elaboration and clarification of the quality of the interaction. If the positive interaction is synthetic, it may be a paranoid reaction, on the part of either patient or analyst, to an expectation of a demand on the part of the other for compliance or flattery. This expectation may be projected or may result from a valid perception. So-called negative transference can also be, on the part of either, simply the literal perception of the degree to which the relationship has become a hostile integration. It is often the expression of the phenomenon described earlier as the paranoid episode.

The pressure of the relationship pushing one of the members toward genuine change explodes and exacerbates this person's central paranoid convictions. These convictions have existed in a continuum since the early relationship with the parents and are exposed in the moment to express the most central fears of the developing relationship. They are "transferred" typically to the agent of exacerbation, to the person who is in the moment being most effective in instigating the move toward growth. Classically, they are transferred by the patient from the parents to the analyst.

These attitudes are chronic and are also subject to acute exacerbation. They become more overt under pressure, but they are necessarily implied in the durable patterns of interpersonal integration. The analyst sometimes tends to overinterpret them as being personally directed against him, on the implicit assumption that reasonable exposure to the therapeutic aspects of his personality should already have enabled the patient to give up patterns of organization (resistance) which he has never given up in any other interpersonal integration.

Counter-transference refers to the therapist's participation in the paranoid and inappropriate aspects of the interaction. By this participation the therapist implements the mutual distorted perceptions. He thus makes it possible for the transference to be explicit and implemented rather than secret and implied. If he understands his own paranoia, the therapist can sidestep certain transference operations of the patient which are not crucial to the immediate therapeutic project, until it is strategic to focus on them. This accommodation on the part of the therapist may be conscious or unconscious. He may be aware of leaving certain attitudes unchallenged, or he may only become conscious of these attitudes at a time when it might soon be appropriate to get to work on them.

We could then define counter-transference, in the technical sense. as the degree to which this accommodation comes unwittingly from the therapist's own unresolved reservations about life, hence acts to reinforce, rather than merely to sidestep and postpone dealing with the patient's maneuvers.

To the extent that the therapist is detached, he functions as therapist under great handicaps. To the extent that he is reactive, he is in danger of stealing the scene from the patient, of demanding the floor, however obscurely, for flooding the interactive field with his own associations, his memories, his necessity for new experience. On the one side, the interaction must constitute a growth experience for the therapist. On the other side, to the extent that it does, it must involve, more obscurely but no less genuinely, growth experiences in the therapist which then evoke his anxiety, resistance and paranoia.

Many therapists are able, without catastrophic loss of self-esteem, to notice their envy, their distance, their feeling of entrapment because of their professional commitment or the money involved, their reactive irritation at the patient's necessity to frustrate their relatively sincere efforts. It is very much harder for the therapist to see his projection onto the patient of his own pervasive resignation. He feels that he is willing for the patient to be the canary in the mine, to move into areas which he himself has been unwilling to explore because of the reorganization that it would involve in his own life. To a certain extent, this is true. He encourages the patient to understand, feel, and move toward things which he cannot conceive of for himself. If he can find someone that he thinks of as like himself, similarly handicapped, who can try and succeed on the basis of the therapist's stated conviction that it should be possible, then it might be possible for the therapist. However, to the extent that he discovers that it would also be possible for himself it may involve more reorganization than he has the courage for. If so, there then must be the tendency to head off the patient from just these changes. That is, the therapist slows the patient to a pace which it is conceivable for him to adopt for himself.

The therapist's overdetermined obtuseness to particular problems of the patient can occasionally involve explosive possibilities. The therapist, in working toward the clarification of some central insight about himself, becomes very perceptive of the peripheral signs of the same problem in the patient, without being able to see the implications for either of them. If the project is current, important, and explosive to the patient, there is a danger that he will put it together first without realizing that the analyst is not ready to see him through it.

The therapist is also subjected to the patient's impressions of who he is, which the therapist may find accurate and disconcerting.

In the communicative interaction between patient and therapist, involving both perception and misperception, reciprocal emotion continuously operates. Any particular emotion in one party tends to set off the obverse or reciprocal emotion in the other. If the therapist is intimidating, the patient tends to be deferential. If the therapist himself is deferential or obsequious, the patient necessarily takes over the structuring of the situation, even though under a handicap. If the therapist is preoccupied and cannot listen, the patient is tongue-tied and cannot talk.

More fundamental than this reciprocal effect are the underlying attitudes that the two tend to hold in common. If the therapist is competitive, the patient tends to meet the challenge. If the therapist is chronically distant and self-protective, the patient will become so, too. If the therapist is detached, objective and cut off from his feelings, the patient has no choice but to be likewise. If the therapist is apathetic and not committed to the integration, the patient also can develop no commitment to it.

The therapist's genuineness and spontaneity in the therapeutic situation are primary instruments for developing the same qualities in the patient. If the therapist experiences and is in good contact with his own emotions, the patient can perceive this and may allow himself some of the same freedom. If the therapist knows what is going on and can handle it appropriately, the patient can afford to be less controlled in his emotional responses. If the therapist is glad to find out what was wrong about a speculation of his about some unhappy aspect of the patient's life, the patient will be able to learn to enjoy correcting his own misperceptions with equal freedom.

The genuineness of the interaction takes precedence. This holds true if the therapist is not indelibly opposed to the patient's getting better, and if he is willing and able to accept some part of the responsibility for this project. One difficulty is that often therapists are secretly ashamed of the person who they think they are, and cannot see how anything therapeutic could come of being really themselves.

The therapist cannot afford to limit himself to those operations which are integrated comfortably in his self-system, but needs all of the resources of his integral personality for the project at hand. If he is essentially committed to the project, he can trust his own unconscious to operate in a constructive direction.

In any interchange between two people, the direction of flow tends to be from the more experienced person to the less experienced, from the one who is better oriented to the immediate topic to the one who is less so. The content may range from an elaboration of what it takes to run a short-order restaurant to an elaboration of some point of psychiatric theory. Thus the direction of flow is back and forth, and it should be.

The therapist should refrain from using a disproportionate amount of therapeutic time for his education about peripheral things. When he is using it to learn more about the details of the patient's life, that is useful to therapy.

A more difficult complication can arise if the therapist is himself desperately in need of a therapist, and the patient happens to be a perceptive person. Here, too, the therapist's awareness of the situation can suffice to keep it within manageable proportions, and some implementing of it is a lesser danger than mutual detachment. More harm is done if the therapist, because of pride or dutiful commitments, has to be unaware of his needs and thus cuts off the patient from expressing even the most obvious and urgent of his speculations about the therapist's problems.

The therapist himself should have enough capacity for action to be able to project himself into whatever action it is urgently necessary for the patient to undertake in order to cope with his problem. The therapist does not have to be able to do exactly everything that the patient, with his different abilities and techniques, is able to do, but only to conceive of himself as being able to do it if he had equal facility with the techniques.

There are certain necessary attitudes. The therapist must have a modicum of tenderness, and must provoke no more anxiety than is necessary for a reasonable rate of progress. The self-image, originally formed as the reflection of the attitudes of significant people, may become in a situation of flux particularly vulnerable to the attitudes of the therapist. It is in this context that the importance of respect of therapist for patient is underlined. Respect refers to an acute sensitivity to the patient's liabilities and assets, to the historical necessity of his having developed his more unfortunate attitudes, and to an affirmation of the patient's wish and capacity to grow.

All of this is necessary, but insufficient for therapy. What the therapist has to offer is his expertness, his grasp of the theory of personality and personality development, his capacity to apply that theory to the data of the patient's life. Insofar as this is an undertaking involving an intimate integration between two people, the personality of the therapist is the instrument for the collecting of data, and for the patient's freedom to use it.

The therapist's capacity for handling anxiety lies in a special category. If the therapist either has so completely dissociated his own capacity for anxiety that he cannot predict when it might occur in his patient, or is so vulnerable to it that he cannot resist the inevitable pull to blow into a panic along with his patient, he is essentially unqualified for his profession. If the therapist cannot evaluate the dangers involved, he can sometimes throw the patient into an unnecessary open psychotic break. If the patient does unexpectedly blow into a panic, the therapist must be able to perceive it, and still be in reasonable shape to handle it. He must be able to repress his surprise, stick to the concrete, keep going until he finds some resolution of the problem, and allow some room for partial repudiation of the insight that caused the trouble. After the particular session, he should figure out what central factor in the patient he had overlooked which made it take him by surprise, and what his problem was about experiencing it in the first place. He should know that it must have been something central to his own problem.

Another process that interferes seriously with the therapist's functioning smoothly in the presence of severe anxiety is the therapist's own low self-esteem. When the patient begins to get better and therefore gets anxious, the therapist has the feeling of, "Why me? Could it be that psychotherapy, as represented by my mother's little boy, could really be facilitating deep changes in this person's capacity to deal with life? This must not be, and therefore it must be, as the patient says, that it is really all very terrible, that nothing good has happened and that nothing good could come of it if it had."

Another error is that therapists persist in derogating their own specific importance to the patient. The therapist functions as a kind of necessary transitional lifeline from the umbilical cord to the consolidation of current relationships. This matters tremendously to the patient. Many therapists feel that it is so unlikely that what they do, in a few hours a week, could be of such importance in the other person's life that they find it necessary to overlook its importance.

II

CONSCIOUSNESS

A) VALIDATION

Validation, confirming direct perception by sharing it, does not imply full agreement or affirmation of the activity concerned. Intertest in and responsiveness to how the situation seems to the other may be enough to clarify it.

If the patient cannot perceive certain aspects of a situation, he tends to fill in the gaps with hallucinations or misperceptions. At the same time, the perceived aspects are seen in distorted perspective. Delusional systems are built on subjective conviction about some aspects of a total experience, and dissociation of others.

The patient's intense investment in the delusional system cannot be reduced simply by focusing on his distortions and omissions. The therapist must first validate the accurate perceptions embedded in the delusional construction.

In many homes all strong feeling of any kind is derogated. Or, particular emotions, of anxiety, grief, despair, love, anger, or self-revulsion, for example, may have been experienced in childhood as either dangerous or not tolerated. Or, a specific feeling such as intense chronic anger interacting with that of the parent may have been promoted to the near exclusion of all other strong emotion.

There are many problems for the therapist in the validation of strong feelings. One is the problem of identifying them. Grief and depression are often clearly expressed in voice, posture, and related psychosomatic phenomena. Joy and euphoria show in the patient's overall carriage and belie his verbal Cassandraic attitudes. Both bewilderment and loneliness are often easy to recognize, even for the patient to acknowledge, if brought to his attention. There is a poor correlation between the subjective conscious experience and the inner reality of love and hate. Anxiety, fear, and anger are also hard to disentangle. Fear and anger are somewhat interchangeable. Anger is one mode of coping with a frightening situation. Fear is the immediate response to the other's anger and is an inevitable element in the fear of retaliation for one's own anger.

In a more chronic and organized form, both fear and anxiety can be integrated into what is referred to as insecurity. In one sense, insecurity refers to the degree of defect in the organization of basic dynamisms, hence the degree of difficulty in operating in the adult world. Its manifest content may appear either in manifest inadequacy or in compensatory preoccupation. Insecurity is also the disguise for the neurotic need for unlimited domination or overweening ambition: "If I am anything less than superlative in position or performance, I will feel degraded." Manifest insecurity can also camouflage fear resulting from one's well-organized and smoothly operating destructive impulses. It expresses and incorporates fear of being found out to be destructive. It is used to alibi oneself and to mislead the audience.

At the core, these emotions—anxiety, anger, fear of being destroyed, and fear of being found out to be destructive—are all very closely related. Nonetheless, it is important to disentangle them. If the analyst of the very angry patient is intimidated into perceiving the "fear of being found out" as the anxiety contingent on faulty organization of the self-system, the patient can never clarify the extent, intensity or manifestations of his central paranoia. On the other hand, if manifest rage is not related to the early components of fear and anxiety which set it off, it remains as an unresolved source of isolation and low self-esteem.

The layering in origin need not coincide exactly with the present layering from manifest content to inner reality. In origin, the impulse to expansion sets off anxiety. It also exacerbates the repressive aspects of the central paranoia, i.e., it activates the lifelong fear of the mother's anger. This fear, deeply incorporated into the self-system, has been the stimulus for organizing a retaliatory operation based in anger. The patient, in whom the retaliatory operation has been broadly elaborated into the adult modes of interpersonal interaction, attempts repeatedly to find expression for it without being either inhibited or destroyed by the recipient's perception of and retaliation for the operation.

Validation is an important lever of the analyst for sorting real from synthetic need, facilitating moves toward growth, and discouraging moves toward deterioration. It facilitates moving a particular experience from the domain of the integral personality into the self-system. This involves active response to, and affirmation of, moves of the integral personality toward self-system expansion. It also sometimes involves invalidation of pervasive security operations by giving them a minimum of attention or affirmation.

Derived emotions and misperceptions expressive of the central paranoia must be related to as intense subjective experiences, sufficiently to get them into awareness and communicated. They must be validated as having been the inevitable response to the original situation. They then can be evaluated as to their appropriateness to the immediate situation. Here, there is a clear distinction between validation and affirmation. This process of sorting out can best be exemplified in respect to growth moves toward needs and satisfactions.

If needs are real, the only way that their satisfaction can become less of a preoccupation is for them to become more satisfied.

Usually one does not move toward satisfaction directly, but searches deviously for the conditions he thinks would make the pursuit of satisfaction possible. Couched in terms of the small child and the old ghosts, these are: 1) to master the fear of danger, 2) to keep the creativeness secret, and 3) to win the fight for power with one or the other parent.

The primary diversionary activity is the full-time preoccupation with the processes of restricting growth. Security operations are, by their nature, insatiable. The opportunity to satisfy them is responded to smoothly and directly, and as if achievement is just about to be reached. Partial achievement opens new vistas in the direction of greater achievement. This is obvious and traditional in the case of people who become cleaner and cleaner, or richer and richer, or more and more important. It is equally true of people who become progressively more respectable or manipulative, or "better adjusted" or well organized and controlled.

The satiation of satisfaction runs a different course. If there is no blocking, a new satisfaction is also approached smoothly and directly. However, if this satisfaction has been repudiated, then the opportunity first is repudiated, and the person feels no response. If the stimulus continues, the need may suddenly be experienced as overwhelming and insatiable. If some satisfaction is experienced, the intensity of longing begins to decrease. As one moves toward adequate supply and the deficit is decreased, more of the need can be experienced. This increased awareness of need and deficit may alternate with periods of unawareness of need. During this period, subjectively, each positive experience seems to intensify rather than decrease the need. This can be discouraging.

There may be apparently insatiable needs for what should be a reasonably satisfying experience. Here the original deficiency was the absence or prohibition of validation of the experience of satisfaction. The persistent defect, then, is based on the inability to experience satisfaction of the need to the extent that the satisfaction is at hand, as, for example, in nymphomania. As long as the awareness of pleasure in the experience is cut off, the deficit can never be made up. This concept can be applied to a wide variety of human functions, from compulsive eating to the compulsive search for the subjective experience of being loved, lovable, and loving.

If one cannot experience fully in the moment the contentment of being loved, the giving and receiving of affection is very much restricted as a basis for further growth. The patient may even cherish secretly the fantasy that his perpetual mild derogation of the loving that he gets is a necessary, if somewhat undignified, maneuver to get an extra ration in order to make up the deficits of the early childhood. In fact, this derogation serves as a perpetual erasing machine, so that the experience of the previous night is never quite a useful base in getting on to the next.

The analytic inquiry, then, must be directed toward what interferes with one's full experience and enjoyment of the satisfaction involved. Why don't you enjoy what you eat? Why don't you know when you are full? What keeps you from feeling subjectively like a member of the group to which objectively you belong? What interferes with your sense of possession of an object which makes it necessary to possess more and more objects? In what way is your enjoyment of the sexual act limited and restricted? What interferes with your experiencing yourself as graceful and attractive?

"Insatiable" needs can also be used in the service of some entirely different maneuver. The apparently insatiable demand for affection may be in fact one aspect of the drive for power. The preoccupation with social acceptability may also serve to obscure and camouflage either the individual's more infantile needs or his real capacities for further growth, and his anxiety about it.

Diversionary activities express some corner of representation of an originally valid need, even though the need is so distorted and obscured that it could never be satisfied in this fashion. For example, the drive for prestige may subsume a deficit in the thoroughly legitimate necessity of feeling like a member of the group. If the deficit could in some way be made up, the diversionary prestige drive would be undercut.

The therapist's function is not the validation either of the pursuit of the "insatiable need" or of the acting-out of the destructive one. Rather, the more fully the experience can be brought into awareness, the more manageable the drive will become. Also, the more the legitimate and ineradicable aspects of the need represented can be efficiently and fully satisfied, the more the neurotic insatiability can be undercut.

The most important functions to validate involve perfectly legitimate, necessary, universal human needs derogated and overlooked, either because of individual distortion or because of social pressure. These may appear in the awareness of the patient as insatiable needs, or they may appear as matters of complete indifference. The person whose personal property was never respected as a young child may maintain throughout life the fiction that things don't matter to him. One of the early turning points in the analysis of a patient, who, before analysis, lived only on corn flakes because of her profound indifference to her bodily needs, occurred when she began eating strawberry shortcake every day for lunch.

The analytic project is very frequently to lure the patient into experiencing the need which he wrote off originally as unattainable. Then it is possible to help him organize his own methods of satisfying it. Most of these needs became blocked early in childhood and are now characterized either by the patient or by the society he lives in as adolescent, childish, or babyish, and, by implication, as bad. Many patients would rather deny how entrenched the deficiencies are than to get ahead with learning to satisfy the derogated need.

B) THE CONTENT OF CONSCIOUSNESS

Each person tends to organize his life so as to perpetuate perception in the original context. Such intent pervades every aspect of human living, subjective and objective, directly interpersonal and less overtly interpersonal, including the content of consciousness.

Resistance to growth, and hence to analysis, is focused on the maintenance of barriers to new perceptions. Since the process of organization of experience is interpersonal and begins at the moment of birth, these original contexts, which tend so strongly to color all subsequent related experience, took place long before the earliest memories. This adds to the difficulty of reorganizing them, since their reconstruction can only be by the process of imaginative reconstruction and analogy.

The stream of consciousness is dominated by the central paranoia a good part of the time. Much of the actual content of consciousness seems to be obsessive and diversionary: counting; grocery lists; scraps of songs; or the replaying in one's head of past, future, or imaginary conversations, plots or interactions. The manifest intent of such mental operations is to obscure the train of thought going on slightly beneath the surface. Even so, all roads lead to Rome. The actual content of the obsessive activity chosen is symbolic either of the paranoid assumptions or of events which were slightly incompatible with the paranoid expectations. In the latter case, these events are being recorded both for the purpose of memory and for later integration and, simultaneously, for the purpose of re-evaluation to fit the *status quo*.

Another large section of the content of consciousness could be referred to as reality-testing, which would be more accurately described as unreality-testing. It is the apparent review of recent events for clarification and extrapolation into future activities. This process is the homework of reconciling the paranoid assumptions with the recent experiential data, and hence implementing the subtle expression and maintenance of the central paranoia in the day-to-day life operations.

The integral personality also has its representation in the waking stream of consciousness in marginal thoughts and perceptions of experiences unexpected and incompatible with the usual expectations. Such thoughts, events, and perceptions are hard to remember and, when remembered, they come with an unreal aura. It often occurs that the person who is being unexpectedly but incontrovertibly friendly is seen or heard as if through a veil or at a great distance.

In the face of this relative domination of the stream of consciousness by the paranoia, each individual organizes certain situations of relative sanity, that is, of relative escape from the paranoia, in accordance with his individual patterns. Some people can be themselves only when alone, others can be themselves only when they are not. Some people develop a talent for establishing a particular kind of rapport with certain other individuals, so that this rapport taps and makes possible in the moment a different level of the stream of consciousness. Some people are more related when they are drunk or half-asleep. Others are more paranoid.

By such devices the integral personality in fact gains considerable representation in the stream of consciousness, as well as considerable implementation in interpersonal relations. Again, such situations are remembered faintly or distantly, or are remembered with some paranoid reconstruction and misinterpretation. It is only if they become inerasably validated and integrated into the stream of consciousness that they set off more active and explosive paranoid repudiations.

At the other end of the spectrum from relatively sane moments is frank reliving of the central paranoia. The most striking example of this is in nightmares. These usually consist of a clear, simple, frank statement of the frightened aspects of infancy and early childhood, with a statement of how the various familial figures appeared to the child at that time. The parallel relationship between the dream and current events of a person's life can be taken to exemplify the strong trend to reinterpret current events in infantile terms.

The extent to which similar dream states take over major sections of daytime life is much less frequently noticed. These appear as trance states in which the inner feeling is the reliving of the childhood terrors no less than the nightmare. Such states tend to be easily unnoticed by both subject and observer. They are most frequently obscured by a state of immobility and paralysis and a conscious experience of a vague bewilderment. The person feels either irresistibly sleepy or is overcome by a leaden fatigue. In either case, the terror inherent in such a state is disguised.

Such states tend to take over in special situations. If the situation is such that the terror is in great danger of taking over in full force, the subjective experience is often disguised by a sense of depersonalization and great distance. The patient literally feels nothing. It is not only that he lives in a world in which there are no people, but rather that there is no world to be depopulated.

If this happens during an analytic session, the intuitive psychotherapist can devise ways of being perceived by the patient who has gotten into such a trance state, and, having been perceived however distantly, of leading the patient back to where he can feel something, to where he can experience being back in the room. It is something like the process of reassuring a frightened child who has waked from a nightmare but cannot quite get awake. It is complicated by the fact that the patient is so frightened that he is not able to experience the terror that is going on. If he could, by virtue of the therapist's maneuvers, become slightly less terrified, he would be able to experience his terror, and therefore to experience other things.

Much of the immediate resistance in the patient, then, comes out of a reluctance to experience fully what he knows is going on, partly because of its inherent unpleasantness and partly because he becomes preoccupied with whether the therapist can get him through this into a quieter state. He is afraid he will spend the rest of the day or the rest of his life reliving his terror.

The intrusion of terror into the therapeutic situation may be triggered by various factors. One is the patient's pull towards mental health, which frees him, however unpleasantly, to drag his most bothersome symptoms into the therapeutic situation. Another factor is the extent to which the patient experiences that this is probably the safest place that he has been in for investigating his inner feelings. These feelings may tend to be set off simply by his walking into the room with a sympathetic person, before the therapist has said anything. Probably a more frequent precipitant is that during the course of investigating the material of which the central paranoia is constructed, such feelings are often re-evoked, and their reliving takes over unnoticed by either patient or therapist. When this happens, so far as the patient is concerned, the therapist has either become non-existent and therefore can not be heard at all, or the therapist is seen as the malevolent counterpart of the patient's feeling of being brutalized. Anything he does from then on is interpreted in this context.

If the style of terror is active, it is more likely to be perceived, but very difficult to escape. In this case, the therapist becomes actively the malevolent counterpart of the dream life, and the patient sets out on uncensored acting-out of his early childhood. The other person in the room is not a contemporary at all, but a mannikin or prop for the inner drama. The activity may be strikingly bizarre and out of context. There may be screaming, inconsolable sobs directed to the wilderness of outer space and not to any particular person. Or, the patient may become as violent and assaultive as any two-year-old in a temper tantrum, acting as if the strangers around him were his natural enemies. This is the context of a considerable variety of the so-called bizarre, repetitive symptoms, obsessive or hysteric, with the re-enactment of particular symptoms in a fugue state, as if one were in a repetitive nightmare.

Such frank, uncensored reliving of the nightmares from childhood is universally experienced as terrifying, and almost everyone attempts to avoid it at all costs. It happens during the waking life, it is somewhat more likely to happen alone, it is somewhat more likely to happen at night, but these are statistical variants. It is accompanied by intense subjective isolation and conviction that one will never be able to find his way out alone, and that there is no one else in the world who will be able to reach him. This conviction also, of course, tends to prove itself out. If neither the patient nor his friend realizes the point at which the fugue state takes over, the first thing that the patient does in the early stages of the fugue state is to treat his friend as if he were a monster to be driven away.

This repetitive experience is frequently the base for recurrent intractable patterns of integrating hostile and restricted partnerships. That is, there is an absolute priority on avoiding simple physical aloneness, an absolute priority on having another human individual there. In this frame of reference the attraction of the hostile integration is that it is somewhat more durable, if less enjoyable, than a friendly one. Each plays on the other's disadvantage to keep him pinned down.

To return to the stream of consciousness: there is a slightly different management of such states by the patient in accordance with whether he thinks of himself as sane or insane, and whether he considers that sanity, in this sense, is constructive or restrictive. If he thinks of himself as sane and values it, he attempts to restrict his stream of consciousness to "rationality" and to exclude both random thoughts from the integral personality and frank formulations of the central paranoia, both of which are experienced as disorganized or disconnected from the external "adjusted" reality. If the self-image is of a person who is insane, there is also the risk that all such intrusions are categorically blocked, because the focus is then on the attempt to appear sane, that is, to maintain consistency at all costs.

There is a third pattern, that of the person who knows his central paranoia well and recognizes it as delusional on an empirical basis but considers it inevitable. His energy is put into trying to keep it in the fantasy realm. He allows time in every day for it to roll on. He may undertake the search for "saner" situations, for situations which are in fact less pervaded by the paranoid system. His anxiety is intensely aroused at being confronted with failures to disregard the central paranoia in the interpersonal relationships. Any such failures must be denied. He cherishes the fiction that he has always known that most of the night and much of the day he lives in a delusional system, but that he can get through life without acting it out with people. He knows that the thoughts persist in his head, and that they are felt with conviction although they are delusional. He knows that the saner thoughts feel unreal, but deludes himself that he has beaten the game by this insight. He cannot face the fact that his interpersonal relations are infiltrated by the subtle, subdued influence of the delusion.

A difficulty in therapy is that trance states are either not remembered at all or are remembered in fragmentary ways. They may be screened by heavy drinking and a blackout, or they may be screened by a blackout without any physiological or pharmacological aids. Such blackout can apply equally to the saner moments of alive relatedness, so that the phenomenon of blackout itself cannot be used to identify the nature of the fugue. If the fugue states are remembered, they are not easily recognized as simple psychotic episodes. They are remembered vaguely, with apology, and are rationalized.

Therapy includes the study of the patient's own devices for self-limiting of the fugue states, of how the patient has in the past brought them to an end, of their usual duration in any case and of his available interpersonal resources for breaking through them. But the primary focus of therapy is to understand fully their significance as dissociated memories. The fugue state, then, is the reliving of fragments of the central paranoia, with temporary organization of the self-system to fit.

Frank breakthrough of material from the integral personality temporarily shatters the organization of the self-system. The therapist must organize the interpersonal field so that this anxiety gradually becomes less debilitating. When the patient is anxious, he has difficulty concentrating. His mind may feel blank, or it may be filled with disconnected fragments of relevant associations.

The goal is to help the patient to learn to sort these experiences into a coherent whole, filling in the gaps. In view of the limited time available in therapy, it is often helpful for the therapist to offer a simple structuring of those elements of the situation of which he feels sure. This can give the patient a framework to start from. It is often easier in his state of confusion for him to correct errors in the therapist's simple structuring than to formulate one himself for joint revision.

If the confusion is somewhat less intense, or the situation less urgent, the therapist may make his best contribution by creating an atmosphere in which the patient can think better. This involves playing down the feeling of urgency. It includes formulating the therapist's expectation that the patient knows what he has in mind and, given time, will be able to sort it out. Usually if the therapist is irritated about either the confusion or the paranoid side effects, the patient will be less able to think clearly. Friendliness, acceptance, and patience are essential. The therapist's active role is alertness to the implications of the patient's associations and unobtrusive responsiveness to them.

It also happens that for some people the quickest way to come out of a catatonic confusion is to pick a fight. The other's irritation then mobilizes the person's paranoia, thereby reinstating self-system functions. The relief then experienced is in the reassurance of being able to think, not in any pleasure in being dealt with irritably. A similar emergency maneuver is to do something dangerous. The evoked fear then mobilizes the self-system.

Each patient has developed before getting into therapy his personal style for weaving his way out of such confusional states. As the therapist begins to understand this style, he can facilitate it and shorten its course. He becomes less impressed by the characteristic defenses which were derived from the fear of disorganization in the power fight with mother. He develops more confidence in the patient's capacity to learn how to formulate wispy thoughts from the integral personality. He gets a glimpse of the patterning of his patient's creative thinking and so can be more alert to the disconnected fragments of it that emerge in the therapy situation.

Both patient and therapist come to respect the particular operations of the patient's integral personality. Neither of them has any great influence over the patterning of this emergent personality. They can only facilitate or discourage its emergence. The patterning is already laid out by the patient's total previous experience and his particular organization of this experience in the integral personality.

Dreams are the reflection of the process of integration that goes on constantly in the integral personality. They embody the descriptive synthesis of the conclusions drawn about a particular incident in its immediate context, its broader context, and its reference to the original integration with the parents. The form of the dream is allegory—the situation to be evaluated is stated in terms of a known one. The core, or continuum, between the various references—the incident being sized up, (i.e., the immediate reference), the incident chosen to symbolize it, the reference to the general current context and the reference to the earliest determinant situations—lies in the implied feeling: joy, fear, anger, disgust, danger, flight, moving ahead, or stepping back.

Dreams are often concerned with formulating the impact of some forbidden impulse from the integral personality on the central paranoia. The message, then, may express a powerful repudiation of the intended action and reaffirmation of the paranoid convictions. Since the self-system is the auditor, it must be stated in terms that communicate some aspect of the message. Such dreams may either contribute to the decision to reject a forward move, or may serve to clarify the irrationality of the fears evoked by it, and thus to facilitate the move.

The immediate function of dreams is to formulate and communicate to the self-system without interrupting the progress of the work being done by the integral personality in integrating new perceptions. Sleep serves the purpose of allowing this integration to go on. Dreaming, then, serves the function of protecting the continuum of sleeping. When dreams wake the dreamer, they have failed in this basic function. When they are entirely unavailable to conscious memory, it is usually because even minimal formulation would involve too much terror. Some people dare to dream only if they are not sleeping alone. The communication may, in this case, be condensed into an occasional nightmare which persists for several hours of a semi-waking state. It can spill over into the waking life in intensified interruptions and intrusions on the stream of consciousness.

In a similar way, fantasies, hypnogogues, marginal thoughts, and ruminative states serve, in the waking state, to communicate directly between the integral personality and the self-system. They embody much of the same material and can be interpreted along much the same lines. They are more subject to censorship and reformulation in terms of the self-system. They tend to be toned down in dramatic impact—to formulate and refer to chronic attitudes and assumptions which are implemented in daily life but are rarely in focal awareness. They often refer to chronic discouragement, intimidation, or encirclement, without including either an intense determination to escape or an intense conviction about the dangers of escape. Daydreams, for example, state the persistent unfulfilled wish which is most urgent, but incorporate in the context a perpetual reminder as to why satisfaction is unavailable.

These symbolic dramas of the waking state, as with dreams and sleep, serve partly to protect the continuity of consciousness. They may be suppressed from easy recall in order to preserve the lifelong organization of the self-system during certain periods of a person's life when even this attenuated message is too explosive. The result is the episodic breaks in the continuum of the waking state. Such episodes may be either catatonic or paranoid or, more usually, show a mixture of the two.

These explosive episodes incorporate their repudiation. The acute, manifest disorganization cannot easily be reinterpreted as a healthy safety valve. The reaction is flight into "sanity" and more intense exclusion of any marginal thought that might lead to a recurrence. The acting-out, expressing the condensed, full-blown statement of the most central paranoia, exposes to public view aspects of the person difficult to incorporate in any operable daytime self-image. They can also, since they are in fact manifest actions in the real world, involve social or legal reprisals which intensify resolution to dissociate the whole dynamism.

III

ORGANIZATION OF THE PRODUCTIVE FUNCTIONS

A) THE SELF-IMAGE

The self-system tends to a logic which must compromise the incompatible personifications of early infancy. This compromise is integrated in late infancy and childhood and is initially in relation to the fusion of the personifications into the perception of the parents as entities.

The resultant construct is first expressed with contemporaries during childhood, in relations of parallel play. It can continue throughout life as the constellation of attitudes called upon to deal with strangers and acquaintances.

The self-image is the condensation of this early logical integration of the self-system and incorporates a naive reflection of parental attitudes. It is often formulated in early memories dating from two and a half to four years. Both the self-image and its origin tend to become obscured by later development. It is also modified by later experience.

However, to the extent that it is obscure it tends to persist in the adult personality. This early resolution of incompatible perceptions tones down the disturbing intensity of the experiences of early infancy. It also tones down the enthusiasm for new experience. Stability takes precedence.

This social personality incorporates those aspects of the security and productive apparatus that are easily conscious. It also expresses reflections of both the productive and the paranoid experience which were at one time in awareness and were later suppressed. At the same time, it maintains the suppression of these experiences from easy awareness. It is highly concerned with how the person will be perceived by others, though he may be very edgy if the other's perception is made explicit.

This system of perceptions and attitudes is often the first presented by the patient in therapy. The patient tries to act as he would like to be perceived by the therapist. The therapist's first task is to organize the relationship so that such a concern for appearances and for consistency will become irrelevant. This process must be started in the initial consultation in therapy.

In undertaking to change the self-image during subsequent therapy, attitudes toward strangers and acquaintances become an important source of data. They throw light on the implicit logic of the self-system.

It is important for this purpose to reconstruct more accurate concepts of the parents, as they were during the patient's early years and as they changed subsequently. When this is done, the sources of later fictions about them, and hence about their attitudes toward the infant, can be identified.

As the self-image gets identified, it begins to expand. One major dimension of expansion is giving up of the thought of oneself as a child in the adult world.

B) THERAPY AND THE PRODUCTIVE FUNCTIONS

The major focus in taking the patient's history is to follow the progress of integration of productive functions with people into the self-system. For the therapist to relate usefully to the productive functions, he must be clear on the point of arrest and on the area at which growth is taking place, that is, the point of fixation.

This involves detailed chronological reconstruction of the relevant history of the patient's life. This process begins with the initial inquiry. This includes the social events and external achievements, the overall patterns of relatedness, and the events of the different interpersonal eras. The process continues throughout therapy. Any major insight or difficulty is referred back to those historical events which contributed to it, positively or negatively. Certain periods of the patient's life can eventually be identified as turning points for growth or discouragement, and these are then reconstructed in more detail. The historical onset of a particular symptom often coincides with the entrenchment of defeat in some area of attempted development. In this reconstruction, the therapist must take an active role.

The patient, in order to maintain his neurosis, is of necessity vague about the sequence of events in his life, the significance of the events, and the connections between them. The therapist provides the theoretical framework and directs the inquiry toward the kind of events that were significant to development. The therapist also must be the one to remember the major outlines of the history and the timing of particular events, because it is especially his responsibility to be alert to possible relationships between them. The patient had to remember them in different connections in order to avoid seeing the causal relationship which he was unable to cope with at the time that the event occurred. The therapist must provide retrograde hypotheses as to what might have been the state of affairs in particular foggy periods in the patient's history, including those of the very early years, in order to evoke confirmatory or dissenting memory.

It is also the therapist's responsibility to grasp and state the possible connections between current events, the present adult character structure, and the detailed events of the history. As the analysis progresses and the overall history becomes clearer, the patient is able to take over a part of this responsibility for reconstructing the events of his life and their connection with current patterns. The obscuring of this data is so crucial to the maintenance of neurosis that the patient will never stumble on the most significant connections without some organizing help from the therapist. This is one of the functions in which the presentation of summaries by the therapist to the patient is useful both to help the patient to get the overall picture, and to force the therapist to put in the time to get it straight himself.

Any therapist who can master the intricacies of higher education has the capacity to master and reorganize the objective data of the patient's life in a useful pattern. The therapist's reluctance to undertake this responsibility lies partly in the amount of work and level of attention that it involves. The patient usually presents a plausible version of the romance of his life, with the unlikely conclusions carefully bolstered and the crucial omissions carefully obscured. He invests considerable energy into taking the auditor along with this construction. The therapist must make a positive effort to extricate himself from the patient's version in order to be free to see the data in its more probable context. It also involves a considerable investment of time and work to fix the details firmly enough in the therapist's memory for him to have them on call when needed.

A more central impediment is the therapist's defense of his residual fogginess about the crucial sequences in his own life. This creates unnoted investment in the thesis that a grasp of the general trends should suffice and that attention to detail is obsessive and redundant. He may find that, in his first serious attempts to reconstruct a patient's history, he needs to keep a separate notebook for his personal associations, in order to keep these out of the way. On the first few life histories he may learn a little about his patient and a great deal about himself. When the therapist can, without anxiety, give a reasonably accurate and detailed chronological history of his own life, he will find it much easier to do the same for his patient.

It is reassuring to the patient and salvaging to his self-esteem to realize that current neurotic attitudes were at one time appropriate and necessary. It is necessary for their re-evaluation that they be identified as originally reactive and not inherent.

The detailed reconstruction of the actual situation from which adult attitudes derived is also necessary to give these attitudes enough substance and background for convincing, explicit formulation. If this is not done, the descriptions of adult behavior tend to be glossed over in the clichés of either the adult social milieu or the particular psychological theory. Remembering the feeling of wetting one's pants on the first day of school is more convincing than any verbal description of shyness, chronic embarrassment, low self-esteem, or dependency.

When separated out from the false logic of the self-image and the bitter logic of the central paranoia, the productive apparatus can easily be revealed to incorporate many gaps and inconsistencies. Its central proposition derives from the discontinuous infantile perception of the Good Mother. On this base, certain functions have already been integrated, and certain needs can be expected to be satisfied. The execution of these functions can be pleasant and not anxious.

Under the pressure of the self-system's attempt to incorporate the incompatible logic of the central paranoia, many otherwise rewarding, productive functions have been expended.

For example, one patient as a child did well in first grade. During second grade his parents quarreled bitterly and considered a divorce. His work understandably deteriorated. The uneasy compromise subsequently worked out by the parents was expressed partly in competition over the child's performance, and this exploitation perpetuated his reading disability. Subsequent attempts to overcome it operate against the handicap that the source of trouble is not understood. If identified, the competence in first grade provides a usable base for later integration.

The first grade experience was at that time firmly a part of the productive function of the self-system. It was later driven underground in a threatening situation which strengthened the central paranoia. In the attempt of the self-system to negotiate a livable arrangement with these particular parents, the function of reading was remembered as "naturally difficult." The parents were remembered as interested and helpful.

When, after a dream, the patient reconstructed the original circumstances, it was clear that the parents had been preoccupied with themselves and irritated with the child for having difficulties that showed. Further, they had played on his difficulty in concentrating in such a way as to intensify his fear of disorganization.

The integral personality had never accepted the conscious fiction that the elements of academic learning were difficult. It continued to organize life in the direction of a style of creative work which would assume an easy facility with reading. It was in the process of implementing this plan, against both manifest anxiety and manifest paranoid explosions, that the dream occurred.

Current adult relationships, loving or hostile, primary or peripheral, constitute another major content of therapy. They provide the basic data for reconstruction of the covert organization and assumptions of the character structure. The central paranoia is explicitly implemented in the hostile aspects of contemporary interactions with people. Moves toward growth from the integral personality constitute attempts to explore forbidden aspects of relatedness with contemporaries. Certain productive functions, antecedent to the zone of fixation, are performed competently and with satisfaction.

The process of therapy consists of identifying those recurrent interpersonal patterns concerning which the patient is unclear, and understanding their implications. One purpose of clarifying the latent content of the personality is to enable the person to reorganize his current relationships to provide a better base for present growth and a better prospect for future growth.

Over and above the specific barriers to growth originally set up, the contemporary culture tends to set rigid rules and sequences as to which patterns of adult relationships are praiseworthy, which can be tolerated, and which evoke serious censure. These proscriptions rarely fit the current growth needs of the individual. At the same time, his capacity for disentangling from the machinery of the conventions depends partly on his security in individual relationships. The therapist can help to bridge the transition.

The patient, sometimes in discouraged conspiracy with the therapist, may try to improve the form of his relationships without the substantial internal reorganizations which would make this improvement valid. A frequent example is the couple who, realizing that they both have a central fear of closeness, stubbornly try to overcome it with each other long after the particular relationship has ceased to be a growing one. The intellectual attempt to solve the central problem of closeness with each other becomes, in broader context, an avoidance of finding integrations where growth would take place spontaneously.

At the same time, once a destructive pattern has been identified, the conscious attempt to change it can serve constructive ends, if this effort is open-minded. It can expose, in the chronic hostile integration, the underlying apathy and unaliveness of the relationship. It can expose the over-determined nature of the hostile patterns and greatly facilitate identifying the underlying paranoid assumptions. Finally, it can make possible, in an experimental context, confrontation with better experiences. The attempt to expose the paranoid interaction may also result in validation and development of the constructive aspects of the relationship. This confrontation can only strengthen the move toward health and weaken the commitment to the paranoid way of life.

In the work of clarifying the patient's character structure, it is important for the therapist to identify where things stand with the basic human growth challenges: Which functions can the patient perform easily and naturally; what can he do fairly adequately, while at the same time having to disown his competence; which performances are by rote and without understanding; what can he not do at all?

The problem of identifying the level of damage lies in two major areas: 1) the level of arrest of systems or functions, and 2) the general level of arrest in respect to the eras of interpersonal development.

The level of damage to functional systems can be very explicit. There are people, relatively at ease in the world of music, color, and texture, who have never quite gotten a grasp of space, time, and the body image. This deficit leaves them feeling chronically disorganized and disoriented, and much energy is devoted to overcoming this inner disorientation.

The person with trouble about time may have memory difficulties with major gaps in his reconstruction of the recent or remote past. He may find it hard to experience in context relationships in which the events over a period of time are of crucial importance. If in the present moment he relates to his friend as if with no clear knowledge of past events between them, or of future expectations, it can keep the interaction on a perpetual seesaw. A momentary interaction is experienced in full force out of the context of the durable relationship. The partner's moment of irritation is experienced as the proof of the malevolence of his character. The moment of abstraction is taken as the revelation of total indifference. The moment of rapport becomes temporarily the basis for unreserved validation of the search for the Good Mother.

Some people have a rather accurate grasp of the major aspects of the nature of interpersonal relationships, and the non-verbal modes of communication. They know the definitions of words and may even be talented at syntax, but they have never seriously learned to use speech for communication—they never really learned to talk. The verbal productions are then subsidiary to the drama of the particular interaction and are, actually, an elaborate mode of acting-out. A story is told for effect rather than in the serious attempt to formulate an inner experience and share it with the other person. Since words are necessary for consensual validation between adults, these people feel perpetually in doubt about their perceptions and perpetually isolated from the people around them. They never put what they really feel into words; it never seriously occurs to them that the other person is doing so, and they are thus limited in the confirmation of their intuitive impressions.

If a defect is of long standing and it has been necessary to compensate for it, the person is almost never clear on what it is that he has difficulty with. The first step is to identify it. It then becomes possible to disentangle it from compensatory efforts, from carefully memorized performance, and from filling in with related functions. It is something like the task of the orthopedist, identifying which muscle groups are paralyzed from watching the overall gait, and studying the tension and overdevelopment of neighboring muscles. A major impediment to identifying an area of dissociated, and consequently malfunctioning, development is that the patient, knowing intuitively that it represents a primitive difficulty, cannot conceive of any other adult's having the patience to see him through it. The defect is usually conceived of as irremediable, even congenital. He may be very resistant to undertaking any common-sense investigation. He has been trying to correct this all his life, and by now he is ready to give up and write it off. This discouragement reinforces the original anxiety.

Alternately the patient may be obsessively preoccupied with mastery. The actual performance may be fair and the supportive functions are usually excellent. The blockage is in the subjective experience of competence. This inability to experience ordinary effectiveness may be disguised either by excessive ambition, or by conspicuous lack of ambition.

Once a deficit has been identified, it is possible to investigate when and how it should have been learned and what interfered with it. The interference may have been a general disorganization of the emotional atmosphere at the time the function was being acquired, as with a patient who experiences unsteadiness on his feet and is easily subject to dizziness relating to a family catastrophe when he was learning to walk.

The original cause of the defect may have been specific to some aspect of the particular function. Some people cannot hear tonal variations because they were afraid to register the anger in mother's voice.

Once the deficit is identified the patient can find opportunities for making it up. This may involve hiring a private tutor for basic arithmetic or organizing some leisure time for sitting in the park. The interest in many art forms and hobbies, music, sports, painting, drawing, often, in part, constitutes an unconscious attempt to fill in perceptual deficits.

Sometimes concern with the function was preserved as a focus of tension, essentially as a marker for some very significant blockage of facility in interaction with peers. Once it has served its analytic purpose, concern with it may disappear.

The overall level of interpersonal relatedness is difficult to identify. It is complicated by the fact that the fixation involves a band rather than a point of experience. The trouble began at birth and progressively, but unevenly, distorted relatedness in later modes. This progressive handicap was reinforced or relieved to some degree by the particular circumstances of learning during subsequent periods. Beyond the point of arrest, there is, for social survival, considerable role-playing and much dissociated experience. It is easier and more acceptable to admit being tone deaf and to become resigned to this defect than to admit that one is unable to make a friend.

People continue trying to create favorable conditions for interpersonal growth. If they are deeply discouraged about it, the only method of progress they can conceive may be pretense. They either develop valid relationships, carefully camouflaged as meaningless, or persevere in meaningless ones and defend them as real.

For widely divergent levels of interpersonal maturity, the overt activities may be very similar. At any emotional level the individual may be married or single, may organize his life to spend time on personal property, crowds, groups, or individuals, or may avoid, to a large degree, any or all of these. The difference is in what they mean to him.

The person's true level of relatedness can sometimes be identified by his value system rather than his interpersonal activities. His conception of the basic reality of the value of other people, of what can be expected of them and how one needs to interact with them, is often a clearer criterion for identifying the level of relatedness than is the patient's current overt interpersonal operation.

The basic reality of infancy is the black and white of Good Mother—Bad Mother. The basic reality of childhood is crime and punishment, penance, redemption, appeasement, and forgiveness. All is referred to the authorities. The basic reality of the juvenile era is cause and effect—you get what you pay for, there are no bargains, nothing is free—but it includes a subjective experience of citizenship, membership, and belonging by right. The basic reality of preadolescence is that people matter, not in groups as available objects, but as particular people, in terms of their particular sense of well-being.

The purpose of clarifying the degree of unrelatedness is to get the underbrush out of the way, to relieve the patient of the burden of maintaining an illusion of intimacy beyond the true level of experience, so that the therapist can relate to and encourage growth processes where they are in effect taking place. Once he has identified where they are not taking place, there is usually no great gain in dwelling on the substituted activities.

Occasionally, it may be necessary in the meantime to give the patient some coaching in simulating in order to secure a social situation for the opportunity for growth. For instance, the patient whose problem is his right to possess objects may have to earn a living in order to get some, and for this he may have to learn some of the rules of getting along with people in the office. Nonetheless, the necessity for pretending in order to find the minimum conditions for new experience is greatly exaggerated.

Picking up at the zone of fixation is the quickest way to get there, and it need not take as long as the patient expects. If one begins at the beginning and takes it in a reasonable order, it can be done with maximum efficiency, minimum reorganization, and minimum anxiety. Isolated pieces of experience which occurred out of order can be fitted into place as soon as the place is ready for them.

An omission in interpersonal development is a liability to later developments, even if the latter met with less disapproval. The patient tries to skip this gap, often making an intense study of the next step in the sequence. He is, figuratively, in hopes of mastering algebra without understanding arithmetic, or calculus without grasping algebra. The patient has an intense investment in misleading himself, and incidentally, his analyst, as to the facts of his problem. He is also resistant to putting the insight into action even after it is clear to him, partly for fear his friends or acquaintances will catch on to who he is. He has probably already selected them partly on the basis of their being at the same spot, and they have probably been trying to tell him much of this "new" insight for years. His relationships cannot get worse by his understanding them. They can only get better.

If the therapist can validate the patient's own moves to identify where he left off and to act on this, there is usually a recognizable upsurge of energy and optimism about analysis and life. Growing up can take place rapidly. Progressive integration of observations thus far maintained in dissociation can proceed.

It is of equal importance to validate productive functions which have already been integrated into the self-system. There are many functions which the patient can perform competently; he can, with pleasure, satisfy known needs without anxiety. These capacities can be easily overlooked in the situation in therapy. Patient and therapist often conspire to focus only on the unhappy or negative aspects of the patient's functioning, and thereby fail to note and consolidate what he is able to do with pleasure. The patient, in the organization of his own life in deference to his central paranoia, often considers that those things that he can do easily are therefore relatively expendable in terms of the emergency in which he conceives himself to be.

Current validation of such functions is needed for their extension into unexplored areas. If the patient is able, with relatively little inhibition, to enjoy food, this can lead to his learning how to enjoy sex. If the patient can enjoy sex, it can lead to his exploring the tender aspects of the sexual exchange. If the patient can enjoy acquaintances, he may be able, then, to learn to be comfortable in the dynamic of group relationships. If he enjoys groups, this provides a base for selecting one of the group as a potential chum.

The exercise of productive functions already integrated into the self-system is not to be derogated because it is not in itself a growth experience. Orientation toward pleasure, toward satisfaction of needs, rather than orientation toward defense in the presumed struggle for survival, points in the direction of integration of new functions for additional satisfaction. The affirmation of satisfactory experiences, then, is relevant to the process of growth. It facilitates the identification of the point at which growth is taking place. In therapy this validation must be pushed to the limit. That is, the therapist must identify the most forward point at which the patient can function competently, without anxiety. In this identification he will contribute to the identification of the zone of fixation, the zone at which growth is currently taking place. He will also contribute to the overall orientation of the patient toward satisfaction rather than toward the expendability of satisfaction in pursuit of safety.

Such validation of already integrated positive experience is inherent in the process of consultation, the process of the taking of a detailed history. It is also inherent in the day-to-day process of keeping track of the patient's current activities. The patient in fact does not sidestep all of his pleasurable activities; if he did, he would find life hardly worth living. He tends rather to participate in them but to play them down in the therapeutic situation in his attempt to focus the therapist's interest on "where the problem lies." This playing down may give both the patient and the therapist a false perspective on the actual content and direction of current life.

There are other experiences which are not currently implemented but which, on the basis of previous competence, the patient is entirely capable of performing and enjoying if he would notice that he has excised them from his current living. These are equally important to identify and validate. They were excised because they were, at some time, in conflict with the paranoid search for living safely. They are in no actual conflict with the necessities of survival and could, if reintegrated, make a significant contribution to the direction of growth and satisfaction.

Thus vaguely obscure but available functions organized to satisfy particular needs can be important to reintegrate into the current patterns of living.

C)   THE ROLES OF SIGNIFICANT FIGURES

Mother, here used in a somewhat generic sense, does not necessarily mean the biological mother—the infant may have been adopted at birth. It also does not mean adjunctive persons, such as the baby nurse, or the grandmother, because they and the infant know that their commitment is more transitory than that of the mothering one. At the same time, to some extent, one must include all significant, surrounding, durable figures of the first three years as, however peripherally, involved in the mothering function, and thus in some sense making up a sort of composite mother from the point of view of the infant.

There tends to be, during the first year, selection by the infant of one adult with whom to form a one-to-one relationship. The patterning of this initial primary integration becomes the point of departure, the base line, for the development of subsequent integrations. Surrounding adults, whether more loving or more hostile, have an important impact in modifying the course of development, but the patterning tends to be determined in the central relationship.

The choice of the particular adult is not in terms of either exposure in time or capacity for love. It lies primarily in the intensity of feeling beamed toward the infant by this mothering one. The determinant of this level of intensity lies in the impact of the infant's birth on the lifeline of the various adults involved. The mothering one, thus, is the one whose lifeline is most affected by the fact of the existence of the infant. In our culture, this is almost always the biological mother. She cannot walk away without intense guilt and defiance, and her defects will be exposed if she accepts the commitment to her child.

There are many well-intentioned mothers who are capable of constructive, durable relationships with friends. Some find their way to loving relationships in their marriages. But the demands of any infant for relatedness far exceed the mother's operating capacity. To the extent that she cannot meet the challenge by growing she must deal with it by dissociation. The degree of this is indicated by the discrepency between the undeveloped creative potential in the infant, and the actual operating performance of the adult.

It is important in therapy to clarify in detail the relationship with the mothering one, in order to release the patient from his commitment to her deficiencies. If the analyst substitutes social generalities or cultural euphemisms, he can miss the specific flavoring of the details of the interaction between the infant and this particular individual with her personal version of the cultural neurosis. The formulation that she was probably no less loving than her contemporaries—she was only an integral representative of them—misses the point.

In analyzing the relationship with the mother, one important element is for the patient to validate his own experience of frustration in coping with her. In order to make the real interaction believable, it is useful to get some understanding of the forces in her life which made responsiveness difficult for her. However, the patient's impulse to dwell on his compassion, before he has freed himself of the damage, avoids clarifying his feelings. The resentment persists, conscious or not, until the original integration is no longer a restraining influence.

There is no clear evidence as to whether or not the newborn infant has already developed some symbiotic emotional synchronization with the mother *in utero*. But the availability of a close integration with one mothering person is crucial to the full development of basic physiological and emotional patterns. On the other hand, if the infant experiences no variety in the available validator—father, nurses, relatives, roomers, older siblings, or family friends—very early in life, we might expect a rigidity and dependence on the particular patterning of the single mothering one which would be particularly hard to broaden in later life.

If the integration with the mothering one is set up, even with supplementary substitutes, and is then discontinued prematurely by death, desertion, or divorce, this discontinuation becomes the most significant fact in the infant's life. In our culture, when the father is persistently remembered as the central early figure, it is almost always because of a gross failure in the integration with the mother. This original failure of integration persists as the primary hostile integration and becomes the core of later troubles. It is most frequently acted out in intensification of the repudiation of the need for tenderness. It is generally expressed in terms of the inadequacies in tenderness of the father. Even when these accusations are fitting, the central accusation is that he was Not Mamma, and no one else would do. The only satisfactory solution would have been for mother to be transformed by the child's need into a more loving person. If father subsequently used the infant as part of the battleground with mother or child and became possessive, controlling, and exploitative, the active, hostile integration with him, superimposed on the original rejection by the mother, can set the central paranoia in an almost unbreakable die.

One of the highly important roles of the father is to fill in where the mother is lacking. After infancy, when the child is more responsive to integrations other than the primary one, the father can exercise much more leverage for growth or inhibition. He may be the only other adult whom the possessive mother cannot exclude as an alternate validator.

As the child moves more into preparation for the adult world—in the juvenile roles, in preadolescent friendships, and in adolescent explorations—the father's influence becomes still more manifest. He operates as a model, or as a limitation, in defining available adult roles. His validation or prohibition of the beginning sexuality of the early adolescent, boy or girl, is often the primary determinant in the course of its development over the next decade. The use he makes of his financial position for facilitating or blocking independence from the family after puberty can throw the balance in the overall choice between growth and deterioration in adult life.

The unrealistic social demands on the mother for producing what she cannot produce for the infant are an important factor contributing to her discouragement in making the best use of what she has. Conversely, the milder demands on the father—the acceptability of limited exposure and moderate tenderness—can make it possible for him to be a less desperate and more affectionate figure in the infant's life.

This places the father as a factor for growth essentially in a supplementary role in the early years. After the primary damage is set, the father may, nonetheless, open the possibility of revision and expansion in adult life.

As the person for whom the effect of the baby on his lifeline is second only to that on the mother's, he is clearly the most important corrective influence available. His awareness of the infant's real dependence, social identification, and legitimate claims on him intensify this. As the small child identifies individuals, he discovers father as a second source of aliveness, of almost equal power to mother's. Father's response to the child's discovery of father's investment in him, and hence of father's importance to him, may be the turning point in determining future directions.

If he is very angry, the father, or any older person living in the household, can have a somewhat disproportionate influence. The primary deficit is in productive functions that fail to be integrated. Nonetheless, the effect on the young child of a father whose wish, however controlled, was that the spontaneity of the baby would be disposed of, once and for all, by the annihilation of the baby, can be crushing. The fear of death from the fluctuations in mood of the mother fuses with the fear of the father's anger and becomes the central preoccupation.

As an inescapable emotional responsibility, the new baby makes less intense demands on father than on mother in our culture. However, the opportunity for reintegrating infantile and childhood experience is the same for both parents. The father can use both the relative freedom of access to this and the relative opportunity of temporary escape to organize the relationship with the child as a progressively expanding growth experience for both of them. If he does, his positive orientation to his own growth in the relationship with the infant and child will be of great influence in helping the child to orient toward growth.

If the father avoids reintegrating particular eras of his own growth in relation to his child, he can find ways to avoid the child during these periods. Such desertion is somewhat less catastrophic than similar periods of the mother's desertion. Because his avoidance of the child's appeal is more socially acceptable, the father has room to be less defensive and therefore less punitive. Also, the father is somewhat less centrally the lifeline of the child's emotional development. Nonetheless, these periods of withdrawal reinforce both the child's repudiation of his need for tenderness and his discouragement about reaching beyond the mother for additional sources.

A primary role of the father, then, is that he is the first and most important step away from the mother in the search for additional validators.

The frequent close relationship of the father in preadolescence and adolescence has some constructive features in the same direction. However, it is more appropriate and useful in the earlier years. For father, it can be that he feels less awkward with the older child, and thus tries, belatedly, to make up for earlier failures. For the older child, this attractive offer can be misleading and diversionary from the more urgent relationships with contemporaries. Sometimes the immature, emotionally isolated father cherishes the fantasy that the daughter will grow up to be the one who understands him and comforts him. The daughter's move into adolescent experience then bursts the bubble and his dismay can precipitate resentful possessiveness. The secret flirtatiousness that sometimes occurs between father and adolescent daughter is frequently motivated by the father's restrictive and possessive impulse. If so, this may haunt her through many later love affairs.

A major lever of influence between father and child is through the father's impact on mother. Father's direct competition with mother, and the unrealistic attempt to become more important than she is, can only create distress and confusion for the child. Affection for the child must incorporate an appreciation of the emotional importance of the mother to the child.

Conversely, irresponsibility in regard to the needs of the child and anger at the demand that he relate to them are often expressed by the father in manufacturing a feud with the mother. The quarrel with the mother may seem to be about her inadequacies as a mother to the child. The impact of it will be to make her more paranoid and less adequate to both husband and child. However denied by father, this style of attack is directed toward interrupting and not toward facilitating the child's development.

In therapy, the importance of understanding the patient's early relationship with mother is clear. Its reconstruction is in terms of the Good and Bad Mother personifications. In the period before memory, father and nurse began to take the role of important additional validators, but the particular interactions with them are often even harder to identify in adult patterns.

After childhood, other adults can usually be sorted out in memory; their importance as sources of tenderness, consistent or interrupted, must be reconstructed. The father's attitude toward the child's physical and intellectual interests and developing relationships with contemporaries is often crucial. The analytic identification of his support or disdain is important.

The analytic management of the adult's current relationship with his parents is no different from the analysis of other adult primary interactions. It is hard to reduce it to these common-sense terms until the course of the past interactions with them has been accurately reconstructed.

D) THERAPY AND THE GOOD MOTHER

In the adult, clear, strong feelings of contentment and satisfaction are always at least partly alienated from the productive functions. In early infancy such feelings were organized in the context of the personification of the Good Mother. Then, in the vicissitudes of growing up, the continuing elements of the Good Mother personification in the adult personality became attached to unattainable, subversive, or self-limiting relationships, and detached from the productive functions.

The adult experience of situations that should evoke positive, clear, satisfying feeling must become reconnected in therapy with its historical roots in the infantile experience with Good Mother. Otherwise, potentially satisfying relationships continue to feel flat and to become ritualistic. In therapy, it may be that many productive functions are expanded and can now be performed with objective success. This success will continue to be experienced as empty and meaningless if it is not reintegrated with the infant's intensity of pleasure in performance and satisfaction.

If this connection is omitted, the patient will also continue on with the perpetual search for intense satisfaction in unrealistic directions, misguided by the original terms of early infancy.

Confrontation with the unproductive aspects of this search is experienced as the demand that the person involved in the search must write off desire and the need for love. Such debunking is usually fought even more bitterly than a common-sense appraisal of the paranoid search for integration with the Bad Mother.

The experience of intense desire and the orientation toward open-ended growth, and the enjoyment of it, are characteristically relegated to the deeply dissociated organization of the integral personality. When such emotion breaks through into the self-system it is with a lack of conviction, and with feelings of eeriness or unreality. It tends to be little acted upon, quickly obliterated.

In the adult, then, identifying the persistent impact of the personification of Good Mother on perceptions of adult relationships is both necessary and elusive. The patient usually defends the choice of relationship made in the service of this search as crucial to growth and as intensely satisfying, and usually later repudiates it as hallucinatory and inappropriate. Such choices have the double character that they are important for identifying and reintegrating obscure but crucial infantile experience, and that they usually constitute a poor base for the search of the adult for tenderness and validation. The intensity of the experience seems to belie its ephemeral quality.

The moment of major breakthrough of material from the integral personality, whether of creativity or of expansion of the capacity for mutual love, is often experienced as or accompanied by infatuation. This infatuation is a self-system activity, incorporating the self-system aspects of the Good Mother. It does not feel unreal. It is experienced at the time with subjective conviction. Its events remain available to conscious memory, even after the initial enthusiasm has died down.

The experience of infatuation is not in itself diversionary, but moves in the direction of a constructive affirmation of the experience of desire. It is important to validate this. The joyful aspects are not destructive. Their validation may be complicated by subsequent developments in the actual relationship. At the same time, it can help to identify personality patterns referring back to the experience of Good Mother and Bad Mother, and to connect them with productive and security functions of the self-system.

This flare-up of self-system perceptions of both Good and Bad Mother during a period of expansion may be partly because of their closeness, in origin, to those aspects of Good Mother that were dissociated into the integral personality. There is a certain tendency for the greater intensity of desire, ecstasy, to remain in the integral personality and to be associated on breakthrough with eeriness.

The reconnecting of present experiences of satisfaction with early experiences of ecstasy and elation is necessary to the active integration of a growth dynamism within the patient. In organizing a growth dynamism, the patient draws strongly on the initial direct interaction with the integral personality of mother, evoked temporarily by the dramatic phenomena of his birth and the needs of his early infancy. Whether mother subsequently succeeded in integrating or repudiating her positive response to his needs, the initial experience nonetheless lies within him.

Because of its ties with the overall growth dynamism of the integral personality, the intrusion of the concept and search for the Good Mother, the glamorizing of particular not-understood, but characteristically intense, attractions to adult individuals who are reminiscent of her, will continue until the growth dynamism is, itself, incorporated into the conscious personality, the self-system.

The patient must integrate his wispy, free-floating concept of the original Good Mother with his actual positive experiences of the good friend in the chumship in order to consolidate a serious and reasonably durable loving relationship. If he does not, the illusory and hallucinatory construction of the Good Mother, incorporating all of the loving aspects and, by definition, excluding all of the anxious, angry or frightened aspects of the Bad Mother, will by comparison vitiate any actual real-life adult experience.

The move to the chumship was originally, as with other major moves toward love, associated with the re-evoked personification of the Good Mother. This was, at the time, toned down in the service of the urgent real necessity to integrate this important experience. Subsequent attempts at adolescent love incorporate more clearly the overlay of the search for the Good Mother. This is partly in order to implement their defeat, before they should become permanent integrations.

The elation and ecstasy associated in infancy with the early experiences of satisfaction, and the original enthusiasm for expansion of experience and growth, are indispensable aspects of human experience and must be reintegrated. This enthusiasm should normally be evoked both by new expansion of the experience of tenderness and by new explorations of growth and creativity.

The competent and effective search for adult, intense love relationships incorporates the original, ecstatic growth experiences of early infancy with the satisfying mother, but does not come into being in these experiences. It develops in the explicit, detailed evolution of the capacity within the individual for experiencing adult love. In such adult love relationships, the concept of friendship implies more, rather than less, intense interaction. The frequent overlay of the search for the Good Mother incorporates a degree of repudiation of the optimal adult interaction. For the adult, the mutual exchange of concern and affection is significantly more rewarding than the unilateral demand for unreserved devotion to one's needs.

The subsequent search for expansion of loving experience can be more intense, but it can never be again intensely exclusive. It has now incorporated the knowledge that to love one person implies some love for others, and to explore the possibilities of love with others will expand the interaction with the one who is more centrally important.

IV

THERAPY AND THE INTEGRAL PERSONALITY

A) THE DYNAMIC OF THERAPY

The dynamic of therapy lies in the patient's moves toward growth. The basic therapeutic maneuver is to guide the outcome of a particular move toward integration rather than toward paranoid repudiation.

This guidance requires the therapist's use of his knowledge of personality. He applies this knowledge to the reconstruction and description of the patient's life, to the management of the relationship with the patient, and to the management of the course of therapy.

Therapy must be directed, in order to take place at all. The patient has already done what he could without direction. Further, the time available for therapy is very short for reorganizing the patterns of a lifetime, and the patient is pressed by the current demands of his life. The therapist must elicit latent moves toward life, and actively invalidate the central paranoia and the security operations. He must investigate areas of universal experience omitted either in the patient's original development, in the present organization of his life, or in his subjective awareness.

In the management of the course of therapy the therapist evaluates the minimum that must change in the patient to make the effort worthwhile, plans what sequence of events might accomplish this, and implements the plan with an open mind as to whether the patient has the interest and capacity to far exceed this minimum. The therapist's security in taking such an active role lies in his intimate knowledge of the patient. If the therapist is clear on what is really going on with the patient, he is not likely to make major and unreconstructable errors.

The therapist must organize the patient's experience to facilitate hope and undercut despair, in order to increase in frequency and intensity the moves toward life. This involves decreasing resistance to growth and handling the anxiety which accompanies the resultant moves. The resistance lies in the organization of the personality to avoid anxiety and is predicated on limitation of awareness, communication, and experience.

Modes of resistance can contain some positive aspects of communication and may involve some attempt to maintain contact in spite of powerful covert disjunctive emotions. Nonetheless, they are primarily distance maneuvers. The frank anxiety is disorganizing of consciousness and communication, and thus involves many disjunctive elements. But its content includes some spontaneous thoughts from the integral personality which have previously been excluded from communication. The sense of isolation concomitant with frank anxiety increases the need for tenderness, and the analyst must relate to this. However, if he does so crudely or abruptly, this unexpected increase in intimacy can make the patient still more anxious, hence more paranoid.

The outcome of a particular growth move is influenced by the availability of tenderness, the historical position of the particular experience in the patient's development, the patient's grasp of the processes of anxiety and of his own usual modes of decreasing it, and the patient's grasp of his own character structure. Another factor is the degree of reorganization of the self-system involved if this particular experience were to remain validated. If the patient really believes this new insight, how many other attitudes toward himself or the world would have to be relinquished?

A parallel factor is the degree of reorganization of the actual patterns of living which would be involved if a particular insight were accepted. Of the many situations to which this applies, the most recurrent is the relinquishing of a particular primary hostile integration, married or unmarried, legal or illegal, which the patient is using as the underpinning of his neurosis. The accumulation of discouragement—the memory of previous repeated attempts at growth in this area, and their repeated failure—can reinforce a feeling of fateful doom in the patient about any new attempt. So also can the degree of actual inexperience, conscious or unconscious, in the area involved. The therapist's contribution lies in reasonable preparation for the particular push to grow, in helping the patient to keep track of what is going on in the course of it, and in using the quality of the relationship between them to get through periods of intense anxiety.

We distinguish between functions with which there has been experience, whether this is available to consciousness or locked in the integral personality, whether accepted or derogated by the patient; and functions with which there has been very little experience at any level. If the experience is essentially new, then the feeling of unfamiliarity will be cumulative with the disorientation of the self-system. On the other hand, there is sometimes a compensatory factor in the lack of experience, whether the lack was rooted in the strength of the original prohibition or in the lack of stimulation. The original prohibition has not, in this case, had to be reinforced by accumulated failures or by constant preoccupation with the active process of repudiation. In particular instances, the therapist may be more impressed than the patient with how much the patient needs to learn. The patient, once brought to tackle the problem and acknowledge the deficit, can sometimes move ahead in mastering the experience with relative freedom, partly because of a limited grasp of the implications.

Real experience, poorly consolidated, may accrue from repetitive moves in the direction of integrating into the relevant situations, each move either camouflaged or dissociated just short of consolidation.

Sometimes there is a reservoir of potential competence, acquired by example from an adult in the early family constellation, which could not be acted on. One example of this is the son who is impelled to be a bum because father was so successful and so competitive with his son. The son has always, at some level, picked up the essential operations which would be necessary for success. He lived with them, he watched them, he knows how they work. He may never have used them overtly, and he is profoundly convinced that they are not for him, but if he is thrust into the situation, he will turn out to be a much more competent business-man, professor or whatever else, than one might expect on the basis of his active experience. The child was apparently cut off from something at which the parent was extremely competent but, if his anxiety about trying it can be sufficiently reduced, it turns out that he has known all along exactly how it ought to be done.

The conscious emphasis on the degree of inexperience comes from the patient. This is often used to deny and disguise the inner competence. It acts as an apology for the expectation of relatively quick mastery once the situation has been integrated. The patient's emphasis on his fear of revealing his inner disorganization is often a more or less direct expression of his fear of exposing the high level of inner organization, to the anger of the parents.

B) DISCOURAGEMENT

One searches for the setting in which a particular experience can take place. This searched-for experience was at first incomplete, then was artistically avoided through a lifetime of evasions. It was vaguely known to be there, just beyond the fingertips. When it comes into being, the moment of this experience often occurs with intense conviction and is then rather quickly toned down or dissipated. It seems to evaporate.

This may be rationalized by the thought that "it has been gotten out of my system"; it has been found, consolidated, and integrated, and one can go on to the next dimension of experience without lingering. Or the thought may simply be that such intense alive and clear experience could only happen in the briefest and most subversive mood.

It is never easy for either patient or therapist to know whether in ending such an experience one is on the move toward expansion or is compulsively committed to the discontinuity of contentment; whether the lifelong conviction that happiness must be subversive, stolen, secretive, and momentary is the dominant paranoid pattern of one's life, implemented by the seeking out of good experience in the less promising, more guilty or more subversive context; or whether the overvaluing of security is used to avoid noticing that much of living is like a shooting star; whether the concept of peace has become contaminated by the search for stabilization.

Each phase anticipates the next phase. Any moment of true and valid clarification and consolidation—any moment in which two people are there at the same time and in the same place without the paranoid static of the past—gives us more knowledge of the past, present, and future. Sometimes the desire to hold on to the moment of clarification beyond this meaning can only result in its repudiation.

Both the patient and therapist tend to oversimplify the complexities of identifying this operation. Archaic fears from infancy add confusion to the sorting of the adult reality. The illusory expectation of catastrophe can operate as a force in either direction: either to amputate an experience long before it has been fully explored, or to try to perpetuate it beyond its usefulness.

There is a tendency, in trying to evaluate what is going on about a particular event, to apply labels: neurotic, competitive, constructive, experimental, friendly, or unfriendly. Such labelling can only obscure the complexities of understanding what is really going on about this particular event.

The demand for, and the fear of, what is often referred to in our culture as "commitment" plays a similar obscuring role. Superficially, the involvement in this concept most frequently takes the form of a preoccupation with the prediction of the unpredictable future, substituted for the simple freedom to be fully present.

An active output of energy is involved in seeing a bright light dimly or in devaluing a satisfying experience. The chronic depression involved in this effort is cumulative with the chronic disappointment of limited participation and the incomplete reward.

The resultant fatigue from the cutoff of satisfaction plus the effort of maintaining dissociation lends an illusion that the expanded life would take too much energy. One is avoiding not the effort but the pain of reliving painful experiences which, however transitory, is the inevitable accompaniment of coming to life. No miserable experience can ever be gotten over until it has been shared, touched, and relived with another person. The transition process of coming alive is always painful. It is not, even in the moment, particularly fatiguing and it results in greater aliveness and more satisfaction. It may be accompanied by an increase in relaxation which, if one is still uneasy about the new breakthrough, can be misinterpreted as an exacerbation of the old fatigue and misused as an excuse for not breaking through the next barrier.

The reawakening must involve some disorganization of the old obsessive structure and some period of reorganization in terms of the expanded scheme of values. Such reorganization may make for temporary inefficiency. It may require episodic withdrawals from responsibility. It may necessitate either intense personal support or temporary privacy. This transitional phase can be mistaken for a period of increased work, since, for the moment, operations and responsibilities which have been handled previously with moderate effort may become extremely irksome. Even so, even the acute transition, the moment of insight, of breakthrough of old needs and new experience, is not in itself more effortful than the hour before of restraining the breakthrough. It may be more exciting, more painful, or both.

C)   DISCOURAGEMENT AND THE CULTURE

In any society, a crucial point of individual freedom is the freedom to grow and to seek out that experience which is next in our moves toward creativeness and tenderness. Existent cultures are ambivalent about the growth of the individual. Social change is urgently needed in the direction of validating the universality of this potential in the individual. In his attempt to move in the direction of much-needed changes in the structure of society, the individual is pressured to formulate and re-evaluate his own value system. We speak of the mediation of the values of the culture through the parents in the person's growing up. These values continue to impact directly and pervasively on the individual as an adult in many ways. The depressive aspects of these values add to the individual's burden of discouragement.

It is therefore important for the therapist to have a working grasp of the impeding demands of the culture on both himself and the patient, if he is to understand the field of interaction in which he is functioning. This is especially so at any particular moment of expansion of the patient's personality.

For example, industrial civilization places an intensely high value on the capacity to function, and is consciously punitive about gaps and deficiencies in this capacity. The fear of social disapproval may be intense after such minor lapses in awareness as losing a key, forgetting the contents of a conversation, forgetting an obvious name or face. Actual retaliation for longer-lasting periods of disorganization is expressed in the contempt for mental patients, and implemented in the almost universally poor quality of mental hospitals.

In this way, while there is some positive value placed on both continuing growth and continuing creativity into the adult years, the necessary accompaniments of such are forbidden. While moments of intense competence are a source of social pride, the citizen dare not cherish the equally necessary periods of ineffectiveness. He must love and enjoy only those parts of himself which are immediately socially useful. More than that, he should maintain himself in such a state that he is able to become socially useful at a moment's notice. The relatedness to the details of external reality takes eternal precedence over the equally pressing private imagery and overwhelming emotion.

The experiencing of accumulated fear, rage, loneliness, doubt, despair, and confusion is attended by guilt, disgust, or shame; is secretive and defensive; and is pared down by each of us to the bare minimum essential for the maintenance of legal sanity.

We start life with a crushing superstructure of naive cultural clichés, passed on mainly by the parents. Any reorganization of these must involve transitional disorganization, both intellectual and emotional. And such disorganization is illegal in cultured homes. This prohibition becomes, then, a central force in the perpetuation of the original clichés.

Thus society operates on the conviction that what it needs from its respective contributing members is a stable level of competence.

This cultural illusion, if it can be seen in the perspective of the conditions for growth, is a suicidal cliché directed ultimately to the destruction of the particular subculture that succeeds in implementing it. Every society needs creativeness and ingenuity, resourcefulness, the capacity, the interest, and the heart to challenge and re-evaluate those assumptions upon which it has been operating, but which at the same time set the limitations of its subsequent development.

The history of cultures parallels the forces of growth and deterioration found in the history of individuals. Certain new, creative, expansive ideas are put into action. Certain false and unrealistic clichés are incorporated into this because of the lack of emancipation from the mother culture. To the extent that these clichés, in the name of security, which is a way of living dangerously, are incorporated, implemented, and enforced by the police power of the state, to exactly this extent any particular culture is slated to be outgrown by any of its children. In this sense, the rise and fall of cultures in the history of civilization is remarkably similar to the rise and fall of individuals in the history of any particular family over several generations.

Civilization, which depends for physical survival on its capacity to change, on its capacity to reorganize its basic assumptions, operates self-destructively in such a way as to inhibit, or, if not successful in that, to annihilate, those individuals who are making a somewhat serious attempt to save it.

This also becomes one of the major impediments to the search for a theory of growth of each individual operating in the context of a particular culture. Out of his own original necessities, the scientist in personality is already more or less swamped, in his personal life, in reliving the destructive experiences of his own childhood. As an adult, he attempts to get some perspective on the realities of the culture in which he lives. He also, in the attempt to re-experience the emotional realities of childhood, tends to project these onto the social structure. The state may be seen as the family, or the stranger as the mother.

There is thus pressure on both patient and therapist to experience the search for growth in secret, subversive, defiant, or rebellious ways. Society to a considerable extent structures it so that this search is interpreted as destructive in a style not too different from the way in which the rigid mother interprets the infant's simple, physiological spontaneity as malevolently organized to attack her at her weakest point.

With this overlay of diversionary factors, personal and cultural, patient and therapist often have a hard time finding out what the patient is trying to do in the direction of organizing even minimal growth, without in the process doing such damage to the social context in which he is operating as to eliminate the possibility of any future useful human interactions.

V

THERAPY AND THE CENTRAL
PARANOIA

A)   SELF-PERPETUATING ORGANIZATION OF THE
     CENTRAL PARANOIA

Much therapy is based on a naive acceptance of the theory of
conditioning as the core of personality development, hence of
reconditioning as the lever of change. The patient's proposition is:
"My mother, who was deficient in her capacity to love, and thus
angry at me for needing it, had red hair. Therefore, it is sensible
for me to avoid redheads in my search for love." The therapist
answers: "Notice that not all redheads are unloving." Such
atomistic re-experiencing of all the regrettable experiences in one's
life that took place by the age of five, let alone by twenty-five,
would make the task of therapy in a finite time entirely unfeasible.

The fact that therapy is feasible is based in the phenomenon of
the active dynamic integral organization of the personality into
dominant motivational trends. In the course of life, the original
catastrophic conditions of the particular infancy (those which
were not catastrophic are not the source of trouble) become organized into an elaborate way of life, the active direction of which
is to maintain the particular feelings appropriate to the original
condition.

The active avoidance of redheads is a minor cog in the wheel.
When the question of love comes up, all potential partners are

382

seen as mothers, and their hair takes on a reddish glow. This distortion leads to maneuvers actively designed for the specific purpose of maintaining itself. Where the core of the integration with mother was a fight, the less hostile woman may be experienced as bland and dull. If she should be noticed to be less stingy than mother, she is then seen as dangerous and exploitative. However, the problem in infancy was to overcome the fear of mother's anger, so the relationship might still progress.

The next attempt is to integrate a relationship incorporating the expectation of the other's malevolence. Such incorporation leads to behavior on the patient's part designed to bring out in the other the particular pattern of betrayal originally most feared. If the mother never listened, intimate thoughts do not come into the patient's head in the presence of someone from whom he wants understanding. If the mother extracted information to use against him, he will project elaborate soap operas to mislead the loved one. If the mother was actively a danger to survival, the patient will compulsively stab at the other's most vulnerable points, so that the other will withdraw and thus be the agent of abandonment, or else retaliate and be disqualified as a source of love.

Such compulsive maneuvers are camouflaged with an intensification of the subjective experience of the search for love.

The whole dynamic operation has become a military machine directed toward a defense of the proposition that the world is the way one once experienced it at mother's breast. As the reality of chronological adulthood makes this proposition more ridiculous, more active effort must go into recreating the conditions which make it plausible. The previously defensive machinery goes into offensive military tactics.

Situations paralleling the original situation must be searched for actively, or imposed forcefully on the material at hand. Alternate positive experience must be erased by an active retrograde reintegration of the data, since the original family constellation is not there to do it. When the family was around, the opportunity for friendships with a different kind of person often never came up. Now such people must be driven away in such a manner as to discredit the impact of finding out that other patterns of relating are available.

These offensive maneuvers are unique to each individual in their specific patterning. Each person tends to take over the patterns exercised at home, either on him or by him, since he is most familiar with these. Superficially, the intent would seem to be to integrate relationships into familiar patterns. If the partner can be provoked to want to murder the patient, the resulting interaction can both obscure and express the patient's own murderous impulses. If such provocation can be stated as the clumsy attempt to integrate a relationship, the camouflage is that much more effective.

Hate is the usual response to a situation that is frightening and inescapable. Intimidation, bribery, exploitation of the other's fears of independence, reproach, which constitute the general pattern of possessiveness, thus guarantee an element of hatefulness in the progress of the relationships. Where this pattern of possessiveness, of professed insecurity masking an insatiable demand for security, is the overall trend, it tends to structure the search for the partner who will be most vulnerable to one's particular methods of victimization.

It may seem even to the perceptive observer that the motive is either to take revenge on the other, or to be victimized by him in order to justify one's own rage. This can even go so far as actual physical harm to the other, or literally driving him crazy. However, this apparent focus on the other as an individual is, again, a cog in the wheel. It seems as if the partner were being punished for his resemblance, real or constructed, to the parent. In fact, he is being attacked for the points of difference, however accurate the type casting in the original choice.

The force behind this elaborate military campaign is to invalidate and erase the alternate positive experience which went on in the initial interaction, before the destructive patterns had time to develop. After the interaction is infused with anger, reproach, or quiet despair, the initial enthusiasm about a broadening experience is either forgotten or looked back on with horror and revulsion. It has ceased to exist in consciousness as a base for the continuing search for further expansive experiences. The specific patterning of such obliteration of experience is highly individual. It involves distance machinery, by either attack or withdrawal, or by developing a penchant for infighting. The focus of discrediting may be on the partner's inability to give, or on one's own. The apparent shocked surprise at the end of the relationship is pervaded with a feeling that one had known it all along. The surprise is really that there is no surprise. Each had unconsciously spotted and then exploited traits in the other which were manipulated in order to prove out the original inner convictions.

When security goals are substituted for satisfactions, when the search for direct satisfaction, for pleasure in people and creativity, has been given up on, one can speak of the individual as then devoting himself to the substitute satisfactions of power, prestige, and exploitation. This formulation is somewhat misleading, however.

It is the intent and goal of the central paranoia to recreate in the patterning of the adult's relationships the specific unpleasant emotional interactions that characterized his situation during infancy. This period represented the most direct and intense interaction with the mothering figures. The central paranoia became the condensation of all the negative and frustrating aspects of this early interaction. It incorporates feelings of fear, rage, disorganization, rejection, and expectation of abandonment. The specific patternings of this interaction are peculiar to each individual's experience and are inherent in the specific interaction in the home at that time. These emotions were in direct reaction to certain emotions in the significant adults, such as rage, rejection, and the wish to escape from interaction with the infant.

As the infant becomes an adult he organizes the patterns of his primary integrations, of interactions with those people who are of particular significance to him, in such a way that they will have the same attitude toward him that his mother had when he was a baby. The person whose central early emotional experience was terror of the chronically and acutely angry mother will make a life work of provoking anger towards him in the people that matter to him. The immediate intent of the fight for power is to make the other rageful and vindictive, and thereby to validate and perpetuate one's own inner feeling of being afraid of the other's rage and vindictiveness. It seems to be an emotional necessity that the partner hate one as much as the mother did. The immediate intent of exploitation is that the person who is exploited should feel exploited and thereby be pulled into either counter-exploitation, chronic resentment, or feeling abandoned. The immediate intent of destructiveness, of the direct attack on the other's needs for satisfaction and growth, is to induce rage and counter-destructiveness against one's own needs. The intent of terrorizing is to provoke counter-terrorizing, thereby validating one's own chronic inner feeling of terror.

It is an almost universal cultural myth that the search for the so-called satisfactions of power, prestige, and exploitation is a regrettable but inevitable undertow in each of us in our search for happiness. There is no satisfaction per se in any of these operations. Plagiarism avoids the satisfaction of creativity. A power fight crushes the satisfaction inherent in intimacy, in the enjoyment of the other person's company. There is no joy as such in prestige. Its attraction lies in the virtuous renunciation of simple interchange with others who are experienced as equals.

Each security maneuver is planned for the purpose of inducing negative, hostile, unfriendly feelings in the opposite partner of the interaction, which feelings will then be most directly beamed at oneself. There is the subjective illusion that the power operation is necessary to avoid an inerasable, chronic, inner terror and keep it out of awareness. In actuality the power maneuver perpetuates the terror by provoking retaliation in the partner, however subtly. The terror, both hallucinated and perpetually recreated by such patterns of interaction, also serves a purpose. It is cherished and perpetuated for its efficacy in refuelling the whole dynamism of the central paranoia.

The power mode is thus not a regrettable by-product, but rather the focus of the total organization of the personality directed toward the active destruction of the shattering experience of love, growth, and creativity. Within its context all positive experiences can be either avoided or eradicated from memory. Terror is tenderly nourished.

It could be relieved by any moment of tenderness and understanding with another person. Such moments as occur must be either overlooked or repudiated by inducing either rage or withdrawal in the other person. The resulting isolation becomes then one of the factors in perpetuating the terror. The individual perpetuates the terror by experiencing himself as isolated and as the object of vengeful retaliation by the people whom he has, covertly or overtly, recently attacked. Each person who has thus been provoked into retaliation or driven away is then enshrined in memory as a significant bulwark to the structure of the central paranoia. Each such memory becomes additional evidence for the thesis that this way of life is inevitable. The dominating theme is: "It is the fate of man to live in an atmosphere of perpetual emotional malnutrition, if not of total starvation."

In the process of therapy directed towards radical change, towards the repudiation and relinquishing of the whole overall style of life just described, the major difficulty lies in getting a broad enough grasp of the total operation. Methods of therapy usually rely extensively on a piecemeal attack, the invalidation of one after another particular paranoid maneuver. The therapist, in defense of his own despair, finds great difficulty in seeing his patient's situation in broad enough scope. He finds additional difficulty in confronting the patient with it. Unless the patient can see it as a whole, he remains snared in its assumptions and is then unable to conceptualize any alternate way of life. He must continue to drive away the mice that are continually gnawing at the ropes that bind him.

One important phase in such therapy is the validation of the continuity of alternate experience: moments and periods of tenderness and its exchange, of joy in exploration, of original work. But the major focus is for the patient to get an operating grasp of the hallucinatory, ephemeral, "self-inflicted injury" quality of those bitter experiences about which he cherishes the feeling that they are inevitable and ineradicable.

The first step in the therapeutic management is to help the patient to identify his manifest current patterns of living and their relationship with his early experiences with tenderness. This exploration must be specific to be of any help. The identification of current operations is impeded by the patient's defensiveness and fears of retaliation or of interference from the therapist.

Resistance is often initially in the form of evasiveness and incomplete or highly censored selection of data, rationalized either by a fear of the analyst or by a wish for his approval. As more data are collected, the patient becomes involved in denials and apologies. As the evidence becomes undeniable, the patient often responds with an intensification of discouragement which becomes, in itself, a form of resistance to change.

The feeling that "It is all so overwhelming, and isn't it hopeless?" is again both hallucinatory re-experiencing of past entrapment and manipulative resistance. It is set off, at this point, not by the overwhelming nature of the confrontation, but by the hopefulness implied in the first grasp of the problem in its totality. Here is the first glimmer that the whole mess is a self-inflicted injury, inappropriate to current reality, and requiring much expenditure of energy for its maintenance. Such a reaction alternates with stolen moments of genuine ecstasy. It is also complicated by more violent explosions of active paranoia, in which the therapist, and/or the partner, are seen as agents of malevolent destruction.

The goal of therapeutic management is to reduce these intense reactions to a pragmatic attitude involving more work and less dramatics. After this it is possible to set up a collaborative investigation of how the recurrent patterns of interpersonal integrations are directed toward the avoidance of new situations and the erasure of new experience.

The reality of the avoidance and denial of just those experiences for which one thinks he has been searching is more important than the reason. The reason why they are expected to be so catastrophic often remains obscure. The subjective expectation is of either literal death or psychological disintegration with a resultant permanent total isolation; one fears that he will lose all capacity for communication, and be returned to the helplessness of infancy, with no one there to rescue him. It refers back to the feeling of nonexistence inherent in discontinuity with mother.

It is a highly unpleasant but not catastrophic state, self-limiting in its duration, dangerous only if the therapist becomes panicked or apathetic, and hence unavailable himself.

The fact of the intense avoidance of intimacy is the core of resistance and must be a central focus of therapy. The patient pleads his wish to change his motivations as in itself constituting change. The conscious wish to find a route to closeness thus becomes a rationalization for the avoidance of it. The therapist must also focus on the individual's characteristic machinery for the destruction of useful experience.

B) THERAPEUTIC DIFFICULTIES

Mother and infant interact emotionally. Here emotions are organized which remain as the core of the adult's attitudes in any context involving intimacy with another. The negative experiences include the intensity, pervasiveness and patterning of the interactive rage, the patterns of aggression, wished or expressed, and of its counterpart, of fear and terror, the mutual evocation of disgust, repulsion and mutual revulsion toward self and the other. Or there may be a blanking out of emotional awareness in the presence of the detachment due to the mother's blocked emotional awareness, so that the mutual experience, the secret bond, is a mutual deadness, a conspiracy not to feel in feelingful situations.

Overt and covert brutality in the mother's style of denial of her tender impulses, of the implementation of rageful ones, establish in the infant appropriate supplications and counter moves.

The quality or absence of mother's response to the intensity of the infant's longing for aliveness, response, food, contact, peace, joy—what is evoked in her by the overwhelming urgency and immediacy of experiencing the need to satisfy biological needs in the timeless world of the newborn—is unique to each mother. The patterning includes such vectors as the mother's tolerance of calm or of strong emotion; her alertness to or denial of genuine emergency; her flatness or sensitivity, alertness or distraction; her relatedness or self-preoccupation, guilt, anger, joy, frustration; self-aggrandizement, self-derogation; feelings of desperation or of adequacy; envy or pleasure in the giving of satisfaction; stinginess or generosity.

The newborn is a small raft on an immense and stormy emotional sea. He can only sink or float and each need is crucial to survival. He has little influence over the direction of the tides. The first reaction is to be swept with them, to respond to rage with terror, to apathy with demand, to stinginess with intensified longing. For purposes of survival, some restraint of these reciprocal responses must begin to be organized immediately.

The demand for satisfaction of needs in the newborn becomes organized into a more pervasive flatness fused with the infantile substitution of apathy for unmanageable terror. The rage at frustration of crucial needs is transmuted into chronic whining, the unsatisfied longing into a quieter despair. Already, the core feelings of the central paranoia have begun to become obscured. As the complexity of consciousness evolves, the infant begins to show more differentiated responses to both areas clear in the mother's awareness as well as to her dissociated areas. Areas in clear awareness, flowing freely, can be integrated with the mother openly. The infant must also orient to the attitudes dominating the mother's central paranoia, however much his response may evoke and clarify them, and hence increase her tension, fear, suspiciousness, revenge, apathy, or despair. With great urgency, but under the handicap of the difficulties of learning in the presence of anxiety, he struggles to learn to avoid provoking response in areas which are strongly dissociated in the mother. Evoking the dissociated in her can precipitate the blank disconnectedness of catatonic panic, which to the infant means total disconnectedness. The search for unhampered and extended relatedness hence may lead to the threat of total cessation of all relatedness.

From infancy on a very large part of the energy of the self-system is devoted to the preservation of these original patterns of emotional integration and the obscuring of their nature from oneself and others. They must be obscured to avoid the mother's resentment at being confronted with her own bitter assumptions about life: her bullying, her fear, or her disinterest. As the child's picture broadens, he gains an additional motive for obscuring—his chronic antagonism, vengeance, or confusion could evoke retaliation from his contemporaries. At the same time, he is compelled to organize his life to perpetuate and to express his own inner emotional reality.

The central paranoia disguises itself by the transfer of blame, usually by blaming the other for one's own limitations in being a friend. However, the diversionary focus of blame can be either onto the other or onto oneself. Either whitewashing or blackwashing oneself in the particular interaction serves the same purpose, to obscure the smooth cooperation for staying apart.

In therapy the mechanisms for the preoccupation with and misplacement of blame must be identified.

The person who is chronically angry feels himself to be a murderer at heart and expects that he will destroy anyone of value to him. He assumes that the other—stranger, acquaintance or partner —will be either frightened or angry if he discovers this attitude, and he hence becomes evasive and distant, or hypocritical and controlling. He repeatedly subtly provokes the other to anger, with the result that he can simultaneously express some of his own feelings and blame them on the other's behavior.

This implied accusation toward the other is unique to each individual. Examples include: "You find me repulsive"; "You, like all people, are sadistic—and my concern is only whether or not you are in a position to vent it on me"; "No one means a thing to you"; "You chronically view people with contempt"; "People disgust you." Each such broad, pervasive expectation of human nature from the other, parallels an equally powerful conviction about oneself, i.e. "I am: dead, fickle, revengeful, helpless, hopeless, defenseless, dangerous, boring, or bored."

Such persistent convictions, however superficially controlled by elaborately studied social roles, inevitably cause an immense amount of static in the process of developing either communication or affection between two people. Thus, partly because of the difficulty of doing anything else, and partly because he operates with most conviction on his own emotional wave length, each person integrates in those situations to which the pervasive emotion is appropriate. Such a partner evokes in him the old infantile longings which are the emotional fountain for the attempt to seek out the other. The difficulty about doing anything else is rooted in the archaic expectation that the free flow of alternate emotional experience would result in the blanking out in which all relatedness would disappear.

One reason why the patient avoids clarifying his central paranoia is that he is afraid he will then control his hostility even less effectively and become even less tolerable to himself or the other. His fear of new experience is cast in the style of his particular security operations, but this casting is diversionary. Centrally, it is the expectation of death and disintegration: of total loss of consciousness and contact, of being paralyzed and unreachable, involved in moving into emotions which would have been entirely unresponded to by mother.

He fears it will throw sudden, clarifying light on his whole way of life—on the multitude of maneuvers devoted to the avoidance of experiencing what could go on, and the restriction of development of spontaneous interaction. If unimplemented by the paranoid interaction, any direct human contact is enlivening to both parties. Then to believe in the validity of the experience would result in immediate re-evaluation of all past interpersonal experience and rewriting of all current operations. No human interaction is entirely devoid of contamination by the private paranoid assumptions. This implied cataclysmic reorganization adds to the archaic expectation of nonexistence in relation to mother.

Another danger lies in an overview of the length of time such reorganization will take if undertaken. Each atom of interaction must be identified, studied and re-experienced—feelings with strangers, acquaintances, friends, lovers, men, women, children, employers, employees, students, teachers, cab drivers, cops—by individuals and by categories, to try to identify the distortion in the interaction which operates to perpetuate the system. "How do I make women be hateful to me so that I can continue to hate them?" "How do I keep people from listening to me so that I can continue to believe they are uninterested?" "What irrelevant issues do I introduce to obscure the friendly interaction?"

The therapist's formulation might be: "Since you see all people as bullies, you think that anyone bigger than you is fair game; and since you see yourself as small and ineffectual, this includes practically all the human race, possibly excepting sick children and frankly psychotic adults."

"How do you guarantee the other will resent you, so that you can continue hating him?"

"You hallucinate the other's anger, so that you can justify your own and then provoke his."

Analysis must be directed toward disrupting the smoothness of these cooperative enterprises.

Such elementary re-education in the alphabet of direct experience can seem overwhelming in its urgency and massiveness. It is, in fact, the normal process of aliveness. Children do not find new experiences, or the expansion of old experiences, dull or burdensome. The effort that it evokes in the adult is an expression of his attempt to control his fear of the unknown.

One of the main tasks of therapy is reorganization of this perceptual pattern. It must be exposed, elaborated, and clarified repeatedly with the patient. The basically antisocial attitudes must become a part of common-sense conversation between patient and analyst, without the kind of sympathy or distaste that would divert the focus from an enthusiastic study of interactions which perpetuate these attitudes. The specificity of the formulation of each individual's central paranoia demands a high level of literary and artistic creativeness on the part of both. As this is attempted, each patient will accuse and reproach the therapist, in terms of just those private clichés which are under scrutiny. The patient intensely experiences the therapist as indifferent, cruel, angry, or betraying. At the same time, any focused attempt on the therapist's part to directly repudiate these accusations is one mode of relating to them and runs the risk of reinforcing them.

The intensity of the patient's resistance comes out of the fact that the main fabric of his life has been so woven out of such concepts that he can conceive of no alternate way of life. In this connection, the patient's vague notion that others live differently carries no inner conviction for him, however he may claim to believe it. Rather, he uses it as an additional subjective, isolating factor, failing to find companionship even in the universality of the subjective dilemma. And his perpetual search for the other with no hostility and no withdrawals leads, repeatedly, to again experiencing each reaching out as a betrayal.

The patient experiences both a great urgency: "I must stop this lifelong operation between today and tomorrow," and an overwhelming lethargy of postponement of change to the perpetual tomorrow. Change is not related to as a common-sense work process, requiring study, observation, experimentation, and consolidation. Human interaction is diverted to the self-absorption of presenting a mirror of one's progress—or lack thereof—with great emphasis on the optimism or despair resultant. Each event, as with the infant, becomes timeless and eternal, thereby obscuring its place in the progression of experience as directed toward reorganizing experience.

Stating, sharing, and validation of the lifelong despair about change is absolutely essential to change. At the same time, this too needs to get into perspective as only one of the underpinnings which perpetuate the paranoia. It tends to be intensified as the patient begins to suspect that change is possible, but that it is a long-range project involving a great deal of hard, inelegant, unglamorous work.

In addition to a frank and relatively cheerful scientific study of his patterns of self-inflicted injury, the patient needs to grasp the manipulative nature of the paranoid emotions. He feels them with intense and overwhelming conviction, a conviction based in its appropriateness in infancy. His persistent demand of the world has been that people have an intense preoccupation with the original conditions of his inner life. He pushes that the experience with the therapist must not only help toward change of experience, but must be in terms of the original trauma. As the therapist tries to introduce different values, the patient feels misunderstood and again perpetuates his isolation.

Memory traces of early infancy do not appear in the adult conscious mind in the form of pictorial memories. They appear as direct sensations and emotional coloration in situations which tend to "remind" us of the situation of the early infant. One such situation is the attempt to relate to a person who has certain characteristics similar to the mother as she was during the first year of her child's life. The direct physical skin-to-skin contact of the sexual act also re-enacts in certain dimensions the physical and emotional interaction between early infant and mother.

The dreams of deep sleep often restate such mood memories in symbolism taken from later experiences. Likewise, the moods of falling asleep or of slow awakening can represent such reliving, with or without more contemporary symbolic content. The infancy mood memory quality of many paranoid perceptions can be identified by their relative inappropriateness to the situation in which they occur.

A young man suddenly experiences the girl in his arms as a piece of clay. So his mother related to him as infant. As the patient puts it: "Holding her—feeling her to be a piece of clay or putty— arms holding a baby. Mother holding me—what she felt when she held me."

Many irritable and frustrated moods, apparently directed toward spoiling the fun for oneself and the other of an otherwise pleasant occasion, restate the original mutual irritation and frustration between mother and infant.

For emotions to get reorganized, they must be recognized as hallucinatory. This happens first in retrospect. The emotion of the night before is experienced as inappropriate to the situation in which it took place. This process tends to arouse the archaic fears of insanity and disorientation and, again, of ultimate isolation.

An important focus in this reorganization is the playing down of the drama of it and the organization of relearning into a prosaic work project. People feel some things which are not there, and do not feel others that are. This insight is a necessary preliminary to change. The re-evaluation of feeling, then, requires the quieter atmosphere of sensitive introspection.

One must temporarily relinquish his attempt to control either his own or the other's inner thoughts, in order to perceive, understand, and re-evaluate them.

To the extent that the self-system is dominated by the paranoid assumptions, the content of the stream of consciousness tends to become organized in the unrealistic polarity of low self-esteem and its grandiose reformulation. Feelings in relation to oneself and others are organized on the axis of attraction-repulsion or destructive-constructive. All experience is siphoned off into these dimensions, thereby constantly reiterating the notion of their relevance. The sense of self, one's needs and one's expansion are expended in this perpetual search. The erasure of the original assaults and indignities to one's self-esteem become the lifework experienced as necessary preconditions to the pursuit of satisfaction.

The other person becomes the mirror to this process. He is perceived in the axis of the Good or Bad Mother, and maneuvers are instigated to pull him into this value system. He seems to become the total repository of one's self-esteem.

In this undertaking, the more interaction develops between two people the stronger the tendency for each to slough off the superficial compensatory functions and to experience directly the negative emotions which constitute the core of the central paranoia—terror, rage, despair, and disgust. Hence the more frantic the attempt becomes to maneuver the partner into both validating and erasing them. The analyst becomes the focus of a demand that he relate in these dimensions rather than to analyze them—that, in an alternating current, he both validate the inevitability of the despair and reassure one that he is not whom mother thought. The patient demands validation of his paranoid perceptions partly as reassurance that his direct perceptions are not crucially distorted.

The impact on the partner is of an exceedingly self-centered operation. All interaction tends to be reduced to its utility in relation to one's own and the other's opinion of oneself, and hence to introduce a degree of unrelatedness by the relative fogging-out of the specificity of the interaction in its own right.

It is necessary to get in touch with the emotions of the central paranoia and their implementation in current interpersonal relations, and to get some degree of acceptance that the resultant antisocial behavior is in the range of human behavior. At the same time it needs to be perceived as self-defeating in the direct sense of evoking hostility and defeating relatedness, and in the way in which it perpetually reaffirms the inevitability and immutability of one's own unsuitability for human interaction.

There is also a constructive aspect to the push to get the accumulated negative and anti-social thoughts and feelings into the open with all comers. This is directed immediately to the necessity to drop off the strain of perpetual play-acting and denial. It involves the wish for acceptance and understanding. It also, in parallel, tries to evoke the opposing demand for punishment and retaliation, so that the inner feelings will be confirmed as impossible of understanding, and the appalled audience will cooperate in the work of maintaining dissociation. Reorganization of the central paranoia involves two interactive processes. One is that it get fully into awareness as day-to-day, obvious, common-sense knowledge about oneself. The other is the searching out of situations which could evoke a different pattern of experience. The latter process immediately evokes operations to reinstitute the paranoid interpretation. The process of rewriting the history of the new event in the old terms takes over.

The experience of intimacy evokes empathy and more self-awareness and thus constitutes an alternate to the restrictive childhood experience. One of the immediate complications in this evolution of new experience is that the first emotions of which one becomes aware are usually those related to the chronic paranoia. The fear or anger or emptiness of which one is chronically unaware are the first feelings to emerge. Therefore it is necessary for these feelings to be solidly in awareness before anything more human can be noted.

In the course of therapy, as some awarenes is approached, there tends to be an exacerbation of malevolent patterns. As with all acting-out this flowering is based on unclarity about its nature. The more clearly the emotion can be experienced, the less is the necessity for dramatization.

The therapist must be alert to the varied symbolic screens behind which growth presses against the restrictive barriers. The experience of sex is one of these. Preoccupation with rage, fear, vindictive revenge, terror, betrayal, and disappointment is an expression of the central paranoia. It focuses on destroying the experience of the exchange of intimacy. Intimacy implies that one is available without reservation to respond to the other person, and the other person is available to respond to the response, without any great overlay of past, present, or future fears. In this situation either's needs are part of the context to which the other one, without reservation, responds. The exchange of tenderness, then, is not only an intent of the move towards growth. It is also that aspect of the interchange between humans necessary for growth.

The emphasis here is on the word exchange. One of the more frequent mechanisms for erasing intimacy lies either in the demand that it should all go towards oneself, or that it should all go away from oneself. People have, in that sense, two formulae. One is, "If I have to give anything, this invalidates the interaction." The other is, "I can feel tender if he is out of the room; it is at the moment that I get a response to my tenderness that I blank out." Both are equally well designed to destroy the central experience of the natural surges in any accumulated growing dynamic interchange.

The experience of sexual interchange frequently becomes a symbol of the quality of the exchange of tenderness. A common example of this is the person who expects that the partner should enjoy sex only with him. That is, if she does not enjoy sex at all it is the proof of her disloyalty, and if she enjoys sex wherever she gets into a sexual situation this is an even more intense and bitter proof of her disloyalty than in the first situation. The fantasy is that one can find a girl free for physical and emotional response with him alone, who learns nothing from this experience that would make it possible for her to be responsive with another person.

The experience which is persistently either avoided or derogated is simple exchange and interaction using known techniques of relating. It is avoided because, if it were validated, it would inevitably implement a more accurate and constructive choice of partners for a wide range of intimacy including the sexual.

Stated another way, the phenomenon of amputated and stereotyped sexual experience in our culture is almost universal. The emphasis for the man is put on the definition that potency is the capacity to respond sexually to any woman with whom he finds himself in bed. The more restrictive standard for the woman is that she is expected to be able to evoke potency in any man that she happens to find herself in bed with, whether or not she herself would expect to have an orgasm. This cultural symbolism expresses the pervasive discouragement about the exchange of tenderness and, at a simpler level, about the empathic interaction of physiologic response.

The experience of sex can evoke any of a variety of emotions available to either person, including such feelings as anger, fear, depression, discouragement. To the extent that sex becomes the symbol of tenderness, therefore the symbol of the original situation which set up the fear of tenderness, it must involve the re-enactment of the original trauma. But strands of the symbolism become obscure. Because of the great cultural investment in the concept of sexual adequacy, the symbolism obscures and frustrates the reality. On the other hand, the dynamic of sex is covertly used to attempt to get important dissociated elements of feeling into awareness. But as the drama is enacted it is difficult to get these elements identified and clearly experienced. In the constructive sense each element left out must involve a denial of lasting experience.

Infancy is the last time people were frankly undressed, open, available, unprotected and with the unconflicted wish for un-hampered physiological interchange with another person. Thus the sexual situation, if fully integrated, could involve the temporary re-experiencing of the most devastating expectations of the inte-grations with mother: the expectations of mother's suicidal and depressed moods, for example, as in the insidious postpartum depression; the expectation of mother's catatonia and withdrawal; the expectation of mother's getting in touch with her own pervasive —and, at that point, irrepressible—fear of aliveness; and, perhaps most powerfully, the expectation of mother's open or hidden moments of destructive rage.

The simplest formulation of many people's behavior in the sexual encounter would be, "Don't hit me."

C) Security operations

Therapy proceeds against layers of resistance. These consist of different patternings of the elements of the security apparatus of the self-system. It is necessary for understanding how the character structure works to understand the primary mechanisms for maintaining dissociation.

A security operation is an active maneuver of the self-system to re-establish the previous equilibrium under the impact of a new experience. It must re-establish, symbolically or in reality, the connectedness with mother, which has been threatened by the impending expansion. It must limit awareness; it must impose some pattern on the external situation; it must protect the restrictiveness of the self-system and express in some way the specific nature of the dissociated breakthrough.

The nature of the symptom is determined by the particular interpersonal situation involved and will represent these four elements in varying degrees.

Security operations have usually been classified in terms of the mode of communication, perhaps because a security operation brought into play in the course of communication between patient and analyst is one of the basic modes of resistance. Thus analysts must be particularly concerned with it. According to this classification, they include selective inattention and psychosomatic phenomena, hysteric, compulsive, and obsessive. Operations that serve exactly identical purposes may involve equally interpersonal maneuvers, momentary or durable, which utilize any or all modes of communication in a versatile way.

The acute paranoid episode accomplishes both the inner repudiation of the data and the physical driving away of the person who set off the trouble. As the patient's paranoid organization becomes more entrenched, he relies more consistently on the malevolent transformation, derogatory processes, projection, and grandiosity for both maintenance and repair of the fortifications. The psychopath takes this one step further, and tries to destroy the person, as well as the experience, if he is challenged by someone to come to life. The graying-out of experience in a pervasive underlying depression with its occasional periods of exacerbation serves the same function as do the more acute episodes of despair and suicide.

This can be acute, temporary maneuvers set off by a particular incident, or can be incorporated into the whole fabric of living, for a lifetime organized to avoid exposure to certain types of growth experience.

There is a continuum from the frank catatonic breakthrough to the smooth social maneuver that avoids a particular interpersonal experience. There is also variation in the effectiveness in avoiding the insight, and in the cost of the operation, per se, to other lines of development. Some security operations involve, by their nature, an almost total preoccupation with their execution. Others may constitute relatively little hindrance except in the specific areas to which they are directed. Some allow for more experience to be accumulated on the side than others.

The limit to flexibility of choice of repressive maneuvers lies in what might be effective in the immediate social situation, in the level of original development of talents in organization and communication, and in the degree of regression.

Since the security operation gets into action under the threat of anxiety, some degree of regression is always involved. One is never quite at his most efficient, though his dedication to the project may compensate to some extent for his shakiness. Because of the regressive aspects, the more intense the anxiety, the more primitive are the security operations that are thrown into the battle to hold the line.

The choice of security operation always has some symbolic reference to the situation in which the particular line of development about to break through was blocked. Sometimes they can be interpreted as if they were dream symbols. Often they represent the reliving and incorporation of an explicit prohibitive maneuver by the parent. They also include elements of the push for growth that they are designed to interfere with.

They interfere with communication, postpone insight, and obviate the necessity for acting on it. It is necessary to identify them and how they operate, and to help the patient understand what their impact is on the environment. Their interpersonal nature is not usually clear in the patient's awareness when he is doing them. He provokes attitudes of the people around him which reinforce his picture of the world without knowing what he did to provoke them. When the patient can experience his provocations as self-defeating moves, not as primary needs, he can attend to more profitable concerns.

The central paranoia is an integration of the denial of fear, the repudiation of the need for tenderness, and the avoidance of disorganization. Misperceptions and misconceptions about oneself and the world, including the self-image, are embodied in these maneuvers.

The preoccupation with overcoming the fear of death expresses itself in avoidance of danger or in seeking it out. Either contributes to obscuring the intensity of the fear and its true origins, by necessitating the perpetual postponement of really experiencing it. The most usual maneuver for suppressing fear is for the patient to make a hostile integration with someone whose inner impulse is to do him in. This may be done by maintaining real or fantasy contact with the original, more dangerous parent, with an adult partner who is similar, or with a rotating series of "promiscuous" contacts, similarly selected, bolstered by the fantasy of remaining essentially unattached. In the presence of danger one suppresses a large part of the fear.

When there is a temporary separation, the suppression is weakened and the fear tends to come through. This is one of the frequent causes of the inability to be physically alone. Even when there is an apparent discontinuing of a hostile integration, there is often a clinging, at least in fantasy, to some sort of continuity. If one accepts the discontinuity, the whole machinery for suppressing the anger in the relationship is disconnected. At this moment the infantile fear comes through.

Thus the hostile integration, though partly expressing and perpetuating the underlying emotion of fear, also obscures it. The hostile integration can be used in retrospect to identify the original childhood fear and ultimately to re-experience the contemporary situation as an inappropriate reminder of it.

The therapeutic maneuver is to push for a temporary discontinuity in the pattern. At this moment, with support and observation, it may be possible to identify the fear of retaliation, which exposes the core of the relationship. This is, then, a new level of insight for reconstructing the original terrors and sorting out their sources and their adult residuals.

With respect to the repudiation of the need for tenderness, the patient is often preoccupied with trying to make up the tenderness which was missed in infancy in the mode in which it should originally have been experienced. However, what facilitates change is the patient's re-experiencing the grief of the infant at the lack of tenderness in the mothering one, not in returning to infantile modes of interaction. Some adult patients search for the adult mothering one who is subject to the pathology of wishing to adopt and raise a thirty-year-old. Each time this Good Mother is found, it is discovered with great surprise that the adopting one had his own self-centered and neurotic investment in the adoption. An alternative is to use the excuse that one's own deprivations were so great, one's needs so insatiable, that there is no use in ever getting started on the project.

Both styles are resistance. The way to make up a deficit is to satisfy it. Mutual sharing can implement a rather rapid making up of considerable deficits. There is as much affection available in the world as anyone has the capacity to integrate with at any particular time, and it is just as possible to progressively expand as to progressively restrict one's capacity for integrating with it. Every interaction of tenderness makes the next experience more accessible, fuller and deeper. Either way, toward expansion or toward restriction, the effects are cumulative.

The dissociation of spontaneity in the first few years involves the delineation of the plausible directions of growth. This is formulated in the self-image. When this image of oneself can be gotten into clear focus and seen in its historical context as the natural and inevitable result of the pressures and prohibitions of infancy, rather than as a reflection of the real subsequent development of actual potentialities, then it begins to reformulate.

One difficulty in undoing the fear of disorganization lies in the fact that the exacerbation of the fear of the threatening one is synchronized with the temporary disorganization of shifting reference points. The machinery for maintaining distance, for organizing perception, and for coping with people perceived as authorities and with contemporaries is temporarily out of commission. The person feels as much at the mercy of the surrounding world as any two-year-old actually is. At the same time, he expects authoritarian retaliation for the move, insight, or growth experience which set off the anxiety.

The patient needs to try to organize his life so as to avoid the old experiences and to expose himself to new ones. Without the comparison with more hopeful patterns of relatedness, no progress can be hoped for. But external reorganization is rarely sufficient.

None of this happens in one great reorganization. Analysis is the process of gradually developing enough of a base in the real contemporary world to gradually relinquish the fallacies of the old one. Each episode of integration is a crisis involving disorganization, loneliness, and the fear of destruction. After each crisis, these feelings fade out of the foreground, and the patient prepares for the next effort.

Full awareness of the emotions of the central paranoia would have been too dangerous for the infant in relation to mother when angry. It is only when the adult is in a position to totally relinquish, if indicated, this original connectedness that it is possible for him to thoroughly re-evaluate them.

D) ANGER AND DESPAIR

Anger has been said to be: a) the response that occurs when someone is not treated with the deference to which he would like to be accustomed; b) the normal response to frustration; c) the normal response to helplessness in the face of sadism and cruelty. In the context of interpersonal theory, anger can best be understood as one of the disguises of anxiety. The presence of anger frequently identifies a situation in which there is a move toward expansion of the personality against resistance. The anger is part of the resistance, part of the paranoid episode implementing the trend towards the paranoid integration, the intent of which is to repudiate the move toward expansion.

Anger is always partly an expression of the problem of authority, of the problem of power, and of being in the power of someone. Anger in the adult world depends on the momentary reliving of the childhood situation of helplessness, entrapment, and lack of choice, in which the other person seems to have the power of life and death. Such inequality between individuals exists only in the relationship of parent and child. Neither parent nor infant can choose whether this is the optimal integration or whether one could find a different partner who would work better. Also concerning deference, the only situation in which it is reasonable for one individual to demand that the other be unilaterally preoccupied with mind reading, adaptation, and responsiveness, is in the relationship between infant and mother.

Thus adult anger is the resultant of an internally determined misperception of the actual adult interaction. It often bursts into awareness at the moment of relinquishing the misperception, and is then focused on the experience of having been taken in. If one had not participated, for his own purposes, in the antecedent conspiracy of overlooking manifest pathology, his experience would be not anger, but relief that a confusing field of interaction had finally been clarified.

Another frequent situation in which anger can seem justified and appropriate in the adult is when he is the object or the captive audience for gross brutality. Here, .again, the trigger is often the neurotic naïveté and wishful fantasy about the world, which predicated the element of surprise in the experience. If the same psychotic, rageful bully were, in fact, a patient in a mental hospital, most people would not experience rage in dealing with him, but rather a resigned prospectus on the amount of work it would take to help him reorganize his reactions.

Another aspect of adult rage is that there is always a manipulative component. Its impact is to intimidate the immediate audience, whether or not this audience is presumably the object of the rage. It thus acts to entrench the central paranoia of the auditor, and to restrict his capacity for creative analysis of the problem at hand.

Rage tends to be presented in socially acceptable formulae. In a culture in which sexual jealousy is socially expected, marital quarrels focus around this issue. If psychological betrayal, or psychopathic lying, is considered to be more of an assault, the partner's neurotic evasiveness and hypocrisy is the more central focus. This legalistic casting of the choice of issues is determined partly by the wish to get real or fantasied public validation for the emotion. More centrally, it plays on the other's conventionality and acceptance of cultural norms of decent behavior. To the extent that the other accepts the prohibitive conventions, he adopts the accusation as his own and adds his own exacerbated self-revulsion to the intensified intimidation, thus reinforcing his own central paranoia. This is the mechanism by which rage works as a manipulative device. Its intent may range from simple diversion from the immediate issues, to organized destruction directed at its object.

There are emergencies in which the conscious manipulative intimidation of the other to perform constructive acts is useful. However, this technique must be weighed against the undertow that the method used, in itself, contributes to the pathology in the other. It is to be noted also, that there are times when the open expression of lifelong controlled rage has a constructive aspect, since it is a pre-condition for the clarification of its infantile origins and of its contemporary distortions of perception.

People sometimes have the subjective experience that they can move into constructive action in a danger situation only on a burst of rage. From this they conclude that the rage is the motor for action. The reality is more complex. The perceived danger, real or distorted, may have evoked parataxic infantile terror which tends to be paralyzing: the rage, by inhibiting the awareness of the terror allows for the mobilizing of action. In fact, in most action situations calling for rapid exercise of complex intuitive judgment, both the experienced terror and rage narrow and limit productive functioning.

There are certain pseudo-dilemmas in the therapeutic management of anger. The progress of closeness between two people risks the exacerbation of the latent anger in each. As the closeness tends to force this latent anger into the open, it may initially act to reinforce the machinery for control of the anger. This increase in control reduces, in its immediate effect, the horizons of self-awareness and the capacity for free interaction with people. This dynamism is the basis for the subjective conviction that staying away from people is a suitable route for ultimately getting closer to them.

Anger is the most efficient medium for maintaining a fixed distance between two people. The validation of a strong emotion tends to evoke rationalizations to justify it. It also involves considerably more freedom in its expression in interpersonal interaction. It involves at least a temporary transfer from the symbols of the past to the individuals in contemporary integrations. In effect, the patient who is in the process of clarifying and validating his own anger is subject to great irritability and poor "control." His preoccupation is that, in the process of trying to be himself, he will either drive away all the people with whom he is somewhat related, or, if not that, will at least lose any chance to win their love.

Anger is not an inherent part of the integral personality, of that part of the person which is organized to reach out for more love and greater satisfactions. It lies at the fulcrum of the self-system and, overt or covert, it must have some impress on all of one's actions. To the extent that it is brought clearly into awareness, the pressure for acting it out overtly decreases.

Thus, even in the moment of transition, the more the underlying chronic anger gets into awareness the less it is a disjunctive force. It must become conscious before one can re-experience the source and disentangle it from current integrations. The acute exacerbation of explosive temper and manifest anger is often a transitional phase.

The patient tries to avoid noticing the degree to which his self-system, his manifest life, is organized around anger at the mother, at the other partner in the primary integrations, and in casual contacts. Avoiding open explosions is also partly in the service of the old power fight. Open anger may put one at something of a disadvantage to one's better controlled opponent.

Chronic anger can be expressed by provoking anger in the other. The one who is provocative can experience a superior calm, or be the innocent victim and thus avoid entirely experiencing his own anger.

Therapists, often timid people, sometimes complicate the therapeutic management of anger. Consciously or unconsciously, they are tempted to maneuver the patient in order to avoid or tone down the full expression of the patient's anger. They are overly quick to interpret, and thereby promote reinstituting control in terms of the irrationality of the emotion, cutting short its full exploration.

The subjective component of rage at oneself includes self-disgust and self-castigation. A detailed analysis of the content shows it to constitute the anticipatory reliving of the expectation of the parents' anger and contempt.

Feelings of being angry with oneself are similar to guilt feelings, in that both are defensive, obsessional preoccupations that derive their force from the residue of self-hatred that carries over from childhood. Self-disgust, self-blame and dyseuphoria are prominent, and are essentially insincere insofar as they are righteously passed off as genuinely directed toward a productive solution. If rooted in self-love, the emotion would not be anger, but a taking stock of the situation accompanied by a preoccupation with the means of righting this attitude in the next similar interaction. Insofar as there is rage with oneself, rather than genuine resolution to take responsibility for one's own actions and life, the rage is always irrational and incorporated parataxically from the eidetic parental authority.

People get into analysis because of the struggle between hope and despair.

Hope represents the integral personality. It is the inner knowledge that one could be more than he is, that life does not have to be as bitter as it has been. Hope is the reflection of the impact of love and freedom on the forces of life in the individual. It is the response to unexpected but convincing validation of unpredicted performance beyond the confines of the self-system.

Despair is the watchdog of the self-system. It comes in a variety of roles—inner, secret convictions that any profound success, creativeness, love, or ecstasy would in some mysterious way prove fatal. The drama of life is played out for the avoidance of intense satisfaction, which avoidance in itself provides sufficient reason for the pull toward death fantasies.

The individual becomes alive, both painfully and joyously, only when experience is intense. It is at the faint twilight glow of hopefulness that the pull toward nonexistence is most intense.

Here again, it is couched subjectively in terms of futility, of fear of losing hope—but it expresses directly the inner compulsion to destroy those sprouts of hope which are beginning to emerge. It is the reflection of the awareness that the only way to destroy the wish to be fully alive is to die.

Despair equally is the active agent in both acute psychosis and culturally acceptable resignation. In psychosis the dramatic repudiation of interpersonal integration, of simple, common-sense humanness, expresses, like the acute suicidal pull, little hope and much skepticism. It incorporates an appeal, however unheard, for rescue.

Resignation lies closer to full acceptance of the empty existence. These patients include the ones who admit to no significant problems, the ones who are always just about to get them worked through, the ones whose problems do not bother them.

The therapeutic management of despair is an analytic continuum. Every therapeutic maneuver must decrease to some degree the totality of the patient's despair about himself and his potential satisfactions. This undercutting of despair operates centrally in the initial interview, but rarely in explicit formulation.

At various later points in therapy, validation of feelings of hopelessness is crucial to change. A danger is that, in the moment of validating, these feelings may strike the patient as eternal truths. Bringing into awareness the patient's obscure lifelong plan to die, rather than to come to life, should be postponed until he can be expected reasonably to use the insight for growth.

With the more chronic form of despair, the difficulty is in getting the desperation into awareness. The patient's life operation is of a wistful sort of cheerfulness. The experience of despair is too intense for him. The therapeutic maneuver here is to move the patient into more satisfying experiences, at which point the lifelong depression will become an active, malevolent force tending to push him back into the restrictive mold. At this point the depression can be interpreted. The interpretation includes:

1) The lifelong accumulation of failures and discouragement and the chronic lack of satisfaction.

2) Conscious despair as a security operation expressed in obsessive preoccupation, or discouraged choices of central or peripheral

relationships, for the purpose of maintaining control, limiting intense positive experiences, ruling out ecstasy and destroying creativeness. In this context, the exacerbation at the moment of growth is dramatic and demonstrable.

3) The pull to the compliance with the death wishes of the reluctant mother. The need to give mother what she needed—a dead hero instead of an alive and embarrassing offspring. This has been symbolized in patients' dreams and psychoanalysts' imaginations as the return to the womb, the pull to the mother sea. In more prosaic terms, it is the wish to give up the battle, to win final peace by total capitulation. It lies in the repetitive experience that each minute satisfaction is at the cost of so much terror that it is in the end not worth the candle. "She would rather see me dead than free, happy, loved, or independent, and she has won! I give up." The giving up may be in the fantasy of returning to the newborn status (the dream of being blind, paralyzed, totally invalided), and so forcing her succor. Or it may be in an even more total defeat—the repudiation of the struggle for physical survival.

The impulse to suicide has also been described as repudiation of unconscious death wishes against the mother. This could be restated: The impulse to suicide contains the unconscious push to destroy the rage which the patient experiences as incomprehensible because he cannot clarify its sources, and to accuse the mother of hers.

4) The preoccupation with hopelessness lies in the attempt to accustom oneself to the fear of death. In the frankly psychotic, the same fantasy is played out in the dream of death and rebirth. "It is only by experiencing death that I can be free of the miserable, lifelong fear of it that paralyzes me in every defiance of the maternal restrictions." A serious impulse to self-destructiveness is in the context of a reproach to the surrounding interpersonal climate. It is usually constructed by the patient as the unavailability of the minimal degree of tenderness necessary for life. If the patient experiences love for, and from, another human being, suicide is unlikely. In the moment of crisis, presence of friends or even acquaintances can be important.

The central therapeutic operation lies in the formulation of the truth, however painful, of what is going on in the patient's life. It lies in the demonstration, not the exhortation, that change is possible, in the context of this particular patient's personal dilemmas, and in terms he can use practically.

Disciplined dissection of an event to clarify the elements involved in it is always more reassuring than reassurance. The truth may be less than the patient's hopes, but it is never worse than his worst fears. Even erroneously unflattering formulations, if they are the therapist's sincere and considered opinion based on the available evidence, are a constructive contribution to the project of clarifying where the trouble lies. They have at least been made available for mutual correction.

VI

GOALS AND LIMITATIONS

PEOPLE GET INTO analysis with various complaints. Some get there because of the unpleasantness of the experience of acute anxiety. Others have become dissatisfied with, and embarrassed about, the futility and discomfort of their pervasive security operations for the avoidance of anxiety. These may include self-terrorizing patterns, suicidal preoccupations, obsessive operations and repetition neuroses. These consume a tremendous amount of energy without producing satisfaction. However, the patient has always considered that his security operations are less painful than the anxiety or fear that they are designed to avoid.

Only if the patient has the expectation, however obscure to him, that the ultimate satisfaction might make it worthwhile to experience the anxiety, is he willing to reorganize seriously his defensive patterns.

Seducing a person into therapy, if he is not genuinely motivated, will come to a discouraged end if the effort is based solely on either his social embarrassment or his friends' dissatisfaction with his neurotic patterns. At the end of the effort, the patient will have proven that therapy also can be of no help to him. This experience of failure can act as a hindrance to the later use of therapy.

Sometimes the friend perceives the inner change slightly before the patient. The patient may maneuver the situation to allow himself to be pushed into getting help, as is sometimes so with the patient who either gets himself hospitalized or gets himself in trouble with the police. In this situation, the therapist may be able to make contact with the patient's positive motivation.

A patient may come into analysis describing a general feeling of emptiness, meaninglessness, and futility. His organization of living provides insufficient basic satisfactions.

If the patient is there primarily because his patterns of living are a nuisance to the people around him, this is an inadequate reason for change unless the fact of being a nuisance is of serious concern to him personally. The patient has an internal and effective motive for change when he has some conviction that he could get more satisfaction out of life. In that event, he gets into analysis because, however unnoticed, he has had some good experiences which either have thrown into relief the futility of his previous investment or have caused him acute anxiety. This anxiety then provides both the excuse and the urgency for getting help. The patient must also have reason to think that analysis might help, either intellectually or by example.

The integration with the therapist is a new and stimulating growth experience in itself, but the dynamic of therapy does not lie in the talents of the therapist. Progressive clarification may augment moves in the patient toward life: new experiences, new emotions, and new insights. But if the initiative remains with the therapist, the process will peter out. Some patients come for therapy dutifully for years, cherishing the fantasy that something can come of therapy on the therapist's motivation.

Unless the patient feels that life without love is not worth living, he will not subject himself to the momentary fear of death that is involved in serious reorganization. In reality, life without love is often prematurely fatal; the patient is subject either to psychosomatic decay or to apparently accidental death. His investment in life is insufficient to maintain life. But if the patient cannot be brought to awareness of the suicidal pull, not much will happen in analysis.

The therapist responds to the patient's growth moves. These moves are associated with anxiety, which serves as a kind of marker. He can also help the patient to organize his life so that the patient will have some experiences with people which will evoke and bring into action a greater investment in life. The therapist can develop the relationship between them into one in which the patient's capacity for affection can be validated.

The patient needs someone who cares if he lives. This person can be the therapist, but if so, his concern with the patient has to be genuine. If his concern is with his pride in his work, the relationship cannot have this dynamic effect. Such genuine concern can only be developed in interaction with the patient's reaching out, both in being available for help and in giving to and sharing with the therapist, at least in the mutual project of understanding the patient.

Primarily the therapist must illustrate that radical change is possible and is not as hopeless as the patient has always thought. He starts with what the patient has integrated and works toward unconscious experience which is pushing for implementation. His task is the concrete demonstration, in terms of the specific current and past data of the patient's life, that therapy is an effective method for improving on the *status quo;* that knowledge, insight, and self-awareness are indispensable tools for getting more satisfaction.

According to early concepts, analysis was thought to be limited, on the basis that there were needs which could not be reintegrated because they were thought to be essentially antisocial. This assumed an acceptance of nineteenth century social conventions as embodying fixed basic reality. A later concept was that the essential human needs were valid and potentially satisfiable within a more fluid concept of social structure, but that the associated processes, the phenomena of distressing anxiety, acting-out of antisocial, disorienting or self-demeaning impulses, were all bad, that reorganization would have to be achieved without these by-products. Supplementing this discouragement was the concept that severe deficit in infantile integration of function could not be made up. This involved a somewhat rigid picture of the facilities available to the adult in areas equivalent to those experiences omitted in infancy and early childhood.

Another excuse for avoiding radical change is respect for accumulated discouragement. The patient claims that analysis is for the young and strong. It is as if he were convinced that if he has invested fifty years in a neurotic structure, he cannot write off a bad investment in order to have more satisfying experiences for the last thirty years of his life. His private rationalization is that this would involve too much loss of face.

Age, indeed, has some advantages. Analysis later in life involves some distressing insight as to alternate interpretations of past behavior which may have seemed necessary at the time, but which, in retrospect, seem to have been grossly inefficient. On the other hand, however derogated and however much out of awareness, the years must have added a considerable accumulation of experience beyond the reference points of the original childhood situation, which is available to be reintegrated into a wise and mature approach to life, if the process of reintegration is seriously undertaken.

The handicaps with increasing years lie in loss of prestige, and in the prejudices of therapeutic facilities and personnel. Loss of face and the inadequacy of social facilities for handling periods of severe disorganization may constitute practical handicaps to therapy at any age. Nonetheless, these difficulties, oriented to the primacy of the social conventions, should not be equated with inherent insolubility of psychiatric problems.

The other deficiency is essentially in the limitations of the theory and of the analyst.

The practical evaluation is whether change is probable in terms of the actual facilities available to a particular person at a particular time. Would analysis involve hospitalization? If so, would the fact of hospitalization be such a handicap to subsequent living that the patient would never get out? Would the conditions actually available for hospitalization be adequate for a resolution of the acute episode, or would this probably result in further deterioration? How long would analysis probably take? Can the patient manage to maintain it? Is there an analyst available who could and would undertake it with a reasonable prospect of success?

Both ideal and minimal goals of therapy should be differentiated from the appropriate time of termination with a particular therapist.

Termination is a practical decision influenced by time, money, and the personalities of the two people involved.

Mutual agreement that an analysis has been completed is always misleading. Under the best of circumstances, the patient may need help at a future time with a later phase of development which could not be explored under the circumstances of the original analysis. It is also not helpful to leave the patient with a vague feeling of being the property of the analyst, with the expectation that if he had received real help from a particular analyst, it would be ungrateful to work on a different aspect of the problem with a different analyst. Not all reasonably competent analysts working with reasonably sincere patients can necessarily solve all important problems. Termination should come when a particular analyst is not making a sufficiently active contribution to the solution of those of the patient's problems which are sufficiently available to analysis to justify the time, money, and effort involved in continuing at that time.

Termination should include a brief review of residual unresolved areas, relating them to the barriers from which they derive. If the patient is not prepared to reorganize his living arrangements to allow for more growth, it should be clarified, without derogation, that this is his choice at the present time. If the patient does not choose to risk major periods of anxiety, this can be noted as the basis for discontinuing. Therapy can be suspended because the patient needs more urgently to acquire new experience and, for this phase of development, therapy may be uneconomical.

The need for therapeutic integration is rooted in the necessary operation of getting through resistance. As the process is genuinely fulfilled, the specific tension of the therapy situation can wither away. If so, self-awareness could go on without outside intervention, and growth situations could be met and integrated without professional support. This would represent the integration of the dynamism for growth.

Therapy is oriented toward accomplishing what is necessary for the patient's subsequent living to tend toward maturity, in its overall pattern, rather than toward deterioration. Discussing one's patterns of living with another person for a few hours or a few hundred hours should result in some improvement of those patterns. If several hundred hours result in no improvement, it could only be the outcome of a centrally hostile integration between the two, which acts to reinforce both central paranoid systems.

A frequent compromise is to settle for improvement in achievement. Since these changes are good for the reputation of the therapist as well as for that of the patient, they are often substituted as the goals of therapy.

Facility in achievement can represent a move from the emotional level of infancy to that of childhood, the period of mastery of the object world. In adult life, it can be necessary for the first economic and physical independence from home. Often, the compromise is at the level of the juvenile values. Comfort in groups and facility in either competitive or cooperative situations are necessary for adult living. It is also rational to take leadership and responsibility in those areas of which one has a superior grasp. But the necessity to be in a position of leadership, the concern with prestige and security, represents a hangover from childhood insecurity and preoccupation with security, and is, in itself, a disguised form of not being one of the group.

Improvement in already initiated functions is not enough. On the other hand, the concept of cure, partly a hangover from the medical origins of the field, is, as applied to growth in personality, manifestly ridiculous. It would constitute a complete distortion of living to devote enough time to analysis to reintegrate all the experiences originally dissociated or held partially out of awareness. The impulse toward growth will be taking place, after analysis, in areas not available to explore at the time of analysis. It frequently happens that what has taken place during the course of the analysis in the direction of the integration of the function of growth has been exceedingly covert and only becomes clear during the years after its termination.

There are some factors which make the positive results of analysis more lasting. One is the capacity of the patient to locate and respond to tenderness. Tenderness is needed for survival and for enthusiasm. It makes a degree of aware anxiety tolerable and is necessary for consensual validation of one's own perceptions.

These needs can be supplied in fairly peripheral contacts as well as in one-to-one partnerships.

The importance of capacity for a loving partnership lies in the fact that the chum is the first person who becomes more important than mother. This experience opens the door for a series of replacements and experiments, each with different frames of reference, capable of nourishing and validating further growth in similar or different directions. Also, if the level of fixation can be identified accurately, real moves, which might otherwise be derogated as regressive and infantile, can be validated. The social front becomes a practical operation, not central to the character structure. One can admit privately that one is frankly learning in areas which have been forbidden, and stop pretending that one already knows.

In practice, this often involves identifying the highest level of achievement during the person's life in a particular line of development. Subsequent loss of function, then, can be understood as a deteriorative process. If this subsequent deterioration is understood and counteracted during analysis, the patient can proceed to develop from the level previously achieved.

The patient must also get some grasp of the normal process of growth, both in general and in himself. He needs to understand that moves take place in or out of awareness, that no growth is possible without anxiety, that his fears are more of success than of failure. He learns his own modes of communication between the integral and the self-systems, his personal ways to find out what is going on in himself, and to supplement these with new ways previously forbidden. He learns the dangers of isolation under stress and develops methods and facilities for seeking out helpful persons in a crisis. He learns the importance of actual experience in living, as against literary, fantasied, or verbal attempts to circumvent it.

The therapist must see the patient through enough growth experiences for the patient to look on further expansion of his personality as possible without the help of the therapist. If new insights are not implemented in new actions, the therapist must not only identify the interference, but must also find ways of getting the patient into action. The fear of pressuring the patient is irrelevant if it is clear between them that the therapist is insisting on exploration and investigation rather than on imposing on the patient a particular way of life.

The patient also needs a specific grasp of his central paranoia. He must be conversant with the origin and patterning of those emotional distortions which became a central reality of his inner life, how he learned to see himself, and how this has limited his growth, his views, his satisfactions and achievements. He needs to know what about affection was originally frightening, and by what means he has succeeded in keeping these central, subjective realities out of awareness.

It is also crucial to reconnect the good experiences of infancy and their adult representations with the adult execution of mastery and the adult search for new experience. The turning point for successful therapy occurs when the patient makes the discovery that experience is more important than the avoidance of anxiety and that love and truth are more important than prestige and security.

The initial complaints of the patient entering therapy must be understood in terms of these goals. The complete removal of anxiety is not possible in modern civilization. It is reasonable to expect that the explosive and disruptive quality of necessary anxiety be reduced to manageable, tolerable proportions.

It is much the same with security operations and with symptoms in the more classical sense. To the extent that one continues to postpone or to attempt to avoid anxiety, security operations come into play. These operations become more flexible and less of a handicap as a result of analysis, but there remains a tendency to fall back on the familiar ones under stress. It is possible for them to be given up as a predominant way of life and to remain primarily as part of the latent equipment for coping with emergencies. They can be expected to be less frequent, of shorter duration, and to be experienced in greater perspective. On the other hand, the symptom serves a purpose. If a headache serves a purpose, half a headache will not do. Symptoms therefore do not lose as much of their intensity as a result of analysis as one might hope.

As for feelings of emptiness and futility, the chronic fatigue of perpetual repression, these are the necessary and legitimate targets for the therapeutic process. If they are not reorganized significantly during the analysis, or subsequently as the result of it, the investment of analysis has been unjustified.

Initially, many patients come to analysis literally struggling with the decisions of life or death—psychosomatic suicide, literal suicide, or suicide by the disruption of the whole fabric of social existence by some dramatically self-destructive compulsion. The first phase of the analysis is frequently directed primarily to undercutting these. When the basic decision to stay alive is more or less clear, the patient often shifts over his lifelong jungle psychology to more everyday problems. The war may be over, but he is still carrying his gun. Preoccupation with security, prestige, property, and status may now be rationalized in terms. of the old framework of life or death. The central paranoia is still intact and the life operations are less dramatically, but no less consistently, directed toward the restriction of experience.

Many patients leave analysis at this point. They do not necessarily go back to the old, dangerous patterns. Sometimes they consolidate their gains, incorporate some useful human experience, and return in two or three years to continue the analytic project. Often they try to stabilize the personality organization on the intermediate basis. Such partial results do not involve inherently any commitment on the patient's part to further growth. And if the patient does not continue to grow, the deteriorative process sets in, quickly or slowly.

VII

STRATEGY

A) INTERPRETATION

The therapist observes and reacts to the immediate interaction between himself and the patient, to the context of the interview and to the general context of the patient's life. He evaluates what is important about all. This involves evaluation of the overall problem of the patient over a considerable span of time.

If a frank catatonic break is imminent, the therapist must judge the patient's intent to head it off. Is the patient trying to find a face-saving device to get into a hospital, or is he trying to find a way to stay out of one? The therapist estimates the degree of entrenchment of the patient's paranoid patterns. If these involve most of the patient's life, how did he get into therapy and what does he think he is doing here? The therapist estimates the intensity of the suicidal pull and the patient's real and obsessive resources to defend himself against it. Is the patient's apparent insecurity the reflection of his operations to control anxiety, or superstructure for the fear that someone will expose the extent and intent of his hostility? Does it disguise and slow down the impulse to love, or is it the disguise for an intense destructiveness which has frightened the patient by the imminence of its expression and of some grasp of the social retaliations involved?

Is the patient focused directly on growth experience and on the anxiety inherent in it, or is he still essentially battling old ghosts in the search for a situation in which concern with growth would be plausible? There is an important difference in management.

In the former case, the daily operations have a modicum of validity and are to be investigated in these terms. In the latter case, the operations exist for a purpose other than their own exploration. They may be in the "constructive" mode of creating an illusion of maturity. They may be in the "destructive" mode of coping with the dangerous problems before becoming involved with the less dramatic ones. It is urgent that all of these be identified. They are detours from direct experience and should not be taken at face value. In the real world, is the patient here to get help in leaving his parents, or to get married, or to get some freedom from the responsibilities to his children? Is he stuck on finding a job, or on arranging and implementing his retirement?

What role does the patient hope the analyst will take in the analysis? Is it useful to go along with it? Which factors in his development is he trying to clarify, what are his difficulties in clarifying, what can he allow the analyst to contribute? The therapist needs to know not only the focal point of interpersonal growth, but what supportive antecedent experiences the patient feels that he needs to fill in before he thinks he can move ahead on it. Is he trying hurriedly to catch up and make use of an unusually good opportunity for growth?

The underlying emotional tone can be expressed in mood, gesture, and psychosomatic difficulties. It is not fully and freely experienced, and it is rarely exactly what the patient would say he is feeling if asked. It may be lifelong, or it may be chronic for the recent past, but the patient is inattentive to its full impact and inured to its peripheral manifestations. He may have been on the verge of tears for months, but literally have never noticed it. The therapist can assume that this emotion is appropriate to the patient's situation as the patient perceives it.

The construction of an overall interpretation is a little like reading between the lines in the newspaper to find out what a particular political crisis is really about. There is a lot of noise, known to be of one bias or another. There are various hypotheses, equally plausible in terms of the provable facts. The data sift down into the unconscious of the analyst and, one day, he knows. A great deal of data which previously had equal weight come into perspective. He has a hunch about what is going on. Having formulated the hunch, he can test, revise, invalidate, or improve it. The therapist then has to draw the patient's attention to just these various points where growth is stymied and get him going on them. In the meantime, however indirectly, the patient is trying to get back to where he left off.

The convention of setting therapy at regular intervals at the convenience of the patient and the therapist is based on the assumption that there is, in the patient, a continuing margin of awareness of and preoccupation with the project at hand. The patient also is searching for the conditions which are necessary for his growth and has a secret plan for securing these conditions. It may be a very discouraged plan, based on acceptance of hurdles which are thoroughly unnecessary.

It is the therapist's business to validate emergent experience without serious regard for either his or the patient's immediate social convenience. However, he should first understand what is the patient's secret plan for gaining access to necessary experience and understand why the patient thinks it has to be so devious. The therapist cannot participate in the patient's doubts and fears, but, before challenging them head-on, he should evaluate just which reorganization would be involved. If the therapist makes a move which seems to the patient to upset the patient's secret plan, the patient is likely to experience it as destructive and disruptive unless it is carefully placed in context. Usually the therapist either disagrees strongly with the plan or tries to guess, formulate, and improve the efficiency of it, eliminating fallacies where he can. There are occasions, however, when the plan seems to be reasonably efficient, and where it would be much more difficult to carry out if it were clearly in awareness. Then the therapist would do well to enter into the silent conspiracy without too much formulating.

Coming into therapy usually represents a major decision to reorganize one's life on a better basis, however much it may be presented in terms of urgent anxiety. If the therapist can be alert to what the patient has in mind, he can be a considerable influence in facilitating this direction. He also has an opportunity to find out what else needs to go on in the patient's life to get it on a reasonable basis.

That the patient has a plan does not relieve the therapist from his responsibility to organize the analysis. It is relevant for him to perceive the patient's plan and to validate what is constructive and usable in it. But his business is to get the patient to work most efficiently on his real problems and to reduce the digressions to a minimum.

In choosing interpretations, there is a certain automatic protection in the fact that consciousness is somewhat limited to that which could conceivably be communicated to the listener. Marginal thoughts and emotional attitudes inappropriate to the patient's present status tend to be screened from the therapist's awareness by the therapist's sense of what is going on in the patient at a particular time. Most of the major complications that arise from uncensored spontaneity, either in emotional or in intellectual self-expression, derive from a misconception on the part of the therapist about the patient. These ill-conceived moves from the therapist include inappropriate anger, inappropriate demands for affection, inappropriate apology, and self-derogation.

Another protection lies in the fact that in a relatively valid relationship, very few errors are irreversible. The therapist almost always has the interval between one session in which something has misfired and the subsequent session to understand what he was overlooking in either himself or his patient.

The data to be formulated include the elaborate clarification and description of the manifest personality, of the self-system in interaction with people. Such formulations are addressed to the integral personality and imply, explicitly or implicitly, a different scheme of values from those on which the described actions are based.

Interpretations include the structuring of the total situation and speculation as to the immediate underlying attitudes in the patient. They include a reconstruction of what must have happened in the patient's history to produce certain results, and concise speculations about the form in which these events might have taken place.

Any repetitive reaching out is based on actual experiences which were at one time, however briefly, clearly in awareness. They are usually remembered, if at all, as failures. The clarification of these early moves can expedite their re-evaluation as a useful point of departure.

An important aspect of this re-evaluation is to identify the original real source of opposition. Those aspects of the memory itself that have been available to consciousness will have been filed away in the wrong slot. Memories will have been used to prove either that the world must let one down or that the patient was incapable of participating. The original difficulty was either in omission of validation, in the anxious response substituted for validation, or in the threat of active destructiveness.

The therapist has the task of identifying a variety of puzzling or eerie experiences as to their meaning in the normal process of growth or deterioration. It is important to understand in what way a particular preoccupation represents the paranoid move to inhibit growth, and in what way it represents an attempt to make up, however repetitively, blocked experience from early childhood.

The therapist should sort out, in interpreting states involving a shift in the level of consciousness—during rumination, illness, acute anxiety, withdrawals, or fugue states—possible attempts to integrate new experiences or to symbolize forgotten experiences. It is also necessary to identify in what way a current misperception is based on the reliving of early childhood experience, as evoked by the challenges to growth in the current situation.

The therapist must organize the situation so that the patient can become more aware of his feelings. The therapist's emotional response, if clear to the therapist, can be used to evoke and clarify the underlying emotional tone in the patient—anger, despair, elation, anxiety, fear, revenge, or whatever. The therapist does not clarify the patient's hostility by setting out to make him angry. If the therapist is sensitive and perceptive of his empathic responsiveness to the patient, the patient is more free to experience his own anger, whether it is projected onto the therapist, or onto some current figure, or, with less obscuring, it is directed toward the original authority figures.

The overall emotion which the patient experiences is always appropriate to his situation as it is perceived by him, but his perception of this situation is limited by the stencil of his previously integrated experience. From the patient's emotion, then, the therapist can read back, for example, that the patient feels or acts as if he had been deserted, attacked, or exploited, in order to identify the patient's private perception of his situation.

The intent of the integral personality is formulated out of the conscious fault, communication, or act, as screened through the self-system. If the growth need, let us say, is to get ahead with the experience of chumship, and if this patient can proceed only in secrecy, then how is this represented in the consciousness of the person who is undertaking it? How does he think about it; how does he perceive it?

Communication operates as if there were in fact two personalities, the self-system and the integral personality, having access to only one telephone line, so that every communiqué, every thought, statement, gesture, act, or even perception must represent some negotiated compromise between the necessities of the two systems. To the extent that they diverge, every communiqué has to be double talk. That is, every act is both the expression and the repression of the conflict between growth and the *status quo*. The way to get at the useful interpretation is to formulate the questions: What is the overall need that the integral personality is trying to work through into the manifest personality? What are the limitations on its getting into awareness? Under these limitations, how will a particular event be represented in consciousness under what varying conditions? From this formulation it is then possible to develop a talent in the reading back of what is probably really going on when, under certain circumstances, the patient says, does, feels, or acts in a given way. This is the main content of interpretation.

When the security functions are operating smoothly and unchallenged, there is a minimum of direct leakage from the content of the integral personality. On the other hand, dreams are the most direct expression of the inner perception, minimally screened by the potential audience, but limited by the capacity of the self-system to perceive and remember them. If there is an interruption in this minimum degree of rapport between the integral personality and the self-system, the patient dreams intensely but can get no grasp of the content of his dreams.

Material from the integral personality will be reformulated in consciousness in terms of the immediate audience, the therapist, as seen by the patient. It tends also to be limited by such other primary integrations as potential audience, as well as by the logic of the self-system. Another limitation is set by the implications for direct action of such a thought, if it should be expressed and validated.

A security operation must re-establish the connectedness with mother; must limit subjective awareness; must impose some pattern on the external situation; and must in some way express the specific nature of the dissociated breakthrough. Interpretation involves the identification of the above elements in any specific security move.

Situations in which frank, even if brief, anxiety occurs comprise another focus for useful interpretation. Anxiety may be identified in brief or longer interruptions in the stream of consciousness, or in the appearance of physiological or mental regression. The anxiety marks the intrusion of a thought from the integral personality into the self-system. The identification of the exact circumstances may facilitate recapturing and integrating this thought.

One important mode of interpretation is in the grasp and formulation of inversions. It occurs with remarkable frequency that the manifest content of the patient's statement is a sort of upside-down of what he is really trying to communicate. For example, the feeling of persistent insecurity is one way of saying that the patient overvalues security. The fear of loneliness is often the patient's way of communicating that he is in danger of driving everyone away because of his fear of closeness. The fear that the recurrence of a particular symptom of anxiety presages a relapse is often a statement of the fear of growing up. Fear of death is in a way fear of life. Fear of failure is often a cover-up for fear of success. Pre-occupation with the danger of being deserted is often a reaction formation against the fear of entrapment. To reformulate such inversions rarely touches directly on dissociated material and may be reassuring to the patient, as well as productive of new material.

The prevalence of the use of inversion is partly based in its usefulness in banishing unacceptable needs by putting them in unacceptable categories; for example, that friendship is homosexual, that longing for the same satisfactions as one sees in one's friend is necessarily hostile to him, or that the ambition to achieve mastery over a process must be competitive.

The mechanism of projection involves perception in another of characteristics unavailable to full awareness in oneself. It also involves a sharp cutoff of awareness of alternate processes in the other. Such alternate processes would otherwise be perceived by the normal exchange of resonant emotion. Direct interpretation of this projection usually meets more resistance than formulation of the simpler inversions, since it is based on a more entrenched dissociation.

Explicit statement of the assumptions on which the logic of therapy is based has some supplementary usefulness in the art of interpretation.

These assumptions include the concepts of psychological determinism, of relatively unlimited human potential for development if unhampered by restrictions, and of the possibility of clear verbal communication of marginal thoughts and feelings. They incorporate a relevant scheme of values; for example, that life is better than death, that self-awareness is more important than the avoidance of anxiety, that satisfaction is better than prestige, that expansion is better than restriction.

The therapist must have a conviction about what matters in life and be willing to take a stand on it. The therapist should be willing and able to clarify his assumptions whenever this would facilitate an interpretation. It is often useful to give the patient a brief elaboration of such concepts as the stages of development, the significance of chumship, the role of promiscuity in the development of sexual adequacy. Such theorizing will not itself change the structuring of the self-system, but it can undercut some of the logical support of the structure which maintains the mechanisms of dissociation.

Among the most important of these challenges to the restrictive assumptions of the self-system are those in the area of the basic conventions within the culture in which both the patient and the therapist operate. Some of the cultural myths that should be questioned include: success in living is in direct proportion to professional success; blood is thicker than water; the fact of legal marriage determines the quality of the interpersonal relationship; living and major decisions can be postponed for the duration of the analysis; sexual experience is only valid if the two people are in love; certain types of character structure have more value than others. The obsessional is often considered better than the hysteric; the hysteric better than the catatonic; the detached better than the dependent.

Simple, clear, concise formulation is also involved in defining the patient's problem and undercutting the patient's misconceptions about it. The clearest example of logical, conscious reorganization of the reference points of the self-system is the presentation of summaries at intervals throughout the course of analysis. The therapist draws together, with foresight and as homework, his impression of the total trend of a particular line of development, either in life or in therapy.

Summaries are necessary to the process of therapy. The patient has a frame of reference based on a variety of assumptions, many of which have a certain validity, some of which are of necessity fallacious, all of which form a relatively integrated whole. If, as is usual, some piece of experience is validated which is incompatible with only one of these assumptions, leaving the others intact, this experience will eventually be erased as extremely unlikely or probably misinterpreted. It is not of sufficient power to reorganize the whole frame of reference. Presenting a summary raises the possibility of, and occasionally accomplishes the substitution of, an entirely different set of assumptions, achieves a new perspective on his style of life, thereby validating not only particular islands of experience but opening the possibility for much new experience.

Many therapists are reluctant to give the patient comprehensive summaries. This reluctance lies partly in the conventional taboo against telling another person frankly face to face either exactly what one has been able to make of his communications, or one's exact impression of him as a person. The therapist is also partly afraid of revealing his ignorance about either the patient in particular or the field in general. He may be afraid of precipitating the patient's anger, often disguised as discouragement.

Any serious attempt to reveal and to share with the patient one's own sincere impression either of his history or of his operations must, basically, contribute to decreasing the patient's discouragement. If he claims to feel discouraged as a result of the summary, it is either an expression of his anger at loss of face or a cover for anxiety stirred by the opportunity for growth or the direct expression of his impulse to derogate the new opportunity and repudiate it as a growth experience.

Therapists also dislike doing unpaid homework. The same amount of time spent in reading a novelist's projection of a fictitious person's possible life experiences would feel more dignified. Time spent understanding and disentangling similar factors in the real person can be even more rewarding for the therapist.

After a summary, the patient almost always shows a spurt of improvement. The therapist also has the opportunity to learn something about how another person has lived, what has been effective in his growth, and what has been destructive to it. Also, by comparing his spontaneous impression with the record of what the patient has actually said to him as recorded in his notes, the therapist has the opportunity to identify some of his own blind spots, in his work and in the processes of living. Presenting a summary also gives the therapist practice in direct communication not contingent upon his eloquence or his capacity to formulate on the spot. He can work it out in advance, write it down, read it if necessary, and expect that what he is going to say will be relevant and comprehensible.

The therapist runs the risk that his summary of the patient's life will be more of a reflection of the therapist's personal problems, or of the therapist's entirely personal scheme of values, than it will be a condensation of the patient's personal problems and a validation of what is relevant in the patient's scheme of values. To a certain degree, this is inevitable; the therapist must operate as a whole person. This process of explicit formulation and consensual validation is a necessary component of the process of the therapist's sorting out his personal bias from the biases of his friends and patients. It will not get clarified without exposing it, and the impulse to keep the personal bias a secret can only serve to maintain it and to perpetuate acting it out.

The therapist has many assets for this project. He has a concept of the stages of development. He has in his notes the dates and ages at which significant events took place. It is a routine job to get these set into chronological sequence. He has some grasp of the difference between the satisfaction of real needs and security operations, and can organize the summary to highlight the distinction. He has made a study of the possible significance of early figures and early events in the patient's life; thus he is in a position to follow their influence throughout the history. Even if there are major deficits in his grasp of theory, the fact is that anyone would be benefited by the opportunity to hear any other serious person's honest impression of what his life is like. The listener cannot fail to stumble on some important omission that the teller has been trying to obscure.

The summary may be organized around some particular recurrent difficulty, around some particular aspect of development, or in terms of the overall history of the interpersonal developmental sequence. The point of departure may be the therapist's impression of an overall operation that is going on during a particular period of analysis, or it may be an attempt to reconstruct the probable events in some poorly remembered period in the patient's history which seems to have been a turning point in the development of later attitudes. It is always somewhat directed toward the simplification of life; it plays down the elaboration of security operations and throws the spotlight on the events relating to growth or deterioration.

The summary must be directed toward what the patient can integrate at the time. The data used are data that are at the time available to awareness. The therapist moves the pieces around to make a reorganized picture of them. He should develop hypotheses to fill in gaps that seem to him to be necessary extrapolations, but which are in fact not currently in the patient's awareness. However, he must label these clearly as hypotheses, and be open-minded if the patient, under this stimulus, comes up with entirely contradictory concrete data. At the end, he must give the patient some opportunity for rebuttal and reaction.

The patient's first reaction to hearing a somewhat extensive review of some aspect of his life is almost always a significant contribution to the data. After registering this contribution, if the timing in a session devoted to a summary is managed so that the session will end at this point, this helps to cut down the patient's erasing the implications by mustering the obsessive operations of the self-system. Reorganization, if it takes place at all, will take place first in the integral personality and will be done best alone and least well in the presence of the person who has precipitated it.

It is clear that this process includes clarifying the central paranoia and its manifold implementations. The patient must notice the least attractive characteristics in himself. He also is confronted with real deficits between pretended levels of functioning and what he would really need to accomplish in order to approximate his own goals. At the same time, he has for the first time in his life some basis for hope that he can change, some glimpse of the resources in his integral personality on which he can expect to draw.

At this crisis, the patient sometimes elaborates intense paranoid accusations against the therapist and pulls out of therapy. Such a possibility is no excuse for the therapist's avoiding clarification of the patient's problems.

One impetus for such withdrawal is the problem of saving face with either the therapist or the patient's immediate associates. In this case, the therapist often gets concerned, in retrospect, with whether his style of confrontation pushed the patient out of therapy in a paranoid episode and thereby made subsequent resolution more difficult. This is a somewhat self-centered orientation. Whether or not mismanagement by the therapist was the immediate agent of the traumatic experience, the therapist cannot be primary. The primary factor has to be that the patient has gotten enmeshed in an illusion seriously out of pace with his real situation. He can see no way out except to transfer blame for the anticipated repetitive failures that he thinks would be inevitable.

Such flight from therapy does not necessarily mark the turning point away from further self-understanding. The patient knows more about himself than he ever has before. He may find this knowledge a useful map for further self-analysis once he is out of the immediate battle with the therapist.

On the other hand, self-analysis is inefficient. To choose self-analysis in preference to therapy is usually intended to slow down the process of growth, with an eye on the point of one's life at which one expects to be too old to enjoy the rewards of the process. Progress may be held back to a rate which is calculated to be significantly slower than the rate of physiological deterioration at the present state of medical knowledge.

B) SEQUENCE IN THERAPY

Therapy takes place in an orderly sequence. Certain things must happen before others can be brought into awareness. There is little knowledge about what determines this sequence, and it is usually handled intuitively by the therapist.

The sequence at the beginning of therapy is relatively clear. Therapy begins with the first contact. If the broad outlines of a constructive integration are then not established, something else is. Later experiences in therapy tend to be fitted into the categories established here.

A brief review of how the patient experiences his difficulties, and a brief inquiry into interpersonal history, constitute the early contacts. Then follows an attempt to get together with the patient on a formulation of the problem, in terms of the patient's inability to handle certain interpersonal situations. This is the initial definition of the problem. This formulation must be at least conceivable to the patient in his present frame of reference.

The patient then generally goes on to tell about his current patterns of living; the therapist, to try to identify these in terms of his knowledge of theory. However, the situation is not really integrated until some spontaneous attitude of the patient's has yielded to a therapeutic maneuver. After this, the patient has some reason to believe that the process might work, and he can begin to make an investment in it.

Several other trends can be traced in studying the sequence of useful therapy.

One is progressive nonparticipation of the therapist in the patient's security operations. This is both overt and covert. This includes the progressive simplification of security operations so that there can be room for something else to happen. The overall direction throughout the analysis is toward progressive closeness. At the beginning the patient tends to adjust to the therapist's security maneuvers, and the therapist to the patient's. A certain minimum of this is necessary for the continuing of therapy.

The therapist must demur about striking distortions of fact, partly because the patient suspects their delusional nature. The therapist must also demur if the patient expresses outrageous and inappropriate low self-esteem. For example, the therapist has to take a stand even initially against a patient's suicidal neglect of medical care. If the therapist ignores such blatant maneuvers, he becomes in effect, a silent participant.

In the early part of therapy, the therapist makes as many adjustments to the patient's neurosis as are compatible with the smooth progress of the work. One of the subsequent processes in therapy is the progressive disentanglement from participation in the patient's neurotic demand that the therapist persist in these adjustments. It is the therapist's responsibility to choose an optimum point in the continuum of closeness-distance at any particular time in analysis and to maintain it gently against the patient's maneuvers, but to move it according to the patient's needs.

Another trend in sequence is progressive sensing through layers of resistance to get to the data leading to the central problems. There is also progressive validation of the durable availability and nonmalevolent character of the therapist, and progressive expansion of the frame of reference of the patient's self-system.

As marginal thoughts are validated and brought into awareness, the next layer, previously dissociated, can become marginal.

The only reason for delay in the identification of points of difficulty in the original chronological development is that it may take the therapist time to get enough data to be reasonably concise about it.

Given this starting point, there is a tendency for the project of therapy to follow roughly these eras of development: 1) handling the psychosomatic, if this is a prominent feature; 2) handling the physical world, separation from mother, physical and financial independence; 3) feeling of community in groups; 4) individual friendships on an equalitarian basis.

This sequence is often camouflaged by the fact that the conditions for growth usually require some capacity for social stability, so that much of the actual initial phase of therapy may lie in helping the patient get into a situation in which he can get to work on the project. Occasionally a patient goes into a hospital as a transition from a tightly restrictive situation to one allowing room for growth. Insight, leading to premature disruption of an integration which is contributing to a necessary preparation for a phase of growth, can set off unnecessary panic.

There are situations in therapy when the therapist feels that the integration of a particular function is crucial to further progress. The patient has been making abortive moves toward it but seems to feel less urgency about it. The therapist may then focus emphatically on consolidation of this particular function. He should not do so unless either the groundwork for its integration is well laid or he can see no other contributory areas that can be consolidated in reasonable time.

A major push from the patient should rarely be discouraged, even if the therapist feels it is premature. However, in this case, he should be prepared for the possibility that the patient will later repudiate major segments of the insight gained. If the undertaking fails, the therapist should be alert to counteract the use of failure by the patient to avoid future efforts.

With a particular patient it may be necessary to wait until there is a basis for enhanced self-esteem through considerable improvement in his interaction with people around him, and hence improvement in how he is seen by them, before it is wise to clarify how mutilated he felt. We have to expect that all of the major underpinnings of the self-system—the self-image, the repudiation of tenderness, and the continuity of childhood terror—will hit the patient with a temporary conviction of truth and reality when they are first brought into focus. The patient has spent his life proving them in his desperate attempt to repudiate them. So although they may be referred to casually in the general structuring of cause and effect, full impact is to be avoided until, within a finite period, they can be restructured as anachronistic.

This applies also to the patient's awareness of his lifelong despair about himself. He needs to validate some experiences contributing to hopefulness and to consolidate some grasp of the general structure of his life patterns before he can cope usefully with full awareness of his feelings of hopelessness.

The orderly sequence in therapy also lies in the limitations in the capacity of the patient to implement a particular insight in real life at the time. The decision as to what must be done first lies in what a particular insight could mean to the patient at this particular moment as he sees the world. Sometimes the question is literally, "What could he do about it?" However, very often all that he might be expected to do about it is to register it vaguely as a possible alternate perception opposing his entrenched convictions about the world. At the same time, perceptions entirely inconceivable to him are of little use. Pushing them will be experienced by the patient as a negation of analysis, as attempts on the part of the therapist to pressure him into pretending a virtue which he has as yet no basis for achieving in reality.

With these precautions in mind, it is nonetheless to be noted that therapists tend to underestimate the capacity of the patient for simple, common-sense action in the real world. Many very disturbed people perform a great variety of everyday functions within the context of their disturbances.

The analyst must be able to conceive of change in general in order to construct a concept of a useful sequence of change in a particular patient. He must also have a specific notion of what symptoms will probably flare up in the process of growth—whether they be depression, catatonia, paranoia, psychosomatic difficulties, or competent detachment. The analyst has less influence over the course of a particular growth episode than he might wish, but a close knowledge of, and a capacity to predict, the patient's patterns may head off the more unfortunate side effects of a particular move.

The analyst must also have some concept of the various specific reorganizations necessary in order for other destructive patterns to fade out in a particular patient. Is a patient's exploitativeness partly continued by her wish for a standard of living which she has never been equipped to provide for herself? Is the man's emotional exploitation of the woman a thin diversion from his wish to relate only to men?

Every symptom reflects some solidly integrated defensive character trait which the patient hopes to hang on to throughout the anticipated reorganization. These character traits in turn derive from and replace other deficiencies in necessary functions. The therapist must understand such problems of organization so that he can partly anticipate the ramifications of a head-on investigation of any particular self-defeating operation.

The concept of orderly sequence does not mean that progress goes on in a straight line. Some aspect of the integral personality takes the leap. The subsequent working through in analysis is the process of dragging everything else up to the new front.

VIII

THERAPY AS RADICAL REORGANIZATION

THERAPY CAN BE defined as a cooperative research project involving the full resources of two persons for the purpose of improving the facilities for satisfaction and expanding experience of one of the two persons, which project can, in fact, implement its purpose only on the basis of expanding the capacities and experience of both. It proceeds by means of elucidating the patient's present and past patterns of living, understanding them in terms of the dynamics of psychological determinism, and reorganizing them accordingly.

In therapy one takes a particular product, an adult personality, and undertakes a factor analysis of the various determinants which went into the final result, and of the ways in which these determinants continue to operate. Analysis, then, must be the process of reading back from any particular adult experience to what might have happened to make this particular outcome likely. The therapeutic operation includes this reading back both in terms of the historical interpersonal events and in terms of current interpersonal events, and their partial representation in the patient's consciousness.

Therapy undertakes to broaden and intensify the patient's awareness to include progressively more and more of his life experiences, past and present. These come to be understood as derivatives of the personality systems and, reciprocally, as having an impact on them. The responsibility of therapy is to help the patient organize and utilize this new knowledge in improving his actual behavior toward greater satisfaction.

438

Therapy is the factor analysis and resynthesis of the relevant data, to put together an overall view, highly documented, of how the drama is acted out, including the supporting emotions and convictions, and of how major areas of simple, direct experience are sidestepped. Analysis can speed up the growth episode, not only by encouragement and interpretation but also by introducing new tools of which the patient was unaware. Such tools for growth include free association, fantasies, new experience, and the getting in touch with, rather than the continued repression of, feelings encapsulated in infancy, such as fear, rage, or intense desire.

The self-system is organized to defend certain important fallacies. It operates to limit perception and to control the experiences to which an individual is subjected in such a way as to maintain certain false impressions of the world and of one's potential role in it. Before the patient comes to analysis, countless experiences have impinged on the organization of the self-system, with some impact on it, and by virtue of these experiences the self-system has expanded in many dimensions. This leaves, in the adult, primarily those fallacies which are crucial to its organization. Areas of growth which were not too strongly forbidden have often been selected as areas of interest and have received considerable energy in their expansion. Minimal compromises have already been worked out for the satisfaction of pseudo-satisfaction or partial satisfaction of primary and ineradicable needs.

For this process one does not need an analyst. The investment of time, money, energy, love, and skill involved in analysis can only be justified if the goal of analysis is to reorganize the basic assumptions of the self-system in order to open up new areas of potential experience for expansion and integration. Piecemeal chipping away at one or another parataxic perception or mode of acting-out can never be enough.

On the other hand, like any revolution, it cannot be accomplished in one major crisis, but proceeds through a series of reorganizations and disorientations, followed by the implementing of the new perspective into the actual process of living. The force causing the explosion always derives from the covert attempt to satisfy forbidden needs. The process of consolidation involves the extension, expansion, and validation of the previously covert experience, so that it can come to be taken for granted as a part of one's view of the world. Analysis, then, cannot be effective without crises, but dramatic confrontation without the adequate antecedent support and subsequent implementation and working through cannot hope to achieve permanent results.

Such radical reorganization is the direction rather than the prescription for what can be achieved. The original reference points could not conceivably be challenged entirely and reorganized fully within the confines of a finite analytic process. The progressive disintegration of the restrictive aspects of the self-system can be expected to go on throughout life. Significant points of progress will re-evoke many of the original childhood fears, and intermediate frames of reference must themselves later be challenged and reorganized by the infinite possibilities of expanding growth. The same applies to the expanding of choices of people with whom to share growth experience.

At best, the process of growth can only be given fresh impetus and a better framework during therapy. Each subsequent move in the direction of progressive capacity for intimacy raises both the original prohibitions and also the intermediate limitations which have been set by the more recent interpersonal experience.

For example, in the juvenile era, contemporaries for the first time take on an importance of somewhat similar magnitude to the importance of the parents, though no one contemporary is of equal individual importance. In integrating a chumship one musters, of necessity, some capacity to break with the values of the juvenile era.

It is illusory to think that in the process of analysis one can, once and for all, reorganize one's basic assumptions and open one's heart to new acquaintances in such a way and to such an extent that subsequent reorganizations or subsequent reopenings would not be necessary or would not involve major anxiety. The original reorganization which can take place during the analysis must be the first of a long series. Successful analysis involves becoming accustomed to revolution.

The central goal of analysis, then, is that the patient will, after analysis, be able to continue to grow on his own. Growth is the process of integrating functions into the personality. Analysis is the process of integrating the function of growth in the patient. This involves both the breaking up of the patterns of organization which inhibit growth and the validation and development of the normal processes of growth. Therapy aims to catalyze the forces of growth and to counteract the factors which lead toward deterioration. Those aspects of the patient's personality in which growth is taking place without obstruction are not the primary concern of the analyst. The analyst's main concern is with those areas in which growth is taking place against resistance. Under these conditions growth proceeds by stops and starts.

When growth has taken place against the resistance of a moderately well-organized self-system, there are islands of experience which occur essentially in opposition to and in spite of the basic assumptions of the self-system. A concept of physics which may be useful in this connection is the concept of the change of phase. If one starts off with a glass of water and drops oil in it in a form which can be emulsified, the result is oil droplets in a continuum of water. If one continues to drop enough oil in it, there comes a point where instead of the continuum being water and the islands being oil, there is a reversal of phase, and suddenly we have an emulsion of water droplets in a continuum of oil.

During the time when the new experiences are taking place, they are perceived as slightly unreal, unlikely, to be rationalized in some mode or another, and usually as rather disconnected from the main thread of life. In between times, they can hardly be remembered. As more of such experiences take place, there comes a time when it is noticed that they have a certain continuity in themselves, and then there is an intermediate period when the person fluctuates back and forth between the two sets of reference points. During the growth phase, the assumptions of the central paranoia, as, for instance, that life must be filled with terror, or that love is for other people, that it could never happen to me, may seem strange and even bizarre. During the periods of retrenchment, these assumptions take on the aspects of essential reality, and periods of joy and euphoria are looked back on as moments of mild insanity, deception, or self-deception. Ultimately, with luck, there is a more durable reversal of phase in which the periods of growth seem to be the natural reality and the periods of despair take on an episodic and discontinuous quality, even though for their duration they may retain a remarkable amount of their original force.

This is the normal course of a particular growth experience. Over a more extended period of time, the patient can develop some perspective on the process itself and can identify his own unique patternings. By now he can come to recognize particular distressing thoughts, moods, and feelings as familiar exacerbations of the lifetime paranoia. He learns to sort out the subjective, intense aliveness of growth experience from the intermingled anxiety or fear, which latter raises the intensity of the experience to painful proportions.

He accepts the fact that the struggle to grow is limited only by the life span. He learns to respect, organize, and utilize his individual conditions and situations for facilitating awareness and integrating new experience, including opportunities and necessary circumstances for intimacy, sharing, and tenderness.

This minimal grasp of general principles and of the individual specificity of the process of growth constitutes the core of the integration of the function of growth, which is the irreducible goal of analysis. It makes possible the acceptance and affirmation of periodic major expansions and reorganizations throughout the life span, as the satisfactions of succeeding levels of development are integrated and consolidated.

After the self-system has begun to form, growth will take place against its restrictions. Because of this, the organization of the personality in the direction of growth will depend on the capacity for formulating and reorganizing the tenets of the central paranoia. It will also require an understanding of the accompanying anxiety and the capacity to manage it. It requires an understanding of the usual vicissitudes in integrating new experience against resistance in order that the setbacks shall not be too discouraging.

However, durable integration of the function of growth depends on the capacity of the patient to organize within himself a dynam-

ism for growth. This dynamism must be oriented toward the progressive expansion of satisfaction and the progressive expansion of experience. It depends for its force on the reintegration of the infantile experiences of desire and satisfaction, of elation and ecstasy, of excitement in exploration.

The adult personality contains the elements for such a dynamism. These are characteristically disconnected in the adult. The experience of intense desire is rarely associated with situations in which its satisfaction is achievable. It is more characteristically displaced to fugue-like re-enactments of the infantile experiences with the Good Mother. Enthusiasm for new experience is often entirely replaced by mild pleasure in the repetition of well-integrated and unchallenging functions.

The field of interaction with the therapist is one of the important areas in which the patient can discover, exercise, and organize his own dynamism for growth. Here, it can come into interaction with the therapist's dynamism for growth and enthusiasm about exploring. As the therapy situation becomes so integrated, this interpersonal field can take on some of its own dynamic. It tends to push itself to its own logical conclusion and to draw both patient and therapist into even further creative explorations.

In the resynthesis of the constructive aspects of the patient's life, his assets, his push to satisfaction, his curiosity, the therapist must identify and relate to the constructive aspects of separate pieces of the patient's behavior and help the patient to connect these with the constructive aspects of apparently widely divergent facets of behavior. The patient's creative capacity for formulating experience, including the experience of his own craziness, will be necessary for a growth dynamism. The honest impulse to be who he feels he is with the people around him, even though this must incorporate his paranoia, can evolve into a more thorough dropping of the pretenses. The relief experienced at the dropping of any pervasive security operation, whether of chronic depression or of acute terror, leaves more room to develop an increasing investment in the search for pleasure.

Likewise the more, progressively, that a chronically unsatisfied need can achieve partial satisfaction, the less will the patient be afraid of the search for more complete satisfaction.

In the infant the original basic dynamism was "desire—integration of function—satisfaction," and this process was characteristically associated with intense pleasure and even with ecstasy. The separate elements of this original dynamism are present in the adult but have been fragmented by the adult organization of personality systems. By a careful search for their elements and an accurate reintegration of the various aspects, the same dynamism can be reintegrated in the adult.

The immediately available base lies in those competencies the exercise of which is currently the base of the productive apparatus of the self-system. A major step is the progressive orientation of the exercise of these functions toward satisfaction, rather than toward the repudiation of satisfaction. The adult personifications of the Good Mother, both in infatuation and in fantasy, can be reconnected, both with the exercise of productive functions and with the drive for individual creativity. As the real capacity for tender exchange and intimacy becomes progressively integrated into the self-system, such good infantile experiences can be reintegrated into this solidly available and intensely satisfying adult experience.

Again, in the progressive development of the capacity for intimacy, each success decreases the discouragement and increases the impetus toward further development. The pleasure that derives from each such experience adds its impetus to the desire-integration-satisfaction dynamism. Both love and creativity become simply ends in themselves, and not subverted to the purpose of counteracting terror. The values of the central paranoia become irrelevant. As the pleasure in love and creativity takes over more of the personality, the horizons of personal satisfaction come to include ever-expanding circles of members of the human race. The grim concept of social responsibility is transformed into pleasure in the privilege of social participation on as wide a base as the person's capacities will permit.

INDEX

INDEX

SHOWCASE
PRESENTS

THE GREEN ARROW

VOLUME ONE

Dan DiDio Senior VP-Executive Editor

Jack Schiff Mort Weisinger George Kashdan Murray Boltinoff Editors-original series

Robert Greenberger Senior Editor-collected edition

Robbin Brosterman Senior Art Director

Paul Levitz President & Publisher

Georg Brewer VP-Design & DC Direct Creative

Richard Bruning Senior VP-Creative Director

Patrick Caldon Executive VP-Finance & Operatio...

Chris Caramalis VP...

John Cunningham VP-Ma...

Terri Cunningham VP-Managing E...

Stephanie Fierman Senior VP-Sales & Marketin...

Alison Gill VP-Manufacturing

Rich Johnson VP-Book Trade Sales

Hank Kanalz VP-General Manager, WildStorm

Lillian Laserson Senior VP & General Counsel

Jim Lee Editorial Director-WildStorm

Paula Lowitt Senior VP-Business & Legal Affairs

David McKillips VP-Advertising & Custom Publishing

John Nee VP-Business Development

Gregory Noveck Senior VP-Creative Affairs

Cheryl Rubin Senior VP-Brand Management

Jeff Trojan VP-Business Development, DC Direct

Bob Wayne VP-Sales

TABLE OF CONTENTS

UNTIL THE 1970'S IT WAS NOT COMMON PRACTICE IN THE COMIC BOOK INDUSTRY TO CREDIT ALL STORIES. IN THE PREPARATION OF THIS COLLECTION WE HAVE USED OUR BEST EFFORTS TO REVIEW ANY SURVIVING RECORDS AND CONSULT ANY AVAILABLE DATABASES AND KNOWLEDGEABLE PARTIES. WE REGRET THE INNATE LIMITATIONS OF THIS PROCESS AND ANY MISSING OR MIS-ASSIGNED ATTRIBUTIONS THAT MAY OCCUR. ANY ADDITIONAL INFORMATION ON CREDITS SHOULD BE DIRECTED TO: SENIOR EDITOR, COLLECTED EDITIONS, C/O DC COMICS.

HERE IS ONE OF US YOU HAVE NOT MET, *GREEN ARROW...* THE *BOWMAN OF BRITAIN!* HE HAS JUST ARRIVED!

GREEN ARROW AND *SPEEDY!* I CAN'T EXPRESS MY DELIGHT, CHAPS! BEEN LOOKING FORWARD TO THIS FOR EVER SO LONG!

SAY! HE CAN JOIN THE FUN RIGHT NOW! HOW ABOUT IT, PAL?... PICK A TARGET AND PROVE YOUR SKILL!

OLÉ! OLÉ! DRAW YOUR ARROWS, GRINGO!

THANK YOU, FRIENDS...THIS GIVES ME THE OPPORTUNITY TO TEST-FIRE MY LATEST SHAFT-- THE *"BIG BEN"!* IT WOULD PLEASE ME IF *GREEN ARROW* DID THE HONORS...

HOLA! A FITTING TRIBUTE TO ZE MASTER ARCHER!

Tick, TOCK- TICK TOCK,

THE CLOCK ON THIS *"BIG BEN"* ARROW HAS BEGUN TO TICK! I SUPPOSE THE HANDS ARE DUE TO STRIKE THE HOUR WHEN THE SHAFT IS RELEASED, BUT WHAT HAPPENS THEN?

FRIGHTFULLY SORRY I CAN'T STAY TO EXPLAIN, OLD BOY! WORKING ON A CASE, YOU KNOW-- IMPORTANT TO THE *"YARD"...*

I'VE GOT A LEAD ON THIS BLIGHTER-- SLIPPERY FELLOW... ESCAPED TO AMERICA... BUT I CAN APPREHEND HIM NOW, IF I HURRY!

I SEE...

REWARD $10,000 FOR THE CAPTURE OF LIMEHOUSE LARKIN WANTED BY THE CROWN AND THE PEOPLE OF THE BRITISH COMMONWEALTH FOR

WELL, I'M SORRY YOU CAN'T STAY TO SEE THIS SHOT! IT'LL BE A LULU, I PROMISE!

SORRY-- GOT TO DASH...

Tick - TOCK TICK

BUT BEFORE THE IMPOSTOR CAN REACH THE EXIT, GREEN ARROW STARTLES HIS COMPANIONS WITH THE MOST SENSATIONAL SHOT OF THE EVENING!

4

DON'T LOOK SO FRIGHTENED, OLD CHAP! YOUR *TIME BOMB ARROW* WON'T GO OFF... I TOOK THE FUSE OUT WHEN I GOT WISE TO YOU!

GREAT SCOTT! THAT'S "COUNTERFEIT" CARSON, THE FORGER WHO SWORE HE'D GET EVEN WITH YOU FOR SENDING HIM TO JAIL!

THE IMPUDENT FOOL! DON'T LET HIM GET AWAY!

LOOK! ZE BRITISH *ARROWCAR!*

HE MUST HAVE STOLEN IT FROM OUR FRIEND WHEN HE TOOK HIS COSTUME!

As the sleek vehicle roars off...

THIS STOP HIM! LAVA *ARROW*-- FROM ISLAND VOLCANO!

As the Polynesian archer's arrow overtakes and penetrates one of the fleeing car's tires...

HUH--? AN ARROW THAT GIVES OFF VOLCANIC HEAT! THE WHEEL IS MELTING!

Unable to drive without a wheel, the culprit takes to his heels--only to meet with another obstacle...

THERE IS QUARRY! JIUJITSU ARROW DELAY FLIGHT UNTIL THIS HUMBLE ONE MAKE CAPTURE!

WHAT IN BLAZES--?

The Japanese arrow grips the fugitive in a judo leg hold and flips him off-balance, but when his pursuer closes in...

YOU CAN TRIP ME, MISTER-- BUT YOU CAN'T HOLD ME!

AIEEEEEE!

5

THE GREEN ARROW

SAVE YOUR ARROWS, **GREEN ARROW**... THIS IS A JOB FOR THE **RED DART**!

ON THE HORIZONS OF CRIME AND CRIME-FIGHTING APPEARS A BIZARRE FIGURE-- DRESSED IN SCARLET, MASKED AND CLOAKED! WERE HIS TINY DARTS--THOSE NEAR-MAGIC MISSILES-- EQUAL TO THE FAMED SHAFTS OF **GREEN ARROW** AND **SPEEDY**, BATTLING BOWMEN OF LAW AND ORDER? THE SURPRISING ANSWER LAY IN THE STRANGE CASE OF...

GREEN ARROW VS. RED DART

NIGHTTIME, AS WOULD-BE WAREHOUSE THIEVES MEET WITH *SURPRISE* OPPOSITION...

BOSS! IT'S **GREEN ARROW** AND **SPEEDY**--FIRIN' THEIR TRICK ARROWS!

WHILE BEHIND A STACK OF CRATES, HIDDEN FROM VIEW...

THIS IS MY LUCKY NIGHT...I GOT A PERFECT BEAD ON THOSE TWO RUN-DOWN ROBIN HOODS!

BUT, BEFORE THE GUNMAN CAN FIRE...

SEEMS I ARRIVED HERE JUST IN TIME!

THE INTRUDER'S TINY DART STRIKES THE GUN...

ZIP!

...AND CARRIES IT TO A NEARBY WALL, PINNING IT THERE!

ALL RIGHT, **SPEEDY**... AFTER THEM!

THUD!

WHEN THE GANGSTERS ARE IN TOW...

WAIT! LET US THANK YOU FOR HELPING US! WHO ARE YOU?

JUST CALL ME THE **RED DART!** I WANTED TO HELP YOU CATCH MUGGSY MILLER... HE'S OUT ON **BUSH PAROLE**, YOU KNOW!

SURE, WE'VE CAUGHT THE MUGGSY MILLER GANG-- BUT WHAT DID HE MEAN BY "BUSH PAROLE"?

THAT'S A WAY OF SAYING MUGGSY ESCAPED FROM PRISON!

TWO NIGHTS LATER, AS ROBBERS RAID A JEWELRY STORE...

THEY SILENCED THE BURGLAR ALARM, BUT DIDN'T REALIZE WE WERE PATROLLING NEARBY!

JEWELRY

2

WELL... I SUPPOSE THAT AS LONG AS YOU'RE ON THE SIDE OF LAW AND ORDER, THERE'S NOT MUCH MORE I CAN SAY!

IT'S BEEN NICE MEETING YOU-- AND I'LL CONTINUE TO DO ALL I CAN TO HELP OUT!

BACK HOME, AS THE ARCHERS RESUME THEIR EVERYDAY IDENTITIES OF WEALTHY OLIVER QUEEN AND HIS YOUNG WARD, ROY HARPER...

HOW LONG DO YOU SUPPOSE THE **RED DART** WILL BE AROUND, OLIVER?

WHO KNOWS, ROY? WE'LL JUST HAVE TO WAIT AND SEE!

MEANWHILE, AT THE **RED DART'S** HEADQUARTERS... WE'RE IN, BOYS! **GREEN ARROW** THINKS THE **RED DART** IS A GREAT GUY-- ALL ON THE SIDE OF LAW AND ORDER... HA, HA, HA!

GREEN ARROW DOESN'T KNOW THAT I'M ACTUALLY JOHN "THE MIDAS" MALLORY-- THE SAME LUG HE SENT TO PRISON THREE YEARS AGO, WHEN HE CAUGHT ME FENCING GOLD FOR SMUGGLERS AND THIEVES!

NOW, MIDAS MALLORY PLAYS THE ROLE OF THE **RED DART**... AND I'M GOING TO ENGINEER THE BIGGEST HAUL THIS TOWN EVER SAW-- THE GOLD BULLION STORED IN **CENTRAL BANK!** LISTEN-- HERE'S MY PLAN...

NEXT EVENING, WHILE THE CITY SLUMBERS...

THE GOLD BULLION LURE IS DRAWING FLIES, **SPEEDY**-- JUST AS WE FIGURED! LET'S GO...

G.A.! LOOK!

CENTRAL BANK

WE'LL HANDLE 'EM! GET BEHIND THE CAR!

THERE THEY COME!

4

LATER, AFTER THE HYPNOTIC SPELL HAS WORN OFF...

I-I FEEL STRANGE-- AS IF I'D BEEN DAYDREAMING... W-WHAT HAPPENED, GREEN ARROW?

I REMEMBER DRAWING AN ARROW-- THEN DROPPING IT-- THAT WAS A HYPNOTIC SUPER-ARROW! WE'VE BEEN IN A TRANCE!

GREAT SCOTT! "COUGAR" CAIN AND HIS MEN--THEY'RE GONE! AND THEY'VE TAKEN THE SUPER-ARROWS! WE'LL BE LIKE CAVEMEN FIGHTING WITH CRUDE WEAPONS AGAINST MEN ARMED WITH AUTOMATIC RIFLES!

YES, SPEEDY. EVEN IF OUR SHAFTS ARE SO OBSOLETE BY COMPARISON, WE'LL FIND THE TARGET WITH STRATEGY!

IN THE DAYS THAT FOLLOW, THE OUTLAWS ROCK THE CITY WITH THEIR STOLEN SUPER-ARROWS.

THAT INVISIBILITY ARROW WORKED GREAT, COUGAR! IT TURNED EVERY-THING WITHIN ITS TARGET TRANSPARENT! THE LAW WON'T SEE ANYTHING TO SHOOT AT WHEN WE TAKE OFF WITH THIS DOUGH!

"COUGAR" CAIN AND HIS MEN-- THEY'VE COMMITTED ANOTHER FANTASTIC THEFT!

THOSE BANK GUARDS WERE TOO SURPRISED TO STOP US WHEN "COUGAR" FIRED THAT ANTI-GRAVITY ARROW AND IT TOWED US INTO SPACE!

DYNAMITE COULDN'T KNOCK DOWN THAT CONCRETE WALL--BUT IT'S A CINCH FOR THE VIBRATION ARROW TO SHATTER IT!

DON'T TALK-- TAKE!

ALARMED BY "COUGAR" CAIN'S DARING RAIDS, AN EMERGENCY MEETING IS HELD BY THE CITY'S OFFICIALS... AND PRESENT AT THE CONFERENCE ARE GREEN ARROW AND SPEEDY!

THE SITUATION IS GRAVE, GENTLEMEN! IF CAIN ISN'T STOPPED, THIS WILL BECOME AN "OPEN CITY" FOR HIM...TO LOOT AT HIS LEISURE!

NOW HE'S BOASTING OF A PARALYSIS ARROW WHICH CAN AFFECT ALL LIVING THINGS IN A MILE-WIDE AREA! HOW CAN WE STOP HIM FROM USING IT TO GRAB THAT GOLD SHIPMENT TOMORROW?

I KNOW THIS DOESN'T LOOK LIKE MUCH OF AN ANSWER TO CAIN'S PARALYSIS ARROW, GENTLEMEN... BUT I'M GOING TO MATCH THIS ANCIENT ARROW FROM THE BATTLE OF HASTINGS AGAINST THAT SUPER-ARROW FROM THE FUTURE!

READER--CAN YOU GUESS THE G.A.'S STRATEGY?

5

THE GREEN ARROW

THE THIEVES CROSSED THE BRIDGE BEFORE IT WAS RAISED--AND ARE ESCAPING!

WE'VE **GOT** TO STOP THEM, **SPEEDY!** THE **CLUES** ARE IN THEIR CAR!

*T*RAGEDY STRIKES WHEN ONE OF OLIVER QUEEN'S FRIENDS IS REPORTED LOST AT SEA, BUT WITH THE TRAGEDY COMES INTRIGUING MYSTERY IN THE FORM OF THE VICTIM'S PERSONAL BELONGINGS... UNUSUAL OBJECTS THAT SEND OLIVER QUEEN AND YOUNG ROY HARPER ONCE MORE ON ADVENTURE'S TRAIL AS **GREEN ARROW** AND **SPEEDY** IN THE CASE OF...

FIVE CLUES TO DANGER

*T*HOUGH PUMMELED UNMERCIFULLY BY A RAGING STORM, THE FAMED **ARROWPLANE** CONTINUES ITS URGENT MISSION...

IT WAS IN THIS AREA, **SPEEDY**-- 50 MILES NORTHEAST OF WHALE ISLAND-- THAT PROF. ANDERSON'S BOAT WAS LAST SEEN!

WE'LL DO ANYTHING WE CAN TO HELP THE COAST GUARD FIND HIM, **G.A.**-- BUT IN **THIS** STORM...**WOW!**

SPEEDY-- DOWN THERE! I THINK I'VE SPOTTED WRECKAGE! TAKE THE CONTROLS WHILE I DESCEND ON THE ROPE LADDER!

BUT AS THE FAMED **ARROWCAR** CLOSES THE GAP... *GREEN ARROW* WAS AFTER US-- BUT HE WAS SHORT-STOPPED! THE BRIDGE IS GOING UP!

SPEEDY-- THE *CATAPULT!* HURRY!

THE POWERFUL CATAPULT HURTLES THE BATTLING BOWMEN UPWARDS...

HAVE YOUR BOW READY, *SPEEDY!* WE WON'T HAVE MUCH TIME TO STOP THEM! FIRE AS SOON AS WE LAND!

AN INSTANT LATER, TWO BOWS TWANG IN UNISON, AND...

THEY GOT OUR TIRES!

SLI-P-P-R-R-T!

ZIP

ZIP

THEN, AS THEY DROP BELOW...

THEY ESCAPED, G. A.!

BUT THEY FORGOT SOMETHING-- WHAT THEY WENT TO THE TROUBLE OF STEALING TONIGHT!

WHY, THEY'RE PROF. ANDERSON'S PERSONAL BELONGINGS WHICH WERE WILLED TO YOU! WHY WOULD THEY WANT THOSE THINGS?

A MORE IMPORTANT QUESTION, *SPEEDY,* IS HOW DID THOSE THIEVES KNOW THESE ARTICLES WERE IN OLIVER QUEEN'S HOME?

WELL, THESE ARE RATHER ODD OBJECTS TO BE WILLED TO SOMEONE-- AN ELEPHANT STATUE, AN ALMANAC, A WATCH, SOME LETTERS AND A HARMONICA!

SO UNUSUAL, IN FACT-- THAT I THINK WE'RE ON TO SOMETHING!

3

SINCE EVERYBODY THINKS YOU ARE DEAD, NO ONE WILL BE LOOKING FOR YOU!

THAT FORMULA BELONGS TO SCIENCE-- NOT TO CRIMINALS!

I WANT THAT FORMULA, AND I WANT IT NOW! I'M BECOMING IMPATIENT, PROFESSOR!

ABRUPTLY...

AND SO ARE WE! LET'S TAKE THEM, SPEEDY!

TWANG

TWANG

I'M GETTING OUT OF HERE!

LET'S USE OUR BOOMERANG ARROWS, SPEEDY, TO ENCIRCLE THEM AND PREVENT THEIR ESCAPE!

ZIP

ZIP

I'VE STILL GOT THE UPPER HAND, GREEN ARROW! LET US GO-- WITH THE PROFESSOR-- OR THIS WHOLE PLACE WILL EXPLODE!

S-S-S-S-S-S

5

BUT FASTER THAN THE BLINK OF AN EYE...

MY **RAIN ARROW** WILL FIX THAT!

SPT-SPT

THEN, OTHER FIGURES ARRIVE...

WE GOT HERE AS FAST AS WE COULD, G.A.! BUT I SEE YOU'VE GOT MATTERS WELL IN HAND. COME ON, CHARACTERS! THE NEXT ARROW YOU'LL SEE WILL BE POINTING TOWARD PRISON!

SO YOU **DID** FIGURE OUT THE CODE, G.A.! I WAS HOPING YOU WOULD!

LET ME SHOW YOU, PROFESSOR...

I ARRANGED THE ITEMS IN THIS ORDER, AND, READING DOWN, THE FIRST LETTER OF EACH WORD SPELLED OUT **WHALE!** IT LED US STRAIGHT HERE TO WHALE ISLAND!

WATCH
HARMONICA
ALMANAC
LETTERS
ELEPHANT

IF YOU WISH, G.A., YOU COULD PLACE THOSE OBJECTS IN YOUR TROPHY ROOM AS A MEMENTO OF THIS CASE!

THAT I'LL DO, PROFESSOR! WE HAVE MANY STRANGE EXHIBITS IN OUR TROPHY ROOM, BUT THIS IS THE FIRST TIME WE'LL DISPLAY A **WHALE!**

THE END.

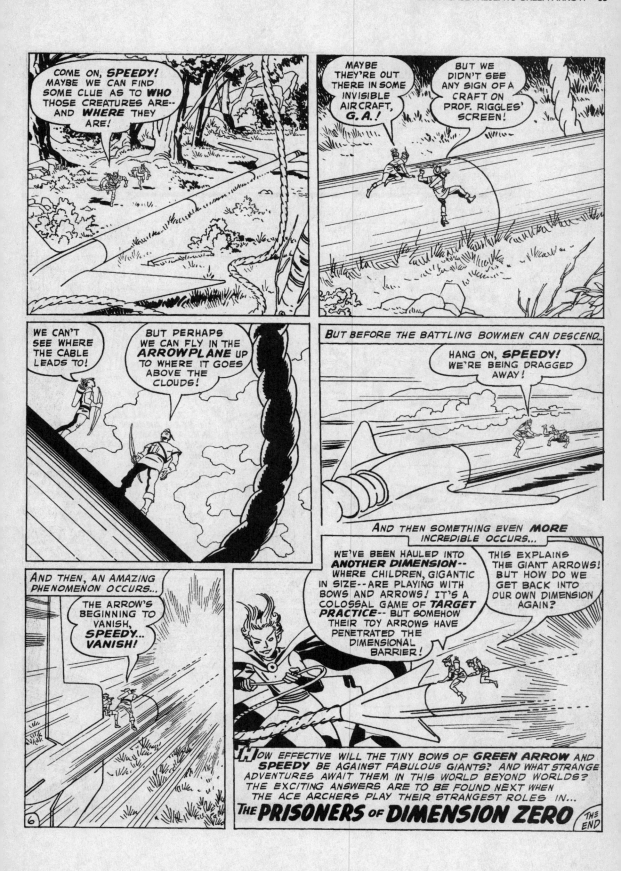

THE GREEN ARROW

G.A.! LOOK! THAT CROOK IS FIRING A *RAY-GUN* AT *XEEN ARROW!*

BUT WHAT CAN WE DO-- WITH OUR SMALL BOWS AND ARROWS?

AWAY FROM OUR TIME, OUR SPACE, OUR *DIMENSION* GO *GREEN ARROW* AND *SPEEDY*-- INTO A STRANGE REALM NO ONE KNEW EXISTED! WHAT GOOD CAN THEIR INCREDIBLY TINY BOWS SERVE THEM HERE, IN THIS, AN ASTOUNDING LAND OF SUPER-GIANTS? FANTASTIC SURPRISES AWAIT THE BATTLING BOWMEN WHEN THEY FIND THEMSELVES... *"PRISONERS of DIMENSION ZERO!"*

ONE DAY, AS A *FLAMING GREEN ARROW* ZIPS OVER THE CITY, FLASHING A FAMILIAR SIGNAL, THE FAMED *ARROWCAR* ROARS FROM THE SECRET *ARROWCAVE*...

WHAT NOW, *GREEN ARROW?*

ANOTHER OF THE MYSTERIOUS, GIANT ARROWS HAS LANDED, *SPEEDY!* WE'RE GOING TO INVESTIGATE IT!

PRESENTLY, WHEN THE ACE ARCHERS ARRIVE AT THE SPOT WHERE THE *COLOSSAL SHAFT* HAS FALLEN...

G.A.! IT'S A *CABLE ARROW!* AND IT'S PULLING US INTO SPACE!

TO WHERE? WHERE ARE WE BEING TAKEN?

AS THE *GREEN ARROW* FACES THE UNKNOWN, IN SPLIT SECONDS PAST ADVENTURES RACE THROUGH HIS MIND, THE WAY A DROWNING MAN REVIEWS HIS PAST LIFE AS HE GOES DOWN FOR THE FINAL TIME...

WILL WE EVER GET *BACK?* WILL WE EVER SEE THE *ARROWCAVE* -- OR THE *ARROWCAR?*

"WILL WE SEE THOSE DAYS AGAIN WHEN KIDS FLOCKED AROUND US IN DEPARTMENT STORES, ASKED FOR OUR AUTOGRAPHS, AND PLAYED WITH THE *GREEN ARROW* TOY ARROW KIT?"

THEN, SUDDENLY, THE FAMED BOWMEN PASS THROUGH A WARP IN SPACE AND EMERGE FROM A TARGET FIRED AT BY GIANT CHILDREN AT PLAY...

HURRY, *SPEEDY!* GET INTO HIDING -- WHERE WE CAN COLLECT OUR WITS!

SOON, UNDER THE COVER OF TOWERING BRUSH...

WHAT'S HAPPENED TO US, G.A.? WHERE *ARE* WE?

THE GIANT ARROWS WHICH LANDED IN OUR WORLD WERE FIRED THROUGH SOME SORT OF DIMENSIONAL "HOLE" BY THESE CHILDREN! WE'RE IN ANOTHER DIMENSION!

ASTOUNDED, THE ACE ARCHERS MAKE THEIR WAY TO THE EDGE OF THE "WOODS," ONLY TO FIND...

WE'RE IN A *ROOF GARDEN* ATOP A HIGH BUILDING -- IN A WORLD OF SUPER-GIANTS!

AND, THOUGH IT EXISTS RIGHT AROUND EARTH, WE COULDN'T SEE IT BEFORE BECAUSE IT'S IN ANOTHER DIMENSION!

SUDDENLY...

A *THIEF!* AND LOOK! HE HAS ROBBED THE BANK!

2

I CAN'T **HEAR** THEM, **SPEEDY!** YET **INSIDE** MY BRAIN I HEAR A VOICE SHOUTING THAT A THIEF IS STEALING SOMETHING! THESE PEOPLE "SPEAK" A **MENTAL** LANGUAGE —A SORT OF "TELEPATHY."

G.A.! LOOK!

THERE HE COMES! **XEEN ARROW!**

XEEN ARROW? HE SEEMS TO BE AN IMITATION OF **YOU**, G.A.!

BUT WHAT AN IMITATION! HE'S AT LEAST A MILE TALL! WE'LL HITCH A RIDE WITH HIM --AND SEE IF HE CAN USE OUR HELP!

THEN, WHEN THE TINY BOWMEN LEAP ONTO THE GIANT ARCHER'S QUIVER...

XEEN ARROW! I'LL FIX YOU!

THE CROOK'S FIRING A **RAY-GUN!** WHAT CAN WE --**OR** EVEN **XEEN ARROW**-- DO WITH BOW AND ARROW?

SPLAT!

THE FABULOUSLY LARGE BOW SUPPLIES THE ANSWER, AS IT FIRES ONE ARROW, WHICH BREAKS INTO **FOUR** MORE!

ONE ARROW IS SHOOTING A "SMOKE SCREEN" AROUND THE THIEF'S EYES--SO HE CAN'T SEE **XEEN ARROW!** THE OTHER GOT THE RAY-GUN!

BUT WHAT WILL THE **OTHER** TWO ARROWS DO?

SPEEDY'S QUESTION IS ANSWERED QUICKLY, AS THE **CHAIN-ARROW** WHIRLS AROUND THE OUTLAW'S UPLIFTED HANDS...

LOOK! IT'S WRAPPING AROUND HIS WRISTS!

AND **THAT** ARROW IS LIKE OUR **ACETYLENE-ARROW!** IT'S "WELDING" THE CHAIN TOGETHER WITH A COLD, HARMLESS HEAT!

3

BUT, THE NEXT MOMENT, THE GIANT ARCHER IN *DIMENSION ZERO* FACES NEW PERIL!...

LOOK, *G.A.!* *XEEN ARROW* IS UNMASKING ONE CROOK, BUT THE OTHER ONE IN THE CAR IS AIMING SOME SORT OF ROCKET GUN AT HIM!

QUICKLY, SPEEDY! WE'VE GOT TO DISTRACT HIM! FIRE YOUR *BALLOON ARROWS!*

THE TINY MISSILES SPEED TOWARD THEIR MARK, EXPANDING AS THEY ZIP THROUGH THE AIR....

SEE! THE ARROWS ARE TAKING IN AIR—GROWING LARGER!

LARGER AND LARGER THEY GROW, UNTIL REACHING THE BURSTING STAGE...

WHAT MAGIC IS THIS? I DIDN'T EVEN SEE *XEEN ARROW* PULL HIS BOWSTRING! YET...

POP!

POW!

POP!

AND IN THAT SPLIT SECOND, *XEEN ARROW* FINDS TIME TO FIRE A HAIL OF *PLASTIC-NET ARROWS!*

LOOK AT *THOSE* FANTASTIC ARROWS! THEY'VE TRAPPED THE SECOND CROOK IN SOME GUMMY SUBSTANCE!

AND WITH THE OUTLAWS OF THIS STRANGE DIMENSION CAPTURED, *XEEN ARROW* SPEEDS AWAY IN HIS VERSION OF AN *ARROWCAR*—UNAWARE OF HIS TWO TINY PASSENGERS...

DEEP INTO THE TOWERING HILLS SPEEDS THE UNUSUAL CAR, AND THEN....

TAKE A LOOK AT *THIS, SPEEDY!* IT'S A SUPER *ARROWCAVE!*

4

A SHORT WHILE LATER, WHEN *XEEN ARROW* ASSUMES HIS OTHER IDENTITY--THAT OF A LEADING SCIENTIST...

WELL, WELL! WHAT HAVE WE *HERE?* TINY PEOPLE-- ARCHERS SUCH AS I! YOU MUST BE THE ONES WHO HELPED ME AT THE ROBBERY! WHERE DO YOU HAIL FROM?

THOUGHT IS OUR ONLY COMMUNICATION WITH YOU! WE CAME FROM ANOTHER DIMENSION -- HAULED HERE BY A GIANT ROPE ARROW!

INDEED! THAT EXPLAINS WHAT HAPPENED TO THE ARROWS OUR CHILDREN USE FROM THEIR *XEEN ARROW* PLAY KITS! THE TOY ARROWS THEY FIRED VANISHED INTO YOUR DIMENSION!

BUT WAIT! YOU WILL BE TRAPPED HERE FOREVER UNLESS WE CAN GET YOU BACK TO YOUR OWN DIMENSION -- WITHIN *SECONDS!* IT'S BECAUSE OF THE *COMET!*

THE *COMET?*

"YES, A WANDERING COMET ORBITED TOO CLOSE TO US AND CAUSED UNUSUAL REACTIONS ON OUR LIGHT-O-SCOPES...

THE COMET DRAWS CLOSER!

OUR LIGHT-O-SCOPES! OBSERVE! AN ALIEN LIGHT IS AFFECTING THEM-- SUCH AS A LIGHT FROM AN UNKNOWN STAR!

THE ALIEN LIGHT MUST'VE BEEN FROM YOUR DIMENSION! TELL ME WHERE YOU ENTERED! I MUST GET YOU BACK THERE AT ONCE -- BEFORE THE COMET DEPARTS --AND CLOSES THE DIMENSIONAL GAP!

WE CAME IN THROUGH A TARGET ON THE ROOF GARDEN!

5

PLACING THE TWO TINY ARCHERS ON THE ARROW, THE HUGE DIMENSIONAL ARCHER FIRES IT, AND...

FAREWELL--FAREWELL! BUT THE ARROW MUST GET THERE QUICKLY-- OR YOU MUST REMAIN HERE FOREVER!

TWANG-G!

OUT OVER THE GREAT, SPRAWLING CITY SPEEDS THE ARROW, WITH ITS MIDGET PASSENGERS...

WILL WE BE IN TIME, *G.A.?* WILL WE?

WE'D BETTER BE! OR...

DIRECTLY INTO THE TARGET'S CENTER IT GOES, AS-- SOMEWHERE FAR OFF --THE WANDERING COMET DEPARTS...

IT CLOSED, *G.A.!* THE DIMENSIONAL GAP *CLOSED!*

AND JUST IN THE NICK OF TIME! DO WE SAY GOODBYE ...TO DIMENSION ZERO?

THEN, ON THE OTHER SIDE OF THE DIMENSIONAL WALL, THE ARCHERS FIRE THEIR *PARACHUTE ARROWS...*

THAT'S IT! NOW WE CAN SETTLE DOWN TO A LANDING!

LATER ON, IN THE *ARROW CAVE'S* TROPHY ROOM...

AND THAT'S IT, *G.A.* -- OUR ONLY MEMENTO OF DIMENSION ZERO! THE GIANT ARROW!

IF WE COLLECT ANY MORE MEMENTOS AS BIG AS THIS, WE'LL HAVE TO HOUSE THEM IN A HANGAR FOR B-29'S!

THE END

THE GREEN ARROW

THE GREEN ARROW

WHAT COULD STOP THIS WEIRD, TRIPLE-THREAT CRIME MECHANISM THAT ROARED THROUGH THE AIR, ROLLED OVER LAND, AND SWAM UNDERWATER? EVEN THE USUALLY UNFAILING BOWS OF **GREEN ARROW** AND **SPEEDY** SEEMED INCAPABLE, THIS TIME, OF PUTTING AN END TO...

THE **MENACE** OF THE **MECHANICAL OCTOPUS**

THIS IS OUR LAST CHANCE, **SPEEDY!** EITHER WE STOP THE OCTOPUS-- OR **WE** GET STOPPED!

HIGH ABOVE THE CITY, IN A TOWERING SKYSCRAPER...

TOMORROW, WE'LL SHIP THESE JEWELS WE'VE INSURED TO OUR MAIN OFFICE... BUT FOR TONIGHT, WE'LL KEEP THEM HERE IN THE SAFE!

SUDDENLY...

GREAT SCOTT! WHAT ON EARTH...?

WHATEVER IT IS, IT'S GOT THE JEWELS!

CRASH!

UPON REACHING A NEARBY RIVER...

A--A SPECIAL COMPARTMENT... IT'S FLIPPING US INSIDE!

SPLASH

"SWIMMING" SWIFTLY, THE OCTOPUS HEADS UPSTREAM, INTO THE DEEP BAY, TOWARD AN UNDERWATER HIDEOUT...

SHORTLY AFTERWARDS, INSIDE...

AND YOU WON'T BE NEEDING THESE BOWS AND ARROWS ANYMORE!

OKAY-- STAY IN THERE!

HA, HA... SO WE'VE FINALLY ELIMINATED THE ONLY MAN WHO MIGHT'VE STOPPED US! WE'LL KEEP *GREEN ARROW* HERE, WHERE HE'LL NEVER BOTHER US AGAIN!

BOSS! LOOK AT THE RADAR SCREEN!

4

SEE THAT BLIP? IT'S THE LINER WE SCHEDULED TO ATTACK--THE LINER WITH THE GOLD BULLION!

OKAY--LET'S GET GOING!

SO ONCE AGAIN, THE GREAT OCTOPUS HEADS OUT TO SEA, TOWARD THE ONCOMING LINER...

AND STARTLED PASSENGERS SOON WITNESS A SCENE OUT OF ANCIENT LEGEND...

LOOK! LOOK AT THAT!

GREAT GUNS! WHAT SORT OF MONSTER IS IT?

WHILE BACK IN THE UNDERWATER HIDEOUT...

THEY'RE GONE--GONE TO LOOT THE LINER... AND WE'RE HELPLESS!

NOT ALTOGETHER, SPEEDY! I WAS ABLE TO HIDE ONE ARROW...THE ACETYLENE ARROW!

WE CAN USE IT TO BURN OUR WAY FREE! BUT DON'T LOOK AT THE LIGHT-- IT'LL HURT YOUR EYES!

SSSSSSSS

PRESENTLY...

NOW WE'LL USE THE AQUA-LUNG ARROW! FIT THE TIP ON YOUR FACE, LIKE THIS, SO WE CAN BREATHE WHILE WE SWIM!

5

THE GREEN ARROW

WHAT'S *THIS?* IT'S THE *ARROWPLANE,* SURROUNDED BY WHOOPING INDIAN BRAVES! YES! LIKE A PAGE OUT OF OLD-WEST HISTORY, *GREEN ARROW* AND YOUNG *SPEEDY* AMAZINGLY FIND THEMSELVES IN A REAL INDIAN-ATTACK ADVENTURE—AND EVEN THEIR WONDER BOWS, USUALLY SO DEPENDABLE, SEEM TO FALTER AS THEY TRY TO FULFILL THE TERMS OF AN IMPOSSIBLE LEGEND AT THE LAST-DITCH SCENE OF...

THE GREEN ARROW'S LAST STAND

THOSE INDIANS ARE ON THE WARPATH AGAIN, G.A.!

LET'S HOPE OUR TRICK ARROWS HOLD OUT! IF NOT—WE'RE ALL GONERS!

IT'S AN INTERRUPTED VACATION IN THE WEST FOR OLIVER QUEEN AND ROY HARPER WHEN A STORM BLOWS UP...

...AND SINCE THE STORM BEGAN, NO WORD HAS BEEN HEARD FROM PROFESSOR HAGEN AND HIS TWO STAFF MEMBERS, LAST SEEN NEAR CROW MOUNTAIN!

CROW MOUNTAIN IS NEARBY, ROY! LET'S SWITCH TO OUR SECRET IDENTITIES OF *GREEN ARROW* AND *SPEEDY* AND INVESTIGATE THE AREA!

TAKING TO THE AIR SHORTLY AFTERWARDS, THE FAMED *ARROWPLANE* DEFIES TEMPESTUOUS WINDS...

THE REPORT SAID PROFESSOR HAGEN AND HIS AIDES WERE IN A CAR!

THEN WE'LL KEEP OVER THE HIGHWAYS—AND MAYBE SPOT THEM!

FINALLY, WITH MOTOR COUGHING, THE *ARROW-PLANE* SETS DOWN IN A VALLEY, JUST AS THE STORM STARTS TO SUBSIDE...

CHG-CHG-CHG

THE TROUBLE DOESN'T SEEM TOO BAD! WE'LL MAKE REPAIRS AT DAWN!

BUT, WHEN THE SUN RISES...

WE MUST BE *MILES* OUT OF THE WAY! I'VE NEVER SEEN THIS VALLEY BEFORE-- AND IT DOESN'T SHOW ON THE MAPS!

MORE IS WRONG WITH THE MOTOR THAN I THOUGHT--IT'LL TAKE A FEW HOURS TO FIX IT!

IN THE NEXT INSTANT, WAR WHOOPS RING THROUGH THE VALLEY, MINGLED WITH THE POUNDING OF HOOFS AND, LIKE A SCENE FROM THE COVERED WAGON ERA, THE *ARROWPLANE* IS SURROUNDED...

HU-U-U-U-HYU

WHAT IN THE WORLD--? AN *INDIAN* ATTACK!

QUICKLY--GET INSIDE THE *ARROWPLANE!*

THEY'RE DESCENDANTS OF THE OLD *SIOUX* TRIBE! THERE WAS A PART OF THE TRIBE THAT WANDERED OFF IN 1880--AND VANISHED! THEY MUST HAVE MIGRATED TO THIS REMOTE VALLEY!

THAT'S NOT OUR WORRY RIGHT NOW!

A DARING BRAVE RIDES TOO CLOSE--ONLY TO BE DRAGGED FROM HIS PONY BY A *ROPE-ARROW!*

AIEEEE!

3

THE FAMED *NET-ARROW* DROPS UPON THREE OTHER BRAVES...

WHAT MAGIC IS *THIS*?

...WHILE THE SHRIEKS OF *SIREN-ARROW*, FOLLOWING QUICKLY, FINISHES THE ROUT...

SHREEEEE

H'YU! RIDE--RIDE! FLEE FROM THE STRANGE ARROWS!

YOU DROVE THEM OFF, *GREEN ARROW!* NOW WE CAN CLEAR OUT OF HERE!

WE'VE STILL GOT TO REPAIR THE *ARROW-PLANE*, AND IF THEY ATTACK AGAIN-- NO TELLING HOW *SOON* WE'LL RUN *OUT* OF ARROWS! THEN WE'RE AT THEIR MERCY!

WHILE AT THE VILLAGE, THE BRAVES GATHER IN AWE AROUND *BIG TURTLE*, THEIR POWERFUL MEDICINE MAN...

NEVER HAVE WE SEEN ARROWS LIKE THOSE-- NEVER BOWMEN SUCH AS THEY!

LET ME ALONE-- *BRAVE TURTLE* MUST THINK!

THE BRAVES FALTER-- AND I ALSO FIND MY POWER *OVER* THEM FALTERING! THEY ADMIRE THIS BOWMAN TOO MUCH!

TRUE! AND IF THEY GET TO TALK TO THE PALE-FACES ABOUT WHAT THE WORLD IS LIKE OUTSIDE THIS VALLEY... YOU WILL LOSE YOUR POWER OVER THEM!

THEY WILL BE TOLD THAT THE INDIAN WARS HAVE LONG CEASED-- AS YOU AND I KNOW! AND THEY WILL WANT TO RETURN TO THE OUTSIDE WORLD!

HERE I AM A *CHIEF*--THERE I WOULD BE *NOTHING!* I WILL STOP THE PALEFACE BOWMAN--AND I THINK I KNOW HOW! I WILL INVENT A TRIBAL *"LEGEND"* TO HELP ME DISPOSE OF THE PALE-FACE ARCHER!

4

LATER, UNDER A PEACE SIGN, *BIG TURTLE* AND THE BRAVES APPROACH THE *ARROWPLANE*...

WE COME TO PAY HOMAGE TO THE GREATEST BOWMAN OUR TRIBE HAS EVER SEEN--THAT IS, IF YOU *ARE* THE GREATEST! HOWEVER, THE TERMS OF THE *LEGEND* SHALL DETERMINE THAT!

LEGEND?

AN ANCIENT LEGEND SAYS THAT THE *GREATEST* OF BOWMEN CAN FIRE ARROWS AT THE *FULL MOON,* AND CAUSE THUNDER TO ROAR ACROSS THE SKIES--AND RAIN TO FALL!

HE'S COME UP WITH A PHONEY LEGEND-- AND IT'S A WHOPPER!

*T*HEN, SPEAKING ALONE WITH *SPEEDY*...

IT'S THE PERIOD OF THE *QUARTER-* MOON, TO BEGIN WITH--AND THERE'S NO RAIN IN THE AIR! AND WHERE'LL WE GET *THUNDER?*

WE'RE GOING TO HAVE TO DO SOMETHING--FAST! THAT MEDICINE MAN IS UP TO SOME REAL, TALL TRICKERY! NOW, LISTEN...

A MOMENT LATER, THE FABLED ARCHER FIRES AN ARROW INTO THE SKY...

TWANG

*A*ND STILL ANOTHER MOMENT LATER, THE ASTON- ISHED BRAVES GAPE IN AWE AT A STAGGERING SIGHT...

LOOK! LOOK THERE! THE *FULL MOON!*

TWANG

YES--AND I MUST FIRE OTHER ARROWS AT THE MOON!

*T*HEN, THUNDER ROARS ACROSS THE HEAVENS... AND...

KWR-OOOM

THUNDER-- AND *RAIN!*

TRULY--HE IS THE GREAT BOWMAN OF THE LEGEND!

5

AS FOR *YOU*, BUSTER, YOU AND YOUR COHORTS BETTER START RIDING-- BECAUSE WHEN I TELL THE TRIBE WHAT THE OUTSIDE WORLD IS LIKE NOWADAYS, THEY'LL TRAMPLE YOU ON THEIR WAY OUT!

*T*HEN...

NICE WORK-- WE PULLED IT OFF IN NEAT FASHION!

YEAH! WE SHOT OFF THE *TWO-STAGE ROCKET ARROW*, WHICH RELEASED THE ILLUMINATED *BALLOON ARROW*, RESEMBLING A FULL MOON! THEN...

... OUR *FIRECRACKER-ARROW* EXPLODED ABOVE THE CLOUDS-- SOUNDING LIKE THUNDER--

AND OUR *DRY-ICE ARROW* BROKE OPEN, SEEDING THE CLOUDS WITH BITS OF DRY ICE, STARTING THE RAIN! YOU KNOW, WE'RE RATHER LUCKY-- EVERYTHING *WORKED!*

6

*A*ND LATER, AS THE CRAFT LIFTS SKYWARD...

THERE GO THE INDIANS LEAVING THE VALLEY! BUT WHERE'S *BIG TURTLE?* DID ANYBODY SEE HIM?

I DON'T THINK *ANYBODY* WILL BE SEEING MUCH OF HIM ANYMORE! HIS LEGEND GOT ALL WET-- THANKS TO *G.A.!*

THE END

THE GREEN ARROW

THE GREEN ARROW

PRESENTLY, ADRIFT IN A LIFE-RAFT...

WE'RE IN A SPOT, **SPEEDY!** THIS RAFT HAS NO PADDLE! WE'RE DRIFTING AWAY FROM THE OTHERS!

WOW! IN THESE HEAVY SEAS, THERE'S NO TELLING **HOW** FAR AWAY WE'LL BE BY THE TIME SEARCH PARTIES ARRIVE!

THE FOLLOWING MORNING...

NO SIGN OF LAND OR SHIPS, **SPEEDY!** JUST SHARKS TRYING TO CHEW UP OUR RAFT!

I'LL GIVE 'EM A TASTE OF DRIFTING **WOOD**, NOT MEN!

I ONCE READ HOW THE NATIVES DEFEND THEMSELVES AGAINST SHARKS!-- THRUSTING AN UPRIGHT STICK BETWEEN THEIR OPEN JAWS!

GOOD IDEA, **SPEEDY!**

TWANG!

I'LL GO NATIVE, TOO! WE'LL GIVE THOSE SHARKS A DIET OF **SEA WATER!**

WHUMP!

A DAY LATER, AS A SWIFT CURRENT WASHES THE LIFE-RAFT ONTO A LONELY BEACH...

I WONDER IF THIS ISLAND'S INHABITED?

IT'S NOT LIKELY, **SPEEDY!** WE'RE FAR FROM SHIPPING LANES! IT'S PROBABLY JUST AN UNKNOWN DOT IN THE PACIFIC!

SUDDENLY...

ATTACK! DRIVE THE YANKEES INTO THE SEA!

GOLLY, **GREEN ARROW!** WE'RE SURROUNDED BY WORLD WAR II JAPANESE SOLDIERS! THEY'RE GOING TO OPEN FIRE!

2

QUICK, *SPEEDY!* A BARRIER OF ARROWS!

RIGHT, *G.A.!*

TWANNGG! ZINNNGGG!

AS THE ATTACK MOMENTARILY WAVERS...

DON'T SHOOT! WE'RE NOT YOUR ENEMIES! WE'RE CASTAWAYS WHO STUMBLED ACROSS THIS ISLE!

LIAR! YOU AMERICAN ENEMY! YOU INVADE *TONGI ISLAND!* THROW DOWN YOUR WEAPONS OR PERISH!

FOR 13 YEARS WE WAIT AMERICAN ATTACK! NOW WE READY TO DESTROY INVADER!--THREE CHEERS FOR THE EMPEROR!

THAT MAJOR ISN'T KIDDING, *SPEEDY!* WE'D BETTER TAKE OFF! USE YOUR RIOT-SMASHING ARROWS!

THE *TEAR-GAS* ARROWS!

SSSTTT! PTTT!

AND THE *SMOKE SCREEN* ARROWS! SOON THEY WON'T SEE *THEMSELVES!*

YIII!

WHOOSSHHH!

BUT AS THE TWINS OF THE TWANGING BOW RACE PAST THE CONFUSED SOLDIERS...

KEEP GOING, *SPEEDY!* I'M RIGHT BEHIND YOU! I---- OOOFFF!

3

A MOMENT LATER...

SURRENDER, GREEN-CLAD YANKEE! YOUR ARROW TRICKERY IS FINISHED! MY PATROLS WILL SOON FLUSH YOUR COMPANION FROM HIDING!

OKAY, MAJOR! BUT THIS IS SOME WEIRD JOKE! THE WAR YOU'RE FIGHTING IS *OVER!* JAPAN LOST WORLD WAR II IN AUGUST, 1945!

LIAR! THE EMPEROR IS SUN GOD! JAPAN *DESTINED* TO WIN! WE RECEIVE LAST ORDERS 13 YEARS AGO-- TO HOLD *TONGI ISLAND* TO LAST MAN! BANZAI! THIS WE *DO!*

HOLY CATS! YOU MEAN YOU'VE *LOST CONTACT* WITH TOKYO HEAD-QUARTERS FOR *13 YEARS?*

SURE! RADIO COMMUNICATIONS DESTROYED BEYOND REPAIR BY TYPHOON IN JUNE, 1945! BUT ORDER IS ORDER! WE HOLD *TONGI* AGAINST YOUR BIG FLEET NOW STEAMING OFF EASTERN POINT!

YOU'RE *SEEING* THINGS, MAJOR! THERE'S NO U.S. FLEET HERE!

BUT AS THE MAJOR LEADS THE GREEN ARROW *TO A LOOKOUT POST...*

Y-YOU'RE RIGHT! BUT THAT FLEET CAN'T BE HERE FOR AN INVASION! JUST *MANEUVERS!*

MAJOR TAYAKO NO FOOL! YOU AND YOUNG COMPANION SNEAK ASHORE AS *SCOUTS!* FLEET WAIT TILL YOU RECONNOITER! THEN ATTACK! BUT I *FIX* THEM!

1945 TYPHOON AND 1950 AMMUNITION EXPLOSION WRECK ALL BIG GUNS! THEREFORE WE NEED *NEW* WEAPONS TO DESTROY FLEET!--MINES! AERIAL BOMBS! TORPEDOES!--WHICH *YOU* MAKE FOR MAJOR TAYAKO, YES?

TO DESTROY THE U.S. FLEET? YOU'RE MAD!

NO! MAJOR TAYAKO *GENIUS!* YOU BUILD *GIANT* ARROW TORPEDOES, BOMBS, MINES! LAUNCH THEM FROM BIG CATAPULT! SINK FLEET OR YOU DIE! YOU UNDERSTAND?

I SURE GET THE POINT! OKAY, MAJOR! IT'S A DEAL!

4

THAT NIGHT AS *SPEEDY* EVADES THE ENEMY PATROLS AND CIRCLES BACK TO CAMP, HE OBSERVES A STRANGE OPERATION...

GREEN ARROW--HELPING THEM! BUILDING GIANT ARROWS FROM SCRAP MATERIALS-- AND MOUNTING 'EM ON IMPROVISED CATAPULT LAUNCHING-PADS!

ALL SET, MAJOR! I'M FIRING THE *TORPEDO-ARROW* FIRST!

TAKING CAREFUL AIM, *GREEN ARROW* RELEASES THE HUGE CROSS-BOW TRIGGER, AND...

I'M FIRING *ALL* THE ARROW-MISSILES AT ONCE, MAJOR!

WHUMPH!

YES! YES! WE MUST ACHIEVE MAXIMUM *SURPRISE!*

MOMENTS LATER, THE *ARROW-BOMB* AND THE *ARROW-MINE* TAKE OFF...

ZIPPPP!

VROOSSHHHH!

PERFECT LAUNCHINGS, ALL OF THEM! THE AMERICAN SHIPS WILL NEVER KNOW WHAT HIT THEM!

BUT THE ISLAND GARRISON IS HELPLESS AGAINST THE DELUGE OF THEIR FOOD-STORES...

GOOD WORK, *SPEEDY!* BUT YOU HAD NOTHING TO FEAR FROM MY GIANT ARROWS! LOOK OUT TO SEA!

SUDDENLY, AS *SPEEDY,* UNABLE TO FATHOM *GREEN ARROW'S* COOPERATION, FIRES A SALVO OF ARROWS...

THEY'VE FIRED THEIR LAST GIANT ARROWS! I'LL FLOOD 'EM WITH COCONUTS -- KNOCKING OUT *BOTH* MEN AND CATAPULTS!

L-LOOK, MAJOR! THE *OTHER* YANKEE RETURNS!

SQUOOOSSHH!

THE TORPEDO *COLLAPSES* ON IMPACT AND RELEASES A HUGE *GREEN ARROW--*TO TELL THE FLEET I'M HERE!

THE MINE IS SOFT METAL THAT RIPS OPEN ON CONTACT, RELEASING A RED DYE-- THE INTERNATIONAL SYMBOL FOR DANGER!

WHEREAS THE *ARROW-BOMB* IS REALLY A THREE-STAGE ROCKET THAT RELEASES THREE SHORT ARROWS-- THREE LONG ARROWS-- THEN THREE MORE SHORT ONES... THE WORLD'S CODE SIGNAL FOR *DISTRESS!*

I-IT'S AN *S.O.S.* FROM THE *GREEN ARROW!*

AS FOR THIS LAST WEAPON, I'LL DEMONSTRATE! IT'S AN *ARROW MACHINE-GUN!* -- GET THE *POINT*, MAJOR TAYAKO?

GULP!

PLUT-PLUT-PLUT-

LATER, AS A JAPANESE NAVAL OBSERVER ASSIGNED TO THE U.S. MANEUVERS CONFIRMS *GREEN ARROW'S* STORY...

SORRY I DIDN'T KNOW WAR WAS OVER, *GREEN ARROW!* I INSIST ON FORMAL SURRENDER-- ESPECIALLY TO SO WORTHY AN OPPONENT!

THANKS, MAJOR TAYAKO! I'M GLAD WORLD WAR II IS FINALLY ENDED--- *EVERYWHERE!*

THE END

THE GREEN ARROW

THE GREEN ARROW

STARTLING HEADLINES... "GREEN ARROW AND SPEEDY REVEALED AS OLIVER QUEEN AND ROY HARPER!" THUS IS THE WORLD'S GREATEST SECRET FINALLY EXPOSED, WHEN THE BATTLING BOWMEN GO INTO ACTION AS...

"THE UNMASKED ARCHERS!"

OUR LAST ARROW! WE'LL FIRE IT TO STOP THE GETAWAY CAR--THEN END OUR CAREERS AS *GREEN ARROW* AND *SPEEDY!*

YES... WITH OUR SECRET IDENTITIES EXPOSED, WE'RE USELESS AGAINST CRIMINALS!

IN SWIFT PURSUIT OF A GANGLAND CAR, *GREEN ARROW* TAKES AIM, AND...

OUR LAST ARROW, *SPEEDY--* OUR LAST CASE!

TWANG!

YES-- FINISHED, G.A.!

WHAT IS BEHIND THIS STRANGE TALK? WHY ARE THE FAMED TWANGING BOWS TO BE SILENCED FOREVER? TO LEARN THE ANSWER, WE MUST GO BACK TO THE PREVIOUS DAY...

... WHEN WEALTHY OLIVER QUEEN WAS LEAVING HIS CLUB...

DON'T FORGET TOMORROW'S MEETING, OLIVER,... WE'LL BE ANNOUNCING THE NEW CLUB OFFICERS!

HA, HA...HOPE I'M NOT SADDLED WITH THE ENTERTAIN-MENT COMMIT-TEE AGAIN!

1

EARLY THE NEXT MORNING, JUST AS OLIVER AND HIS YOUNG WARD, ROY HARPER, FINISHED BREAKFAST...

LOOK, OLIVER... THE **ARROW-SIGNAL!**

YES-- OUR CALL FROM POLICE HEADQUARTERS! INTO YOUR COSTUME, ROY...

BUT A MOMENT LATER, AS THEIR EYES CAUGHT THE MORNING NEWSPAPER...

ROY! LOOK AT THAT HEADLINE!

IT'S ALL HERE IN BLACK AND WHITE... OUR SECRET IS OUT!

WH-WHAT'LL WE DO NOW?

POST-HERALD

BULLETIN:
GREEN ARROW AND SPEEDY IDENTIFIED!

TWO AND THE SAME!

IN THE MOST SENSATIONAL NEWSBEAT OF THE YEAR IT WAS REVEALED THAT...

PLAYBOY OLIVER QUEEN AND WARD, ROY HARPER...

GREEN ARROW AND SPE...

ONLY ONE THING WE **CAN** DO... ANSWER THIS LAST CALL, THEN HANG UP OUR CRIME-FIGHTING COSTUMES FOR GOOD!

*SHORTLY, AS THE SLEEK **ARROWCAR** SPED THEM TO THE SITE OF A BANK HOLDUP...*

THE CROOKS USED SMOKE BOMBS! QUICKLY, **SPEEDY--** THE **FAN-ARROWS!**

WHIR-RRRRR

WHIR-R-RRa

ZIPPING ABOUT, THE AMAZING ARROWS QUICKLY DISPELLED THE FUMES...

IT'S **GREEN ARROW!** HE'S CLEARED THE SMOKE AWAY WITH TRICK ARROWS!

LET'S RUN FOR IT! MAYBE WE CAN LOSE HIM IN TRAFFIC!

2

BUT AS THE CAR ROARED AWAY...

WE SHOULDN'T HAVE TROUBLE TRAILING THEM WITH A *FOUNTAIN-PEN ARROW!*

THE UNIQUE MISSILE, WRAPPING ITSELF AROUND THE CAR'S REAR BUMPER, WITH ITS POINT DRAGGING THE STREET, LEFT A FLOWING TRAIL OF INK...

...WHICH LED THE ARCHERS TO THIS GRIM MOMENT, WHEN THEY FIRED THEIR LAST SHAFT!

WHREEEEEEEE

BLAST IT--AN ARROW GOT OUR TIRE...AND HERE COMES A SQUAD CAR! WE'RE TRAPPED!

AND AS THE POLICE TOOK OVER...

LET'S GO, *SPEEDY*... THE COMMISSIONER MUST BE WAITING TO SEE US!

YES.... BY NOW, EVERYBODY IN TOWN HAS SEEN THE PAPERS--AND KNOWS WHO WE ARE!

SOON AFTER, IN POLICE HEADQUARTERS...

WELL, COMMISSIONER, HAVE YOU ANYTHING TO SAY TO US BEFORE WE HANG UP THESE COSTUMES FOREVER?

OLIVER QUEEN! ROY HARPER! WHAT ARE YOU DOING IN THOSE OUTFITS? GOING TO A MASKED BALL? --HA, HA!

IT'S A NICE JOKE, OLIVER--BUT REALLY, I'VE GOT WORK TO DO NOW! MAYBE I'LL SEE YOU LATER AT THE CLUB! I WANT TO BE IN ON THE INITIATION OF THE NEW CLUB OFFICERS!

INITIATION? HMM...YES-- MAYBE WE *WILL* SEE YOU LATER!

3

TOWARD NIGHTFALL, WHEN THE UNMASKED ARCHERS RETURN TO THEIR SECRET *ARROW-CAVE*...

NO WONDER THE COMMISSIONER THOUGHT THIS WAS A JOKE! LOOK AT ALL THESE OTHER NEWSPAPERS...NOT A SINGLE STORY ABOUT OUR REAL IDENTITIES! WE'VE BEEN *FOOLED!*

FOOLED? HOW?

I'M CERTAIN THAT I'VE BEEN ELECTED TO A NEW POST AT THE CLUB--AND THE NEWSPAPER "REVEALING" OUR TRUE IDENTITIES WAS A GAG-- PART OF MY *INITIATION!*

OOOOH! AND WE'VE ALREADY UNMASKED--JUST BECAUSE A TRICK NEWSPAPER WAS PRINTED UP!

BUT, WAIT--THEY DIDN'T BELIEVE US! MAYBE IT'S OKAY!

I WISH IT WERE THAT SIMPLE... BUT NOW THEY MIGHT BEGIN GETTING IDEAS--SO WE'VE GOT TO FIGURE OUT A WAY OF *CONVINCING* THEM THAT IT *WAS* A JOKE! THE QUESTION IS-- *HOW?*

LET'S SEE...WE'VE GOT THE *BOXING-GLOVE ARROW,* THE *FOUNTAIN-PEN ARROW*--THE *BALLOON ARROW,* WHICH WE CAN MAKE INTO ANY SHAPE, AND...

R-R-RING!

THE PHONE'S RINGING UPSTAIRS IN THE HOUSE! WE'D BETTER TAKE IT ON THE EXTENSION!

HELLO, OLIVER--THIS IS THE COMMISSIONER! IF YOU STILL WANT TO PLAY *GREEN ARROW*... HA, HA... THERE'S A ROBBERY GOING ON! *GREEN ARROW* WILL BE THERE IN PERSON-- TO WATCH YOUR ACT!

ER... SURE, COMMISSIONER! SOME JOKE!

AFTER GETTING THE FACTS...

ROBBERY? WHERE?

A MODERN "JESSE JAMES" HAS HELD UP A TRAIN CARRYING GOLD BULLION! WE CAN BE OUT THERE IN A FEW MINUTES-- AND SO WILL THE POLICE!

4

AS THEY SPEED TO THE SCENE...

HERE'S OUR PROBLEM... WE'VE GOT TO CONVINCE THE POLICE THAT WE'RE STILL CARRYING OUT OUR "JOKE"-- WHILE, AT THE SAME TIME, WE STOP THOSE CROOKS!

I GET IT... IN REAL ACTION, WE'LL HAVE TO LOOK BAD WITH OUR BOWS--YET BE GOOD ENOUGH TO NAIL THE TRAIN ROBBERS! THIS WILL BE TOUGH!

UPON REACHING THE TRAIN...

I DON'T SEE **GREEN ARROW** YET, OLIVER-- BUT MY MEN HAVE THE CROOKS HEMMED IN AROUND THE TRAIN! CARE TO TRY YOUR SKILL ON THE BOW?

ALL RIGHT-- HERE'S MY CHANCE TO... UH... **PROVE** THAT I'M **GREEN ARROW!**

JUST THEN, HOWEVER...

GREAT GUNS! THERE ARE THE CROOKS! THEY'VE GOT THE CONDUCTOR--HOLDING HIM AS HOSTAGE!

GREEN ARROW WILL FIX THAT! WATCH!

TWANG!

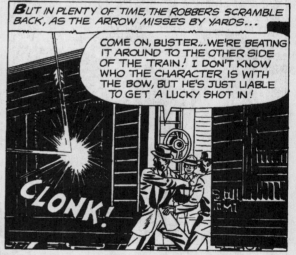

BUT IN PLENTY OF TIME, THE ROBBERS SCRAMBLE BACK, AS THE ARROW MISSES BY YARDS...

COME ON, BUSTER...WE'RE BEATING IT AROUND TO THE OTHER SIDE OF THE TRAIN! I DON'T KNOW WHO THE CHARACTER IS WITH THE BOW, BUT HE'S JUST LIABLE TO GET A LUCKY SHOT IN!

CLONK!

GOOD TRY, OLIVER--BUT DON'T WORRY... MY MEN WILL HANDLE THEM, EVEN IF **GREEN ARROW** DOESN'T APPEAR!

BUT NOBODY SEES THE ARROW THAT "MISSED", AS IT SUDDENLY PIN-WHEELS UP INTO THE AIR...

5

...AND THEN, ON THE OTHER SIDE OF THE TRAIN, OPENS UP INTO A NET!

HEY! WHAT HAPPENED? WE'RE CAGED!

WE GOT A BREAK! GREEN ARROW APPEARED--FROM NOWHERE--AND NAILED THEM WITH HIS FAMOUS NET ARROW!

SEE? DIDN'T I TELL YOU I WAS GREEN ARROW?

NOW-NOW, OLIVER, YOU SIMPLY WON'T GET AWAY WITH IT! THERE GOES GREEN ARROW, HE'S FLYING OFF IN HIS ARROW-PLANE!

OH... HA, HA... I DIDN'T NOTICE! I GUESS THAT'S THE END OF MY LITTLE... ER... JOKE, COMMISSIONER!

BUT HOW IS THIS ILLUSION POSSIBLE?

THE ANSWER LIES BACK IN THE ARROW-CAVE, AWHILE LATER...

WHILE YOU HELD THE COMMISSIONER'S ATTENTION, I WAS ABLE TO GET THE BALLOON-ARROW OFF IN TIME! I SHAPED IT LIKE THE ARROW-PLANE, AND THAT'S EXACTLY WHAT IT LOOKED LIKE UP AGAINST THE MOON!

AND IT TOOK US OFF THE HOOK, SPEEDY... A REAL BIG HOOK! WE'RE BACK IN BUSINESS AGAIN--AS USUAL!

THE END

THE GREEN ARROW

THE GREEN ARROW

Did you ever wonder how GREEN ARROW came to be? Were you ever curious about how he became such a fabulous archer-- and how he invented his thousand-and-one different trick arrows? If so, then join SPEEDY, his young companion, in a surprising adventure that turns back the calendar to reveal...

THIS IS THE GREEN ARROW TODAY...

I'LL FIRE THE FLARE ARROWS, SPEEDY-- LIGHTING UP THE STREET! THEN THE LAW CAN MOVE IN ON THOSE CROOKS!

AND THIS WAS THE ORIGIN OF THE GREEN ARROW...

THERE'S ONLY ONE WAY TO STOP THESE MUTINEERS-- THAT'S BY USING TRICK ARROWS! BUT I'VE NEVER USED THEM LIKE THIS BEFORE! WILL THEY WORK?

THE GREEN ARROW'S FIRST CASE

AS OLIVER QUEEN AND HIS YOUNG WARD, ROY HARPER, WATCH A MORNING TV NEWSCAST...

...AND FOR THE FIRST TIME SINCE REMOTE STARFISH ISLAND WAS OBSERVED BY PLANE, AN EXPEDITION IS SAILING TO EXPLORE IT!

GREAT GUNS!

AND HERE IS AN AERIAL SHOT OF THE ISLAND...

OLIVER! WHAT'S WRONG?

GET INTO YOUR COSTUME, ROY! WE'VE GOT TO GET TO THAT ISLAND BEFORE THE EXPEDITION DOES! IF NOT, MY SECRET IDENTITY WILL BE DISCOVERED!

WITHIN MOMENTS, OLIVER AND ROY HAVE SWITCHED TO THEIR FAMED COSTUMES AS *GREEN ARROW* AND *SPEEDY,* AND AS THEY SOAR AWAY IN THEIR *ARROWPLANE...*

IT WAS ON THAT SAME *STARFISH ISLAND* THAT THE *GREEN ARROW* WAS BORN!

THAT'S RIGHT-- I REMEMBER NOW... YOU TOLD ME SOMETHING ABOUT IT A LONG TIME AGO!

YES, UP TO THEN, I WAS JUST OLIVER QUEEN, WEALTHY PLAYBOY, WORLD TRAVELER! I WAS ON A VOYAGE IN THE SOUTH SEAS...

"ONE NIGHT, I ACCIDENTALLY FELL OFF THE SHIP AND MY SHOUTS FOR RESCUE WENT UNHEARD... BY MORNING, I'D DRIFTED FAR OFF COURSE! THEN I SPOTTED IT..."

AN *ISLAND!* LUCK'S WITH ME!

"I CRAWLED ASHORE AND SLEPT FOR SEVERAL HOURS, THEN BEGAN TO EXPLORE THE ISLAND..."

NOT MUCH CHANCE OF A SHIP ANCHORING HERE! EVEN IF ONE DID COME BY, THE SHOALS WOULD PREVENT IT FROM SAILING IN!

"I FOUND A CAVERN, WHICH PROVIDED SHELTER, AND WAS ABLE TO START A FIRE BY STRIKING STONES TOGETHER..."

YES-- I'VE GOT A "HOME" AND A FIRE! AND I'VE FOUND FRESH SPRING WATER! BUT I NEED FOOD!

"I FASHIONED ARROWS MUCH AFTER THE METHOD USED BY INDIANS-- I CHIPPED STONE TO GET ARROWHEADS, AND TIED THEM TO SHAFTS WITH THIN, STRONG VINES..."

NOW I'D BETTER GET IN SOME PRACTICE, BEFORE I LOSE MY FIRST ARROW WITH MY FIRST SHOT!

2

"I DREW A TARGET ON THE HILLSIDE--AND YOU SHOULD'VE SEEN THAT FIRST SHOT!"

TWANG

DRAT! I OVERSHOT IT!

"HOURS AND DAYS OF PRACTICE TAUGHT ME THAT THE FARTHER AWAY A TARGET IS, THE HIGHER YOU AIM--AND I SOON BEGAN TO LEARN HOW TO ALLOW FOR WINDAGE..."

NOW--I'M BEGINNING TO GET THE HANG OF THIS BUSINESS!

"IT WAS A BITTER LESSON WHEN I SHOT MY FIRST FISH--IT SWAM AWAY WITH MY ONLY ARROW!"

WHAT'S THE GOOD OF NAILING A FISH--UNLESS YOU CAN BRING IT IN!

"I THEN DECIDED TO IMPROVISE A 'FISHING LINE,' SO I ATTACHED A STRONG VINE TO THE ARROW'S SHAFT--AND THUS, FOR THE FIRST TIME, THE *ROPE ARROW* WAS USED..."

THAT'S MORE LIKE IT! BUT I THINK I CAN DEVISE EVEN A BETTER METHOD OF CATCHING *MORE* FISH!

"I LATER FASHIONED A NET FROM MORE VINES, AND PUT IT INSIDE AN ARROW SHAFT, WHICH I HAD HOLLOWED OUT..."

WHEN FIRED, THE NET SHOULD FALL FREE FROM THE ARROW! LET'S SEE IF IT WORKS...

IT OPENED--AND THE NET IS FALLING! SO FAR-- SO GOOD!

3

WHAT A CATCH!

"MY SUCCESS WITH THESE TRICK SHAFTS WAS SO ENCOURAGING I QUICKLY REALIZED INGENIOUS ARROWS COULD BE USED FOR ALMOST *ANY* PURPOSE -- SO IT BECAME A CHALLENGE TO INVENT NEW ONES..."

THERE MUST BE A MEANS OF GETTING COCONUTS WITH AN ARROW! HMM -- I THINK I HAVE IT...

"I FITTED AN ARROWHEAD TO A SHAFT SO THAT IT TURNED FREELY..."

NOW, IF I HAD A RUBBER BAND, I COULD RIG UP A TRICKY LITTLE "MOTOR"...

"I REMOVED THE ELASTIC BAND FROM MY *SOCKS*, ATTACHED IT TO SMALL HOOKS ON THE ARROWHEAD, AND THEN WRAPPED IT AROUND THE SHAFT..."

NOW -- FOR THE COCONUT TEST!

"I FIRED, AND AS THE ARROW SPED TOWARD THE COCONUT, THE ELASTIC UNWOUND ITSELF, SPINNING THE ARROWHEAD IN THE PROCESS..."

ANOTHER VARIATION OF THE *ROPE ARROW* -- BUT I'LL CALL THIS ONE THE *DRILL ARROW!*

TWANC

"SUCCESS! THE REVOLVING ARROWHEAD DRILLED THROUGH THE TOUGH SHELL--"

COCONUT FOR SUPPER!

④

"THEN, TO CAMOUFLAGE MYSELF WHILE HUNTING SMALL GAME, I COVERED MYSELF WITH A GREEN-LEAF SUIT, AND YOU MIGHT SAY THIS WAS THE ORIGIN OF THE *GREEN ARROW* COSTUME..."

"IMPORTANT DURING ALL THIS WAS THE FACT THAT EACH DAY I WENT INTO THE CAVERN AND CHISELED A SORT OF RECORD ON THE STONE WALL..."

THERE--I'VE NOTED MY NAME, THE DATE I CAME HERE--MY PROGRESS WITH THE BOW AND ARROW--AND ALL ABOUT MY GREEN COSTUME! AN ISLAND DIARY!

"AND THEN ONE EVENING, MY LONELY ROBINSON CRUSOE EXISTENCE WAS BROKEN BY A SHOT!"

A SHIP! A SHIP! IT'S A COMMERCIAL FREIGHTER! THAT WAS ONE OF THE DECK GUNS I HEARD! MAYBE THEY'RE SIGNALING TO SEE IF ANYBODY IS ASHORE!

"SINCE THE SHIP WAS ANCHORED, I DIDN'T TAKE TIME TO START A FIRE--WHICH WOULD BE THE ONLY SIGNAL TO ATTRACT THEM IN THE DESCENDING DARKNESS--BUT INSTEAD, COMMENCED SWIMMING OUT TO THEM..."

FREEDOM AT LAST!

"I EASILY MADE MY WAY UP THE GREAT ANCHOR CHAIN--ONLY TO LEARN WITH DISAPPOINTMENT THAT PART OF THE CREW HAD MUTINIED!"

YOU WON'T GET AWAY WITH THIS, CARTER! YOU KNOW THE PENALTY FOR MUTINY!

HA! BUT BY THE TIME WE UNLOAD THIS CARGO, AND SELL IT, YOU WON'T BE AROUND TO REPORT ANY-THING, CAPTAIN!

"MY AIM WAS TO QUELL THE MUTINY! I FIRST RUBBED MY FACE WITH ANCHOR-CHAIN GREASE, SO THE DECK LIGHTS WOULDN'T REFLECT AGAINST THE WHITENESS OF MY FACE--AND THAT BECAME THE FIRST TIME I EVER WORE A MASK'..."

NOW--TO MAKE MY MOVE!

5

"BUT THE DECK-WATCH SPOTTED ME AND SHOUTED AN ALARM..."

THERE'S A GUY LOOSE ON DECK! GET HIM!

MY *DRILL ARROW*-- THE SAME ONE THAT PENETRATED THE COCONUT SHELLS! NOW, IF IT'LL DO THE SAME WITH THAT *OIL BARREL*...

"I SHOT MY ARROW AND A JAGGED HOLE APPEARED IN THE THIN METAL OF THE BARREL, AND..."

LOOK OUT! HE'S FLOODED THE DECK WITH OIL!

BLAST IT! I CAN'T STAY ON MY FEET!

"QUICKLY, I BROUGHT THE *NET ARROW* INTO PLAY..."

HUH? WHAT HAPPENED?

WE'RE CAUGHT IN A NET!

THAT GREEN COSTUME-- THE BOW AND THE TRICKY ARROWS-- AND THAT MASK OF GREASE! JUST WHO *ARE* YOU?

"I KNEW THEN, IN THAT SPLIT-SECOND, THAT MY EXISTENCE ON THE ISLAND COULD NOW SERVE A USEFUL PURPOSE! WHEN I RETURNED TO CIVILIZATION--I WOULD FIGHT CRIME WITH MY TRICK ARROWS! FROM THEN ON I WOULD BECOME TWO PEOPLE-- OLIVER QUEEN AND..."

THE *GREEN ARROW!* YES, THAT'S IT--JUST CALL ME *THE GREEN ARROW!*

SO YOU SAILED HOME AND BECAME THE *GREEN ARROW* AFTER THAT! *BUT*-- EVERYTHING IS WRITTEN DOWN ON THE CAVERN WALL!

EXACTLY--AND IF THE EXPEDITION SEES IT, THEY'LL KNOW MY SECRET IDENTITY! AND RIGHT NOW-- THEY'RE GOING ASHORE!

BUT, A MOMENT LATER...

TOO LATE! THEY'RE ENTERING THE CAVERN! IN ANOTHER MINUTE THEY'LL KNOW THE WHOLE STORY!

NO, WAIT! THEY'RE CARRYING A GEIGER COUNTER--TO CHECK FOR RADIATION! THIS ISLAND MIGHT'VE RECENTLY BEEN SPRAYED BY NUCLEAR FALL-OUT! THAT GIVES ME AN IDEA!

6

THE GREEN ARROW

Why, suddenly, should the underworld no longer fear the usually awesome bows of **GREEN ARROW** and **SPEEDY?** Why do criminals suddenly laugh tauntingly as again and again they thwart the amazing arrows of the battling bowmen? Indeed, what phenomenon could possibly explain...

THE ARROWS THAT FAILED

TAKE YOUR TIME... DON'T WORRY ABOUT **GREEN ARROW** AND **SPEEDY!** THEY COULDN'T HIT THE SIDE OF A BARN!

H-HE'S RIGHT... OUR ARROWS REFUSE TO STAY ON COURSE!

IN THIS HIDEOUT, ON THIS PARTICULAR DAY, ONE OF GANGLAND'S MOST UNUSUAL CRIME PATTERNS IS BEING DRAWN UP...

BOSS, WE DON'T GET IT! DID YOU SAY THOSE TWO BOTTLES OF STUFF CAN STOP **GREEN ARROW?**

PRECISELY! THIS CHEMICAL FORMULA I'VE STOLEN FROM THE STATE LAB WILL PUT AN END TO HIS BOTHERSOME BOW! YOU'LL SEE -- TOMORROW -- AFTER THE CELEBRATION!

DAILY FANFARE
CITY TO HONOR FAMOUS ARCHERS
GREEN ARROW AND SPEEDY IN TIC...

THE NEXT DAY, AFTER A BIG PARADE, AS THE FAMED ARCHERS SIGN AUTOGRAPHS FOR FANS AND WELL-WISHERS...

ALL RIGHT -- MOVE IN... NOW'S OUR CHANCE!

TOWN HALL

UNNOTICED BY EITHER THE BOWMAN OR ANY OF THE THRONG, TWO PAIRS OF HANDS QUICKLY GO TO WORK...

THAT EVENING, WHEN *GREEN ARROW* AND *SPEEDY* RESUME THEIR EVERYDAY IDENTITIES OF WEALTHY OLIVER QUEEN AND HIS YOUNG WARD, ROY HARPER...

WHEW! I SURE APPRECIATED THE GOOD TIME THE FOLKS SHOWED US, OLIVER--BUT I GET LESS TIRED FIGHTING CROOKS THAN SIGNING ALL THOSE AUTOGRAPHS!

WELL, THEN-- GET BACK INTO YOUR COSTUME, ROY! THERE'S THE *ARROW-SIGNAL* FROM POLICE HEADQUARTERS... WE'RE NEEDED IN TOWN!

IN A FEW SHORT MOMENTS, THE FAMED *ARROWCAR* WHEELS INTO A DARK, TWISTING STREET...

THERE THEY ARE, BOSS--*GREEN ARROW* AND THE KID!

NO NEED TO RUSH, BOYS... THEY CAN'T HARM US!

TOWN ASSAYING OFFICE

THEY CAME OUT THE TOWN ASSAYING OFFICE, *G.A.*!

YES--LOADED DOWN WITH PRECIOUS MINERALS... BUT WE'RE IN TIME TO STOP THEM! FIRE THE *SMOKE ARROWS*--THEN WE'LL CLOSE IN!

BUT AMAZINGLY, JUST BEFORE THE ARROWS REACH THEIR MARKS...

YOU WERE RIGHT, BOSS... THEY SUDDENLY TURNED UPWARDS! I NEVER SAW *GREEN ARROW* MISS BEFORE!

SURE--JUST AS I PREDICTED! NOW, LET'S CLEAR OUT OF HERE!

GOLLY-- WHAT WENT WRONG, *G.A.*?

BEATS ME, *SPEEDY!* ANYWAY, WE CAN'T GO AFTER THOSE CROOKS... WE MUST RETRIEVE OUR *SMOKE ARROWS* BEFORE THEY FALL ON SOME INNOCENT PARTY!

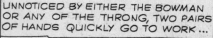

AGAIN THEY ESCAPED! WHY ARE OUR ARROWS *MISSING*?

WHIIEE...

YOU'VE GOT ME, *SPEEDY*! NOW WE'VE GOT TO DO SOMETHING ABOUT THE ARROWS AGAIN! THOSE SIRENS WILL LEAD THE POLICE ASTRAY!

INSTANTLY, TWO *BAZOOKA ARROWS* STREAK UP TO SILENCE THE WAILING SHAFTS...

BLAM! WHROOM!

SEE?... OUR AIM WAS GOOD AGAIN WITH THE *BAZOOKA ARROWS*! WHY DID WE MISS THE CAR WITH OUR *SIREN ARROWS*?

AS IN THE CASE OF THE *SMOKE ARROWS* YESTERDAY, I CAN'T EXPLAIN IT, *SPEEDY*!

LATER, AT THE GANG'S HIDEOUT...

THAT FORMULA WORKS WONDERS, BOSS! WE CAN PULL JOBS RIGHT UNDER THE *GREEN ARROW'S* BOW-- AND HE'S HELPLESS! HOW DOES IT *WORK*?

I'LL GIVE YOU A SIMPLE DEMONSTRATION!

IT'S A NEW CHEMICAL CONCOCTION THAT CONTAINS NEGATIVE AND POSITIVE IONS! BRING NEGATIVE AND POSITIVE CLOSE TOGETHER, AND THEY ACT LIKE MAGNETS, ATTRACTING ONE ANOTHER...

...BUT COAT AN OBJECT, SUCH AS THIS BALL, WITH THE *NEGATIVE* FLUID, THEN ROLL IT NEAR THE *NEGATIVE* SUBSTANCE...

...AND THE REVERSE EFFECT OCCURS! AS WITH MAGNETS, *UNLIKE* CHARGES ATTRACT, *LIKE* CHARGES REPEL!

I SEE... SO OUR CAR IS COVERED WITH ONE FLUID-- AND WE DUMPED THE SAME FLUID IN THE QUIVERS! ANY ARROW THEY FIRE AT US IS REPELLED, JUST LIKE THAT BALL!

4

THUS, ON THE FOLLOWING NIGHT, AS RAIN FALLS ON THE CITY...

WHATTA YOU KNOW? OUR BIG BRAVE ARCHERS AGAIN... ⸚HA, HA⸚

YEAH...THEY CAN SHOOT ARROWS TILL THEY LOOK SILLY-- BUT THEY CAN'T TOUCH US! EVEN THE RAIN DOESN'T WASH THE CHEMICAL OFF THE CAR!

BUT THE RAIN MAKES *ANOTHER* DIFFERENCE--IN THE BATTLING BOWMEN'S WEAPONS...

GREAT GUNS! WE TRIED CHANGING BOWS--AND FORGOT TO WAX THE NEW STRINGS!

THE BOWSTRINGS ARE USELESS! QUICKLY, *SPEEDY*-- GET THE INNER TUBE FROM THE SPARE TIRE... WE'LL *IMPROVISE* A BOW!

SOON...

A MAKESHIFT BOW-- AND A MAKESHIFT ARROW TO FIT IT... A *ONE-WAY STREET SIGN!* HERE GOES...

THE UNIQUE "ARROW" IS RELEASED, AND...

HEY! HE SPEARED THE LOOT!

NO TIME TO GO AFTER IT, EITHER! JUMP IN...WE GOTTA SCRAM OUT OF HERE!

ZI-I-I-I-P!

AND AS THE CAR VANISHES INTO WINDING STREETS...

WE RECOVERED THE BANK MONEY-- EVEN THOUGH THEY GOT AWAY! OUR AIM IS GETTING BETTER!

FUNNY, THOUGH... WE DIDN'T EVEN USE ONE OF OUR OWN ARROWS-- AND FOR THE FIRST TIME SINCE WE WENT AFTER THIS GANG, WE SCORED A HIT! HMM-- I THINK WE'LL VISIT THE POLICE LAB!

ON THE NEXT NIGHT, AS THE GANG PARKS IN FRONT OF A JEWELRY STORE...

LOOK--UP THERE! THE *ARROW-PLANE* AND *GREEN ARROW!* GET BACK IN THE CAR-- WHERE HIS ARROWS CAN'T TOUCH US!

JEWELRY

5

BUT AN ARROW DOES STRIKE THE CAR... AN *OVEN ARROW!*

YOW! THE CAR'S GETTING AS HOT AS A FURNACE! GET OUT!

WHAT HAPPENED? WHAT WENT WRONG?

SHORTLY AFTERWARDS...

LUCKILY, IT RAINED LAST NIGHT-- OR WE MIGHT NEVER HAVE FOUND THE ANSWER TO WHY OUR ARROWS WERE FAILING!

YES-- WHEN WE SUCCEEDED WITH AN *IMPROVISED* BOW AND ARROW, THAT WAS THE TIP-OFF!

WE FIGURED SOME CURIOUS PHENOMENON WAS OCCURING EVERY TIME WE SHOT AN ARROW AT THE CAR, SO WE HAD THE POLICE LAB GO OVER OUR ARROWS AND THE QUIVER! THERE WERE DEFINITE TRACES OF A NEW CHEMICAL THAT HAD BEEN STOLEN FROM THE STATE LAB!

AND WE PROMPTLY LEARNED OF THE "ATTRACTION" AND "REPELLENT" QUALITIES OF THE CHEMICAL-- AND THUS SOLVED THE MYSTERY!

AND SO, WHEN THE GANG IS LED AWAY...

WELL, ALL THE LOOT HAS BEEN RECOVERED... AND THE GANG WILL WIND UP BEHIND BARS! MEANWHILE WE'D BETTER GET UNDER COVER... IT'S STARTING TO RAIN AGAIN!

NOW-NOW, *SPEEDY*... LET'S NOT FORGET-- IT WAS RAIN THAT HELPED SOLVE THIS CASE! WE CAN STAND A LITTLE MORE OF IT!

THE END

THE GREEN ARROW

INSIDE THE WINDOW BELOW, THE CONCENTRATED BEAM STRIKES A SAFE...

SSSSSS

...AS TWO MEN COME RACING IN!

IT WORKED! THE HEAT WAS SO TERRIFIC, IT BURNED RIGHT THROUGH THE SAFE'S DOOR!

CLANG-CLANG-CLANG!

YEAH--BUT IT ALSO SET OFF A BURGLAR ALARM! WE'D BETTER HURRY!

WITHIN SECONDS, AT THE MANSION OF WEALTHY OLIVER QUEEN AND HIS YOUNG WARD ROY HARPER...

THE **ARROW** SIGNAL, OLIVER!

POLICE HEADQUARTERS' CALL FOR **GREEN ARROW** AND **SPEEDY**! LET'S GO, ROY... INTO OUR COSTUMES!

IN THEIR COLORFUL CRIME-FIGHTING GARB, THE TWO SPEED SWIFTLY TO THE SCENE OF THE CRIME!

THE FASTEST WAY UP IS WITH THE **ARROWCAR'S** CATAPULT!

AND WHEN THEY LAND SAFELY ON THE ROOFTOP...

THE CROOKS ARE ESCAPING BY HELICOPTER!

NOT YET...

TWANG TWANG!

②

AS ONE SHAFT CUTS THROUGH THE ROPE LADDER..

TELL THE GUYS IN THE 'COPTER TO DO SOMETHING BEFORE **GREEN ARROW** AND **SPEEDY** NAIL US!

A THIRD ARROW STRIKES THE SATCHEL OF LOOT, AND...

YOW! THERE GOES THE MONEY FROM THE SAFE!

BUT FROM THE HELICOPTER, MORE LENSES COME INTO PLAY!

THIS'LL STOP 'EM!

L-LOOK, G.A.! THE 'COPTER SEEMS TO HAVE BROKEN INTO PIECES!

ANOTHER STUNT WITH LENSES...THERE'S NO SOLID TARGET TO AIM AT!

BY THE TIME **GREEN ARROW** AND **SPEEDY** CAN OVERCOME THE ILLUSION...

THEY LOWERED A SECOND ROPE LADDER...ALL OF THEM ESCAPED!

WELL... AT LEAST WE CAN RETURN THE STOLEN MONEY!

THEY SURE PRODUCED A WEAPON WHICH CRIMELAND'S NEVER USED BEFORE!

YES...AND I WONDER WHAT **OTHER** TRICK LENSES THEY'VE GOT UP THEIR SLEEVES?

THE ANSWER COMES THE FOLLOWING DAY, AS AN ARMORED CAR PULLS UP IN FRONT OF A JEWELRY STORE...

Bartiers JEWELRY

AT THAT MOMENT, "WORKMEN" ABOVE BEGIN LOWERING A SHEET OF "PLATE GLASS"...

AND, TIMED WITH THE LOWERING OF THE GLASS, TWO FIGURES MOVE OUT FROM A DOORWAY...

OKAY, PETIE-- LET'S GO! GET YOUR MASK READY!

THIS SHOULD CAUSE A NICE LITTLE PANIC!

YEAH... AND BRING US PLENTY OF LOOT, TOO!

ABRUPTLY, A FANTASTIC SIGHT GREETS THE ASTOUNDED GUARDS...

EEYOW! G-GIANT CREATURES!

4

IT WORKED! THAT DISTORTION LENS CONFUSED 'EM LONG ENOUGH FOR *US* TO TAKE OVER! GRAB THE LOOT!

THE GANG TAKES OFF WITH ITS HAUL, JUST AS GREEN ARROW AND SPEEDY REACH THE SCENE...

THERE THEY GO--TO THE ROOF!

WE'RE GOING UP AFTER THEM!

BUT AS THE TWINS OF THE TWANGING BOW REACH THE ROOFTOP...

THEY'RE BEING PICKED UP BY A DIRIGIBLE THIS TIME!

MAYBE WE'VE STILL GOT TIME TO STOP THEM!

INSIDE THE DIRIGIBLE'S GONDOLA, HOWEVER...

TURN THE PRISM LENS ON THEM ... THEN WATCH THE FUN!

THE PRISMS CAPTURE BRILLIANT COLORS FROM THE SUN'S RAYS, AND...

THOSE COLORS! THEY'RE SO BRIGHT, I CAN HARDLY *SEE!*

ONLY ONE CHANCE, *SPEEDY...*OUR *INK-ARROWS!* HURRY, BEFORE WE'RE TOO DAZZLED TO SEE ANYTHING!

5

NEXT MOMENT...

TWANG!-
TWANG!

FIRE AWAY!

AT ONCE, THE BIG LENS IS BLOTTED OUT...

THAT'S IT... THE LENS IS USELESS NOW! BRING DOWN THE DIRIGIBLE!

THWACK! THUD!

AND AS A SWIFT FLURRY OF ARROWS PUNCTURES THE CRAFT...

THAT ENDS HER FLIGHT! FIRE THE **SIREN-ARROW**-- AND SUMMON THE POLICE!

WHREEEEEEE

HERE GOES...

6

AND SO, AWHILE LATER, WITH THE CROOKS IN TOW...

YOU CAN COME OUT FROM BEHIND THE LENS, **GREEN ARROW!** WE DON'T HAVE TO MAGNIFY YOU AND **SPEEDY** TO MAKE YOU BIG!

YES--COME TO THINK OF IT, WE'VE HAD OUR FILL OF LENSES FOR A LONG, LONG TIME!

THE END

THE GREEN ARROW

ONE DAY IN SMALLVILLE AS CLARK KENT, SECRETLY **SUPERBOY,** WALKS HOME WITH HIS CLASSMATE, LANA LANG...

WANT TO COME TO THE LIBRARY WITH ME TONIGHT AND WORK ON THE PAGEANT **SMALLVILLE HIGH SCHOOL** IS PLANNING, CLARK?

NO, THANKS, LANA. I'VE BEEN LOOKING FORWARD TO PUTTERING ABOUT IN MY WORKSHOP.

THAT EVENING, IN A HIDDEN BASEMENT ROOM OF THE KENT HOME...

IF LANA ONLY KNEW THAT THIS IS **SUPERBOY'S** WORKSHOP-- WHERE I CONDUCT SUPER-EXPERIMENTS AS A HOBBY IN THE LITTLE FREE TIME I HAVE! HMM... I CAN'T WAIT TILL I PERFECT THIS **TIME MACHINE** GADGET I'VE BEEN TINKERING WITH FOR YEARS!

SUPERBOY ROBOT

IF I CAN EVER GET IT TO WORK, I'LL BE ABLE TO TUNE IN ON THE PAST AND OBSERVE PAST CRIMES WHILE THEY WERE BEING COMMITTED! I'D RECOGNIZE THE CRIMINALS, AND PICK THEM UP EASILY!

CLARION

NO TRACE OF QUARTER MILLION IN GOLD HIJACKED TWO MONTHS AGO!

FOR HOURS THE **BOY OF STEEL** IS ABSORBED IN HIS WORK, AND THEN...

I'VE SUCCEEDED! I'VE TUNED INTO AN ARCHERY CONTEST IN THE PAST AND... WH-WHAT? TH-THAT CAR... AND THE 1959 DATE ON THE LICENSE PLATE! I'VE TUNED INTO THE **FUTURE!** SOME ATMOSPHERIC DISTURBANCE MUST HAVE ENABLED MY MACHINE TO BREAK THROUGH THE TIME BARRIER!

COME ONE! COME ALL! PUBLIC EXHIBIT OF GREEN ARROW'S SKILL PROCEEDS TO CHARITY

1959 B2G-Z

FASCINATED, **SUPERBOY** WATCHES WHAT, TO HIM, IS AN AMAZING SCENE IN THE FUTURE!...

WATCH CLOSELY, LADIES AND GENTLEMEN, AS THE FAMED **GREEN ARROW** SNARES THIS CROOK! THE "CROOK", BY THE WAY, IS A POLICE OFFICER WHO VOLUNTEERED FOR THE TASK. **G.A.** KNOWS THAT IF THE SATCHEL IS DROPPED, STOLEN ANTIQUES WITHIN MAY BE SHATTERED, SO HE FITS A BOOMERANG ARROW TO HIS BOW AND...

RARE ANTIQUES

②

INCREDIBLE! WHY SHOULD A CRIME-FIGHTER OF THE FUTURE USE SUCH OBSOLETE WEAPONS AS THE BOW AND ARROW?

OUT OF *HIS* HANDS INTO *MINE* WITHOUT DANGER OF ITS FALLING WHEN I GET *HIM!* THE RARE ANTIQUES ARE SAFE, BUT AS FOR THAT CROOK...

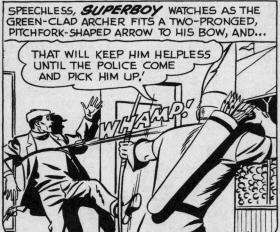

SPEECHLESS, *SUPERBOY* WATCHES AS THE GREEN-CLAD ARCHER FITS A TWO-PRONGED, PITCHFORK-SHAPED ARROW TO HIS BOW, AND...

THAT WILL KEEP HIM HELPLESS UNTIL THE POLICE COME AND PICK HIM UP!

WHAMP!

IN A MOMENT OUR ASSISTANTS WILL START A FIRE ON THAT ROOF-TOP-- AND JUST WATCH HOW *G.A.* PUTS OUT THE FLAMES!

HMM... WITH THOSE INGENIOUS ARROWS AND HIS INCREDIBLE MARKSMANSHIP, NO WONDER HE'S SUCH A GREAT CRIME-FIGHTER!

INSTANTS LATER, *SUPERBOY* PEEKS INTO THE FUTURE TO SEE THE *EMERALD ARCHER* CONFRONT ANOTHER EMERGENCY...

I'LL AIM THESE ARROWS AT THAT WATER TANK, AND WHEN I SINK THEM AT JUST THE RIGHT ANGLE...

...EACH ONE BECOMES A SPOUT SPRAYING A STREAM OF WATER INTO THE BLAZE!

3

GOSH, *I* WAS HOPING TO GO AS WYATT EARP. I'VE GOT A ROBIN HOOD COSTUME I WORE AT A MASQUERADE PARTY LAST YEAR. DO YOU THINK MAYBE YOU'D...

GO AS ROBIN HOOD? OKAY. I ALWAYS ADMIRED HIM... BUT I'VE NEVER BEEN MUCH INTERESTED IN ARCHERY!

HA, HA! HE'LL BE INTERESTED SOON ENOUGH! IT'S STRANGE HOW FATE WORKS! THIS PAGEANT WILL PROBABLY BE HIS FIRST STEP TOWARD FAME AS THE WORLD'S MOST RENOWNED ARCHER!

DAYS LATER, THE STREETS OF SMALLVILLE TEEM WITH TEEN-AGERS DRESSED AS FLORENCE NIGHTINGALE, DANIEL BOONE, CLEOPATRA AND OTHER FAMOUS FIGURES AS THE *HISTORY PAGEANT* BEGINS...

I FEEL A LITTLE STRANGE IN THIS WESTERN SHERIFF'S OUTFIT... BUT WHAT DIFFERENCE SO LONG AS IT PUTS A BOW IN OLIVER'S HANDS! NOW, I'LL HAVE TO FIGURE OUT WAYS FOR HIM TO USE IT... SO HE DISCOVERS WHAT AMAZING HIDDEN TALENTS HE HAS!

SUDDENLY, AS CLARK AND OLIVER WALK OFF TOGETHER...

LOOK! THAT MAN ON HORSEBACK!

HE'S NOT A KID FROM THE PAGEANT! HE LOOKS REAL... LIKE THE REAL...

YEP! LIKE THE REAL JESSE JAMES! THAT'S WHO I AM, PARDNER... THE GREATEST OUTLAW OF ALL TIME COMIN' FROM THE PAST TO GET THE GREATEST SHERIFF... *YOU*, WYATT EARP!

THOSE ARE *REAL* BULLETS, TOO! WH-WHAT CAN IT MEAN?

KRAK!

⑤

...I SEE THE HORSE RIGHT ABOVE ME WITH MY X-RAY VISION! IT WON'T BE SCARED BECAUSE IT DOESN'T SUSPECT I'M HERE!

A MOMENT AFTERWARD, STRONG HANDS OF STEEL REACH UP, AND...

L-LOOK! SUPERBOY'S STOPPING THAT RUNAWAY HORSE DEAD IN ITS TRACKS... FROM UNDER THE GROUND!

LATER, WHEN SUPERBOY RETURNS TO CAPTURE THE BANDIT JESSE JAMES...

HUH? OLIVER IS GETTING JAMES WITH HIS BOW ALL RIGHT... BUT NOT THE WAY THE GREEN ARROW OF THE FUTURE WILL USE A BOW! HE'S CLEVER AND COURAGEOUS... BUT HE SURE HAS A LONG WAY TO GO SO FAR AS MARKSMANSHIP IS CONCERNED!

SOON, AS THE BOY OF STEEL FLIES TO THE POLICE WITH A DESPERADO FROM THE PAST...

HE LOOKS LIKE JESSE JAMES COME TO LIFE, BUT I CAN'T BELIEVE THAT! I WONDER WHO HE IS AND WHY HE RISKED JAIL TO SABOTAGE OUR PAGEANT! HMM... MAYBE THE POLICE WILL GET IT OUT OF HIM!

THAT'S ONLY PART OF THE PUZZLE, SUPERBOY! SOON, AS CLARK RETURNS TO JOIN OLIVER...

AVAUNT, WRETCHED KING! I, SIR MODRED, WOUNDED THEE FATALLY 1,000 YEARS AGO, AND I WILL AGAIN... NOW!

GREAT SCOTT! THAT LAD IS DRESSED AS KING ARTHUR AND HE'S BEING ATTACKED BY SIR MODRED, THE KNIGHT WHO WAS ARTHUR'S LEGENDARY ENEMY... JUST AS WYATT EARP WAS ATTACKED BY JESSE JAMES!

MAYBE OLIVER WASN'T ACCURATE BECAUSE HE WAS SO INEXPERIENCED! I'LL BORROW THAT ROPE FROM MY CLASSMATE, TOM HAWKES, WHO IS DRESSED AS HOUDINI, THE GREATEST ESCAPE ARTIST OF ALL TIME, AND GIVE OLIVER A CHANCE TO MEET THIS EMERGENCY AS THE GREEN ARROW MIGHT!

7

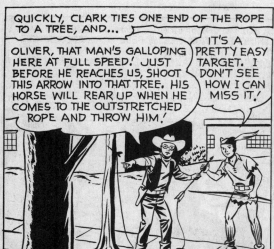

QUICKLY, CLARK TIES ONE END OF THE ROPE TO A TREE, AND...

OLIVER, THAT MAN'S GALLOPING HERE AT FULL SPEED! JUST BEFORE HE REACHES US, SHOOT THIS ARROW INTO THAT TREE. HIS HORSE WILL REAR UP WHEN HE COMES TO THE OUTSTRETCHED ROPE AND THROW HIM!

IT'S A PRETTY EASY TARGET. I DON'T SEE HOW I CAN MISS IT!

HOWEVER, A MOMENT LATER...

I...I MISSED AGAIN! GOSH, I'M *TERRIBLE* WITH A BOW AND ARROW!

HE SURE IS... ESPECIALLY FOR SOMEONE DESTINED TO BE THE GREATEST ARCHER OF ALL TIME. HMM... ALL THAT MILK BEING WASTED... I'D BETTER SWITCH TO *SUPERBOY* AGAIN!

MILK JIM BRAND'S BAR-L

SHORTLY, CLARK VANISHES, AND THE *BOY OF STEEL* APPEARS ON THE SCENE...

SAP FROM THIS TREE SHOULD MAKE A QUICK TEMPORARY REPAIR JOB!

MY SUPER-BREATH AIMED AT THAT HOLE WILL CREATE ENOUGH PRESSURE TO KEEP THE MILK FROM FLOWING OUT, AND MEANWHILE I'LL SQUIRT ENOUGH SAP OVER IT TO MAKE A GOOD STRONG PATCH!

MILK JIM BRAND'S BAR-L

THANKS, *SUPERBOY.* I'M KIND OF NEW IN TOWN, SO I GUESS YOU DON'T KNOW ME! JIM BRAND'S THE NAME... I OWN THE *BRAND B RANCH*... AND ANY TIME I CAN DO *YOU* A GOOD TURN, LET ME KNOW!

I SURE WILL, MR. BRAND!

SHORTLY AFTERWARD...

I DIDN'T HAVE TO WORRY ABOUT OLIVER AFTER ALL. HE'S MADE A SLINGSHOT OUT OF HIS BOWSTRING, AND IT LOOKS AS IF HE'LL MANAGE SIR MODRED PRETTY WELL BY HIMSELF!

8

HE *HAS*! HOW IRONICAL! YOUNG OLIVER QUEEN USES THAT BOW CLEVERLY IN EVERY WAY EXCEPT THE ONE WHICH WILL MAKE HIM FAMOUS SOME DAY IN THE FUTURE!

GOOD SHOT!

YES, BUT I'M CAUSING NOTHING BUT TROUBLE WITH THIS THING. HERE, *SUPERBOY*...TAKE IT FROM ME! I NEVER WANT TO *SEE* A BOW AND ARROW AGAIN!

HMM... I CAN'T LET HIM FEEL THAT DISCOURAGED! WHAT CAN I... WAIT! LITTLE FRED WINTERS FROM MY SCIENCE CLASS WHO IS DRESSED AS BENJAMIN FRANKLIN HAS LOST HIS KITE!

LET ME HAVE IT, OLIVER. I'LL SHOW YOU WHAT CAN BE DONE WITH A BOW AND ARROW... AND I WON'T USE SUPER-POWERS, EITHER... JUST ORDINARY SKILL IN ARCHERY!

THE *BOY OF STEEL* TAKES CAREFUL AIM, AND...

THAT WAS TERRIFIC! YOU PINNED THE STRING TO THE POLE SO THE KITE WOULDN'T FLY AWAY! NOW I CAN GET IT BACK IN TIME TO DEMONSTRATE HOW FRANKLIN USED A KITE TO CAPTURE A LIGHTNING BOLT!

HMM...IT'S STILL OUT OF HIS REACH, SO A FEW MORE ARROWS, AND...

QUICKLY *SUPERBOY* FIRES A VOLLEY OF ARROWS, AND THEN...

Y-YOU MADE A LADDER OF ARROWS FOR HIM TO CLIMB! YOUR MARKSMANSHIP IS INCREDIBLE!

NO BETTER THAN *YOURS* WILL BE SOME DAY!

WHY DON'T *YOU* PRACTICE ARCHERY, OLIVER?

IF YOU WORK ENOUGH, I BET YOU COULD ACQUIRE AS MUCH SKILL! I'LL BE BACK AFTER I TAKE *HIM* TO JAIL!

GOSH, IT WOULD BE WONDERFUL TO SHOOT THE WAY HE DOES, BUT {SIGH} I KNOW IT'S JUST AN IMPOSSIBLE DREAM FOR *ME*!

9

MEANWHILE, SHORTLY AFTER *SUPERBOY* LEAVES...

ON GUARD, THOU GREEN-CLAD OUTLAW! I HAVE SPANNED TIME TO GET THEE!

THE SHERIFF OF NOTTINGHAM... ROBIN HOOD'S LIFELONG ENEMY! HE... HE'S COME OUT OF THE PAST TO GET *ME!* HOW CAN I...HMM... "HOUDINI" HAS MORE ROPE! I'LL BORROW SOME!

INSTANTS LATER...

THIS DOESN'T REQUIRE ANY SKILL. I'LL MERELY SHOOT SO THAT THE ARROW FALLS ON THE OTHER SIDE OF THAT BRANCH!

AND WHEN THE SHERIFF THUNDERS DOWN ON OLIVER...

NOW, SWINGING BACK AND FORTH ON THIS ROPE, I CAN SHAKE ALL THE APPLES DOWN ON HIM!

KLUNK! KLUNK! KLUNK!

SOON, WHEN *SUPERBOY* RETURNS, AND HEARS WHAT HAPPENED...

HE USED A BOW AND ARROW, ALL RIGHT... BUT NOT THE WAY HE'LL USE THEM IN YEARS TO COME WHEN HE'LL BE THE FAMED *GREEN ARROW!* I GUESS MY HUNCH WAS WRONG. WHAT TOOK PLACE TODAY CERTAINLY WON'T INSPIRE OLIVER ON HIS ARCHERY CAREER!

BUT EVEN AS OLIVER AND *SUPERBOY* TALK...

L-LOOK! ANOTHER ONE! THAT'S A CHEROKEE CHIEF ATTACKING WILL BRONSON, WHO IS DRESSED AS DANIEL BOONE! THINK YOU COULD STOP HIM WITH AN ARROW, OLIVER?

I DON'T KNOW. I'M SO BAD THAT...WAIT... I HAVE AN IDEA! I *WILL* STOP HIM WITH AN ARROW!

IF HE'S GOING TO USE AN ARROW, WHY IS HE BORROWING THAT HORN FILLED WITH GUNPOWDER?

10

I DON'T GET IT. HE'S STREWING THE GROUND WITH ARROWS, AND POURING GUNPOWDER OVER EACH ONE!

AN INSTANT LATER... SEE? WHEN THAT INDIAN'S HORSE TRAMPLES ON THE *FLINT ARROWHEADS* HIS IRON SHOES WILL STRIKE A SPARK, IGNITING THE GUNPOWDER! THE EXPLOSION WILL SCARE THE HORSE INTO BUCKING, AND UPSET THE RIDER, MAKING HIM DROP HIS FIRE BRAND!

HOWEVER...

HUH? THE IRON SHOE HIT THE ARROWHEAD... BUT NO SPARKS OR EXPLOSION RESULTED!

A CLEVER IDEA, OLIVER, BUT IT FAILED TO WORK. WELL, IT'S TIME *SUPERBOY* TOOK A HAND.

... I MEAN...ER...GIVE A HAND! A LIGHT TAP AND ANOTHER INDIAN BITES THE AIR!

MOMENTS LATER...

I'LL TAKE THESE CHARACTERS TO THE POLICE, AND SEE IF THEY'VE LEARNED ANYTHING ABOUT THE MYSTERIOUS HORSEMEN FROM THE PAST.

GOSH, ISN'T HE WONDERFUL! HOW I WISH THAT SOME DAY I COULD FIGHT CRIME AND HELP PEOPLE THE WAY HE DOES... BUT I ⸨SIGH!⸩ HAVE NO SPECIAL TALENTS OR POWERS!

PRESENTLY, AT POLICE HEADQUARTERS...

JIM BRAND JUST TOLD ME THAT THESE MEN ARE COWHANDS FROM *HIS* RANCH, *SUPERBOY*. THEY READ ABOUT THE *HISTORY PAGEANT*, AND DECIDED TO GO ON A SPREE AND SCARE THE TOWN!

IT WAS ALL A BAD JOKE! I'LL PAY FOR WHATEVER DAMAGE THEY DID!

11

THEY HURT **ME** MORE THAN ANYONE ELSE. THE HORSES THEY USED WERE THOROUGHBREDS I'D SOLD AND PROMISED TO SHIP ABROAD. THE BOAT SAILED TWO HOURS AGO WITHOUT THEM. WOULD **YOU** FLY THEM TO THE **S.S. GOLDEN LEOPARD** FOR ME, **SUPERBOY?**

WHY--ER... SURE!

THE **BOY OF STEEL** CONSTRUCTS A HUGE, PORTABLE, WOODEN CORRAL, AND...

THERE'S THE SHIP BELOW! I'LL FLY DOWN!

LATER, AFTER **SUPERBOY** HAS GONE...

HA, HA! THE **BOSS** SURE WAS SMART! IF WE HAD TRIED TO SHIP THESE HORSES PAST THE CUSTOMS OFFICE, THEY WOULD NEVER HAVE PASSED THE INSPECTORS. THEY'RE PRETTY STRICT EVER SINCE WE HIJACKED THAT QUARTER OF A MILLION IN GOLD MONTHS AGO!

SHORTLY... MELTING THE GOLD DOWN AND SHAPING IT INTO HORSESHOES WHICH WE PAINTED BLACK WAS CLEVER... BUT EVEN MORE BRILLIANT WAS THE BOSS' IDEA OF SABOTAGING THAT PAGEANT! BLAMING HIS MEN FOR PREVENTING HIM FROM PUTTING THESE HORSES ON THE SHIP GAVE HIM A PERFECT EXCUSE TO ASK **SUPERBOY** TO BRING THEM ABOARD!

MEANWHILE... I'LL SAVE THIS ARROW OLIVER USED TO KEEP IN MY TROPHY ROOM. HE DOESN'T KNOW IT, BUT WHEN IT WAS STEPPED ON AND FAILED TO PRODUCE A SPARK, HE SOLVED HIS **FIRST** BIG CRIMINAL CASE WITH AN **ARROW!**

LATER... HUH? WE GOT FULL STEAM AHEAD AND WE'RE GOING **BACKWARDS!**

RIGHT TO JAIL! MY MICROSCOPIC VISION DETECTED FLECKS OF GOLD WHERE THAT INDIAN CHIEF'S HORSE STEPPED ON AN ARROW! I SUSPECTED THIS SCHEME AS SOON AS I HEARD BRAND'S STORY!

S.S. GOLDEN LEOPARD

12

SOON, AT THE SMALLVILLE POLICE STATION...

THAT HORSE HIT THE FLINT HEAD OF THE ARROW WITH JUST THE RIGHT FORCE TO CREATE A SPARK, SO WHEN NO SPARK RESULTED, I SUSPECTED THAT ITS SHOES WERE MADE OF **SOFTER** METAL-- LIKE **GOLD**--WHICH WOULDN'T PRODUCE SPARKS. YOUNG OLIVER QUEEN REALLY SOLVED THE CASE!

THAT EVENING, IN **SUPERBOY'S** WORKSHOP...

WHATEVER ATMOSPHERIC DISTURBANCE ENABLED ME TO TUNE IN ON THE FUTURE HAS PASSED! I CAN'T GET A THING. HMM... MAYBE IN A FEW YEARS, AFTER LOTS MORE TINKERING, I'LL FINALLY GET THIS MACHINE TO WORK!

AND, NEXT MORNING, WHEN OLIVER PAYS CLARK A VISIT...

I'M AFRAID I WAS A DISGRACE AS ROBIN HOOD. I MUST BE THE WORLD'S WORST ARCHER!

LITTLE DOES HE KNOW THAT ONE DAY HE'LL BE THE WORLD'S **BEST!** HOWEVER, I GUESS HE'LL GET HIS TRAINING SOMEWHERE ELSE. **I** WASN'T SELECTED BY FATE TO GIVE IT TO HIM!

AND NOW THAT YOU'VE SEEN HOW THE *young* GREEN ARROW STRUGGLED WITH A BOW AND ARROW YEARS AGO, TURN THE PAGE TO SEE THE AMAZING SKILL WITH WHICH HE HANDLES ARROWS *Today!*

THE GREEN ARROW

THE GREEN ARROW

NOW THAT YOU HAVE PEERED INTO *GREEN ARROW'S* PAST DAYS WITH *SUPERBOY*, WE SPAN A CHASM OF YEARS TO AN UNUSUAL SCENE IN THE PRESENT— WHERE INDIANS HAVE *RIFLES*, AND SOLDIERS HAVE *BOWS* AND *ARROWS!* AND WE STAND ON THE THRESHOLD OF A SURPRISING ADVENTURE, AS THE FAMED BOWMAN AND HIS YOUNG PAL, *SPEEDY*, BRAVE THE DANGERS OF THE BADLANDS WITH...

"The ARROW PLATOON!"

PLATOON! READY-- AIM-- *FIRE!*

WE'RE THE FIRST U.S. ARMY TROOPS EVER TO BE ARMED WITH A BOW AND ARROW!

THEIR *FIRST* ARROW SHOTS! REMINDS ME OF THAT DAY LONG AGO, WHEN *SUPERBOY* SHOWED ME HOW TO DO IT!

TWANG! TWANG! TWANG! TWANG!

YES, THE YOUTHFUL ARCHER KNOWN TO *SUPERBOY* HAS GROWN UP TO BECOME WEALTHY OLIVER QUEEN, WHO -- ON THIS DAY, WITH HIS YOUNG WARD, ROY HARPER-- DRAWS A FAMILIAR COSTUME FROM A CLOSET...

THERE'S NO GREEN ARROW SIGNAL IN THE SKY, OLIVER! WHAT ARE WE UP TO NOW?

ROY, WE'VE GOT A VERY SPECIAL MISSION ON OUR HANDS TODAY!

THE COSTUME SOON REPLACES QUEEN'S EVERYDAY CLOTHES, AND HE BECOMES THE *GREEN ARROW*, AND ROY HARPER BECOMES *SPEEDY*, TO FORM THE FAMED *BATTLING BOWMEN* DUET!

WE HAVE A *LOT* OF NEW ARROWS TO TEST, AND NEED A LARGE UNPOPULATED TARGET AREA! THE *DESERT* IS JUST THE PLACE!

GOOD, *G.A.!* OUR *ARROWPLANE* CAN GET US THERE IN A FEW HOURS!

MUCH LATER, OVER THE DESERT, THE SLEEK AIRCRAFT IS BATTERED BY A FIERCE SANDSTORM...

WE MIGHT'VE AVOIDED THE STORM WITH RADIO WARNINGS, BUT OUR RADIO WENT DEAD! NOW--WE'LL HAVE TO CRASH-LAND... MOTOR'S CONKING OUT...

CHG-CHG CHG!

SAND GOT INTO THE MOTOR AND RUINED SOME OF THE FINE PARTS, *SPEEDY!* WE'LL WAIT TILL THE STORM LETS UP, THEN WALK TO *STAR CITY* FOR NEW PARTS!

WOW, G.A. THAT'S A 20-MILE HIKE! BUT I GUESS WE'LL HAVE TO HOOF IT!

AN HOUR PASSES, AND WHEN THE STORM DIES...

LOOK, GUYS! IT'S THE *ARROWPLANE*--AND THE *GREEN ARROW* AND *SPEEDY!*

FOUR SOLDIERS! WHAT ARE *THEY* DOING OUT HERE?

WE'RE PRIVATES KOVACS, BRANDT, MARTIN AND COHEN-- FROM FORT SANDS! WE WERE LOOKING FOR SOME CONS WHO ESCAPED FROM FEDERAL PRISON. BUT LOST OUR WEAPONS, CANTEENS AND FOOD RATIONS WHEN OUR DESERT JEEP ROLLED INTO A QUICKSAND PATCH!

WE'RE IN THE SAME SPOT--BUT WE *DO* HAVE WEAPONS! *BOWS AND ARROWS!*

AND A MOMENT LATER...

BUT WE'RE *RIFLEMEN*-- NOT ARCHERS!

NEITHER WAS ROBIN HOOD--UNTIL HE TRIED! HERE--YOU MIGHT AS WELL LEARN HOW TO FIRE THEM!

THIS REMINDS ME OF THE TIME *SUPERBOY* TRIED TO INTEREST ME IN ARROWS, WHEN I WAS A BOY BACK IN SMALLVILLE!

2

FIRST, THE "ARROW PLATOON" LEARNS BASIC ARCHERY...

THERE! YOU'RE ALREADY HITTING TARGETS!

AND I THOUGHT THE INDIANS HAD A MONOPOLY ON THESE THINGS!

TWANG! ZIP! ZIP! TWANG!

LATER, DINNER ON THE WING--A DESERT FOWL--IS BROUGHT DOWN...

ZIP!

EVEN WATER IS AVAILABLE IN THE DESERT--WITH ACCURATE ARROWS...

ALL RIGHT, PRIVATE KOVACS--IT'S YOUR CHANCE TO USE THE *ROPE ARROW!* WE'VE GOT FOOD--AND THE PULP OF THAT CACTUS UP THERE WILL SUPPLY US WITH WATER... IF YOU CAN BRING IT DOWN!

NEXT MOMENT... PERFECT SHOT! YOU'VE GOT IT "HOOKED!" PULL IT DOWN! WE'LL HEAD FOR A GHOST TOWN WE SPOTTED FROM THE *ARROWPLANE,* AND MAKE CAMP!

THEN WE CAN CONTINUE SEARCHING FOR THE CONVICTS!

③

BUT MEANWHILE, THE ESCAPED CONVICTS ARE MET BY OTHER MEN...

GOOD JOB, BOYS! THE TIMING WAS PERFECT!

WE'VE GOT HORSES--FOOD--WATER! *AND,* WE'VE GOT SOMETHING TOO... PLENTY OF AMMO FOR SOME BANK-CRACKING JOBS!

WE'RE GOING TO BECOME *INDIANS*-- AND NO ONE WILL SPOT US! WE SWIPED THIS STUFF FROM A RESERVATION--WAR PAINT AND ALL!

IT IS SOMETIME LATER WHEN THE *ARROW PLATOON* PLODS INTO THE GHOST TOWN...

THIS *IS* A GHOST TOWN! NOTHING MOVING BUT TUMBLEWEEDS, KICKING UP A LITTLE DUST!

BUT I'LL BET THERE WAS A *LOT* OF DUST KICKED UP HERE WHEN THIS PLACE WAS ALIVE!

AFTER "DINNER", MORE PRACTICE...

TWANG-G!

TIP!

SOLDIERS, YOU'LL NEED MORE PRACTICE ON THE *TRICK* ARROWS, SUCH AS THE *BOXING GLOVE ARROW*, THE *BALLOON ARROW*, *NET ARROW* AND *RING ARROW*!

THE UNIQUE ARROWS LOOP SKYWARD, DROPPING DOWN NEAR A DISTANT LIVERY STABLE.

HEY! LOOK OUT THERE! SOMEONE'S SHOOTING ARROWS! WE'VE BEEN SPOTTED!

ARROWS WON'T WORK AGAINST *GUNS*! COME ON! WE'LL BLAST WHOEVER IT IS RIGHT OUT OF TOWN!

THE NEXT MOMENT, THE GHOST TOWN SPRINGS ALIVE, AS FIGURES GALLOP FORTH FROM WITHIN THE STABLE...

WHU-U! WHU! WHU!

SCATTER FOR COVER! THOSE AREN'T GHOSTS, WE'VE AROUSED! WE'RE ATTACKED BY *INDIANS*!

④

THIS IS A TWIST! WE'RE ATTACKED BY *INDIANS* USING *RIFLES*, WHILE *WE* USE BOWS AND ARROWS!

AND NOW WATCH-- AS THEY GALLOP UP!

TIP!

THE FAMED *ROPE-ARROW* ZIPS ACROSS TO THE FAR SIDE OF THE STREET, AND...

LOOK OUT! YI-I-I-I.!

CRASH!

THOSE ARROWS ARE MORE DANGEROUS THAN BULLETS! AND WE THOUGHT THOSE SOLDIERS WOULD BE A CINCH TO DRIVE OUT OF HERE!

ZIP! BANG!

ZIP! POW!

YEAH, BUT WE DIDN'T FIGURE ON THE *GREEN ARROW* BEING WITH THEM--NOT WHEN WE FIRST SPOTTED THEM!

WE WERE GOING TO HIDE OUT HERE, BUT WE CAN'T NOW! LET'S GET THE HORSES AND SCRAM!

CRACK!

THEY'RE STARTIN' TO RUN!

PLATOON-- OPEN FIRE!

BOWS TWANG, AND ARROWS LEAP FORTH--CURLING INTO SPHERES...

TUMBLEWEED ARROWS! THEY'VE STIRRED UP DUST--GIVING US A "SMOKE SCREEN"! RUSH 'EM!

C-CAN'T SEE YOU, PETIE!

5

BOSS! THE SOLDIERS ARE GOING AFTER PETIE AND SLICKER!

YEAH? WE'LL SEE ABOUT THAT! BLAST 'EM WITH RAPID FIRE!

BUT BEFORE THE RIFLES CAN SPEAK, THE BATTLING BOWMEN SPOT THEM, AND...

YI! I DIDN'T GET A CHANCE TO FIRE EVEN ONE SHOT!

ZIP!

ZIP!

THE PACK HORSE IS STILL WITH US -- OUT IN BACK! THE *GRENADES* WE GOT TO BLAST BANKS AND SAFES ARE OUT THERE! *GET 'EM!* WE DON'T NEED RIFLES!

HA! HA! WE'LL MAKE TOOTH-PICKS OUT OF THEIR BOWS AND ARROWS!

WHILE OUTSIDE, PETIE AND SLICKER MAKE A BREAK, BUT...

THEY'RE GETTING AWAY!

NOT YET! WE'LL USE THE *GREEN ARROW'S* TRICK SHOTS...

...THE *BOXING GLOVE ARROW*... AND THE *BOOMERANG ARROW!*

BIFF

CLONK!

GREEN ARROW! LOOK! THEY'RE LOBBING GRENADES!

BUT IN A SPLIT SECOND THE BATTLING BOWMEN DRAW AND RELEASE TWO ARROWS...

THE *ACETYLENE ARROWS!*

THE INCENDIARY SHAFTS MEET THE GRENADES IN MID-AIR, FUSE WITH THEM, AND...

RUN! *RUN!* OUT THE BACK! HURRY!

... THE GRENADES ARE CARRIED SKYWARD, TO EXPLODE HARMLESSLY...

KWHAMMA!

VROOM!

HOTEL

THEN, IN THE SILENCE THAT FOLLOWS...

THEY WERE GOING TO MAKE THIS GHOST TOWN THEIR HEADQUARTERS, *G.A.*--BUT THEY DIDN'T HAVE A *GHOST* OF A CHANCE, EH?

NO, SIR-- NOT WHEN THEY TANGLED WITH THE *ARROW PLATOON!*

THE END

THE GREEN ARROW

AS THE UNUSUAL ARROWS, MADE OF A SPECIAL ALLOY, SPEED THROUGH THE AIR, FRICTION CAUSES THEM TO BEND...

...UNTIL THEY FORM A COMPLETE CIRCLE -- OR *HOOP*...

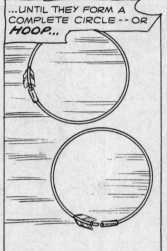

...AND, BEFORE THE ASTONISHED CROOKS CAN MOVE...

BLAST IT! WE'RE NAILED!

A SHORT WHILE LATER, AFTER THE CROOKS ARE TURNED OVER TO THE POLICE...

A CIRCUS, *SPEEDY!* DO YOU FEEL LIKE A LITTLE RELAXATION?

I SURE DO! AND LOOK-- THEY'VE GOT A *GREEN ERROR* CLOWN! SOUNDS LIKE A TAKE-OFF ON YOU!

FEATURING THE AMAZING CIRCUS GREEN ERROR CLOWN

PARKING THE *ARROWCAR* IN A SECLUDED ALLEY-WAY, THE FAMED ARCHERS SWITCH TO THEIR EVERYDAY IDENTITIES OF OLIVER QUEEN AND HIS YOUNG WARD, ROY HARPER...

AND NOW PRESENTING-- THE WORLD-FAMOUS *GREEN ERROR!*

the ERRORCAR

FIRING A "ROPE-LADDER ARROW" INTO A POST, THE *GREEN ERROR* COMMENCES TO CLIMB UP, BUT...

HA, HA, HA! THAT CLOWN'S A RIOT!

KWUMP!

AND NOW-- THE *GREEN ERROR* FIRES HIS FAMED *BOOMERANG ARROW!* WATCH IT, FOLKS*!* *WATCH IT!*

TWANG-G!

YES, THE ARROW *DOES* BOOMERANG, AND ON THE RETURN TRIP, ITS HEAD BLOSSOMS INTO A BOXING GLOVE, AND...

WHACK!

HA, HA, HA THIS IS RICH! HO, HO

HE'S REALLY GOOD, ISN'T HE, OLIVER?

MORE FUN THAN A BARREL OF MONKEYS!

SUDDENLY, WITHOUT ANNOUNCEMENT, THE LIGHTS GO OUT...

HELP! THE GATE RECEIPTS ARE BEING STOLEN!

THIS ISN'T PART OF THE ACT, ROY*!* LET'S SEE WHAT'S UP!

THEN, IN THE SHADOWS OUTSIDE ...

CAN'T WE GET RID OF THESE SILLY ANIMAL COSTUMES NOW, BOSS?

THESE "SILLY" COSTUMES MADE IT POSSIBLE FOR US TO GET THE LOOT*!* BESIDES, THERE'S NO TIME TO SWITCH NOW*!*

SET A LOAD OF THIS -- THE *ARROWCAR* -- THE REAL *ARROWCAR!* HEY*!* THAT MEANS *GREEN ARROW'S* AROUND SOMEWHERE!

OKAY-- FIX IT SO HE CAN'T FOLLOW US! HURRY!

MOMENTS LATER, WHEN OLIVER AND ROY, CLAD AGAIN IN THEIR FAMED GREEN GARB, REACH THE *ARROWCAR*...

SOMETHING'S WRONG, *SPEEDY!* I CAN'T GET THE MOTOR STARTED*!*

AND THERE GOES THE GETAWAY CAR! WHAT'LL WE DO?

R-R-R-R-R-R-R!

3

FOR A BRIEF MOMENT, THE ARCHERS ARE THWARTED...BUT THEN...

WAIT A MINUTE, *SPEEDY!* LOOK THERE!

THE *ERRORCAR!* WHAT'S *IT* DOING OUT HERE?

WE BRING THE *ERRORCAR* OUT HERE BETWEEN SHOWS TO DRIVE AROUND AND EXHIBIT TO PEOPLE! IT'S GOOD PUBLICITY! BUT NOW-- WITH THAT HOLD-UP...

I THINK WE CAN USE YOUR CAR, MY FRIEND! AND MAYBE WE CAN DO SOMETHING ABOUT THAT HOLD-UP!

THERE THEY GO-- AROUND THAT CORNER UP AHEAD!

THE *ERRORCAR* WILL NEVER CATCH THEM-- BUT I'VE GOT AN IDEA! IT'S GOT A CATAPULT-- JUST LIKE THE *ARROWCAR*-- SO WE'LL GO UP ABOVE, TO THE ROOFS, WHERE WE CAN SPOT THEM EASILY!

The ERRORCAR

CLOMP-CLOMP!

THEN, AS THE CATAPULT SWITCH IS PRESSED...

GREAT GUNS!.... THE SWITCH RELEASED A JET OF COMPRESSED WATER!

WOW! THIS IS SOME "CATAPULT"!

WHROOSH!

DON'T COMPLAIN, *SPEEDY*-- IT *WORKS!*

QUICKLY--LET'S USE OUR *FLARE ARROWS!* WE'LL TURN THE NIGHT STREETS INTO DAYLIGHT!

4

BUT INSTEAD OF **FLARES** POPPING FROM THE ARROWS, AS EXPECTED...

WAIT A MINUTE! WHAT KIND OF ARROWS ARE **THOSE?**

BY THUNDER! WE MISTAKENLY SWITCHED OUR QUIVERS FOR THOSE CRAZY ONES USED BY **THE GREEN ERROR** CLOWN!

IN THE NEXT INSTANT, AS THE WEIRD BALLOONS FLOAT DOWN TO THE STREET...

LOOK OUT, LOUIE! WHAT ARE **THOSE** THINGS?

KA-BLAM!

COME ON! WE'LL HAVE TO MAKE IT ON FOOT!

NO, WAIT! THERE'S THE **GREEN ARROW--** AND THE KID!

AS THE CRIMINALS GO FOR THEIR GUNS, TWO MORE OF THE **GREEN ERROR'S** ZANY SHAFTS ARE FIRED, AND...

YIII-II! LOOK OUT!

ROPE ARROWS, SPEEDY! LET'S TRY **THEM** AND SEE WHAT HAPPENS! MAYBE WE CAN TIE UP THE CROOKS WITH THEM!

TWANG!

TWANG!

BUT...

HEY! LOOK WHAT'S HAPPENED!

THE ARROWS ARE DRAGGING US WITH THEM!

5

THE GREEN ARROW

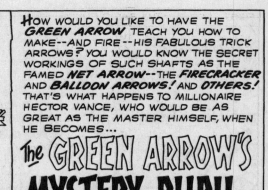

HOW WOULD YOU LIKE TO HAVE THE **GREEN ARROW** TEACH YOU HOW TO MAKE--AND FIRE--HIS FABULOUS TRICK ARROWS? YOU WOULD KNOW THE SECRET WORKINGS OF SUCH SHAFTS AS THE FAMED **NET ARROW**--THE **FIRECRACKER** AND **BALLOON ARROWS**! AND **OTHERS**! THAT'S WHAT HAPPENS TO MILLIONAIRE HECTOR VANCE, WHO WOULD BE AS GREAT AS THE MASTER HIMSELF, WHEN HE BECOMES...

THE GREEN ARROW'S MYSTERY PUPIL

IT'S A SHOWDOWN-- LIKE IN THE OLD WEST GUN DUELS! IT'S THE **GREEN ARROW** AGAINST VANCE--HIS PUPIL! BUT WHEREAS THE **GREEN ARROW** WON'T KILL--VANCE **WILL!**

AN URGENT AD APPEARS IN THE PAPERS ONE DAY, ATTRACTING THE ATTENTION OF TWO FAMILIAR FIGURES...

PERSONAL ADS

GREEN ARROW:
PLEASE CONTACT BOX 41-X
IMMEDIATELY!
IMPORTANT!

PRESENTLY, IN RESPONSE TO THE AD...

I SUMMONED YOU, **GREEN ARROW**, BECAUSE I HAVE A ONE MILLION DOLLAR CHECK FOR YOUR FAVORITE CHARITIES... **IF** YOU'LL TRAIN ME TO BE A MASTER ARCHER LIKE YOURSELF!

I NEVER REFUSE MONEY FOR CHARITIES! I ACCEPT YOUR OFFER!

SOON...

I'VE ALREADY ARRANGED A TARGET GROUNDS! WE CAN COMMENCE AT ONCE!

OF COURSE!

OTHER STARTLING ARROWS ARE PRODUCED FROM THE MAGIC QUIVER...

THE **RAIN ARROW!** IT SERVES MANY PURPOSES-- SUCH AS EXTINGUISHING SMALL FIRES...

AND THE **SMOKE-ARROW**-- USED MANY TIMES AS A SMOKE SCREEN!

AND THE **SUN-ARROW!** ITS BLINDING BRILLIANCE IS CAUSED BY ILLUMINATING GASES RELEASED WITHIN THE SHAFT!

THEN ONE DAY AFTER WEEKS OF TRAINING...

YOU'VE BECOME AN EXCELLENT BOWMAN, MR. VANCE ... AND YOU'VE MASTERED THE TRICK ARROWS! BUT I HAVE A LAST FEW WORDS OF ADVICE!

IN ACTION, I KEEP THE SUN TO MY BACK-- AND IN THE OTHER MAN'S EYES! IT'S A BIG ADVANTAGE! AND REMEMBER-- I **NEVER HAVE,** AND **NEVER WILL,** DRAW AN ARROW TO KILL ANYBODY!

I'LL REMEMBER THAT, **GREEN ARROW!**

LATER THAT EVENING, AS THE BATTLING BOWMEN ASSUME THEIR EVERYDAY IDENTITIES-- AS OLIVER QUEEN AND YOUNG ROY HARPER...

IT'S SURE GOOD TO BE HOME AGAIN!

BUT NOT FOR LONG, OLIVER! IN A FEW MINUTES WE'LL HAVE TO CHANGE BACK INTO OUR COSTUMES AND MAKE OUR REGULAR PATROL!

MEANWHILE, IN TOWN, ARMORED CAR GUARDS FIND THEMSELVES SUDDENLY ENVELOPED IN THICK SMOKE AS THEY ATTEMPT TO MAKE A PAYROLL DELIVERY...

CAN'T SEE! WHAT HAPPENED?

AN ARROW ZIPPED IN-- MADE A LITTLE EXPLOSION-- AND, POW!, ALL THIS SMOKE!

3

THEN, THE AWED CROWD SEES A STARTLING FIGURE RACE FROM THE SCENE AND SWING AWAY-- A *CRIMSON* ARCHER!

THERE'S THE GUY! HE'S GETTING AWAY WITH THE PAYROLL!

HO, HO! INDEED I AM, OAFS!

MEANWHILE, PROWLING THE STREETS ON THEIR REGULAR PATROL IN THEIR SLEEK *ARROWCAR,* THE BATTLING BOWMEN SPOT THE CRIMSON FIGURE, AND...

GREAT GUNS! THE CROOK WHO ROBBED THAT ARMORED TRUCK IS AN *ARCHER!*

WE'LL SEE HOW GOOD HE IS! THE CATAPULT CAN LAUNCH US ALL THE WAY UP!

THE MODERN WILLIAM TELLS LAND ON THE ROOF, AND...

MAYBE A CUFF FROM THE *BOXING GLOVE ARROW* WILL WIPE AWAY THAT "VICTORY" SMILE ON HIS FACE!

TWANG

BUT, AMAZINGLY...

HA! I'LL COUNTER WITH ONE OF THE SAME!

BIFF !

STILL ANOTHER ARROW IS FIRED SKYWARD, WITH ROPE LADDER TRAILING, AND AS THE BATTLING BOWMEN GAPE...

HA, HA, HA! HOW DO YOU LIKE *THIS* FOR A *BALLOON ARROW* EXTRAORDINAIRE?

GREAT SCOTT! HE'S ESCAPING-- BY USING TRICK ARROWS LIKE OURS!

THAT CRIMSON ARCHER USED TRICK ARROWS-- LIKE OURS! COULD HE HAVE BEEN HECTOR VANCE?

H'MMM-- BUT WHY WOULD A *MULTI-MILLIONAIRE* STEAL A FEW THOUSAND DOLLARS? BUT WHOEVER IT WAS, I THINK WE'LL SEE MORE OF HIM!

4

BUT LATER, THE BALLOON ARROW LANDS AT VANCE'S ESTATE, AND...

A DREAM COMES TRUE! ONCE I WAS HERB VRANEY-- KINGPIN RACKETEER!

YEAH! THAT WAS BEFORE THE *GREEN ARROW* CAUGHT UP WITH YOU IN NEARBY GLENVILLE-- AND SMASHED YOUR GANG!

HE SENT ME TO PRISON-- RUINED ME! BUT SINCE, I'VE HAD MY FACE CHANGED BY PLASTIC SURGERY, AND I'VE DUG UP THE MILLIONS IN LOOT I HID BEFORE I WENT TO JAIL! I VOWED TO SMASH THE *GREEN ARROW*-- WITH HIS OWN WEAPONS!

I KNOW THE *GREEN ARROW'S* ONE WEAKNESS! AND WHEN I'VE TAUNTED HIM ENOUGH AS THE *CRIMSON ARCHER*, I'LL MEET HIM IN THE SAME TOWN OF *GLENVILLE*-- AND DESTROY HIM!

THEN YOU'LL BE KINGPIN AGAIN, BOSS-- THE *REAL* KINGPIN!

NIGHT AFTER NIGHT, THE CRIMSON FIGURE UNLEASHES HIS ARROWS- FOR-CRIME ON THE CITY...

THEN, ONE MORNING, TOWARD DAWN, THE *CRIMSON ARCHER'S* TRAIL LEADS INTO THE SUBURBS-- TO THE TOWN OF GLENVILLE ...

WHAT'S HE UP TO NOW, G.A.? WHY IS HE JUST STANDING THERE?

YOU'VE BEEN TRYING TO CATCH ME, *GREEN ARROW!* NOW'S YOUR CHANCE! I'M YOUR OLD PUPIL, HECTOR VANCE-- REMEMBER ME?

EARLY RISERS STARE FROM WINDOWS AT THE SCENE IN THE STREET...

WE'LL STAGE A *DUEL* -- A MODERN VERSION OF THE OLD WEST GUNFIGHTS IN THE STREET! IT'LL DECIDE WHICH OF US IS THE GREATEST ARCHER!

GREAT GUNS! A BATTLE OF BOWS!

THEN, SUDDENLY, THE SLANTING RAYS OF THE RISING SUN FALL SQUARELY IN *GREEN ARROW'S* EYES...

CAN'T SEE--NOT AGAINST THE BRILLIANCE OF THE SUN!

HA! I'M USING AN OLD TRICK OF YOURS, *GREEN ARROW*-- KEEPING THE SUN AT MY BACK!

FIRST, I'LL LOOSEN YOU UP A LITTLE BIT WITH SOME NEAR MISSES! AND REMEMBER-- *YOU* WON'T KILL...BUT *I* WILL!

IF I COULD ONLY GET A BETTER *LOOK* AT HIM! BUT I CAN'T-- THE SUN'S IN MY EYES!

ZING

ROLLING ON THE GROUND, THE FAMED BOWMAN DRAWS AN ARROW--FIRES...

GOT TO USE THE *SMOKE-ARROW!*

THE SMOKE-TRAILING ARROW LOOPS HIGH AND FAR, FORMING A GREAT *CLOUD* BETWEEN THE SUN AND THE TWO BOWMEN...

I THOUGHT *G.A.* WAS A GONER--FOR A MOMENT! BUT HE JUST BLOTTED OUT THE SUN--SO HE COULD SEE! *NOW* LOOK!

ZIP!

TWANG! ZIP!

NEXT, FLAME ARROWS BUILD A FIERY WALL AROUND THE *GREEN ARROW...*

THE *CRIMSON ARCHER'S* GOT HIM RINGED IN WITH FLAMES!

BUT THE FAMILIAR *RAIN-ARROW* IS FIRED, AND...

YOU AVOIDED THE HOTFOOT TREATMENT, *GREEN ARROW!* BUT WAIT'LL YOU SEE WHAT'S NEXT...

TWANG!

GUN-ARROWS! AND THEY FIRE *REAL BULLETS!*

I'LL SAY THEY DO!

K-POW! K-POW!

6

THIS CAN'T GO ON! HE INTENDS TO KILL ME -- AND I'VE VOWED NEVER TO KILL ANYBODY WITH MY BOW AND ARROWS! HE'S GOT A BIG ADVANTAGE, UNLESS...

AND SUDDENLY, ONCE AGAIN, THE SKY LIGHTS UP BRILLIANTLY-- BUT THIS TIME, CAUSED BY...

EH? WHAT HAPPENED? NOW *I* CAN'T SEE!

YOU FORGOT ABOUT MY *OWN* SUN-- THE *SUN-ARROW!*

YOU FINALLY NAILED HIM-- WITH THE *ROPE-ARROW!*

HE DIDN'T MAKE THE MISTAKE OF FACING THE SUN -- SO I "MADE" THE MISTAKE FOR HIM!

LATER IN THE DAY, WHEN THE *CRIMSON ARCHER* IS TAKEN IN TO HEADQUARTERS...

SO IT WAS HECTOR VANCE ALL THE TIME -- OR, RATHER, HERB VRANEY! FINGERPRINTS GAVE HIM AWAY!

YES--AND THE OLD LOOT, PLUS THE NEW, HAS BEEN RECOVERED! HE'S PUT AWAY NOW-- FOR GOOD! MY "PUPIL" WILL NEVER FIRE ANOTHER ARROW!

THE END

THE GREEN ARROW

THE GREEN ARROW

WHY WOULD **GREEN ARROW**, FAMED BATTLER AGAINST CRIME, SUDDENLY TURN AGAINST HIS FAITHFUL YOUNG ASSISTANT, **SPEEDY?** AFTER HAVING SHARED COUNTLESS ADVENTURES TOGETHER, WHY SHOULD THE **GREEN ARROW** WANT TO BREAK UP THEIR TEAM? YET, INCREDIBLE AS IT SEEMS, THIS IS WHAT HAPPENS WHEN **SPEEDY** WATCHES ANOTHER YOUTH TAKE OVER HIS CHERISHED JOB BIT BY BIT IN...

GREEN ARROW'S NEW PARTNER!

EXCELLENT, LAD! YOU FOLLOWED MY INSTRUCTIONS AND MADE A PERFECT **ARROW-SLED!** YOU'LL SOON BE ABLE TO DO EVERYTHING **SPEEDY** DOES!

THERE'S NO DOUBT OF IT NOW! **GREEN ARROW** IS GROOMING THIS BOY TO REPLACE ME! I'M THROUGH AS THE **GREEN ARROW'S** PAL!

ONE NIGHT, AT OLIVER QUEEN'S HOME, AS HIS WARD, ROY HARPER, STUDIES FOR AN EXAM...

I'M GOING OUT FOR A WALK, ROY! YOU STAY HERE AND FINISH YOUR HOMEWORK!

OKAY, OLIVER! PLEASE BRING BACK SOME ICE-CREAM!

BUT, MINUTES LATER, AS ROY LETS SOME AIR INTO THE STUFFY ROOM...

GR-GREAT SCOTT! OLIVER'S SWITCHED INTO HIS **GREEN ARROW** COSTUME!....AND HE'S DRIVING THE **ARROW-CAR!** HE'S UP TO SOMETHING! I'LL CHANGE INTO MY **SPEEDY** UNIFORM AND FOLLOW HIM!

AFTER SEARCHING THE TOWN FOR AN HOUR, **SPEEDY** FINALLY SPIES **GREEN ARROW** NEAR THE RIVER...

HE'S WITH SOMEBODY! A YOUNG BOY...IN COSTUME! HE'S TEACHING HIM HOW TO FIRE ARROWS!

I'VE DEMONSTRATED MY **ICE-CUTTING** ARROW! NOW **YOU** TRY IT! AIM TOWARD THAT FIREBOAT STUCK IN THE ICE-FLOE!

THAT'S IT, LAD! LET THE ARROW CHOP OUT A PATH FOR THE BOAT!

RIPP!

CRUMPPP!

THE FIREBOAT'S FREE! GOOD WORK, SON! SOON YOU'LL BE IN **SPEEDY'S** CLASS AS AN ARCHER! NOW GO HOME AND REST! I'LL SEE YOU TOMORROW NIGHT!

WHO IS THAT BOY? WHY IS **GREEN ARROW** TRAINING HIM BEHIND MY BACK? I--I CAN'T UNDERSTAND IT!

A HALF HOUR LATER, AT OLIVER QUEEN'S APARTMENT...

SORRY I WAS GONE SO LONG, ROY! I MET AN OLD COLLEGE CHUM! HERE'S THE ICE-CREAM YOU WANTED!

THANKS!

HE'S LYING TO ME! HE'S COVERING UP! HE DOESN'T WANT ME TO KNOW WHAT HE WAS DOING! WHY? WHAT IS **GREEN ARROW** HIDING FROM ME?

THE NEXT DAY, AS A DEEPLY WORRIED ROY HARPER WALKS ACROSS THE SCHOOL CAMPUS...

LOOK! THERE GOES THE NEW BASKETBALL COACH! HE JUST REPLACED OLD MAN KINGDON! **NOW** WE'LL WIN GAMES!

R-REPLACEMENT? ...GASP!...W-WHY, THAT'S IT! THAT'S WHY **GREEN ARROW** MUST BE TRAINING THAT BOY! HE--HE'S THINKING...GULP! OF **REPLACING** ME!

THAT NIGHT, AT OLIVER QUEEN'S APARTMENT, AS THE HEARTSICK LAD WATCHES HIS GUARDIAN GET DRESSED...

HE'S GETTING READY TO LEAVE! BUT THIS TIME I'LL TRAIL HIM FROM THE MOMENT HE WALKS OUT THAT DOOR!

ER...ROY! I'M GOING OUT TO PLAY BRIDGE TONIGHT WITH SOME FRIENDS! DON'T WAIT UP FOR ME!

2

SHORTLY AFTER, AS ROY, HAVING SWITCHED TO *SPEEDY,* TRAILS *GREEN ARROW* WITH WILDLY THUMPING HEART...

HMMM... *GREEN ARROW'S* SURE PLAYING *BRIDGE* TONIGHT!-- WITH A BRAND-NEW PARTNER!

TONIGHT I'LL SHOW YOU HOW TO USE THE *AVALANCHE ARROW!*

THE ARROW PICKS UP SNOW AS IT SKITTERS DOWN THE SURFACE OF THE SLOPE... TILL IT CREATES A SNOW-SLIDE!

WHAT ABOUT THE *HARPOON ARROW?* YOU SAID I COULD FIRE ONE TONIGHT!

OKAY! YOU CAN! THERE'S A DERELICT FLOATING IN THE HARBOR! HIT IT FROM THE SHORE! COME ON!

SHORTLY AFTER, AS THE UNKNOWN TRAINEE TAKES AIM...

GOOD SHOT! A FEW MORE PRACTICE SESSIONS AND YOU'LL BE *TREMENDOUS!*-- NOW WE'LL PULL THE DEBRIS TO SHORE WHERE IT WON'T DO ANY HARM!

TWANG!

I--I DON'T UNDERSTAND! WHERE DID I FAIL *GREEN ARROW*? BUT IF HE DOESN'T WANT ME ANYMORE... I-I *MUST'VE* LET HIM DOWN! BUT *WHEN? HOW?* WHY DOESN'T HE *TELL* ME?

LATER THAT NIGHT, AT THE HOUSE...

GOSH, ROY, YOU LOOK SICK! WHAT'S WRONG WITH YOU LATELY, LAD? YOU WALK AROUND AS IF YOU HAD LOST YOUR BEST FRIEND!

HOW TRUE! I *HAVE LOST* HIM! OLIVER DOESN'T TRUST ME OR WANT ME ANYMORE! BUT WHY DOESN'T HE STOP *PRETENDING*? WHY ISN'T HE HONEST ENOUGH TO TELL ME I'M FIRED?

3

A FEW NIGHTS LATER, AS ROY WAITS IMPATIENTLY FOR OLIVER TO COME HOME...

THE *ARROW FLARE!* THERE'S TROUBLE SOMEWHERE IN THE CITY! I'LL SWITCH TO *SPEEDY* AND INVESTIGATE!

BUT A FEW BLOCKS AWAY, AS A POLICEMAN STOPS SPEEDY...

A FUNNY THING JUST HAPPENED, *SPEEDY!* I SAW *GREEN ARROW* FIRE A DISTRESS FLARE... THEN RUN AWAY!

H-HE *DID?*

G-GOOD HEAVENS! THAT FLARE MIGHT'VE BEEN A *DECOY* TO GET ME OUT OF THE HOUSE! I'M GOING BACK TO MAKE SURE!

BUT AS SPEEDY REACHES THE HOUSE, HE MAKES A STUNNING DISCOVERY!

GREEN ARROW'S BROUGHT HIM *HERE!* TO OUR WORK-SHOP! NOW THERE'S NO DOUBT OF IT! I'M *OUT* AND THIS NEW KID IS *IN!*

HERE'S HOW YOU MAKE AN *ARROW SLED!* WATCH CAREFULLY!

SOON AFTER, AS THE VISITOR REMOVES HIS MASK...

THAT'S WONDERFUL, *G.A.!* THE SHAFTS BECOME RUNNERS! NOW I'LL BE ABLE TO BUILD ONE LIKE IT!

G-GREAT SCOTT! I KNOW THAT KID! HE'S IN MY CLASS AT SCHOOL! HIS NAME IS JERRY HALLECK! HOW DID *GREEN ARROW* HAPPEN TO PICK *HIM* AS MY REPLACEMENT?

IN SCHOOL, THE NEXT DAY, AS SPEEDY RESENTFULLY WATCHES HIS RIVAL...

WHAT'S *HE* GOT THAT I HAVEN'T GOT... THAT *GREEN ARROW* SHOULD FAVOR HIM OVER ME?

STUDENTS, WE'VE DISCUSSED THE HISTORY OF ARCHERY FROM THE STONE AGE FLINT ARROW TO ROBIN HOOD'S LONG BOW! NOW LET'S TURN TO MODERN TIMES... AND THE *GREEN ARROW!*

ROY HARPER, CAN YOU NAME ANY OF *GREEN ARROW'S* INGENIOUS ARROWS?

S-SORRY! I DON'T KNOW ANYTHING ABOUT *GREEN ARROW OR* HIS ARROWS!

IF I REVEAL HOW MUCH I *DO* KNOW, SOMEBODY MIGHT BECOME SUSPICIOUS... AND STUMBLE ACROSS MY *SPEEDY* IDENTITY!

I KNOW, TEACHER!

I CAN NAME **ALL** HIS ARROWS! THE SMOKE-SCREEN ARROW! THE BOXING-GLOVE ARROW! THE LARIAT ARROW! THE HANDCUFF ARROW!

SURE! **HE** CAN TALK! NOBODY'LL SUSPECT **HIM**! BUT ONCE HE'S **REPLACED** ME, HE WON'T BLAB SO FREELY! HE...

WAIT! I'VE BEEN TORTURING MYSELF FOR **DAYS**! WELL, I'M **NOT** CONTINUING THIS WAY! IF **GREEN ARROW'S** THROUGH WITH ME, LET HIM TELL ME TO MY FACE! I'M HAVING A SHOWDOWN WITH **G.A.** THIS AFTERNOON!

BUT THAT AFTERNOON, IN THE CITY PARK...

NICE WORK, LAD! THAT'S PERFECT USE OF THE **BUZZ-SAW ARROW!** THIS TIMBER IS CONDEMNED! BUT **ANY** TREE CAN BE CUT DOWN IN THE SAME WAY!

HELP! **HELP!** MY CHILD!

CONDEMNED TREES PARK DEPT.

WHIZZZ! BUZZZZzz!

G-GOOD GRIEF, LAD! A YOUNGSTER'S RUN INTO THE MOOSE CAGE AT THE ZOO! HE'LL BE TRAMPLED!

WAIT, **GREEN ARROW!** YOU DESIGNED AN ARROW FOR SUCH EMERGENCIES! THE **ANTLER ARROW!**

WHEN MOOSE BATTLE ONE ANOTHER, THEY LOCK ANTLERS! THESE **METAL** ANTLERS SPREAD OUT LIKE A PARASOL FROM THE FLYING ARROW!

THE FLYING ANTLERS SLAM INTO THE ATTACKING BEAST'S HORNS, STOPPING HIM IN HIS TRACKS BY THE SHEER FORCE OF THE SHOT!

I- I DON'T KNOW WHERE THAT ARROW CAME FROM! BUT IT SURE SAVED **THIS** KID'S LIFE!

PWACKKK!

BUT AT THAT MOMENT, ON THE HILL, ANOTHER TENSE DRAMA IS ACTED OUT, AS **SPEEDY** STEPS OUT OF COVER...

S-**SPEEDY!** WHAT ON EARTH ARE **YOU** DOING HERE?

I **FOLLOWED** YOU! I'VE SEEN YOU TRAIN THIS BOY! I KNOW YOU'RE GROOMING HIM TO REPLACE ME! BUT WHY MUST YOU DO IT BEHIND MY BACK?

5

FOR *THEIR* PURPOSES--IMPRACTICAL... BUT FOR MINE--*VERY* PRACTICAL! HA, HA!" I MAKE IT A BUSINESS BUYING UP INVENTIONS WHICH NOBODY ELSE WANTS!

WELL!... THIS SURE TURNED OUT TO BE MY LUCKY DAY!

NEXT EVENING, IN THE HOME OF WEALTHY OLIVER QUEEN AND HIS YOUNG WARD, ROY HARPER...

THERE'S THE *ARROW-SIGNAL*, OLIVER... THE POLICE NEED OUR ASSISTANCE SOMEWHERE!

RIGHT... ROLL OUT THE *ARROWCAR*, ROY!

MOMENTS LATER, AS *GREEN ARROW* AND *SPEEDY*, THE PAIR STREAKS TOWARD A FANTASTIC RENDEZVOUS...

GREAT SCOTT! WHAT IN THE WORLD IS THAT THING, *SPEEDY*?

IT LOOKS LIKE A SMALL, ARTIFICIAL *MOON*, G.A!

SUDDENLY, FROM SMALL "CRATERS" ON THE MOON, SQUIRTS A GREEN-COLORED MIST, THROUGH THE OPENED DOOR OF A JEWELRY STORE...

JEWELER

SHWEEE!

...FOLLOWED BY MASKED BANDITS!

THE SLEEPING GAS KNOCKED EVERYONE OUT, TEMPORARILY! COME ON--WE'LL MAKE OUR HAUL!

EWELER

BUT, AT THAT MOMENT, AS THE **ARROWCAR'S** CATAPULT HURLS TWO FIGURES SKYWARD...

THE GREEN MIST WAS EJECTED BY SOME POWERFUL FAN ACTION! WE'LL GIVE IT SOMETHING ELSE TO EJECT-- WITH OUR **SMOKE ARROWS!**

THUS, WHEN THE CROOKS EMERGE WITH THEIR LOOT...

HEY! WHAT WENT WRONG?

DON'T KNOW! I CAN'T SEE!

BUT JUST AS THE BATTLING BOWMEN PREPARE TO WRAP UP THE GANG...

SOMEONE DROPPED A **NET** ON US!

IT LOOKS MORE LIKE A **STEEL COBWEB!**

WHOMP!

GREAT SCOTT! LOOK UP THERE, **SPEEDY!**

S-SOME SORT OF **MECHANICAL SPIDER**... IT'S SPUN A TRAP AROUND US AND THE POLICE!

AND THERE GO THE "MOON CROOKS"!

YES... THE "SPIDER" IS IN LEAGUE WITH THE "MOON"! WE'VE GOT TO FREE OURSELVES-- AND FIND A WAY OF FOLLOWING THEM!

REMOVING AN **ACETYLENE ARROW** FROM HIS QUIVER, THE FAMED ARCHER IGNITES IT, AND...

IF THE WEB IS ELECTRICALLY CONTROLLED, WE MIGHT BE ABLE TO BURN A SHORT IN IT!

IT WORKED... THE STEEL HAS LOST ITS TENSION! COME ON ... WE'RE GOING TO INVESTIGATE THAT "SPIDER"!

THE USELESS WEB DROPS, JUST AS...

LEAP, SPEEDY!

AND SHORTLY AFTERWARD, AS THE ROBOT CREATURE STRIDES TOWARD THE SEA...

IT'S GOING IN THE SAME DIRECTION TAKEN BY THE MECHANICAL MOON! GUESS IT'S HEADED FOR THE GANG'S HIDEOUT!

BUT INSIDE THE MONSTROUS MACHINE...

LITTLE DO THOSE ANNOYING ARCHERS REALIZE THAT I HAVE WATCHED THEIR EVERY MOVE ON THIS TELEVISION SCREEN! HA, HA! ARE THEY IN FOR A SURPRISE!

SOON, AT A FANTASTIC, CLOUD-SHROUDED ISLAND...

THAT BIG BUILDING MUST BE OUR DESTINATION, SPEEDY!

LET'S HOPE WE CAN CLIMB OFF THIS THING BEFORE WE'RE SEEN!

BUT ONCE INSIDE...

I HOPE YOU ENJOYED THE RIDE, *GREEN ARROW!* DON'T TRY ANYTHING... OUR GUNS ARE TRAINED ON YOU!

MY NAME IS *MR. VENTION*, AND I ASSUME YOU CAME TO SEE OUR WONDERFUL LAYOUT HERE! I'LL SHOW YOU EVERYTHING, GLADLY...

THEN...

YOU SEE, THESE ARE MODELS OF USELESS INVENTIONS...USELESS TO OTHERS, THAT IS--BUT VERY USEFUL TO *ME!*

TO COMMIT CRIMES, I GATHER!

EXACTLY! HERE YOU SEE MY PRIZE *BLACKOUT SAUCER--* AND THE *MAGNETIC BOOMERANG,* WHICH CAN ATTRACT AND DESTROY OPPOSING AIRCRAFT! AH-- WHAT POWER THEY WILL BRING ME!

AND WHAT A SURPRISE *I'LL* GIVE YOU, IF I CAN JUST EDGE MYSELF A LITTLE CLOSER TO THE CONTROL BOARD OF THE FULL-SIZE SAUCER!

DEFTLY, THE ACE ARCHER MANEUVERS HIS *ELECTRO-ARROW* NEAR THE SAUCER'S CONTROLS, AND...

HE'S MADE THE *BLACKOUT LIGHT* COME ON! GRAB HIM!

I'VE WORKED MY ROPES LOOSE, *SPEEDY!* NOW'S OUR CHANCE!

AND AS THE PAIR BURSTS FORTH FROM THE BLACK-LIGHT RAY...

THE *MAGNETIC BOOMERANG* PROBABLY WORKS ON A *GYROSCOPE* PRINCIPLE, *SPEEDY!* YOUR *GRENADE ARROW* SHOULD BREAK THROUGH...

ZIP!

WH-POOM

...AND MY ARROW WILL ENTER THE BREAKTHROUGH, UPSET THE GYROSCOPE, AND --*LOOK OUT!* THE THING HAS STARTED UP!

WHRR-R-R

SPINNING CRAZILY, THE HUGE MACHINE'S POWERFUL MAGNETS CLUTCH AND HOLD THE OTHER INVENTIONS... *CARRYING* THEM!

WHRR-R-R

GET OUT OF HERE! THE *BOOMERANG'S* GONE BERSERK!

CRASHING THROUGH LIKE AN UNSTOPPABLE BATTERING RAM, THE RUNAWAY BOOMERANG HEADS INTO THE SEA ...

KWHPAM

THEN, WITH A PAIR OF *NET ARROWS,* THE ARCHERS EASILY SNARE THE ENTIRE GANG...

QUITE A HAUL!

YES--AND THEY'LL KEEP UNTIL THE LAW ARRIVES! THEY'LL HAVE PLENTY OF TIME TO DREAM ABOUT INVENTIONS NOW-- IN JAIL!

THE END

THE GREEN ARROW

THE GREEN ARROW

INCREDIBLE AS IT MAY SOUND, FOR THE FIRST TIME IN HIS LIFE, FEAR HAUNTS **GREEN ARROW**! THE STEEL-NERVED MASTER OF THE BOW, WHO HAD LAUGHED AT SUPERSTITION, NOW QUIVERS WITH ALARM WHENEVER HE GOES INTO ACTION! WILL THE TERRIBLE PROPHECY OF AN OLD SORCERER COME TRUE? IS **GREEN ARROW** DESTINED TO BECOME A KILLER? EVERY TICK OF THE CLOCK BRINGS THE AMAZING ARCHER CLOSER TO A FANTASTIC FATE WHEN HE BECOMES A VICTIM OF...

THE CURSE OF THE WIZARD'S ARROW!

FIRE, **GREEN ARROW**! FIRE YOUR ARROW BEFORE THE THIEVES GET AWAY!

FIRE IF YOU **DARE**! I AM MERLIN, THE GREATEST WIZARD WHO EVER LIVED... AND I PREDICT THAT WITHIN 24 HOURS YOU WILL SLAY YOUR BEST FRIEND WITH AN ARROW!

ONE DAY, AS **OLIVER QUEEN** AND HIS YOUNG WARD, **ROY HARPER**, ATTEND AN OUTDOOR CHARITY BAZAAR...

GOOD H-HEAVENS, OLIVER! YOU COULDN'T HIT THE SIDE OF A BARN!

YOU'RE RIGHT, L.T.! I GUESS I'LL NEVER BECOME A **GREEN ARROW**!

HA, HA! IF L.T. MORSE ONLY KNEW THAT OLIVER **IS GREEN ARROW**, THE WORLD'S GREATEST BOWMAN!

ARCHERY RANGE $1⁰⁰

TWANGG!

PRESENTLY, AS OLIVER AND ROY TOUR THE OTHER BOOTHS...

LOOK, OLIVER! THERE'S ONE MORE BOOTH! MAYBE THIS CHAP WHO CALLS HIMSELF "MERLIN" CAN READ YOUR FORTUNE!

ALL RIGHT, ROY... FOR THE FUN OF IT! BUT YOU KNOW I DON'T BELIEVE IN THE SUPERNATURAL!

MERLIN THE WIZARD I READ YOUR FORTUNE!

FOR INSTANCE, SUSPENDING THOSE OBJECTS IN THE AIR IS JUST MECHANICAL TRICKERY OR ILLUSION!

I HEARD YOU, OLIVER QUEEN! EVIDENTLY YOU DOUBT MY POWERS!

GOSH, OLIVER! H-HE KNOWS YOUR NAME!

I KNOW *EVERYTHING*, ROY HARPER! OF COURSE, THE SCOFFING MR. QUEEN WILL SAY I KNEW YOUR NAMES *BEFORE* I ADDRESSED YOU! HOWEVER, WILL YOU DARE TO HAVE YOUR FUTURE READ, MR. QUEEN?

WHY NOT? IT'S ALL IN FUN! READ ON, "MERLIN"!

THERE'S NOTHING HUMOROUS ABOUT *THIS* PREDICTION! *WITHIN 24 HOURS, YOU WILL KILL YOUR BEST FRIEND WITH AN ARROW!*

AN--AN ARROW? GREAT SCOTT, OLIVER! H-HE KNOWS WHO...

NONSENSE, ROY! IT MUST BE JUST A WILD GUESS!

TEN MINUTES LATER...

WHEW! THAT MERLIN WAS SOMETHING, OLIVER! WHEN HE MENTIONED THAT ARROW...

FORGET IT, ROY! FOR HIS PREDICTION TO COME TRUE, I'D HAVE TO KILL *YOU*, MY BEST FRIEND! IT'S JUST TOO RIDICULOUS FOR WORDS!

THAT NIGHT, AT OLIVER QUEEN'S HOUSE, AS THE TWINS OF THE TWANGING BOW RESPOND TO A FLARE SIGNAL...

I'VE SET UP THE DUMMIES SO THEIR SHADOWS FALL ACROSS THE WINDOW SHADE! NOBODY WILL SUSPECT WE'VE GONE OUT AS *GREEN ARROW* AND *SPEEDY*!

OKAY, *SPEEDY*! I'VE CONTACTED POLICE HEADQUARTERS! THIEVES JUST ROBBED A SUBWAY PAYROLL CAR AT BLAIR AVENUE STATION! THAT'S IN OUR DISTRICT!

SHORTLY AFTER, AS THE ARROW-CAR ARRIVES AT THE STATION...

GREEN ARROW! THE THIEVES ARE RUNNING INTO THE DOWNTOWN TUNNEL!

WE'LL FIND 'EM! COME ON, *SPEEDY*!

SUBWAY

2

SOON, IN THE SUBWAY TUNNEL...
L-LOOK! THE *GREEN ARROW'S* AFTER US! BLAST HIM!

THEY'RE PLAYING FOR KEEPS, *SPEEDY!* WE'D BETTER RUSH 'EM... AFTER I BLIND THEIR AIM!

BANG!   BANG!

AS THE *GREEN ARROW'S* SHAFT STREAKS ALONG THE THIRD RAIL, CREATING EYE-BLINDING SPARKS...

YEOOWWW! I-I CAN'T SEE!

THAT SHOWER OF DAZZLING SPARKS DID THE JOB, *SPEEDY!* THEY'LL BE SEEING *SPOTS,* INSTEAD OF US! CLOSE IN!

SSSSTT!   WWISSSTT!

BUT BEFORE THE FAMED ARCHERS CAN TAKE FIVE STEPS, AN ALARMING ROAR FILLS THE TUNNEL!

AN *EXPRESS TRAIN!* *GASP!* W-WE'RE IN ITS PATH... WITH NO PLACE TO HIDE!

IF I CAN HIT THE SIGNAL TRIPPER WITH AN ARROW, *IT'LL FLASH A RED* SIGNAL! HERE GOES!

RRRRRR!

BUT JUST AS *GREEN ARROW* RELEASES THE ARROW, HE TRIPS, AND...

YIIIII!

G-GOOD *HEAVENS!* MY AIM WAS *DEFLECTED!* T-THE ARROW...

RIP-P!   ZIP!

T-THE PREDICTION! *MERLIN'S* PREDICTION! THAT I WOULD *KILL* MY BEST FRIEND WITHIN *24 HOURS!* I-I CAME WITHIN AN INCH OF *DOING* IT!

D-DON'T STAND THERE, G.A.! THE TRAIN'S COMING ON! FIRE ANOTHER ARROW! *FIRE IT!*

RRRRRR!

BUT AS *GREEN ARROW* TREMBLINGLY RETREATS...

S-SOMETHING'S HAPPENED TO *GREEN ARROW!* I'VE GOT TO HIT THAT TRIPPER MYSELF!

BLANGG!

WHIZZ!

3

TEN MINUTES LATER, AS THE THIEVES ARE LED AWAY...

I- I FROZE UP, **SPEEDY**, BECAUSE I WAS AFRAID TO FIRE ANOTHER ARROW! ANOTHER "ACCIDENT" MIGHT MAKE MERLIN'S PROPHECY COME TRUE! I'VE GOT TO FIND MERLIN AS SOON AS I CHANGE MY COSTUME!

THAT NIGHT, AT L.T. MORSE'S ESTATE...

BUT, OLIVER... I **NEVER** HIRED A FORTUNE-TELLER! SEE FOR YOURSELF! YOU MUST'VE **IMAGINED** THAT MERLIN BOOTH!

THEN... GOOD HEAVENS!... MERLIN COULD ONLY HAVE APPEARED AND DISAPPEARED SO MYSTERIOUSLY IF HE WERE A **TRUE** WIZARD!

LATER, AS MORSE'S VISITORS LEAVE...

THEN Y-YOU BELIEVE KING ARTHUR'S COURT SORCERER STILL LIVES? THAT MERLIN APPEARED TO TELL YOUR FORTUNE? THAT HIS PREDICTION WILL COME TRUE?

I KNOW IT SOUNDS CRAZY, **SPEEDY!** BUT I DON'T KNOW WHAT **ELSE** TO BELIEVE! ANYWAY, I CAN'T TAKE A CHANCE! YOUR LIFE MEANS TOO MUCH TO ME!

THEREFORE... I WON'T FIRE ANOTHER ARROW UNTIL THE 24 HOURS ARE OVER! FOR THE NEXT 14 HOURS I'LL PATROL THE CITY **WITHOUT** MY BOW! FOR THE FIRST TIME IN HISTORY, **GREEN ARROW** WILL USE ANOTHER WEAPON TO FIGHT CRIME!

SHORTLY AFTER, AT THE ARROW-CAVE, AS THE TWO FRIENDS SWITCH BACK INTO THEIR UNIFORMS...

HERE'S WHAT I'LL USE UNTIL NOON TOMORROW! A **RIFLE**... THAT WILL FIRE **NO BULLETS!** JUST SHOTS THAT WILL HAVE THE SAME EFFECT AS MY SPECIAL ARROWS! I'LL BE MATCHING YOUR ARROWS, SHOT FOR SHOT!

AT 2 A.M. THAT NIGHT, AS THE AMAZING BOWMEN SPY A ROOF TOP PROWLER...

THE THIEF'S CROSSING TO THE NEXT BUILDING! **I'VE** GOT TO USE A ROPE ARROW, G.A.! BUT YOURS IS **BACK** IN THE ARROW CAVE. I STILL DON'T KNOW HOW YOU'LL MATCH MY SHOT!

WATCH ME! FIRE YOUR ROPE ARROW!

4

FINALLY, AT 12:59 THAT AFTERNOON, AT THE ARROW-CAVE...

WELL, *SPEEDY*, IT'S ALL OVER! IT'S *MORE* THAN 24 HOURS SINCE MERLIN MADE THAT TERRIBLE PREDICTION! NOW I CAN PUT MY RIFLE AWAY... *PERMANENTLY!*

GREEN ARROW! L-LOOK OUT! IT'S *SLIDING!*

AS THE RIFLE HITS THE CAVE FLOOR SHARPLY...

TWANGG!

BANG!

G-GREAT GUNS, G.A.! THE RIFLE WENT OFF! THE SHOT CUT THE TAUT STRING ON YOUR CROSS-BOW!

M-MY ARROW! *IT'S ENTERED THE CHEST OF ONE OF THE EXTRA DUMMIES* WE USE AT MY HOUSE WHENEVER WE GO OUT ON PATROL!

WAIT, *GREEN ARROW!* I-I MUST CHECK THE *TIME!* GIVE ME T- THE PHONE!

PLUT-T-T!

AS THE OPERATOR TELLS *SPEEDY* THE CORRECT TIME...

G.A., Y-YOU WON'T BELIEVE THIS... BUT WE FORGOT TO TURN THE CLOCK BACK LAST NIGHT FOR *DAYLIGHT SAVING TIME!* OUR WATCHES ARE *AN HOUR SLOW!*

SO IT WAS ONLY 11:59 A.M. ... AND JUST AS MERLIN PROPHESIED, ONE OF MY ARROWS ENTERED THE HEART OF MY *BEST FRIEND!* WAS IT A *COINCIDENCE?* OR DID MERLIN REALLY READ MY FUTURE? I GUESS I'LL NEVER KNOW!

The End

JUST THEN, AS AN ATTENDANT ENTERS FROM ANOTHER ROOM...

OKAY--GET THE OTHER PAINTINGS... AND LET'S MOVE OUT FAST!

WHAT--? HARRY AND FRANK HAVE THEIR HANDS UP--AND I HEAR VOICES FROM UNSEEN PEOPLE! I'D BETTER SOUND THE ALARM!

MOMENTS LATER, IN THE MANSION OF WEALTHY OLIVER QUEEN AND HIS YOUNG WARD, ROY HARPER...

LOOK, OLIVER... THE ARROW-SIGNAL!

SOMEONE NEEDS OUR ASSISTANCE... LET'S GO, ROY!

IN THEIR SECRET ARROW-CAVE, BENEATH THE HOUSE, THE PAIR SWIFTLY DONS THE MASKS AND COSTUMES OF GREEN ARROW AND SPEEDY, ACE ARCHERS OF LAW AND ORDER!

AND AS POLICE RADIO DIRECTS THEM TO THE SCENE OF THE CRIME...

GREEN ARROW! THANK GOODNESS YOU'RE HERE! THREE CROOKS GOT OUT WITH $50,000 WORTH OF PAINTINGS! THEY'RE OVER THERE!

THERE? WHERE?

GUNFIRE--OVER BY THE BRICK WALL! GET BACK!

YES--BUT WHERE'S THE MAN WHO FIRED? I CAN'T SEE HIM!

BLAM!

I DON'T KNOW WHAT THEIR TRICK IS, SPEEDY--BUT LET 'EM HAVE A FEW ARROWS! MAYBE THAT'LL FLUSH THEM OUT!

MUSEUM OF AR

TWANG! TWANG!

2

BUT UP THE STREET, NEAR A PARK, AN "EMPTY" CAR SUDDENLY STARTS INTO MOTION...

IT'S HEADING INTO THE PARK... AND *LOOK* AT IT, *SPEEDY!*

IT'S TURNING GREEN... *GREENER!*

NOW THE CAR'S GONE FROM SIGHT! IT TURNED THE SAME COLOR AS THE FOLIAGE!

≶WHEW!≶ WE CERTAINLY DIDN'T HAVE MUCH OF A CHANCE WITH OUR ARROWS! DO YOU THINK THIS WAS AN OPTICAL ILLUSION?

I DON'T KNOW *WHAT* TO THINK, *SPEEDY!* FOR THE MOMENT, WE'RE STYMIED!

SHORTLY, IN THE THIEVES' HEADQUARTERS...

WE GOT 'EM, BOSS-- WITHOUT A HITCH... 50 G'S WORTH!

NATURALLY! I PLAN WITHOUT MARGIN FOR ERROR! WHERE ARE YOUR SPRAY GUNS?

HA, HA... YOU'VE SURE EARNED YOUR MONICKER, BOSS-- THE *CAMOUFLAGE KING!* THESE CONTRAPTIONS OF YOURS WORK SHEER MAGIC! WE HAD OURSELVES BLENDING WITH ANY BACKGROUND THAT TURNED UP!

AND THAT MADE YOU VIRTUALLY INVISIBLE TO OTHERS!

THAT'S THE BEAUTY OF IT! *ANY* BACKGROUND CAN BE IMITATED INSTANTLY, WITH MY SPECIAL CHEMICAL SPRAY PAINT! WATCH THIS SIMPLE DEMONSTRATION...

3

EACH PAINT CARTRIDGE SPINS AROUND AT YOUR SLIGHTEST TOUCH, SPRAYING YOURSELF--OR A TARGET-- WITH ANY COLOR OR DESIGN YOU WANT!

THAT'S JUST HOW WE DID IT, BOSS! EVEN *GREEN ARROW* WAS FOOLED!

EH? *GREEN ARROW?* WAS *HE* THERE?

YEAH-- BUT DID *HE* LOOK STUPID! HE WAS FIRIN' ARROWS WHERE WE *WEREN'T!*

BUT DON'T BE MISLED...HE'S CLEVER ENOUGH TO FIGURE THIS OUT! I'D BETTER GO WITH YOU ON THE CIRCUS JOB, IN ORDER TO GIVE *GREEN ARROW* A SPECIAL SURPRISE, IF HE SHOWS UP!

INDEED, THE FABLED ARCHER HAS ALREADY TUMBLED TO THE ASTONISHING STUNT OF THE CAMOUFLAGE KING, BACK AT THE ARROW CAVE...

SEE HOW PERFECTLY MY HAND, PAINTED WITH WHITE AND RED SQUARES, FITS INTO THE CHECKERBOARD BACKGROUND?

AMAZING! YOU HAVE TO SEARCH TO FIND IT!

SO *THAT'S* WHAT WE'RE UP AGAINST--A *CAMOUFLAGE GANG!* BUT HOW CAN OUR ARROWS STOP SOMETHING WE CAN'T EVEN SEE?

THAT'S WHAT WE'RE GOING TO WORK ON, *SPEEDY*--EVEN IF IT TAKES MOST OF THE NIGHT! WE'VE *GOT* TO COME UP WITH AN ANSWER!

NEXT DAY, AT THE CIRCUS...

BOSS! OUTSIDE--THE CASHIER SEEMS TO BE IN TROUBLE...BUT I DIDN'T *SEE* ANYTHING--OR ANYBODY AT HER CAGE!

HUH?

SEE?...SOMETHING FUNNY *IS* GOING ON! HER HANDS ARE UP!

I DON'T KNOW WHAT'S GOING ON--BUT I'M FIRING AN ALARM WITH MY SIGNAL-CANE!

WHROOSH!

ADMISSIONS

4

AS A SCREAMING FLARE ROCKETS SKYWARD FROM THE CANE...

BLAST IT! THAT'S A SIGNAL FOR THE LAW! I'D BETTER JOIN THE OTHER GUYS AND SCRAM!

PATROLLING THE CITY IN THEIR *ARROWCAR,* THE BATTLING BOWMEN ALSO SEE THE ROCKET-FLARE, AND...

GREEN ARROW! CROOKS--*INVISIBLE CROOKS*-- GOT ALL THE GATE RECEIPTS! I DON'T KNOW WHERE THEY ARE NOW!

JUST AS I FIGURED! THIS TIME, WE'LL SPLIT UP, *SPEEDY*... SEARCH FOR THEM SEPARATELY!

SCREECH!

BUT AS *SPEEDY* STALKS BY A TENT...

HUH?

WHROOSH!

WHILE NOT FAR DISTANT...

NOW, ARROW-MAN, YOU CAN'T SEE THE KID--AND HE CAN'T SEE YOU... HA, HA!

WHROOSH!

AND AS THE "INVISIBLE" FIGURES RACE AWAY...

NOW'S OUR CHANCE!

G.A.! WHERE ARE YOU?

OVER HERE--BUT I CAN'T SEE YOU!

SPEEDY AND I CAN'T FIRE-- WE MIGHT HIT EACH OTHER-- BUT...

ALL RIGHT, *SPEEDY*-- FIRE THE *MIDNIGHT ARROWS!* AND FIRE THEM *ABOVE THE EXIT GATES!*

AS THE UNIQUE NEW SHAFTS ARC HIGH ABOVE THE EXIT GATES, RELEASING A DELUGE OF BLACK LIQUID...

BOSS! LOOK WHAT'S HAPPENIN'!

ONE OF *GREEN ARROW'S* TRICKS! HURRY-- PUT THE SPRAY GUNS TO WORK!

5

BUT BEFORE THE GANG CAN ACT...

WHERE ARE THEY? I CAN'T SEE THEM!

AS A *BOLO ARROW* HAMSTRINGS THE CROOKS MOMENTARILY...

BLAST IT! IF WE COULD ONLY *SEE* THOSE ARCHERS!

IT WAS OUR OWN DOING! WE MADE THEM INVISIBLE, SO THEY'D WORRY ABOUT HITTING EACH OTHER--AND WOULDN'T FIRE! BUT NOW...

ABRUPTLY, A *NET ARROW* OPENS UP. AND...

EEYOW!

THIS IS IT-- CURTAINS!

AND SO, WITH THE GANG COMPLETELY HELPLESS...

YOU SURE TURNED THE TABLES ON THEM, *GREEN ARROW!*

YES...THE *CAMOUFLAGE KING* AND HIS MEN WILL HAVE A TOUGH TIME CAMOUFLAGING THEMSELVES NOW--BEHIND BARS!

THE END

THE GREEN ARROW

THE GREEN ARROW

SOMETIMES, IN A NIGHTMARE, *GREEN ARROW'S* YOUNG ASSISTANT, *SPEEDY*, RELIVES THE HEART-BREAKING EVENTS OF HIS FIRST MEETING WITH THE WORLD'S MOST FAMOUS BOWMAN! SHUDDERINGLY, *SPEEDY* RECALLS HIS FANTASTIC FAILURE WITH THE BOW... WHEN HE WAS NOT *SPEEDY*, BUT AN EAGER LAD NAMED ROY HARPER, WITH A DREAM OF GLORY HE COULD NOT MATCH WITH HIS SKILL! LET US GO BACK IN TIME, THEN, TO THAT SHOCKING PERIOD IN ROY HARPER'S LIFE WHEN HE WAS...

"The WORLD'S WORST ARCHER!"

HOW THE *GREEN ARROW* MET SPEEDY!

GULP! ...MISSED AGAIN! W-WHAT'S *HAPPENED* TO ME? I-I CAN'T HIT ANYWHERE NEAR THE TARGET!

P-PLEASE GIVE ME ANOTHER CHANCE, *GREEN ARROW!* I KNOW I CAN HIT THAT GOAL POST!

SORRY, SON! TAKE IT FROM ME... YOU'RE WASTING YOUR TIME AND MINE! YOU'LL *NEVER* MAKE AN ARCHER!

CLANK!

ONE DAY, SEVERAL YEARS AGO, AS AN EAGER YOUTH VISITS THE GREENVILLE FAIR GROUNDS...

ALL ABOARD THE SPACE PLATFORM! GET A BIRD'S EYE VIEW OF THE FAIR! ONLY 25¢ A RIDE!

I'VE GOT TO FIND *GREEN ARROW!* I'LL SAVE TIME IF I SPOT HIM FROM THAT AERIAL PLATFORM!

HERE'S A QUARTER! ONE RIDE, PLEASE!

RIDE the SPACE PLATFO

MAGNETIC SPACE PLATFORM 25¢

MINUTES LATER, CARRIED ALOFT BY THE MAGNETIC LIFT...

THERE HE IS, GIVING A DEMONSTRATION FOR CHARITY! WHAT A WONDERFUL PERSON THE *GREEN ARROW* IS! WHEN HE'S NOT FIGHTING CRIME, HE'S RAISING FUNDS FOR THE NEEDY!

PRESENTLY, AS THE BOY JOINS THE CROWD OF ADMIRERS IN FRONT OF A SHOOTING GALLERY...

L-LOOK! *GREEN ARROW* IS SNUFFING OUT ALL THE CANDLES WITH A VOLLEY OF ARROWS!

WHAT SHARP-SHOOTING!

WHIZZZ! WHIZZZ!

NEXT, THE *GREEN ARROW* ENTERS A BASEBALL BATTING-CAGE...

ADJUST THAT BASEBALL-PITCHING MACHINE TO ITS MAXIMUM SPEED! FIRE A DOZEN BALLS AT ME, ONE RIGHT AFTER THE OTHER!

OKAY, *GREEN ARROW!* HERE THEY COME!

ZIP! ZIP!

I'LL THREAD ALL THE BASEBALLS SIMULTANEOUSLY WITH MY UNBREAKABLE PLASTIC *NEEDLE* ARROW!

GASP! I-IT'S INCREDIBLE! THE ARROW IS PIERCING ALL TWELVE BASEBALLS IN A ROW, MAKING A HORSEHIDE NECKLACE!

ZIPP! ZIPP!

LATER, AS *GREEN ARROW* GOOD-NATUREDLY SIGNS AUTOGRAPHS, ONE AUTOGRAPH ALBUM, STRANGELY, BEARS A WRITTEN MESSAGE...

PLEASE TEST ME, *GREEN ARROW!* MY LIFE'S DREAM IS TO AID YOU IN YOUR WORK!

ROY HARPER, EH? VERY WELL, ROY! LET'S GO TO THE ARCHERY RANGE!

Please, Green Arrow, I'd like to be your assistant. I've been trained as an archer for years!

Roy Harper

SHORTLY, AT THE CARNIVAL ARCHERY RANGE...

HITTING THAT BULL'S-EYE SHOULD BE EASY! I MISSED! GASP!...T-THE ARROW'S *CURVING AWAY!*

YES! IT'S HEADING FOR THAT WATER TOWER!

TWANG!

②

I-IT MUST BE THE WIND, *GREEN ARROW!* A SUDDEN GUST OF WIND DEFLECTED THE SHAFT!

WE'LL SOON FIND OUT, LAD! I'LL TRY A SHOT!

CLANGG!

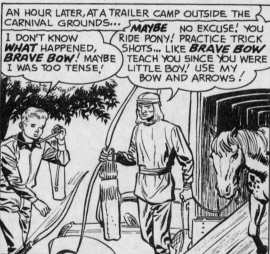

THAT NIGHT, AT THE CAMPFIRE...

THIS BUCKSKIN COSTUME YOU MADE FOR ME FITS PERFECTLY, *BRAVE BOW!*

GOOD! YOU TAKE THIS FLINT ARROW! *BRAVE BOW* KILL HIS FIRST BUFFALO WITH IT! IT WILL BRING *YOU* LUCK, TOO, WHILE YOU TRAIL *GREEN ARROW* TO HELP HIM FIGHT EVIL ONES!

IF YOU PROVE WORTHY IN BATTLE, *GREEN ARROW* MAY BE GLAD TO BECOME YOUR NEW GUARDIAN! ME GETTING TOO OLD! ME SOON GO TO HAPPY HUNTING GROUNDS! WHEN YOUR FATHER DIE MANY YEARS AGO, ME PROMISE TO RAISE HIS INFANT SON!

"YOU SEE, YOUR FATHER RISKED HIS LIFE TO SAVE MINE IN A BIG FOREST FIRE..."

ME NEVER FORGET THIS, RANGER HARPER! ME RETURN NOBLE DEED SOME DAY!

WHEN YOUR FATHER DIE IN AVALANCHE, ME KEEP MY PROMISE! ME RAISE YOU LIKE OWN SON! NOW *BRAVE BOW* TOO OLD! MUST PASS YOU ON TO ANOTHER MAN! THAT IS WHY WE SEEK OUT *GREEN ARROW!*

LOOK, CHIEF! THE POLICE FIRED A FLARE ARROW TO SUMMON THE *GREEN ARROW* NEAR THE FAIR GROUNDS! HE *CAN* USE MY HELP *NOW!* I MUST GO!

THUS, SHORTLY AFTER, AS THE *ARROW CAR* RACES UP TO THE WAX MUSEUM, AND *GREEN ARROW* DASHES INSIDE...

I HAD A FEELING THE *GREEN ARROW* WOULD SHOW UP! BUT WE'RE *PREPARED* FOR HIM! HEAVE THAT SMOKE-BOMB, ED!

FFFSSSTT!

YOU BET! THIS IS *ONE* STICKUP THE *GREEN ARROW WON'T* BUST UP!

WAX MUSEUM

SECONDS LATER, AS THE THIEVES REACH THE STREET...

C'MON, BOYS! *GREEN ARROW* IS STUMBLING AROUND IN THE SMOKE! NOTHING CAN STOP US NOW!

SNAP!

SNAP!

THAT REAL BOW AND ARROW IN THE WAX STATUE OF ROBIN HOOD CAN STOP 'EM... IF *BRAVE BOW'S* FLINT ARROW FLIES STRAIGHT!

AN INSTANT LATER, THE STATUE'S BOW TWANGS, AND...

EEEOOWW! I-I'M PINNED TO THE CAR!

PLANNNGG!

ZIPPP!

THEN, AS THE WAX FIGURE FALLS, FEET FIRST...

OHHH!

SOCKK!

CRACKK!

I--I DID IT! MY TRICK SHOT KAYOED ALL THE HOLDUP MEN. GREEN ARROW WILL BE IMPRESSED HOW I USED "ROBIN HOOD" TO DO HIS JOB!

BUT WHEN GREEN ARROW STAGGERS OUT...

DON'T KID ME, BOY! YOU HAVEN'T THE SKILL FOR SUCH A TRICK SHOT! THE WIND ACCIDENTALLY SNAPPED THE STATUE'S WIRES! LUCK DID THE REST!

B-BUT...

IT'S NO USE! IN THAT SMOKE-FILLED ROOM, GREEN ARROW DIDN'T SEE ME FIRE THE ARROW! I'LL RETRIEVE IT AND TELL BRAVE BOW WHAT HAPPENED!

LATER, AT THE TRAILER CAMP...

HERE'S YOUR FLINT ARROW! IT BROUGHT ME BAD LUCK!

DESPAIR NOT, LAD! TOMORROW YOU WILL SEE GREEN ARROW AT THE CARNIVAL'S TRACK AND FIELD GAMES! ASK FOR ANOTHER TRIAL AND YOU WILL SURELY SUCCEED!

AT THE TRACK MEET, THE NEXT DAY...

ROY HARPER WINS THE 100-YARD DASH IN RECORD TIME!

ROY HARPER LEAVES THE FIELD BEHIND IN HIGH HURDLES!

ROY HARPER SETS A NEW RECORD IN THE MILE!

AS GREEN ARROW PRESENTS THE VICTORY CUP TO THE AMAZINGLY SPEEDY WINNER...

THANKS FOR THE CUP, GREEN ARROW! BUT I'D APPRECIATE EVEN MORE ANOTHER CHANCE TO PROVE MY ARCHERY SKILL!

ALL RIGHT, ROY! GET DRESSED AND MEET ME HERE AFTER THE CROWD LEAVES!

HOURS LATER, IN THE DESERTED FAIR ARENA, AS ROY FIRES AN ARROW AT A WOODEN GOAL POST...

I MISSED AGAIN! I-IT'S CURVING OFF! IT'S HITTING THAT METAL RAILING! WHAT'S WRONG WITH ME?

LOOK, LAD! YOU'RE WASTING YOUR TIME AND MINE! TAKE IT FROM ME... YOU'LL NEVER MAKE AN ARCHER! GO HOME AND FORGET ABOUT BECOMING MY ASSISTANT!

KLANKKK!

AS ROY SADLY RETURNS TO THE TRAILER CAMP...

I'M NO GOOD, BRAVE BOW! I DISGRACED MYSELF WITH GREEN ARROW! HOWEVER GOOD I AM IN PRACTICE, IN A REAL TEST, I'M A FAILURE!

DON'T GIVE UP! ME GIVE YOU FLINTARROW! IT CUT THE STATUE'S WIRES... AND WILL HELP YOU AGAIN! NOW REST AND AWAIT YOUR NEXT CHANCE!

THE "NEXT CHANCE" ARRIVES A FEW HOURS LATER, AS A SIREN ARROW ECHOES THROUGH THE HILLS...

LOOK! ARROW CAR CHASES BAD MAN'S CAR INTO CANYON! TAKE MY FLINT ARROW! FOLLOW THEM!

IT MUST BE THAT CHARITY FUND ROBBERY WE HEARD ABOUT OVER THE RADIO! THE THIEVES SPLIT UP! THEY'RE GOING TO RENDEZVOUS IN THE HILLS!

WHEEE!

BANG! BANG!

PRESENTLY, AS THE ARROW CAR MEETS AN IMPASSABLE OBSTACLE...

MY CATAPULT SEAT GIVES ME A SHORT-CUT UPWARDS! THEY'RE GETTING TOGETHER ON THE CLIFF-TOP!

I-IT'S GREEN ARROW! OUR BARRICADE DIDN'T STOP HIM! GRENADE HIM!

WHOOSH!

BUT AS GREEN ARROW HITS THE DIRT, HE LETS FLY A SPECIAL ARROW!

MY PAINT-BRUSH ARROW WILL SWAB THEIR WINDSHIELD WITH PAINT! THEY CAN'T TRAVEL WITHOUT VISIBILITY!

BLAMM!

SPLOSHH!

BUT THE CAR DOES PROCEED DOWN THE CANYON ROAD... WITH ONLY ROY HARPER BETWEEN THE THIEVES AND THEIR GETAWAY!

I CAN'T GET A CLEAR SHOT FROM THE OVERHANGING CLIFF! I'LL CLING TO THIS BUSH BY MY LEGS, AS IF I WERE RIDING BRAVE BOW'S PONY!

SHORTLY, AS BOTH *GREEN ARROW* AND ROY REACH THE STOPPED CAR...

LOOK AT THE REAR TIRE, *GREEN ARROW!* IT'S FLAT! I PUNCTURED IT WITH MY FLINT ARROW!

THERE'S NO ARROW STICKING OUT OF THAT TIRE! I DIDN'T SEE YOU DO ANYTHING! THE PUNCTURE WAS PROBABLY CAUSED BY A SHARP STONE! MY PAINT SMEAR STOPPED 'EM!

HARDLY! THE WINDOW-WASHERS WIPED THE WINDSHIELD CLEAN! BUT I GIVE UP! KEEP MY BOW AND ARROWS AS A SOUVENIR OF MY FAILURE! I'M GOING BACK TO THE TRAILER CAMP! BY NIGHTFALL, I'LL BE FAR AWAY!

BUT AN HOUR LATER, AS THE ARROW CAR TOWS A RATTLING VEHICLE INTO THE CAMP...

LOOK, ROY! I DUG THIS FLINT ARROWHEAD OUT OF THE TIRE TUBE. IT PROVES YOU *DID* STOP THE CAR! THE SHAFT BROKE OFF! TELL ME, DID YOU EVER GO FOR A RIDE ON THE CARNIVAL'S SPACE PLATFORM?

YES! YESTERDAY, BEFORE I SAW YOU FOR THE FIRST TIME! W-WHY?

TRAILER CAMP — PER DAY

⑦

A *HUGE MAGNET RAISES THAT PLATFORM!* IT *MAGNETIZED* ALL YOUR METAL ARROWS SO THAT YOUR *NON-FLINT* TEST ARROWS ALL HEADED FOR METAL OBJECTS, WIDE OF THE TARGET! AND BECAUSE YOU'RE AS QUICK WITH THE BOW AS ON A TRACK, HEREAFTER I'M CALLING YOU *SPEEDY... PARTNER!*

AHA! ME TELL YOU *FLINT ARROW* BRING YOU LUCK... *SPEEDY!*

THE END

THE GREEN ARROW

THE GREEN ARROW

THERE COMES A DAY IN THE LIFE OF ANY BOY WHEN HE WANTS TO BUY SOMETHING OUT OF REACH OF HIS ALLOWANCE... AND DECIDES TO WORK PART-TIME TO PURCHASE IT! WHEN THAT DAY COMES TO **GREEN ARROW'S** BRILLIANT YOUNG PAL, **SPEEDY**, THE BOY ARCHER NEVER DREAMS THAT A SINISTER PERIL LURKS IN HIS CLEVER ADVERTISING SLOGAN...

HAVE ARROW--WILL TRAVEL!

I HAVE TO HAND IT TO **SPEEDY**! HE'S WILLING TO **WORK** FOR THE THINGS HE WANTS TO BUY!

HA! HA! THAT FOOLISH KID WILL LIVE TO REGRET THE DAY HE DECIDED TO PERFORM ODD JOBS!

SPEEDY, FAMED ASSISTANT TO **GREEN ARROW**, WANTS WORK! JOB MUST INVOLVE ARCHERY SKILL! NO REASONABLE REQUEST TURNED DOWN! **HAVE ARROW-- WILL TRAVEL!**

ONE NIGHT, AT OLIVER QUEEN'S HOUSE, AS HE RELAXES WITH HIS YOUNG WARD, ROY HARPER...

OLIVER, COULD YOU INCREASE MY ALLOWANCE?

NO, ROY! I GIVE YOU ENOUGH MONEY EACH WEEK TO TAKE CARE OF YOUR NEEDS! I DON'T WANT TO SPOIL YOU!

I KNOW, OLIVER... AND I'M NOT COMPLAINING! IT'S JUST THAT... WELL, I'D LIKE TO BUY A SAILBOAT, AND I CAN'T SAVE UP FOR ONE ON MY PRESENT ALLOWANCE!

I SEE! WELL, WHEN I WAS A BOY...

WAIT! I KNOW WHAT YOU'RE GOING TO SAY...AND YOU'RE **RIGHT!** IF I WANT EXTRA MONEY, I SHOULD EARN IT MYSELF! AND I **WILL!** I HAVE A GREAT IDEA, OLIVER! HERE'S WHAT I'LL DO...

THE FOLLOWING DAY, AN AD APPEARS IN THE DAILY NEWSPAPER...

HAVE ARROW...WILL TRAVEL! **SPEEDY,** FAMED YOUNG ASSISTANT TO **GREEN ARROW,** WANTS PART-TIME WORK! JOB MUST INVOLVE ARCHERY SKILL! NO REASONABLE REQUEST REFUSED! LOW RATES! WRITE: P.O. BOX 893-K, CENTRAL CITY

AND THE DAY AFTER THAT, AS ROY RECEIVES A FLOOD OF RESPONSES...

GOSH, ROY! THERE'S ENOUGH WORK HERE FOR A YEAR! HOW WILL YOU FIND TIME FOR OUR **REGULAR** DUTIES AND PATROLS?

DUTY COMES FIRST! THIS STUFF...SECOND! IF ANY EMERGENCY ARISES, FIRE A SIGNAL ARROW! I'LL ALWAYS BE DRESSED AS **SPEEDY** AND I'LL COME RUNNING! NOW I'M OFF FOR MY FIRST JOB!

AN HOUR LATER, AT THE OUTSKIRTS OF TOWN...

AS I WROTE YOU, **SPEEDY,** MY FACTORY CHIMNEY MUST BE CLEANED OUT! IT'S STUFFED WITH SOOT!

I PREPARED EXACTLY THE ARROW YOU NEED! THE BLADE-WIDTH OF THIS FAN ATTACHMENT WILL JUST FIT THE DIAMETER OF YOUR CHIMNEY!

PRESENTLY, INSIDE THE FACTORY...

NOW, IF I'VE AIMED DIRECTLY UP THE CHIMNEY'S CENTER, THE ADDED ROCKET POWER OF THIS **CHIMNEY-SWEEP ARROW** SHOULD DRIVE THE WHIRLING BLADES TO THE CHIMNEY TOP... PUSHING OUT ALL THE SOOT!

A MINUTE LATER...

WHOOOSH! SKKKKT!

LOOK! THE SOOT IS COMING OUT! THAT KID IS A GENIUS WITH AN ARROW!

LATER THAT DAY, AS *SPEEDY* KEEPS ANOTHER APPOINTMENT...

HE'S THE RAREST MONKEY IN MY PET SHOP, *SPEEDY!* THE *A.S.P.C.A.* TRIED TO CATCH HIM FOR ME YESTERDAY, BUT FAILED! THAT'S WHY I WROTE TO YOU!

DON'T WORRY! I'LL COLLAR HIM BEFORE HE REACHES THOSE HIGH TENSION WIRES!

TWANNNGG!

H-HE'S BEING *STRAPPED* INTO SOMETHING!

RIGHT! THAT "SOMETHING" IS A *PARACHUTE* ARROW! HE'LL BE DOWN IN A MINUTE!

LAST STOP FOR *SPEEDY* THAT AFTERNOON IS A MANUFACTURER OF BULL'S-EYE TARGETS...

I'VE RECEIVED COMPLAINTS ABOUT THE STRENGTH OF MY STRAW TARGETS, *SPEEDY...* SO I WANT YOU TO TEST THE RESISTANCE POINTS OF NEW TARGET MATERIALS! THIS TARGET IS MADE OF CLAY!

HMM... CHANCES ARE MY ARROW WILL *ALMOST* PASS *THROUGH* IT!

YOU SURE KNOW YOUR TARGETS, *SPEEDY!* THAT'S WHY I HIRED YOU! WELL, I'LL NEED A DAY TO STUDY THE RESULTS! THEN I'LL PREPARE A NEW TARGET FOR TOMORROW... SAME TIME!

THHUUCKK!

BUT WHEN *SPEEDY* LEAVES...

WHO'D GUESS THIS TARGET RESTED ON A *SCALE?* THE ADDITION OF *SPEEDY'S* ARROW RECORDED THE ARROW'S *EXACT WEIGHT!*

GREAT! NOW I'LL BLOW UP THE PICTURES OF *SPEEDY'S* ARROW IN FLIGHT... FOR MICROSCOPIC EXAMINATION!

THAT NIGHT AS OLIVER QUEEN SWITCHES TO *GREEN ARROW* TO PATROL THE CITY...

WHEW! YOU'RE REALLY MAKING MONEY! KEEP UP THIS WAY, *SPEEDY,* AND YOU'LL *HAVE* THAT BOAT!

IT'S MY STEADY CUSTOMERS I'M COUNTING ON, G.A.! TOMORROW I'M TESTING A *HARD WAX* TARGET FOR THAT SAME TARGET MANUFACTURER!

HAVE ARROW-WILL TRAVEL FUND

3

BUT THE NEXT DAY, AS **SPEEDY** CONTINUES TO FILL SOME OTHER ODD JOBS...

I DECLARE, **SPEEDY!** YOUR **BUZZ-SAW ARROW** CUTS MORE TIMBER IN TEN MINUTES THAN A CREW OF LUMBERJACKS CAN FELL IN A DAY! YOU SURE EARNED YOUR $10 FEE!

IT'S A COMPACT, COLLAPSIBLE GADGET, TOO! THE BUZZ-SAW ATTACHMENT FOLDS RIGHT INTO MY QUIVER!

LATER, AS **SPEEDY** BECOMES A **GOLF CADDY**...

YOU CERTAINLY SOLVED THIS 10TH HOLE MENACE FOR US, J.T.! BEFORE YOU HIRED **SPEEDY**, WE LOST DOZENS OF GOLF BALLS OVER THE FENCE!

YEP! **SPEEDY'S NET ARROWS** STOP EVERY BALL FROM LANDING IN THE RIVER!

PLUTTT!

AT 6 P.M., AS **SPEEDY** KEEPS HIS SUNSET RENDEZVOUS...

YOUR WAX TARGET AFFORDS MORE RESISTANCE, SIR! THIS TIME MY ARROW **DIDN'T** PENETRATE THE BACK OF THE TARGET!

QUITE RIGHT, **SPEEDY!** I'LL MEASURE ALL THE EFFECTS TONIGHT! REPORT TOMORROW AT 6 P.M. AS USUAL!

FWWWTTTT!

HOWEVER, WHEN **SPEEDY** IS GONE...

SPEEDY'S ARROW MADE A PERFECT IMPRESSION IN THE WAX... AND POURING PLASTER INTO THE HOLE GIVES US AN **EXACT DUPLICATE** OF HIS SHAFT!

WONDERFUL! TONIGHT WE'LL MAKE **OUR** VERSION OF **SPEEDY'S** ARROW!

PLASTER   WATER

THAT NIGHT, AT THE MANUFACTURER'S LABORATORY...

IT'S DONE! THE EXACT COPY IN WEIGHT, SIZE AND APPEARANCE OF **SPEEDY'S** SHAFT... EXCEPT THAT **THIS** ARROW CONTAINS A SPECIAL **TIME BOMB** THAT WILL EXPLODE AT MIDNIGHT TOMORROW!

GOOD! ALL THAT REMAINS NOW IS TO SWITCH ARROWS WITH THE KID!

T.N.T

AT THE SAME TIME, ELSEWHERE, AS THE EMERALD ARCHER AND HIS YOUNG ASSISTANT RETURN FROM A MISSION...

GOSH! THE MONEY IS SURE PILING UP! I CAN PRACTICALLY HEAR THE SLAP OF SAILS IN THE OCEAN BREEZE! BUT I WOULDN'T KEEP THIS CASH HERE! I'D BANK IT!

YOU'RE RIGHT, **GREEN ARROW!** I'LL DEPOSIT IT TOMORROW... BETWEEN ODD JOBS!

HAVE ARROW... WILL TRAVEL FUND

AT SUNDOWN, AS SPEEDY WINDS UP HIS DAY WITH HIS REGULAR 6 P.M. APPOINTMENT...

GOOD HEAVENS! A BEE DEFLECTED MY AIM! I-I MISSED THE BULL'S-EYE!

THAT'S OKAY! YOUR ARROW TESTED THE RESISTANCE OF THIS WOOD TARGET! THAT'S ALL I WANT!

BZZZZZZ!

HMM... THE KID DROPPED HIS WALLET!

SPEEDY'S WALLET

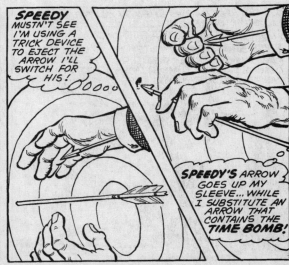

SPEEDY MUSTN'T SEE I'M USING A TRICK DEVICE TO EJECT THE ARROW I'LL SWITCH FOR HIS!

SPEEDY'S ARROW GOES UP MY SLEEVE... WHILE I SUBSTITUTE AN ARROW THAT CONTAINS THE TIME BOMB!

HOWEVER, WHEN SPEEDY LEAVES...

...!GASP!: M-MY WALLET'S MISSING! I WAS SO BUSY TODAY, I FORGOT TO GO TO THE BANK! NOW I'VE LOST ALL THE MONEY I MADE!

THAT EVENING...

WELL, BOSS... IT'S MIDNIGHT! YOUR TIME BOMB ARROW SHOULD GO OFF IN SPEEDY'S QUIVER... AND IF HE'S ON HIS USUAL PATROL WITH THE GREEN ARROW, WE SHOULD BE RID OF THOSE PESTS FOREVER!

THERE'S THE EXPLOSION! NEAR THE WATERFRONT, WHERE GREEN ARROW AND SPEEDY MUST BE PATROLLING IN THEIR ARROW CAR! LET'S GO!

BARRROOM!

BUT AS THE TWO MEN REACH THE DOCKS...

AS WE EXPECTED, THESE CROOKS POSING AS TARGET MANUFACTURERS CAME TO "THE SCENE OF THE CRIME"! USE YOUR LARIAT ARROW, SPEEDY!

≡GASP!≡ T-THEY'RE ALIVE! THEY MUST'VE DISCOVERED I SWITCHED ARROWS AND THEY BLEW UP OUR TIME-BOMB ARROW! B-BUT HOW DID THEY FIND OUT?

PRESENTLY, AS SPEEDY EXPLAINS...

SINCE I ALWAYS HIT THE BULL'S-EYE, YOU PREPARED AN ARROW WITH RED BULL'S-EYE PAINT ON ITS TIP SO I WOULDN'T SUSPECT A SWITCH! BUT I MISSED THE LAST BULL'S-EYE! SO THERE SHOULD'VE BEEN WHITE PAINT ON THE TIP YOU PULLED OUT OF THE TARGET! CONCLUSION? YOU BOMBED YOURSELF INTO PRISON!

5

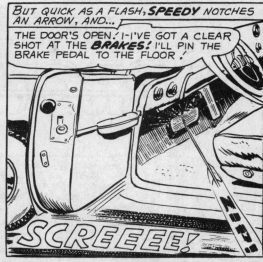

POLICE ANNALS DON'T LIST A MORE DARING AND BRAZEN CRIMINAL, WHOSE BOLD, DAYLIGHT ROBBERIES DEFIED BOTH THE LAW AND THOSE BATTLING BOWMEN, *GREEN ARROW* AND *SPEEDY!* HE LAUGHED AT BULLETS AND ARROWS, FIRED IN VAIN AT HIM, AND DEFIED THE FAMED ARCHERS TO MEET THE...

CHALLENGE of the PHANTOM BANDIT

¡HA, HA! WHY DON'T YOU GRAB ME? I JUST ROBBED A BANK... I *DARE* YOU TO GRAB ME!

THAT CAR-- GOING RIGHT *THROUGH* HIM!

NOONTIME, IN A LARGE CITY'S MOST PROMINENT BANK...

YOU'VE GOT INCREDIBLE NERVE, DEMANDING I OPEN THE SAFE FOR YOU! MY EMPLOYEES CAN SEE YOUR EVERY MOVE!

STOP STALLING-- AND DO AS I SAY!

AS SOON AS YOU STEP OUT OF HERE, YOU'LL BE CAUGHT-- OR SHOT!

WE'LL SEE ABOUT THAT--AFTER I'VE SPRAYED THIS STUFF ON MYSELF AND THE LOOT!

SSSS

AS THE CONFIDENT CROOK STEPS FROM THE OFFICE...

NOW FOR THE ALARM! HMPH-- THE *NERVE* OF THAT CHARACTER!

CLANG! CLANG!

OKAY, YOU... *HUH?* MY HAND WENT RIGHT *THROUGH* HIM!

THAT'S THE STORY, MAC. YOU GUYS CAN'T LAY A HAND ON ME ...GET IT? ⸘HA, HA⸘

IN THAT CASE, I'LL JUST GRAB-- OH, NO! TH-THE SAME THING HAPPENED WITH THE LOOT SATCHEL!

AND, OUTSIDE, AS TWO PATROLMEN TRY TO HALT THE INCREDIBLE THIEF...

JUMPING BLUE BLAZES! OUR SLUGS WENT RIGHT THROUGH HIM-- *HARMLESSLY!*

BANG!

BANG!

MEANWHILE, WHERE THE GUNFIRE HAS ATTRACTED TWO OTHER FIGURES--*GREEN ARROW* AND *SPEEDY,* FAMED BATTLING BOWMEN...

THERE GOES THE THIEF, *SPEEDY!* I'LL STOP HIM EASILY, WITH AN *ARROWLINE!*

TWANG

ZIPPING AROUND ITS TARGET. THE TRICK ARROW LOOPS OVER THE TAUT ROPE, FORMING A KNOT...

OKAY-- HE'S "HOOKED"... HAUL HIM IN, *G.A.!*

2

BUT AS THE ARCHERS PULL ON THE ROPE...

G.A.! D-DID YOU SEE WHAT I SAW?

YES--BUT I DON'T BELIEVE IT!

HA, HA! NOTHING CAN TOUCH ME-- NOT EVEN GREEN ARROW!

BEEEP!

BEEEP!

SCREE

HE'S GONE NOW-- LOST IN THE CROWD SOMEWHERE!

WHEW! I'VE NEVER SEEN ANYTHING LIKE THIS BEFORE!

IT'S FANTASTIC... HE SEEMED TO BE AN ORDINARY, NORMAL PERSON-- YET THERE'S NO SUBSTANCE TO HIM!

GOLLY... WITH A PHANTOM CROOK ON THE PROWL, WE SURE HAVE OUR CHORES CUT OUT FOR US!

LATER, IN A ROOMING HOUSE ON THE OUTSKIRTS OF TOWN...

NOW TO SPRAY THE "COUNTER-CHEMICAL" ON ME, AND RETURN TO NORMAL! AFTER ALL, IT TAKES A SOLID MAN TO DO SOME SOLID SPENDING... HA, HA, HA!

SSSS

YESSIR, JOEY SANDERS, YOU'RE A REAL LUCKY BIRD! NO MORE GRUBBING--NO MORE PICKING POCKETS! YOU'RE THE BIGGEST OF THE BIG TIME NOW-- REAL BIG!

WHEN I KNOCKED OFF THAT CHEMIST AND STOLE HIS TWO SECRET INVENTIONS, I NEVER DREAMED THAT ONE WAS A LIQUID THAT COULD MAKE MY BODY LOSE ITS SUBSTANCE...

3

...AND THE OTHER WAS THE ANTIDOTE THAT WOULD RETURN ME TO NORMAL! IN NO TIME AT ALL, I'LL BE A MILLIONAIRE!

INDEED, ALMOST OVERNIGHT, THE *PHANTOM BANDIT* BECOMES ONE OF THE MOST FEARED NAMES IN THE ANNALS OF CRIMEDOM...

IT'S NO USE, *SPEEDY!* THE *BOXING GLOVE ARROWS* WENT RIGHT THROUGH HIM!

HOW I'VE WAITED FOR THIS DAY-- TO MAKE SAPS OUT OF THOSE BOW-AND-ARROW BOYS!

ZIP!

ZIP!

HOPING SCIENCE MIGHT HAVE AN ANSWER, THE ARCHERS VISIT THE CITY'S LEADING LABORATORY...

THE *PHANTOM BANDIT* IS OBVIOUSLY THE MAN WHO KILLED PROFESSOR ZORN-- AND STOLE HIS HIGHLY SECRET FORMULA!

AH... SO IT'S ALL WORKED WITH A CHEMICAL, EH?

YES--BUT ZORN NEVER REVEALED *WHAT* THE CHEMICAL WAS... HENCE WE CAN'T FORMULATE A CHEMICAL COUNTER-AGENT TO HELP YOU!

I SEE... THEN OUR JOB IS TO FIND SOME WAY TO *GET BACK* THAT CHEMICAL!

LATER, WHEN THE ARCHERS ASSUME THEIR EVERYDAY IDENTITIES--AS WEALTHY OLIVER QUEEN AND HIS YOUNG WARD, ROY HARPER...

AND THERE, SEE?... WE HAVE A *BUNNY!*

I KNOW YOU ENJOY THAT MAGIC ACT, ROY, BUT WE'VE GOT TO CONCENTRATE ON THE *'PHANTOM BANDIT!'*

OF COURSE, OLIVER... I'LL TURN IT OFF!

WAIT! *THAT'S* IT!

...NATURALLY, SOME TRICKS ARE WORKED BY *POWER OF SUGGESTION,* A VALUABLE AID TO MAGICIANS!

WHAT DO YOU MEAN, "THAT'S IT"?

POWER OF SUGGESTION... IF IT CAN HELP MAGICIANS-- IT MIGHT ALSO HELP *US*! LET'S GO!

SHORTLY AFTERWARDS, IN THE CENTER OF TOWN, WHERE *THE PHANTOM BANDIT* STRIKES AGAIN...

WE CAN'T STOP THAT GUY! EVERYTHING PASSES *THROUGH* HIM!

WE'RE ALSO STUMPED! BULLETS WON'T HURT HIM -- AND MIGHT HIT INNOCENT PEOPLE!

ANKS RMORED CAR, INC.

NATIONAL BANK

NATIONAL BANK

CAN'T WE TRY ANYTHING?

YES -- WE'RE TRYING A NEW ANGLE! WE'RE GOING TO *FOLLOW* THE BANDIT AND WAIT FOR HIM TO TAKE THE COUNTER-AGENT THAT WILL RETURN HIM TO NORMAL! *THEN* WE DO OUR BIT!

WITHIN AN HOUR, THE CRIMINAL ENTERS HIS FLAT, AND...

BACK TO MY SOLID SELF *AGAIN*! BROTHER -- THIS IS REALLY IT! I'M STANDING THE LAW ON ITS HEAD!

SSSS

AND AT THAT VERY INSTANT...

HUH? A *ROPE ARROW* -- FIRED THROUGH THE WINDOW!

ZIP!

NOW I GIVE A TUG, AND HE'S OUR... NO! THE ROPE PASSED THROUGH HIM AGAIN!

WHAT--? IT *COULDN'T* HAVE PASSED THROUGH ME! I'VE USED THE COUNTER-CHEMICAL TO MAKE ME SOLID!

5

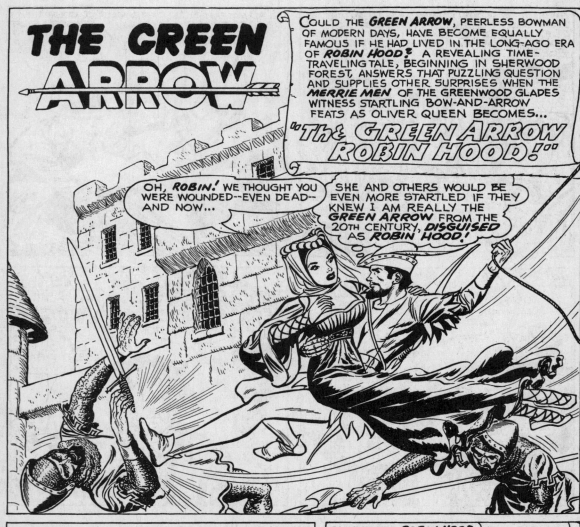

THE GREEN ARROW

COULD THE *GREEN ARROW*, PEERLESS BOWMAN OF MODERN DAYS, HAVE BECOME EQUALLY FAMOUS IF HE HAD LIVED IN THE LONG-AGO ERA OF *ROBIN HOOD*? A REVEALING TIME-TRAVELING TALE, BEGINNING IN SHERWOOD FOREST, ANSWERS THAT PUZZLING QUESTION AND SUPPLIES OTHER SURPRISES WHEN THE *MERRIE MEN* OF THE GREENWOOD GLADES WITNESS STARTLING BOW-AND-ARROW FEATS AS OLIVER QUEEN BECOMES...

"The GREEN ARROW ROBIN HOOD!"

OH, *ROBIN!* WE THOUGHT YOU WERE WOUNDED--EVEN DEAD--AND NOW...

SHE AND OTHERS WOULD BE EVEN MORE STARTLED IF THEY KNEW I AM REALLY THE *GREEN ARROW* FROM THE 20TH CENTURY, *DISGUISED* AS *ROBIN HOOD!*

ON VACATION IN ENGLAND, OLIVER QUEEN AND YOUNG ROY HARPER JOIN OTHERS AT A QUAINT MUSEUM IN SHERWOOD FOREST...

THE *ROBIN HOOD CARRIAGE* WILL BE HERE SOON! WE'LL VISIT THE LEAFY GLENS ONCE ROAMED BY HIM AND HIS MERRIE MEN, THEN TOUR THE MUSEUM ITSELF!

ROBIN HOOD MUSEUM.

OLIVER, DID *ROBIN HOOD* HAVE A SECRET IDENTITY, SUCH AS WE HAVE WHEN WE SWITCH TO *GREEN ARROW* AND *SPEEDY?*

IT IS SAID HE HAD SEVERAL-- AND, ACTUALLY, HIS MISSIONS WERE SOMETHING LIKE OURS! FOR INSTANCE...

"WHEN TROUBLE LOOMS, WE SWITCH FROM OLIVER QUEEN AND ROY HARPER TO PLAY OUR SECRET ROLES AS THE *GREEN ARROW* AND *SPEEDY...*"

"AND MANY TIMES OUR BOWS AND ARROWS HAVE SOLVED PROBLEMS FOR PEOPLE AND PUT DOWN CRIME..."

THE *ARROWCAR'S* CATAPULT HAS HURLED US RIGHT UP TO WHERE THE ESCAPING CROOKS ARE!

AND THE *NET ARROW* WILL HOLD THEM TILL THE POLICE ARRIVE!

ZIP!

BUT, LOOK--THE CARRIAGE IS HERE NOW! YOU JOIN THE OTHERS FOR THE RIDE, ROY! I'M GOING TO STROLL THROUGH THE FOREST!

Robin Hood Coach

AS THE CLOMP OF HOOFS FADES IN THE DISTANCE, OLIVER QUEEN MEANDERS INTO THE GREEN LEAFI-NESS OF THE FOREST, COMING UPON A CAVE...

WELL, WHAT HAVE WE *HERE*? DON'T TELL ME *THIS* IS A HOLD-OVER FROM *ROBIN HOOD'S* DAY!

WELL, I'LL BE--! THIS CAVE WAS USED AS A WAR-TIME ATOMIC LAB, AS THESE OLD PAPERS INDICATE! HMMM...WONDER WHAT ALL THIS MACHINERY WAS FOR...

BUT AS HE TOUCHES ONE OF THE CONTROLS...

WHAT HAPPENED? I MERELY TOUCHED ONE OF THE CONTROLS--AND IT'S LIKE A LIGHTNING STORM HAS HIT THE PLACE!

CR-RACKLE!

2

THEN... HOLD, STRANGER--NOT ANOTHER STEP! IF YOU BE AN AGENT OF THE NOTTINGHAM SHERIFF OR EVIL PRINCE JOHN, THIS ARROW WILL... WILL...

GREAT SCOTT! A WOUNDED MAN, DRESSED EXACTLY LIKE *ROBIN HOOD!*

BUT THE HANDS THAT HOLD THE BOW AND ARROW FALTER, AND...

AH, STRANGER! YOU KNOW ME--YET YOU LIFT NOT A FINGER AGAINST ME! YOU MUST BE A FRIEND!

I'LL CALL THE MUSEUM ATTENDANT AND GET HELP! I'M OLIVER QUEEN, A VISITOR HERE!

IT IS MY *MERRIE MEN* WHO NEED HELP, STRANGER! TODAY, THE SHERIFF LEADS PRINCE JOHN'S MEN AGAINST MY FOREST HIDING PLACE! WARN THEM!

THUNDERATIONS! HE MUST BE PART OF THE MUSEUM PAGEANTRY! HE'S HURT, AND IS DELIRIOUS!

THERE IS LITTLE TIME TO WARN THEM! PRITHEE, TAKE MY LONG BOW AND ARROWS! THEY WILL IDENTIFY YOU TO *LITTLE JOHN!*

OF COURSE! I'LL CALL A DOCTOR FROM THE MUSEUM!

THEN, FOR THE FIRST TIME, QUEEN TAKES STARTLING NOTICE OF THE REST OF THE CAVE...

WAIT A MINUTE! THE LAB EQUIPMENT IS GONE! THE CAVE IS BARE!

YOU WILL GET HELP AT THE BLUE DOLFIN INN! I WORK THERE AS A *CHIMNEY SWEEP!* 'TIS A SECRET IDENTITY I USE TO OBTAIN INFORMATION IN TOWN!

STUNNED, QUEEN RACES FROM THE CAVE, TO SURVEY THE FOREST AROUND HIM...

THE CARRIAGE TRAIL--THE MUSEUM--IT'S ALL GONE! IS THE MAN IN THE CAVE REALLY *ROBIN HOOD?* HAVE I SOMEHOW BEEN FLUNG *BACK IN TIME?*

THE ATOMIC MACHINE I TOUCHED! IT MUST'VE HELD *RETAINED ENERGY*-- AND WHEN IT WAS RELEASED, IT SOMEHOW OPENED UP A WARP IN TIME! I'M ACTUALLY *IN THE PAST* NOW!

③

STUNNED AT FIRST BY THE INCREDIBLE TRUTH, OLIVER SOON COLLECTS HIS WITS AND STOPS AT AN ABANDONED FARMHOUSE...

I'VE CROSSED AN ABYSS OF TIME, INTO AN ERA OF CENTURIES AGO! BUT IF I AM TO HELP *ROBIN HOOD'S* MEN, I'D BETTER DISGUISE MYSELF AS *ROBIN* HIMSELF!

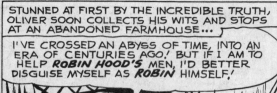

THIS IS A *REAL* TWIST! I MUST CARRY OUT A ROLE SIMILAR TO MY OWN IN 1959 -- PLAYING *ROBIN HOOD AND* HIS SECRET IDENTITY AS A CHIMNEY SWEEP!

BUT THEN, AS THE DISGUISED *GREEN ARROW* STARTS UP A FOOT TRAIL...

I'M ALREADY TOO LATE! THE CAMP OF THE *MERRIE MEN* IS UNDER ATTACK -- AND I HAVEN'T GOT ANY TRICK ARROWS, UNLESS... UNLESS I FASHION SOME -- AND *FAST!*

IN SHORT MOMENTS, THE 20TH CENTURY ARCHER, WITH HIS IMPROVISED TRICK ARROWS, TAKES HIS POSITION IN A LOFTY TREE...

FIRST OFF, I'LL LET THE ATTACKERS GET A GLIMPSE OF SOME *FALCON ARROWS* -- TWO FEATHERS TO SERVE AS "WINGS", AND THORNS TO SERVE AS "TALONS"!

GLIDING SWIFTLY DOWN ON THEIR FEATHERED "WINGS", THE THORNY "TALONS" OF THE UNUSUAL ARROWS SNATCH BOWS FROM THE HANDS OF THE ATTACKERS...

BY PRINCE JOHN'S BEARD! WHAT MANNER OF ARROWS ARE *THESE?*

ZIP!

ZIP!

QUICKLY! REACH OUR HORSES! WE'LL RIDE FOR MORE HELP!

A SMALL BRANCH CONNECTED WITH TWO BENT ARROWS-- THIS SHOULD MAKE A GOOD *YOKE ARROW!*

ZIP!

AND WITH THE "YOKE LOOPS" SETTLED OVER THEIR HEADS, THE TWO FLEEING FIGURES ARE PINNED TO THE GROUND!

HO, HO! SEE WHAT GOOD *ROBIN* HAS DONE!

AND IN QUICK SUCCESSION, A VOLLEY OF *JOUSTING ARROWS* ROUTS THE OTHER ATTACKERS...

BEGONE FROM THIS ACCURSED PLACE! *ROBIN HOOD'S* ARROWS ARE LIKE MAGIC!

POW!

BIFF!

WHAM!

THEN, WHEN THE FOREST IS QUIET AGAIN...

YOU WERE WOUNDED ONLY THIS MORNING, GOOD ROBIN! BUT ALL IS WELL NOW--EXCEPT THAT THE SHERIFF TOOK *MAID MARIAN* TO HOLD AS HOSTAGE IN NOTTINGHAM!

THEN I RIDE, LITTLE JOHN! WAIT HERE FOR WORD FROM ME!

AND LATE THAT NIGHT, BEHIND THE BLUE DOLFIN INN...

NOW--I'VE SWITCHED TO *ROBIN HOOD'S* OTHER IDENTITY--THAT OF THE CHIMNEY SWEEP! BY MORNING, I SHOULD KNOW WHERE *MAID MARIAN* IS HELD...

UPON THE APPROACH OF DAWN, A FIGURE PACES NERVOUSLY IN NOTTINGHAM CASTLE...

OH!

WHROOSH!

I AM HERE, FAIR **MAID MARIAN**, AT THE REQUEST OF **ROBIN HOOD**--TO TAKE YOU FROM THIS PLACE!

INDEED! HOW CAN A MERE CHIMNEY SWEEP SAVE ME WHEN GOOD **ROBIN**, EVEN WITHOUT HIS WOUNDS, COULD NOT ESCAPE THESE GROUNDS?

THE COURTYARD IS LITTERED WITH THE SHERIFF'S MEN-- AND THEY ALSO STAND ON YONDER WALL!

BUT I SHALL ASCEND TO THE ROOF, PULL YOU UP BY THE LONG BROOM-- THEN PRESENT YOU WITH A SURPRISE! HURRY!

MOMENTS LATER...

ROBIN! BUT YOU WERE WOUNDED!

I AM BETTER NOW, **MAID MARIAN!** MY FRIEND THE CHIMNEY SWEEP MADE THIS MUCH OF OUR ESCAPE POSSIBLE--I SHALL DO THE REST!

THAT WAS SURE A QUICK CHANGE INTO THE COSTUME I CARRIED IN THE BAG!

AFTER FIRING A PREPARED **VINE ARROW** TO A DISTANT TURRET, THE DISGUISED **GREEN ARROW** SWINGS ACROSS THE COURTYARD...

HO! THE SHERIFF'S MEN **STILL** LITTER THE COURT-YARD--ON THEIR FACES!

AND UPON REACHING THE FAR WALL...

QUICKLY! TO THE HORSES BELOW! I'LL CONFUSE THE WALL GUARDS WITH **SOOT ARROWS**--MADE FROM SOOT I GATHERED FROM THE CHIMNEY!

AND SOON AFTERWARDS, AS THEY APPROACH SHERWOOD FOREST...

JOIN **LITTLE JOHN** AND THE OTHERS--I SHALL SOON RETURN...

THEN, UPON HIS RETURN TO THE CAVE...

I FEEL MUCH BETTER THIS DAY, STRANGER--BUT *YOU*, HOW DID YOU FARE?

ALL IS WELL WITH YOUR MEN AND *MAID MARIAN*, ROBIN! I MUST TELL YOU--QUICKLY--WHAT OCCURRED...

AFTER HIS STORY IS FINISHED, OLIVER BIDS FAREWELL TO *ROBIN HOOD*--THEN ENTERS THE CAVE...

NOW WHAT'S HAPPENING? AM I BEING RETURNED TO THE 20TH CENTURY?

CR-RACKLE!

THEN, ONCE MORE THE CAVE RESUMES ITS ORIGINAL APPEARANCE...

STRANGE! FOR A MOMENT I THOUGHT I WAS BACK IN *ROBIN HOOD'S* DAY! I THOUGHT ATOMIC ENERGY SOMEHOW HAD OPENED UP A TIME WARP--BUT HERE I AM, AS BEFORE! A DREAM, PERHAPS...

YET, WHEN LATER HE REJOINS THE OTHERS AT THE MUSEUM...

AMAZING, OLIVER! ACCORDING TO THIS EXHIBIT, *ROBIN HOOD ALSO* HAD TRICK ARROWS!

YES, I KNOW! BUT HOW DID I KNOW? HAVE I READ ABOUT THEM IN THE PAST, OR...?

OX-YOKE ARROW

ROBIN HOOD'S TRICK ARROWS

FALCON ARROWS

JOUSTING ARROW

500T ARROW

THE END

THE GREEN ARROW

THE GREEN ARROW

COUNTLESS SNOWS AGO, HE HAD RULED THE TRIBE, A JUST AND MIGHTY CHIEF...AND IT WAS SAID, AFTER HIS PASSING, THAT HE WOULD ONE DAY RETURN--IF THE TRIBE AGAIN NEEDED HIM! BUT FATE TOOK A HAND, TURNING LEGEND INTO REALITY, FORCING THE GALLANT GREEN ARROW, ACE ARCHER OF JUSTICE, TO BECOME...

ALIAS CHIEF MAGIC BOW

SEIZE THE TRICKSTER! HE MEANS TO DO YOU HARM!

GOLLY!... GREEN ARROW ENTERED THE VILLAGE DISGUISED AS AN INDIAN -- ONLY TO BE *ATTACKED!*

VACATIONING IN THE WEST, WEALTHY OLIVER QUEEN AND HIS YOUNG WARD, ROY HARPER, DRIVE OVER REMOTE, WINDING TRAILS, WHEN SUDDENLY...

OLIVER! DID YOU HEAR A CRY FROM THE FOREST?

YES... GET OUT OUR FIGHTING TOGS, ROY!

SWIFTLY, THEY DON A PAIR OF FAMED, COLORFUL COSTUMES...

THAT CRY SOUNDED LIKE TROUBLE...LIKE A JOB FOR-- *GREEN ARROW* AND *SPEEDY!*

INDEED, AT THIS VERY MOMENT, IN A NEARBY GULLEY...

IT IS HOPELESS... THE BEAR IS ALMOST UPON ME!

BUT ALL AT ONCE, FROM UP ABOVE, TWO ARROWS COME STREAKING DOWN...

TWANG!
TWANG!
TWANG!

ONE OF THEM--THE FIRECRACKER ARROW-- EMITS A SERIES OF SHARP EXPLOSIONS...

CR-RACK
BAM!
CR-RACK!
POW!

...WHILE THE SECOND SHAFT--THE ROPE ARROW-- QUICKLY ENCIRCLES THE STARTLED BEAST, BINDING IT!

QUICKLY--I'LL HELP YOU OUT! THAT BEAR WILL FREE HIMSELF SOON--AND WE WANT TO BE FAR AWAY WHEN HE DOES!

SHORTLY AFTERWARD...

I THINK WE'RE SAFE NOW!

YOUR ARROWS--THEY WERE PURE MAGIC... LIKE THOSE OF CHIEF MAGIC BOW HIMSELF!

HA, HA! I AM KNOWN MERELY AS GREEN ARROW-- AND MY YOUNG FRIEND IS CALLED SPEEDY! BUT WHO ARE YOU-- AND THIS CHIEF MAGIC BOW?

I AM PRINCESS MORNING MOON... AND MAGIC BOW WAS MY TRIBE'S FIRST CHIEF, COUNTLESS MOONS AGO!

2

"IT IS SAID THAT HIS MIGHTY BOW PROTECTED THE PEOPLE BY PERFORMING WONDERS--EVEN AGAINST THE GREAT WIND SPIRIT..."

DEPART, ANGRY WIND! ENTER NOT INTO MY REALM, TO SWEEP AWAY MY VILLAGE! *DEPART!*

NOW, OUR LEGENDS SAY THAT *MAGIC BOW* WILL RETURN TO US, SHOULD OUR TRIBE EVER BE TROUBLED AGAIN! OF COURSE--IT IS ONLY LEGEND...

AND *IS* YOUR TRIBE TROUBLED NOW?

AH, YES--EVER SINCE MY FATHER, *CHIEF GREAT TREE*, DIED! THE EVIL CH'WAKA, OUR MEDICINE MAN, HAS BEEN CONTROLLING THE TRIBE WITH HIS BLACK MAGIC!

BLACK MAGIC?

CH'WAKA CALLS IT MAGIC, BUT I KNOW HE IS ONLY A CUNNING TRICKSTER! YET, IF HE IS NOT EXPOSED, HE WILL SOON BE NAMED OUR CHIEF!

HMM... *SPEEDY* AND I WERE ON VACATION-- BUT I THINK YOUR TRIBE COULD REALLY USE A *CHIEF MAGIC BOW!* LISTEN-- HERE'S MY PLAN...

LATER, AFTER SEPARATING FROM THE PRINCESS...

FIRST, I'VE GOT TO DISGUISE MY FACE--AND DYE MY HAIR SO THAT I RESEMBLE AN INDIAN! *BERRY STAIN* SHOULD DO THE TRICK!

SOON...

ARE YOU SURE YOU WANT ME TO HIDE OUT- SIDE THE VILLAGE, *G.A.*?

YES, *SPEEDY*... DON'T SHOW YOUR- SELF--EVEN IF I'M IN TROUBLE! I'LL NEED YOU IN THE BACKGROUND!

IT IS NOT LONG AFTERWARDS THAT AN IMPOSING FIGURE STRIDES BOLDLY INTO THE VILLAGE...

EH? WHO IS THIS STRANGER?

HE LOOKS LIKE THE ONE SO OFTEN DESCRIBED IN OUR LEGENDS, CH'WAKA! CAN HE BE--*CHIEF MAGIC BOW?*

3

CHIEF *MAGIC BOW?* BAH-- I NEVER BELIEVED THAT OLD LEGEND... I ONLY USE IT TO ADVANCE MY OWN SCHEMES! IF THE PRINCESS IS UP TO SOME TRICK, I MUST STOP IT NOW...

SEIZE THAT IMPOSTOR... HE IS NOT *MAGIC BOW!*

GOLLY!... THEY'RE ATTACKING *G.A.*--AND I CAN'T INTERFERE TO HELP, OR I'LL SPOIL OUR PLANS!

BUT BREAKING CLEAR, THE DISGUISED ARCHER HURRIEDLY FITS ONE *FIRE ARROW* AFTER ANOTHER INTO HIS BOW, AND...

AIEEE! HIS ARROWS BURST INTO A FLAMING BARRIER AROUND US! HIS BOW IS TRULY MAGIC!

BUT I WISH TO DO YOU NO HARM... I COME AS A FRIEND!

TWANG!

TWANG

WAIT! ONLY ONE MAN COULD DO THAT! I HAVE HEARD OF THE GREAT PALEFACE ARCHER, *GREEN ARROW,* AND HIS MARVELOUS TRICK ARROWS! BUT I CAN FOIL EVEN HIM!

MAKING HIS WAY INTO A TEPEE, CH'WAKA SOON EMERGES WITH A CHALLENGE...

LONG HAS THE TRIBE KEPT *CHIEF MAGIC BOW'S* QUIVER OF ARROWS! SURELY, YOU CAN PERFORM YOUR FEATS WITH YOUR VERY OWN WEAPONS!

GREAT SCOTT! HE'S FORCING ME TO USE *ORDINARY ARROWS!* WHAT'LL I DO?

SWITCHING QUIVERS, THE ACE ARCHER RECEIVES HIS FIRST TASK...

MAGIC BOW COULD FIRE AN ARROW INTO THE BRUSH-- AND RECALL IT AT WILL! CAN *YOU* DO THE SAME?

WITH MY *BOOMERANG ARROW,* THAT WOULD BE A CINCH! BUT WITH AN *ORDINARY* SHAFT--WELL.... THERE'S ONLY ONE CHANCE...

SLOWLY, THE SKIES DARKEN--LIGHTNING FLASHES, FOLLOWED BY...

RAIN! THE ARROWS OF MAGIC BOW ARE WITH US AGAIN! HE PERFORMED THE WONDERS OF THE LEGEND!

THUS BEATEN, THE MEDICINE MAN RESORTS TO ANOTHER FORM OF "TRICKERY"...

THE IMPOSTOR WILL SOON RUIN ME... MAKE YOUR FIRST ARROW COUNT!

INDEED--IT WILL BE ANOTHER "ARROW FROM THE SKY" THAT SLAYS THE STRANGER!

BUT, ABRUPTLY...

AIEEE!

CR-RACK

FOOLS! NOW LOOK WHAT YOU HAVE DONE!

AS THE OTHERS GATHER AROUND...

NOW YOU KNOW HOW CH'WAKA'S "ARROWS FROM THE SKY" APPEARED! THOSE TWO BRAVES FIRED THEM FROM THE TREE!

CH'WAKA PROMISED US MANY THINGS, TO HELP HIM CARRY OUT FEATS OF MAGIC! WE WILL TELL ALL!

AND, FINALLY, BEYOND THE VILLAGE...

TELL THE TRIBE THERE WAS NO REAL MAGIC ON MY PART, EITHER! FROM HIDING, SPEEDY HEARD ALL, AND FIRED TRICK ARROWS ACCORDINGLY! HIS SMOKE ARROW MADE THE "NIGHT"...

AND MY ELECTRIC ARROW AND RAIN ARROWS CAUSED THE STORM!

BUT THE TRICK ARROWS DID WORK WONDERFUL "MAGIC"... SO YOU MIGHT SAY THE LEGEND CAME TRUE-- IN A WAY!

HMM...IN A WAY, YES-- YOU MIGHT SAY SO!

THE END

THE GREEN ARROW

MUCH OF *GREEN ARROW'S* SUCCESS AGAINST LAW-BREAKERS DEPENDS ON THE INGENUITY AND VARIETY OF HIS TRICK ARROWS! BUT WHAT WOULD HAPPEN IF THE ARROWS THE *EMERALD ARCHER* NEEDS IN AN EMERGENCY SUDDENLY TURN OUT TO BE SILLY AND WORTHLESS? HOW COULD HE SHOOT HIS WAY OUT OF A PERILOUS TRAP IF ALL HE CAN USE IN SELF-DEFENSE ARE...

THE AMATEUR ARROWS!

THESE WELL-MEANING CAMP KIDS MADE A PACK OF CRUDE, CHILDISH ARROWS TO HELP *SPEEDY* AND ME FIGHT CRIMINALS! BUT HEAVEN HELP US IF WE REALLY HAD TO DEPEND ON SUCH SILLY SHAFTS AS A *DOUGHNUT* ARROW AND A *PEPPERMINT STICK* ARROW!

ONE DAY, OUT WEST, AS GREEN ARROW AND SPEEDY ARRIVE AT THE GREEN ARROW CAMP FOR BOYS...

HOWDY, *GREEN ARROW!* THE CAMP KIDS ARE SO EXCITED ABOUT YOUR VISIT HERE TODAY, THEY AIN'T SLEPT ALL NIGHT!

BLESS THEIR HEARTS! WE WON'T KEEP THEM ANY LONGER!

IT'S A GREAT THING YOU'RE DOIN', *G.A.,* GIVIN' ALL YOUR REWARD MONEY AN' SUCH TO CAMPS FOR UNDER-PRIVILEGED BOYS!

THAT'S WHAT MONEY IS FOR, HANK!... MAKING PEOPLE HAPPY! ALL KIDS -- RICH *OR* POOR -- SHOULD HAVE A CHANCE TO ENJOY THE GREAT OUTDOORS!

*AS THE **ARROW CAR** ENTERS THE CORRAL, AN ENTHUSIASTIC ROAR OF WELCOME GREETS ITS OCCUPANTS!*

HURRAY FOR *GREEN ARROW!*

WELCOME, *GREEN ARROW!*

THREE CHEERS FOR *GREEN ARROW* AND *SPEEDY!*

LOOK AT THAT HUGE EFFIGY OF YOU THE KIDS BUILT! THEY SURE IDOLIZE YOU, *GREEN ARROW!*

*SHORTLY, AFTER **GREEN ARROW** THANKS THE YOUNG CAMPERS FOR THEIR WARM RECEPTION...*

AN' NOW, *GREEN ARROW*, HERE'S THE MOST FITTIN' GIFT THE CAMPERS COULD THINK OF GIVIN' YOU! SPECIAL ARROWS THEY INVENTED TO HELP YOU FIGHT CRIME!

EACH OF US BOYS MADE HIS OWN ARROW DURING THE ARTS AND CRAFTS PERIOD!

ER...THANKS! THANKS A LOT, BOYS! UH...THEY'RE JUST WHAT I NEED!

ARE YOU KIDDING, *GREEN ARROW?* LOOK HOW CRUDELY CONSTRUCTED THEY ARE! HALF OF THEM WON'T EVEN FIT INTO OUR QUIVERS!

GOSH, *GREEN ARROW!* AREN'T YOU GOING TO TRY THEM OUT-- TO SEE HOW THEY WORK?

OH, SURE! *SURE!* ER...*THIS* ARROW! WHAT DOES IT DO?

IT'S A *BABY-ARROW!* I FIGURED SOMETIME YOU MIGHT BE AT THE SCENE OF AN EXPLOSION OR A FIRE WHERE EVERYBODY GETS FRIGHTENED, ESPECIALLY BABIES!

"SO WHEN YOU SEE A FRIGHTENED BABY, YOU JUST FIRE MY BABY-RATTLE ARROW! IT HITS THE GROUND WHERE THE BABY CAN REACH OUT FOR IT! THE BABY PLAYS WITH IT AND HE ISN'T SCARED ANY MORE!

2

ER...THAT'S VERY CLEVER AND...UH...USEFUL! AND *THIS* ARROW?

THAT'S *MY* ARROW! I CALL IT A *BAIT ARROW!* SUPPOSE YOU'RE IN THE JUNGLE! A TIGER IS COMING AT YOU! SO YOU FIRE THE *BAIT ARROW!* GO AHEAD! *FIRE IT!*

TWWANNGG!

SEE HOW IT WORKS? AS IT FLIES THROUGH THE AIR, IT MAKES A SOUND LIKE A DUCK! JUST LIKE IT'S ATTRACTING THE ATTENTION OF THAT DOG, THE *TIGER* FORGETS ALL ABOUT YOU AND CHASES AFTER THE DUCK! THEN YOU MAKE YOUR ESCAPE!

QUACKK!

QUACKK!

WOOF! WOOF!

I-I SEE! VERY GOOD! NOW HERE'S ANOTHER ARROW WITH A BIG HUNK OF *BUBBLE GUM* ON IT!

THAT WAS *MY* IDEA! FIRE IT! YOU'LL SEE WHAT IT DOES!

AS GREEN ARROW OBLIGINGLY FIRES THE BUBBLE-GUM ARROW...

THE SHAFT IS HOLLOW SO AIR CAN PASS INTO THE GUM AND BLOW IT UP INTO A BIG BUBBLE!

"THE WAY I SEE IT, THE *BUBBLE-GUM ARROW* GUMS UP CROOKS! FOR INSTANCE, YOU FIRE IT AT A PROWLER! HE GETS CAUGHT INSIDE THE STICKY BUBBLE, LIKE A FLY TRAPPED ON FLY PAPER!"

AS AN HOUR PASSES, DURING WHICH TIME THE G.A. TRIES OUT SUCH OTHER HOPELESS ARROWS AS A DOUGHNUT ARROW AND A PEPPERMINT STICK ARROW...

RUN FOR COVER, KIDS! IT'S A SUDDEN SUN SHOWER!

NO, *GREEN ARROW!* NOBODY HAS TO GET WET! FIRE MY *UMBRELLA ARROW!* STRAIGHT UP IN THE AIR!

3.

AS GREEN ARROW FOLLOWS THE BOY'S SUGGESTION...

WHOOOSH!

THE SPOKES INSIDE THE SHAFT SPREAD OUT AT THE TOP LIKE AN UMBRELLA OPENING UP! THEN YOU HOLD THE UMBRELLA IN PLACE BY HOLDING ONTO THAT CORD ATTACHED TO THE HANDLE!

VERY SMART, SON! ANY ARCHER WHO DOESN'T CARRY YOUR ARROW IS... ER... ALL WET!

LATER, AS THE SHOWER STOPS AND THE SENIOR KIDS PILE INTO THE CAMP STAGECOACH...

CLIMB ABOARD, *SPEEDY!* THE SENIOR CAMPERS EXPECT US TO DRIVE THEM INTO TOWN TO COMPETE IN A PET CONTEST AGAINST OTHER DUDE RANCH CAMPS!

I KNOW!

THE BOYS ARE TAKING ALONG THEIR MICE, RABBITS, HAMSTERS AND OTHER PETS! ALSO THEIR *ARROWS!* I GUESS THEY FIGURE WE CAN'T DO WITHOUT THE SILLY THINGS!

TUSH, *SPEEDY!* THE BOYS MEAN WELL... EVEN IF THEIR ARROWS *ARE* CHILDISH INVENTIONS, USELESS FOR OUR PURPOSE!

TEN MINUTES LATER, AS THE STAGECOACH RUMBLES ALONG A BADLANDS ROAD...

HELP! H-HELP ME! I'M HURT!

TAKE THE REINS, *SPEEDY!* SOMEONE'S IN TROUBLE! BETTER STILL... STOP THE COACH! WE MAY HAVE TO TAKE THIS MAN INTO TOWN!

BUT AS THE COACH STOPS...

THROW YORE HANDS UP, *GREEN ARROW!* REMEMBER ME? I'M ONE OF THE FOUR TERRIS BROTHERS YOU SENT TO PRISON! AN HOUR AGO, BY A LUCKY ACCIDENT, WE ALL ESCAPED WHEN A PRISON VAN, CARRYING JUST THE FOUR OF US, OVERTURNED!

WE REMEMBERED ONCE HIDIN' SOME RIFLES NEAR HERE! SO WE DUG 'EM UP AN' THEN SCATTERED... EACH MAN FOR HIMSELF! NOW FOLLOW ORDERS! THROW YOUR QUIVERS IN THE WATER!

DO AS HE SAYS, *SPEEDY!* WE CAN ALWAYS MAKE NEW ARROWS!

THEN, AS THE YOUNGSTERS FRIGHTENEDLY PILE OUT OF THE STAGECOACH...

AN IDEA JUST HIT ME! I WONDER IF THAT *BABY-RATTLE ARROW* ONE OF THE KIDS MADE COULDN'T DO SOME GOOD! HMM... WE'LL SOON FIND OUT!

IN ONE SWIFT MOTION, THE BABY-RATTLE ARROW IS NOTCHED AND FIRED!

¿GASP!¿ THE SOUND OF A *RATTLESNAKE!* R-RIGHT NEAR ME! IT'S GETTIN' READY TO STRIKE! I...

RATTLE! RATTLE-RATTLE!

QUICK, *SPEEDY!* IN HIS FEAR OF GETTING BITTEN BY A RATTLESNAKE, THAT OUTLAW STUMBLED AND TRIPPED DOWN THE INCLINE! GRAB HIM!

OKAY! BUT I-I'M STILL SHOCKED! THAT KID'S SILLY *BABY-RATTLE ARROW* HELPED US!

YIIIIII!

BUT AS THE INJURED MAN IS TAKEN ABOARD THE STAGECOACH...

HOLD IT! STOP TYIN' UP MY BROTHER... AN GIT OFF THAT COACH! I AIN'T KIDDIN'! *GREEN ARROW!*

IT'S ED TERRIS, THE SECOND OLDEST BROTHER!

BETTER LISTEN TO HIM, *SPEEDY!* HE MIGHT HURT SOMEBODY WITH A WILD SHOT!

BUT AS SPEEDY CROSSES IN FRONT OF GREEN ARROW TO DESCEND...

FOR TWO SECONDS, SPEEDY'S BODY WILL SCREEN MY MOVEMENTS! ENOUGH TIME FOR ME TO FIRE A *BAIT ARROW!*

5

SECONDS LATER, AS THE "DUCK" ARROW FLIES PAST AN AMAZED DESPERADO...

HUH?!

QUACK! QUACK!

NOW IF THE ARROW'S QUACKING SOUNDS ONLY ATTRACT THE ATTENTION OF THAT FLOCK OF WILD DUCKS I SPOTTED A MINUTE AGO!

HONK! HONK! HONK! QUACK!

YEEOOWW!

IT DID! HA, HA! DECOYED BY THE BAIT ARROW, THE FLOCK IS USING ED TERRIS AS A LANDING FIELD! HE'S BEING SWAMPED WITH DUCKS!

PRESENTLY, AS ED JOINS HIS EVIL BROTHER ON TOP OF THE STAGECOACH...

FROM NOW ON, WE'LL CALL THAT SHAFT A DEAD DUCK ARROW! ED SURE WAS FINISHED THE SECOND THAT MECHANICAL DUCK LANDED NEAR HIM!

SPEAKING OF FINISHES, WE'RE NOT FINISHED WITH THE TERRIS BOYS! LOOK UP THERE!

I SEE YOU SPOTTED ME, GREEN ARROW! WELL, I SPOT SOMETHIN,' TOO! YOU AN' SPEEDY GOT NO QUIVERS! THAT LEAVES YOU UNARMED, EH?

YEP, BUD! WE'RE SURE AT YOUR MERCY! ALL I'VE GOT IS THIS HARMLESS, BUBBLE-GUM ARROW! -- THE SILLY INVENTION OF AN EIGHT-YEAR-OLD BOY! WATCH THE BUBBLE GET BIGGER!

I'M WATCHIN'! AN' I'M LAUGHIN' SO LOUD I'M FIT TO BUST! HA! HA! HA!

BUT SUDDENLY, AS THE HUGE BUBBLE BURSTS AGAINST A POINTED ROCK...

⋛GASP!⋚ I-IT'S EXPLODIN' LIKE GUN SHOTS!...LOUD ENOUGH TO BUST MY EARDRUMS! ...⋛GROAN!⋚ T-THE SOUND KEEPS REPEATIN' AN' REPEATIN'...

PWAM! PWAMMM! PWAMM!

6

AS THE ECHOES OF THE BURST BUBBLE MULTIPLY DEAFENINGLY...

BAM! BWAMM! BWAMM! PWAMM!

I- I CAN'T STAND IT ANY MORE!... ;GASP!:... I GIVE UP!

POOR BUD TERRIS! HE DIDN'T REALIZE HE WAS STANDING ON THE EDGE OF *ECHO VALLEY!* BUT *I* DID! I DELIBERATELY AIMED THE GIANT BUBBLE AT THE JUTTING ROCKS... AND WAITED FOR THE *ECHO EXPLOSIONS!*

LATER, AS *GREEN ARROW'S* DESTINATION COMES INTO VIEW...

LOOK DOWN THERE, *G.A.!* IT'S THE LAST OF THE TERRIS BROTHERS! HE'LL REACH TOWN BEFORE WE CAN SOUND THE ALARM... AND HE'LL ESCAPE!

MAYBE *NOT, SPEEDY!* I HAVE AN IDEA-- HAND ME THE *UMBRELLA ARROW* AND THAT CAGE WHICH CONTAINS ONE OF THE KIDS PETS!

PRESENTLY, AS *GREEN ARROW* WAITS FOR THE *UMBRELLA ARROW* TO DRIFT TOWARD THE ESCAPING CONVICT...

NOW I'LL SEVER THE STRING I TIED AROUND ONE OF THE PET CAGES! ANY ONE OF THE OTHER ARROWS, IF FIRED WITH SUFFICIENT FORCE, CAN SNAP THE STRING!

AS THE STRING SNAPS, RELEASING THE CAGE...

THE - THE CAGE BROKE! NOW I'M *DONE FOR!* ...;GASP!:...

CRACK!

AN HOUR LATER, IN TOWN...

WE GOT HIM, *GREEN ARROW!* SOON AS YOU TOLD US OF THE ODOR HE WAS GIVING OFF, WE SPREAD THE ALARM! HE WAS A CINCH TO CATCH!

THAT'S WHAT I FIGURED! A FRIGHTENED *SKUNK* GIVES OFF A MIGHTY DISTINCTIVE SMELL... AND I KNEW THE BAD ODOR WOULD CLING TO THIS TWO-LEGGED SKUNK SO HE COULD BE TRACED EASILY!

THAT NIGHT, AS THE WIZARD ARCHERS LEAVE CAMP TO RETURN TO *STAR CITY* IN THEIR EVERYDAY IDENTITYS AS OLIVER QUEEN AND ROY HARPER...

THIS TIME *SPEEDY* AND I WANT TO CHEER *YOU!* WE CAN PAY YOUR ARROWS NO HIGHER COMPLIMENT THAN TO SAY THAT THEY WILL *ALL* GO INTO OUR PERMANENT ARSENAL FOR FIGHTING CRIME!

AMEN!

The End.

THE GREEN ARROW

ONE DAY, A STRANGE CURSE FALLS UPON THE ARROWS STAR CITY'S FAMOUS ARCHERS USE TO FIGHT CRIME! SUDDENLY, MYSTERIOUSLY, EACH ARROW FIRED BY GREEN ARROW OR SPEEDY DISAPPEARS, AS IF THE TWINS OF THE TWANGING BOW WERE BEING PUNISHED BY UNSEEN FORCES FOR THEIR HEROIC DEEDS! WHAT GHOSTLY POWER CAUSES THIS FRIGHTENING PHENOMENON? WE DEFY YOU TO SOLVE...

THE CASE OF THE VANISHING ARROWS!

STOP FIRING, GREEN ARROW! IT'S HAPPENING AGAIN! OUR ARROWS ARE DISAPPEARING! WHAT'S MAKING THEM VANISH?

I-I DON'T KNOW, SPEEDY! BUT IF THESE EERIE OCCURRENCES KEEP UP, I'M GOING TO START BELIEVING IN THE SUPERNATURAL!

ONE AFTERNOON, AS FAMED BOWMEN GREEN ARROW AND SPEEDY VISIT STAR CITY'S POLICE HEADQUARTERS...

LOOK, SPEEDY! HERE'S A LETTER FROM THE RENOWNED BATMAN, CONGRATULATING US ON CAPTURING THE "RED DART" GANG!

THAT'S GREAT! LET ME SEE IT!

LOOK AT THE BAT INSIGNIA ON BATMAN'S STATIONERY! SAY, G.A., WHY DON'T YOU HAVE A TRADEMARK THAT STANDS FOR GREEN ARROW?

A SPECIAL TRADEMARK FOR MYSELF, EH? NOT A BAD IDEA, SPEEDY! HMMM...

THAT NIGHT, AT THE ARROW CAVE, AS THE AMAZING ARCHERS ASSUME THEIR EVERYDAY IDENTITIES OF OLIVER QUEEN AND ROY HARPER...

I'M TAKING YOUR ADVICE, ROY! WE'RE GOING TO HAVE A TRADE-MARK! FROM NOW ON I'M PUTTING **GREEN STONE ARROW HEADS** ON THE TIPS OF ALL OUR SHAFTS!

=GASP!= **GREEN ARROWS!** IT'S A NATURAL, OLIVER! WHERE'S MORE ROCK? I'LL HELP YOU SHAPE MORE ARROW HEADS!

A MINUTE LATER, OUTSIDE THE CAVE... I FOUND THESE PIECES OF GREEN ROCK JUST OUTSIDE OUR CAVE ENTRANCE! THEY MAKE PERFECT ARROW HEADS!

I'LL GATHER ALL THERE IS! I'M SO EXCITED! FROM NOW ON, **GREEN ARROW HEADS** WILL BE OUR INSIGNIA!

THE FOLLOWING MORNING, AS **GREEN ARROW** AND **SPEEDY** PATROL STAR CITY IN THEIR SLEEK **ARROW CAR**...

LOOK AT THE CROWD AROUND THE NEW STAR BUILDING! IT'S THE TOWN'S TALLEST SKYSCRAPER, AND TODAY IT'S OPEN FOR INSPECTION!

YES, **SPEEDY!** NEWSPAPER-MEN FROM ALL OVER THE COUNTRY ARE HERE TO REPORT ON IT!

STAR BUILDING

WE WERE ALSO INVITED TO ATTEND THE PENTHOUSE CELE-BRATION! BUT WE CAN'T GO BECAUSE IT CONFLICTS WITH OUR REGULAR PATROL!

WE'LL SEE IT ANOTHER TIME! THE BUILDING ISN'T COMPLETELY FINISHED ANYWAY! I READ THAT ONLY ONE ELEVATOR IS RUNNING!

SHORTLY AFTER, ON THE WATERFRONT...

HELP! POLICE!

LOOK, G.A.! IT'S A STICKUP! HERE'S THE FIRST CHANCE TO USE OUR NEW **GREEN ARROW HEADS**

RIGHT! I'LL FIRE MY **GLUE ARROW!**

GLUE

WHISSSTT!

IT RELEASES A TRAIL OF STICKY, CEMENT-LIKE GLUE ALONG THE CROOK'S WOULD-BE ESCAPE ROUTE!

MOMENTS LATER, AS THE THIEF HITS THE TRAIL...

GASP! I-I CAN'T RUN! I'M CAUGHT LIKE A FLY ON FLYPAPER!

WELL, *SPEEDY!* THERE'S *ONE* CROOK WHO'S *"STUCK"* WITH US!

RIGHT! NOW YOU GET HIM AND I'LL GET THE ARROW BACK!

BUT AS *SPEEDY* REACHES FOR THE SHAFT...

GASP!...I-IT'S TAKING OFF BY ITSELF!

WHOOOOSH!

AS THE ARROW ARCS HIGH ABOVE THE RIVER...

GREEN ARROW! DID YOU SEE WHAT I SAW?

IT S-SEEMS AS IF THE ARROW FLEW AWAY UNDER ITS OWN POWER AND LANDED IN THE MIDDLE OF THE RIVER!

W-WHAT CAUSED IT TO DO THAT?

I DON'T KNOW, *SPEEDY*... EXCEPT THAT IT WAS NOTHING *ORDINARY!* LET'S FORGET ABOUT IT BEFORE WE START BELIEVING IN THE *SUPERNATURAL!*

AN HOUR LATER, AS THE ARROWCAR CRUISES PAST THE CITY ZOO...

LOOK OUT! A TIGER'S BROKEN LOOSE! RUN FOR YOUR LIVES!

WE'VE GOT TO CAPTURE THAT BEAST BEFORE HE KILLS SOMEBODY! I'LL FIRE MY *TRAP ARROW!*

GROWRR!

TO ZOO

AS THE LIGHT, TENSILE STEEL SPOKES FAN OUT IN ALL DIRECTIONS FROM THE SHAFT...

THE TIGER MIGHT TEAR AN ORDINARY NET TO RIBBONS, BUT NOT THIS *ALL-STEEL* WEB!

TWANGG!

3

SHORTLY, AS THE TIGER LASHES ABOUT FRENZIEDLY WITHIN THE SPIDERY TRAP...

SPEEDY, TELL THE ZOO-KEEPERS TO HURRY UP WITH A CAGE! I'LL USE MY LARIAT ARROW TO LASSO HIM!

ZIPP!

BUT PRESENTLY, AS THE RAGING CAT IS CARTED OFF AND THE FAMED ARCHERS REACH OUT TO RETRIEVE THEIR ARROWS...

GREAT GUNS!... :GASP!: GET BACK, SPEEDY!

OUR ARROWS... THEY'RE MELTING INTO BLOBS OF NOTHING!

ZZZZT!

SSSST!

SECONDS LATER, AS THE BAFFLED BOWMEN GASP IN AMAZEMENT...

THE ARROWS VANISHED RIGHT BEFORE OUR EYES!... WHAT'S GOING ON, G.A.? WHAT'S HAPPENING TO OUR ARROWS?

I-I KNOW AS MUCH AS YOU DO, SPEEDY! THE ARROWS DISAPPEARED AS IF BY MAGIC! WHY OR HOW... I CAN'T EVEN GUESS!

AS THE PUZZLED ARCHERS RESUME THEIR PATROL...

G.A., I KNOW THIS DOESN'T MAKE SENSE, BUT OUR ARROWS ONLY BEGAN VANISHING AFTER WE CHOSE OUR GREEN ARROW INSIGNIA!

I THOUGHT OF THAT, TOO, SPEEDY! BUT PICKING A TRADEMARK DOESN'T EXPLAIN THE WEIRD DISAPPEARANCES! FRANKLY, I'M STUMPED!

LATER THAT DAY, IN THE BANKING DISTRICT OF STAR CITY...

IT'S LUCKY WE DIDN'T ATTEND THE STAR BUILDING CELEBRATION! THERE'S AN ARMORED CAR ROBBERY! THE THIEVES ARE ESCAPING WITH BAGS OF MONEY!

WE'LL THUMB-TACK THOSE THUGS TO THAT FENCE!

TWANGGG!

4

FIVE SECONDS LATER... GET USED TO CAPTIVITY, BOYS! YOU'LL BE BEHIND BARS A LONG TIME!

GASP! I-IT'S *GREEN ARROW!* WE'RE SUNK!

BUT AFTER THE THIEVES ARE TURNED OVER TO THE POLICE...

I'LL COLLECT THESE ARROWS SO WE CAN USE THEM AGAIN AND... HOLY SMOKES!

WHOOSH!

WHOOSHH!

AGAIN THE ARROWS ARE TAKING OFF BY THEM-SELVES! THIS TIME THEY'RE HEADING STRAIGHT UP, AT TERRIFIC SPEED! G.A. THIS IS MORE MYSTERIOUS THAN *FLYING SAUCERS!*

ZINGG!

ZING!

ZING!

SOON, AS THE FRICTION GENERATED BY THE TERRIFIC UPWARD SPEED MELTS THE ARROWS...

PFFT!

PWISTT!

WELL, THEY'RE GONE FOREVER LIKE OUR OTHER ARROWS! IF THIS KEEPS UP, G.A., WE'LL RUN OUT OF SHAFTS BEFORE THE DAY'S OVER!

I-IT DEFIES IMAGINATION, *SPEEDY!* GULP! NOTHING LIKE THIS EVER HAPPENED TO US BEFORE!

5

LATER, AS THE DAY WEARS ON AND MORE ARROWS VANISH AFTER BEING FIRED...

I'M DOWN TO MY LAST FEW ARROWS, G.A.!

ME, TOO!... WAIT! A POLICE RADIO SIGNAL IS COMING IN!

ATTENTION ALL CARS! LOOK FOR MASKED BANDIT SPEEDING AWAY IN STRIPPED-DOWN HOT ROD! WANTED FOR ROBBERY!

THERE HE GOES! THE FOOL! HE'S RACING AT OVER 100 MILES AN HOUR!

WE MUST STOP HIM BEFORE HE CRASHES INTO A CAR! WE'VE STILL GOT OUR *SAW ARROWS!* LET'S SAW HIS MOTOR OFF!

WHOOOOSHH!!

MOMENTS LATER, AS THE TWO JAGGED-TOOTHED SHAFTS OVERTAKE THE CAREENING HEAP...

WATCH! IN A SECOND, HIS MOTOR WILL ROLL OFF INTO THE ROAD!

H-HEY!

ZZZZ!

AS THE MOTORLESS ROD GRINDS TO A DEAD STOP...

I PUT CUFFS ON THE BUZZARD, G.A., WITH A HANDCUFF ARROW! WATCH HIM WHILE I RECOVER OUR SAW ARROWS!

WHUMPPP!

BUT AS **SPEEDY** *REACHES FOR THE SHAFTS...*

NOW THE ARROWS ARE BORING HOLES INTO THE GROUND! THEY'RE DISAPPEARING FROM SIGHT!

WE'RE JINXED, **SPEEDY!** THOSE ARROWS MUST HAVE A CURSE ON THEM! I-I DON'T KNOW WHAT ELSE TO THINK!

HOURS LATER, AS THE DAY ENDS...

SOME HEROES WE ARE, **SPEEDY!** WE LOST EACH OF OUR ARROWS! WE'RE LEFT WITH EMPTY QUIVERS AND NO EXPLANATION!

G-GREEN ARROW! LOOK! IT'S SUPERMAN!

HI, *GREEN ARROW!* HI, *SPEEDY!* SORRY ABOUT YOUR ARROWS... BUT I *HAD* TO GET RID OF THEM! I DIDN'T DESTROY THEM ALL AT ONCE BECAUSE I KNOW YOU NEEDED THE ARROWS TO ACCOMPLISH YOUR PATROL JOB!

BUT WHAT WAS WRONG WITH OUR ARROWS?

AS FATE HAD IT, IN MAKING YOUR ARROW HEADS, YOU MUST'VE PICKED UP FRAGMENTS OF A *GREEN KRYPTONITE METEOR* THAT LANDED ON EARTH YEARS AGO! AS YOU KNOW, KRYPTONITE IS A DEADLY MENACE TO ME...

"IF ANY OF YOUR KRYPTONITE ARROWS FELL INTO CRIMINAL HANDS, THERE WAS A CHANCE IT COULD BE USED AGAINST ME! SO I SIMPLY FOLLOWED YOU AROUND TOWN WITH MY TELESCOPIC VISION AND DESTROYED THE ARROWS WITH MY *X-RAY VISION* AND *SUPER-BREATH*..."

PRESENTLY, AS *SUPERMAN* FLIES OFF...

SO *THAT'S* WHY OUR ARROWS VANISHED! BUT *WHERE* WAS *SUPERMAN* WHILE HE MADE THE ARROWS DISAPPEAR?

TOO BAD I COULDN'T TELL *GREEN ARROW* AND *SPEEDY* THE *WHOLE* TRUTH! BUT IT WOULD MEAN REVEALING MY SECRET IDENTITY OF *CLARK KENT*, THE METROPOLIS *DAILY PLANET* REPORTER...

"...I SPENT THE DAY IN THE TOWER OF THE STAR BUILDING! THE ONE ELEVATOR TO THE TOWER HAD BROKEN DOWN! IF I FLEW OFF, I'D REVEAL THAT CLARK KENT WAS *SUPERMAN!* SO I HAD TO DESTROY THE KRYPTONITE ARROWS BY REMOTE CONTROL FROM MY BIRD'S-EYE VIEW OF THE CITY!"

THAT NIGHT, OUTSIDE THE ARROW CAVE...

WE'LL GET RID OF ALL THE REMAINING KRYPTONITE FRAGMENTS SO NOBODY WILL EVER USE THEM AGAINST *SUPERMAN!*

RIGHT, G.A.! AND THIS TIME THE ONLY THING THAT'LL VANISH WHEN WE FIRE OUR *GREEN ARROWS* IS ANY CRIMINAL'S CHANCE OF ESCAPING US!

-THE END

THE GREEN ARROW

He was a little man -- with strange, fascinating little weapons -- but his crimes were big! Indeed, what started as mere child's play, for GREEN ARROW and SPEEDY, turned into a game of high stakes, when the emerald archers pitted their shafts against...

THE MIGHTY MR. MINIATURE

HIGH ABOVE THE STREET, A TINY PLANE BUZZES NEAR BULLETPROOF WINDOWS...

NOW, BY RELEASING SOME OF THE SPECIAL GAS FROM MY LITTLE BALLOONS, I CAN DESCEND TO THE WINDOW-- AND ENTER!

INSIDE, MINUTES LATER...

NOT A MOMENT TOO SOON GETTING THE SAFE OPENED! I HEAR RUNNING FOOTSTEPS...NO DOUBT THE BUILDING GUARDS!

ACME SAFE

NO MATTER, THOUGH...I'LL MEET THEM WITH A *TANK ATTACK!*

NEXT INSTANT...

HA, HA! NOW TO MAKE MY ESCAPE!

EEYOW! WHERE'D THOSE THINGS COME FROM?

POP!

POW!

POW!

POW!

AN ALARM GOES OUT, BRINGING TWO COLORFUL FIGURES TO THE RESCUE, JUST AS...

G.A.!... LOOK!

THAT'S OUR CROOK, *SPEEDY*-- AND THAT'S HOW HE GOT IN AND OUT OF THE OFFICE BELOW!

LET'S DEFLATE THOSE BALLOONS...HE'LL DROP RIGHT ON THE ROOF!

SOUNDS EASY ENOUGH...

2

BUT JUST AS THEY TAUTEN THEIR BOW-STRINGS...

GREEN ARROW AND SPEEDY, EH? I'M READY FOR THEIR ARROWS, TOO...

...WITH A GUIDED MISSILE ANTI-ARROW ATTACK! ⸘HA, HA, HA⸘

BAM!

VROOM!

KWHOOM!

NOW TO SEND MY CAMOUFLAGE PLANE INTO ACTION!

GREAT GUNS! HE'S LAYING DOWN A SMOKESCREEN!

WE CAN'T SEE TO FIRE ANOTHER ARROW!

⸘HA, HA⸘ YOU'LL SOON LEARN YOU CAN'T TANGLE WITH--MR. MINIATURE!

BACK HOME, AS THE ARCHERS RESUME THEIR EVERYDAY IDENTITIES OF WEALTHY OLIVER QUEEN AND HIS YOUNG WARD, ROY HARPER...

MR. MINIATURE HE CALLS HIMSELF, EH?

YES, ROY...AND I'M SURE WE HAVEN'T SEEN THE LAST OF HIM!

INDEED, THE VERY NEXT MORNING...

OLIVER!...THE ARROW-SIGNAL! THE POLICE NEED AN ASSIST FROM US!

GET OUT THE COSTUMES, ROY, WHILE I CHECK THE CALLS ON OUR POLICE RADIO!

MINUTES LATER, AS THE *ARROWCAR* GOES STREAKING FROM THE CITY...

JUST AS WE FIGURED-- *MR. MINIATURE* HAS STRUCK AGAIN!

THIS TIME AT THE AIRPORT, EH? WONDER WHAT HE'S PLANNING?

THE ANSWER LIES WITH A CROUCHING FIGURE, SOME DISTANCE AHEAD...

HA, HA! I KNEW THOSE BUMBLING BOWMEN WOULD SOON SHOW UP! SO WHILE MY TINY PLANES KEEP THE AIRPORT IN A TIZZY...

...I'LL MEET *THIS* EMERGENCY WITH THE CARS I DESIGNED ESPECIALLY FOR RUBBER TIRES!

AND AS THE *ARROWCAR'S* WHEELS HIT THE LITTLE SPIKED VEHICLES...

TWO BLOWOUTS! CAR'S OUT OF CONTROL!

BAM! BAM! BAM!

ONLY ONE THING CAN SAVE US... THE *CATAPULT!*

YES... WE CAN CATCH ONTO THOSE TREE LIMBS!

CONFOUND IT! I FORGOT ABOUT THEIR BLASTED *CATAPULT SEATS!* I'LL HAVE TO CANCEL MY OTHER PLANS-- GET AWAY FROM HERE... FAST!

4

AS THE MIDGET AUTO SPEEDS TOWARD A NEARBY HILLSIDE, CAMOUFLAGED DOORS OPEN, AND...

THERE MUST BE OTHER WAYS TO GET RID OF THOSE PESKY ARCHERS! I'LL STAY IN HIDING FOR AWHILE-- THEN TRY SOMETHING ELSE!

BUT AS THE DOORS START TO CLOSE, TWO BOWS TWANG, AND...

WE SPOTTED THAT ENTRANCE IN THE NICK OF TIME, SPEEDY! OUR ARROWS WILL PREVENT THE DOORS FROM CLOSING!

ZIP!

ZIP!

SO THIS IS MR. MINIATURE'S STRONGHOLD!

AN ACTUAL CITY -- IN MINIATURE! HE MUST DESIGN AND STORE ALL HIS WEAPONS HERE!

SUDDENLY...

HA, HA! YOU TRACKED ME DOWN, GREEN ARROW, BUT I'M READY FOR YOU!

TINY PARATROOPS-- FIRING LITTLE GUNS AT US!

BANG!

BANG!

BANG!

INSTANTLY, THE ACE ARCHERS MOVE INTO ACTION...

WE'LL JUST TACK THOSE PARATROOPS AGAINST THE FAR WALL-- OUT OF THE WAY!

ZIP!

TWANG!

TWANG!

WHILE UP ABOVE...

HA, HA! THEY FORGOT I CAN EXPLODE THEIR EVERY ARROW WITH MY GUIDED MISSILES!

5

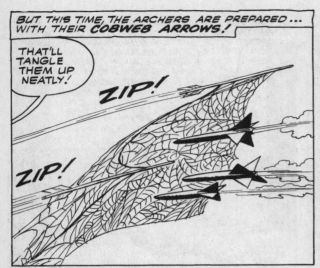

THE GREEN ARROW

ON THEIR HOME GROUNDS IN STAR CITY, THE GREEN ARROW AND SPEEDY CAN BATTLE A HOST OF ENEMIES AND EMERGE TRIUMPHANT! BUT AWAY FROM DRY LAND, TRANSPLANTED TO A WATERY WORLD FULL OF UNFAMILIAR PERILS, THE FAMED TWINS OF THE TWANGING BOW ARE AT A TOTAL DISADVANTAGE! THEY MUST FACE THE PERILS OF FIERCE SEA CREATURES AND A CUNNING MASTER CRIMINAL WHEN THEY BECOME...

The UNDERWATER ARCHERS!

THAT WHALE SEES US! IT'S COMING FOR US! WHY DOESN'T GREEN ARROW DO SOMETHING?

GASP!... T-THERE ISN'T AN ARROW IN MY QUIVER THAT CAN STOP THAT MONSTER! WE'RE GONERS UNLESS I CAN IMPROVISE SOME DEFENSE IN THE NEXT FEW SECONDS!

AT THE SECRET HIDEOUT OF HORACE KATES, ALIAS THE WIZARD, SHORTLY AFTER HIS ESCAPE FROM PRISON...

B-BOSS!... GASP! YOU ESCAPED FROM PRISON LIKE YOU PROMISED!

NATURALLY! MY PLANS ALWAYS WORK... EXCEPT WHEN THAT BLASTED GREEN ARROW INTERFERES!

BOYS, I'VE BEEN DOING MUCH THINKING IN PRISON! I'LL INVENT NO MORE "FLYING BUZZ-SAWS" OR "DRILL-MOBILES" THAT BREAK THROUGH BANK WALLS! FROM NOW ON I'LL CONFINE MY CRIMES TO THE SEA, WHERE I'LL BE SAFE FROM GREEN ARROW!

A FEW DAYS LATER, AS OLIVER QUEEN AND HIS YOUNG WARD, ROY HARPER, BUSY THEMSELVES IN THEIR SECRET *ARROW CAVE*...

THE PRISON GRAPEVINE SAYS THAT, TO AVOID *GREEN ARROW,* THE WIZARD HAS MADE THE *SEA* HIS NEW HUNTING GROUNDS!

IF WE'RE GOING TO RECAPTURE HIM, WE MUST CONSTRUCT EQUIPMENT SO THAT WE CAN BREATHE AND OPERATE UNDER-WATER! HELMETS, OXYGEN TANKS, FINS...

...AND UNDER-WATER BOWS AND ARROWS! OUR PRESENT WEAPONS ARE USELESS, OLIVER!

DAYS LATER, AS THE NEW MATERIAL IS TESTED...

ANOTHER BULL'S-EYE! THESE SPECIAL BOWS, EQUIPPED WITH A PNEUMATIC AIR-PRESSURE DEVICE, FIRE AN ARROW AS ACCURATELY *UNDER WATER* AS ABOVE IT!

WHOOOSH!

THAT NIGHT, THE BIGGEST PIECE OF NEW EQUIP-MENT ENJOYS ITS MAIDEN VOYAGE!

OUR NEW *ARROW-BOAT* HAS PASSED EVERY TEST FOR SPEED AND SEA-WORTHINESS, ROY! WE'RE NOW READY TO BEGIN OUR SEARCH FOR THE WIZARD!

FINE! I'M RARING TO GO!

THE NEXT MORNING, AS OLIVER AND ROY SWITCH TO THEIR FAMED IDENTITIES AS *GREEN ARROW* AND *SPEEDY*...

I RECEIVED AN UNDERWORLD TIP THAT *THE WIZARD* VOWED TO STEAL A SHIPMENT OF GOLD BULLION FROM THAT FREIGHTER NOW LEAVING STAR CITY HARBOR! WE'LL TRAIL IT AND SEE WHAT HAPPENS!

BUT AS DAYS PASS UNEVENTFULLY..

≈YAWN!≈ ...THAT TIP MUST'VE BEEN A PHONY, *G.A.!* HERE WE ARE, IN WHALING WATERS, AND NO SIGN OF *THE WIZARD* YET!

WAIT, *SPEEDY!* SOMETHING FUNNY IS HAPPENING! A WHALE IS SWIM-MING RIGHT UP TO THE FREIGHTER!

SHORTLY...

WHIZZ! WHOOOZZZ!

I GET IT, *G.A.!* LIKE A DONKEY PURSUES A CARROT TIED IN FRONT OF ITS NOSE, THESE WHALES WILL RACE AFTER OUR BAIT ARROWS AND DRAG US TO SHORE!

NOT *US* ALONE, *SPEEDY!* THEY'RE ALSO DRAGGING THE MECHANICAL WHALE BACKWARDS! IN A FEW MINUTES, *THE WIZARD'S* "SEA CREATURE" WILL BE BEACHED!

AT THE SAME TIME, INSIDE THE "WHALE"...

W-WE'RE HELPLESS, WIZARD! UNLUCKILY, THE HARPOON KAYOED OUR ELECTRONIC CONTROL SYSTEM! WE CAN'T OPEN THE HATCHES OR FIRE OUR GUNS!

I'M PREPARED FOR SUCH A CRISIS! COME THIS WAY!

THUS, BEFORE LAND IS REACHED...

TOO BAD ONLY THREE OF US CAN ESCAPE IN MY EMERGENCY UNDERWATER ROCKET! BUT I PROMISE TO AVENGE THIS DEFEAT!

YOU'D BETTER, *WIZARD!* WE MISSED GETTIN' CAPTURED BY INCHES!

LATER, ON THE ISLAND...

THE COAST GUARD'S IN CONTROL NOW, *G.A.!* BUT *THE WIZARD* ESCAPED!

HE'LL TURN UP AGAIN, *SPEEDY!* WHEN HE DOES, *WE'LL* BE THERE TO KEEP HIM COMPANY!

A WEEK LATER, AS A PRISON SHIP ACCIDENTALLY STRIKES AN ICEBERG IN THE FOG...

W- WE'RE SINKING FAST! SEND OUT AN S.O.S.! FREE ALL THE PRISONERS AND GIVE THEM LIFE-JACKETS! IT'S EVERY MAN FOR HIMSELF!

YES, CAPTAIN! I'LL GIVE THE ORDER TO ABANDON SHIP!

4

AS THE ARCHERS REACH THEIR DESTINATION, THE CAR'S *CATAPULT* SENDS THEM STREAKING UPWARD...

WE'LL STATION OURSELVES UP HERE, TO PREVENT A ROOFTOP ESCAPE-- WHILE THE POLICE CLOSE IN BELOW!

BUT INSIDE THE STORE...

AH...*GREEN ARROW* JUST ARRIVED, AND IS AWAITING US ON THE ROOF! PRESS THE *INFLATION BUTTONS* ON YOUR BELTS, AND WE'LL BE READY FOR THE BIG SURPRISE!

FINE WATCH

NEXT INSTANT...

GREAT GUNS, *G.A.!* LOOK AT THAT!.... *HUMAN BALLOONS!*

THAT'S A NEW ONE, *SPEEDY!* GET OUT YOUR *ROPE ARROW*...WE'LL HAUL 'EM IN!

BUT JUST AS THE BATTLING BOWMEN RELEASE THE TRICKY MISSILES...

HA, HA! LET ME INTRODUCE MY *PNEUMATIC GUN!* YES-- JUST AN *AIR* GUN...BUT WITH A BLAST THAT CAN TURN BACK YOUR ARROWS!

WRRRR

AND WHEN HE TURNS THE MACHINE TO AN EVEN HIGHER SPEED...

HANG ON, *SPEEDY!* THAT WIND IS AS STRONG AS A TYPHOON!

WHRRRR

SHORTLY...

THE BLAST OF AIR HAS STOPPED... BUT WHERE ARE THE FLOATING CROOKS?

NOT A SIGN OF THEM ANYWHERE!

INDEED, NOT A SIGN OF THE UNIQUE CROOKS -- BECAUSE THEY'VE ASCENDED ABOVE THE CLOUDS, TO BE TAKEN ABOARD A HOVERING BLIMP...

AND AS THEY DEFLATE THEIR UNIQUE SUITS...

AND SO WE'VE SEEN, GENTLEMEN, HOW THE SCIENCE OF *PNEUMATICS* -- THE MECHANISM OF PLAIN AIR -- CAN BAFFLE CRIMELAND'S GREATEST NEMESIS! AND THIS IS ONLY THE BEGINNING, AS THOSE ARCHERS WILL SOON DISCOVER!

NEXT DAY, AS HUGE SUMS OF CASH ARE TRANSFERRED FROM A BANK TO AN ARMORED CAR --

--THREE CABLES SUDDENLY DESCEND, AND...

HEY! WHERE'D THOSE THINGS COME FROM?

BY THE TIME *GREEN ARROW* AND *SPEEDY* ARRIVE...

THERE GOES THE MONEY! IT'S BEING HAULED AWAY BY *BLIMPS!*

TOO LATE TO STOP THEM -- SO LET'S *JOIN* THEM, SPEEDY!

3

AND AS THEY MAKE THEIR WAY INTO THE SUSPENDED VEHICLE...

SNAP!

WE'RE *REALLY* INSIDE NOW-- *LOCKED* IN!

FAR OUT TO THE COUNTRY SPEED THE BLIMPS, UNTIL...

G.A.!... LOOK BELOW! A HIDEOUT IN THE HILLS!

THEY'VE THROWN OFF BALLAST... WE'RE DESCENDING!

DOWN WE GO-- WITH A BUMP...

ARMORED CAR SERVICE

CLOMP

SHORTLY AFTERWARDS, THE DOOR IS FORCED OPEN-- BUT BEFORE THE ARCHERS CAN MOVE INTO ACTION...

WHROOSH

G.A.! WHAT'S HAPPENING?

"HA, HA," THEY DIDN'T THINK WE SAW THEM "HITCH" A RIDE ON THE ARMORED CAR! BUT NOW THEY'RE TRAPPED IN THE *VACUUM MACHINE,* AND WE CAN HAUL THEM UP TO THE *PNEUMATIC PRISON!*

AND AS THE ARCHERS ARE EJECTED INTO A ROOM OF THE CASTLE-HIDEOUT...

WHY... WE'RE HOVERING IN *MID-AIR!*

EXACTLY... IN MY *PNEUMATIC PRISON,* WHICH NEEDS NO BARS! FORCES OF AIR, FROM THE TOP, BOTTOM AND SIDES, KEEP YOU IN PERFECT BALANCE!

4

YOU'LL FIND YOU'RE POWERLESS TO ESCAPE-- AS IF HELD BY THE STRONGEST BONDS! JUST THINK, *GREEN ARROW,* YOU WERE BEATEN-- AND NOW YOU ARE HELD PRISONER BY PLAIN *AIR!* PONDER THAT IN MY ABSENCE!

SOON...

THEY'LL KEEP UNTIL OUR RETURN! LET'S GET GOING... WE HAVE OTHER JOBS TO PERFORM IN TOWN!

WHILE IN THE STRANGE PRISON CELL...

AN ARROW! HOW DID YOU HIDE IT IN YOUR SHIRT?

WHEN THE VACUUM CONTRAPTION WAS DRAWING US FROM THE ARMORED CAR, I GRABBED IT AUTOMATICALLY--LIKE A MAN CLUTCHING FOR A STRAW! WE MIGHT BE IN LUCK ... IT'S THE *JET ARROW!*

ITS JET IS USUALLY TRIGGERED BY THE BOWSTRING, BUT I CAN SET IT OFF BY HAND! GRAB HOLD OF ME -- AND HANG ON! HERE GOES...

THE JET POWER OFFSETS THE BALANCING AIR FORCES-- *AND UP WE GO!*

YES -- UP AND *OUT!*

5

THE GREEN ARROW

ONE AMAZING NIGHT, THE *GREEN ARROW* FINDS HIMSELF IN AN AMAZING WORLD...NONE OTHER THAN ENGLAND AT THE TIME OF *KING ARTHUR!* HOW THE FAMED ARCHER AND HIS YOUNG ASSISTANT, *SPEEDY*, ARE WELCOMED AT CAMELOT CASTLE, AND THEIR ASTONISHING ADVENTURES AS TWENTIETH CENTURY KNIGHTS OF KING ARTHUR'S SIXTH CENTURY ROUND TABLE, IS THE INCREDIBLE TALE OF...

The GREEN ARROW in KING ARTHUR'S COURT!

THE STRANGERS ARE ENEMY SPIES! DESTROY THEIR WAR CHARIOT WITH OUR CATAPULTS!

WE'VE GOT CATAPULTS OF OUR OWN, *SPEEDY*... OUR *CATAPULT SEATS*...THAT CAN SAVE US *TEMPORARILY!* BUT DON'T ASK ME HOW WE'LL ESCAPE KING ARTHUR'S KNIGHTS *LATER!*

ONE NIGHT, AT OLIVER QUEEN'S HOUSE, AS HIS YOUNG WARD, ROY HARPER, PREPARES A BOOK REPORT FOR SCHOOL...

MANY INTERESTING FACTS ARE IN THIS OLD ALMANAC, ROY! LIKE THE DATES OF ANCIENT EARTHQUAKES, ECLIPSES, AND...

NEVER MIND THAT, OLIVER! PLEASE HELP ME ORGANIZE MY NOTES ON MARK TWAIN'S NOVELS!

SO FAR I'VE COMPLETED THE BIOGRAPHY OF MARK TWAIN AND WRITTEN A SYNOPSIS OF THREE OF HIS BOOKS...*TOM SAWYER, HUCKLEBERRY FINN* AND *THE PRINCE AND THE PAUPER!*

NOW YOU WANT TO KNOW WHAT TO HIGHLIGHT FROM HIS OTHER WORKS FOR YOUR MARK TWAIN BOOK REPORT! OKAY! LET ME REFRESH MY MEMORY!

PRESENTLY, HOWEVER, AS THE BRAVE BOWMEN AWAKE...

ROUSE THYSELVES, KNAVES! WHAT DOST THOU NEAR CAMELOT IN THIS STRANGE CHARIOT?

ONE SIDE, GUARD! THEY ARE ENEMIES WHO CAME TO SPY ON KING ARTHUR'S DOMAIN!

GASP! *GREEN ARROW!* WHERE... WHAT...

STAND BACK, MAN! SIR SAGRAMOR CHARGETH AT THE STRANGERS!

G.A.! LOOK!

I'M AWAKE NOW, *SPEEDY!* I-I SEE HIM! I DON'T KNOW *WHO* HE IS OR *WHERE* WE ARE BUT I CAN'T ASK QUESTIONS NOW!

TWANG!

THIS *TELESCOPED* ARROW KEEPS LENGTHENING AS IT FLIES! IT'S THE NEAREST TO A *LANCE ARROW* I'VE GOT!

OOF!

ZOUNDS! *GASP!* ... THE STRANGER UN-HORSED SIR SAGRAMOR! ATTACK HIM!

PLUNK!

WE'RE TAKING OFF, *SPEEDY!* THOSE CHARACTERS ARE CARRYING THIS JOKE, IF IT *IS* A JOKE, TOO FAR!

I DON'T UNDERSTAND WHAT'S HAPPENING, *G.A.!* THE LAST THING I REMEMBER WAS BEING GASSED BY THOSE CROOKS! THE FULLER CASTLE WAS A RUIN...

RRRRR

THIS CASTLE IS JUST LIKE IT, BUT IT'S *PERFECT!* AND LOOK AT ALL THESE PEOPLE, DRESSED IN ARMOR! WHERE'D *THEY* COME FROM?

HALT, KNAVES! WHAT WITCHCRAFT IS THIS? WHAT ENGINE OF DESTRUCTION DOST THOU DRIVE THAT ROARETH LIKE THUNDER?

ART THOU FRIEND OR FOE? SPEAK, MAGICIANS! I, KING ARTHUR OF CAMELOT, COMMAND THEE!

K-KING ARTHUR? CAMELOT? *GASP!* LOOK, MISTER... WE'RE FROM STAR CITY, U.S.A.! THIS IS THE YEAR *1960*...ER... *ISN'T* IT?

WAIT! I HEARD THE STRANGERS SPEAK! THERE IS NO U.S.A. AND THIS IS *JUNE 21, 528!* O'KING, THESE ARE WICKED SORCERERS! *PUT THEM TO DEATH!*

HMMM... MERLIN, MY COURT MAGICIAN, IS THE MOST POWERFUL WIZARD IN ENGLAND! WE MUST HEED HIS ADVICE!

APPARENTLY THESE FOOLS ARE IN DEADLY FEAR OF MAGIC!

HOLD! *MY MAGIC IS MORE POWERFUL EVEN THAN MERLIN'S.'* I WILL BLOT OUT THE NOONDAY SUN WITH AN ARROW.' WATCH!

ARE YOU CRAZY, *G.A.?* NO TRICK ARROW CAN CAUSE AN ECLIPSE!

TUSH, *SPEEDY!* I KNOW NO ARROW CAN MOVE THE MOON IN FRONT OF THE SUN! BUT REMEMBER THAT *ALMANAC?* I RECALL FROM READING IT THAT AN ECLIPSE OCCURRED *AT NOON ON JUNE 21, 528!* LOOK! IT'S GROWING DARK ALREADY!

PRESENTLY...

IT'S A TOTAL ECLIPSE!‡GASP!‡ *SPEEDY...* THIS PROVES WE'RE *ACTUALLY LIVING IN 528 A.D.!* WE'VE BEEN SENT BACK IN TIME!

‡GASP!‡ 'TIS FANTASTIC MAGIC, STRANGER! THOU HAST REPLACED MERLIN AS THE FIRST MAGICIAN IN ENGLAND!

BUT AS THE SUN REAPPEARS...

SHOW US MORE MARVELS, STRANGER!

FOOLS! HE CAN'T PRODUCE MARVELS! HE'S A FRAUD!

FRAUD, AM I? VERY WELL, MERLIN! I'LL BLAST YOUR TOWER INTO RUBBLE WITH A SINGLE ARROW!

‡GASP!‡ 'TIS BLACK MAGIC! M-MY TOWER IS DESTROYED!

THANKS TO MY *DYNAMITE ARROW!* SPEEDY AND I WILL BE SAFE HERE AS LONG AS THESE IGNORANT BARBARIANS REGARD TWENTIETH CENTURY INVENTIONS AS FANTASTIC MAGIC!

BARRROOM!

4

LATER THAT DAY...

I HEREBY APPOINT THE STRANGERS TO MY *ROUND TABLE* SO THEY MAY DISTINGUISH THEMSELVES IN *KNIGHT ERRANTRY* AS WELL AS MAGIC!

PSSST... *G.A.!* LOOK AT SIR SAGRAMOR AND MERLIN! THEY'RE GLARING AT US!

THEY HATE US! THEY'LL DO ANYTHING TO GET RID OF US!

FORGET 'EM, *SPEEDY!* THE KING GAVE US AN ASSIGNMENT! TO PROTECT A VILLAGE MENACED BY A PACK OF WILD BOARS!

AS THE NEW KNIGHTS DRIVE INTO THE COUNTRYSIDE...

THERE THEY ARE! WHAT UGLY CRITTERS! IF THEY TOUCH THOSE VILLAGERS WITH THEIR RAZOR-SHARP TUSKS...

THEY WON'T, *SPEEDY!* FIRE YOUR *FIRE-CRACKER ARROWS!*

A MINUTE LATER...

HMMPPHH! STAMPEDING WILD BOARS BACK TO THE FOREST IS HARDLY MY IDEA OF CHIVALROUS ACTION!

ALL IN THE DAY'S WORK, "SIR *SPEEDY*"! KNIGHTS CAN'T RESCUE DAMSELS IN DISTRESS *EVERY* TIME!

POP!

CRACK!

POP!

BUT ON THEIR NEXT MISSION...

THIS IS THE CASTLE OF THE WICKED AND BEAUTIFUL QUEEN MORGAN LE FAY! HER DUNGEON IS FULL OF ENSLAVED PRISONERS!

OKAY! LET'S BLAST AN EXIT IN THE DUNGEON WALL WITH OUR ROCKET ARROWS!

TWANGG!

FORWARD, MEN! DESTROY THE INVADERS OF MY CASTLE!

HERE COME THE QUEEN'S KNIGHTS! I'LL TAKE CARE OF 'EM, *SPEEDY,* WHILE YOU FIRE A *PULLEY ARROW* TO THE PRISONERS WHO STREAM OUT OF THE DUNGEON!

BWAMM!

5

MOMENTS AFTER... THAT BUZZ-SAW ARROW I FIRED SHOULD DO ITS JOB BEFORE QUEEN MORGAN LE FAY'S KNIGHTS CROSS THE DRAWBRIDGE!

BZZZZZ!

SHORTLY... ≥GASP!≥ THIS IS BLACK MAGIC! THE B-BRIDGE IS CUT IN TWO!

YIIIIIII!

NOW SPEEDY CAN FIRE A LINE TO THE CAPTIVES IN THE DUNGEON!

CRACKKK!

PLOP!

A HALF HOUR LATER...

WE DID IT, G.A.! WE'VE EMPTIED THE DUNGEON OF THE QUEEN'S SLAVES!

YES, SPEEDY! IF WE MUST LIVE IN THE PAST, AT LEAST WE CAN FIGHT INJUSTICE AND TYRANNY, JUST AS WE DID IN STAR CITY!

BUT THAT EVENING, AT KING ARTHUR'S COURT...

HAIL OUR NEW KNIGHTS! THEY'VE BEGUN TO OUT-SHINE THE REST OF US IN GOOD DEEDS!

≥GASP!≥ G.A.! SIR SAGRAMOR...

NOT FOR LONG! I'LL SETTLE MY SCORE WITH GREEN ARROW NOW!

AS THE EMERALD ARCHER QUICKLY NOTCHES AN ARROW...

MY BACK MAY BE TURNED TO SIR SAGRAMOR...BUT MY EYES CAN SEE HIM IN THESE POLISHED SHIELDS! HENCE... AN OVER-THE-SHOULDER SHOT WITH A LASSO ARROW!

TWANNGGG!

≥GASP!≥... SPARE MY LIFE, GREEN ARROW!

I'LL PASS SENTENCE UPON THEE, ROGUE! THOU ART BANISHED FOREVER FROM CAMELOT FOR ALLOWING THY ENVY OF GREEN ARROW TO TURN THEE INTO AN UNCHIVALROUS KNAVE!

6

THAT NIGHT, AS THE TWINS OF THE TWANGING BOW RETIRE...

IT WAS THOUGHTFUL OF THIS OLD WOMAN TO BRING US MILK BEFORE WE GO TO BED! I... ≶GASP!≷... G.A.! T-THE ROOM IS TURNING AROUND!

≶GASP!≷... I-I'M GETTING DIZZY TOO! T-THERE'S SOMETHING IN THIS MILK!

AYE! A MAGICAL POTION THAT I, MERLIN, IN THE DISGUISE OF AN OLD WOMAN, PREPARED TO MAKE THEE FALL INTO AN ENCHANTED SLEEP! HA! HA! NOW I AM RID OF THY RIVALRY FOREVER!

≶GASP!≷ MERLIN! Y-YOU REVENGED YOURSELF... OHHHHHH!

SOON AFTER, HOWEVER, AS THE SLEEPERS AWAKE...

≶GASP!≷ WE'RE BACK IN FULLER'S CASTLE! G.A.! THOSE CROOKS... THEY'RE STILL HERE! THEY FELL ASLEEP, TOO!

THANKS TO THE FAN ARROW I FIRED BEFORE I PASSED OUT, SPEEDY! THE FAN BLEW THE GAS THEIR WAY, TOO! NOW WE CAN JAIL 'EM!

THE NEXT DAY, AT OLIVER QUEEN'S HOUSE...

SPEEDY, I'M RECORDING OUR EXPERIENCE IN THE GREEN ARROW'S CASE FILE! BUT I-I CAN'T DECIDE! DID WE DREAM EVERYTHING? LIKE THAT ECLIPSE...?

≶GASP!≷ ECLIPSE? THAT'S FUNNY! THE HERO IN "A CONNECTICUT YANKEE IN KING ARTHUR'S COURT" ALSO SAVED HIS LIFE BY PRETENDING TO CAUSE AN ECLIPSE!

HE ALSO RECALLED AN ALMANAC DATE OF AN ECLIPSE ON JUNE 21, 528! THEN, HE, TOO, "MAGICALLY" DYNAMITED MERLIN'S TOWER... AND DEFEATED SIR SAGRAMOR WITH A LASSO!

≶GASP!≷ THEN WE ACTUALLY RELIVED MARK TWAIN'S STORY!

HMM.... WE WERE READING "CONNECTICUT YANKEE" BEFORE WE PURSUED THOSE THIEVES! AFTER BEING GASSED, MAYBE WE RELIVED THE INCIDENTS OF THE BOOK BY DREAMING THE SAME DREAM!

...OF GREEN ARROW IN KING ARTHUR'S COURT, EH? ≶SIGH!≷... ALAS, SPEEDY, WE'LL NEVER REALLY KNOW!

A CONNECTICUT YANKEE IN KING ARTHUR'S COURT

The End.

A QUIVER-FULL OF TRICK ARROWS, A BRAIN-FULL OF CLEVER IDEAS, A BODY FULL OF STEEL-SPRING STRENGTH, AND A HEART FULL OF LOVE FOR FREEDOM AND JUSTICE... THIS IS THE AMAZING COMIC BOOK HERO *THE WIZARD ARCHER!* BUT ARE THE FEATS HE PERFORMS WITH HIS BOW AND ARROW TOO INCREDIBLE? FOR THE ANSWER, SEE WHAT HAPPENS WHEN THE *GREEN ARROW* AND *SPEEDY* HAVE TO DECIDE ON THE SKILL OF...

The COMIC BOOK ARCHER!

GREEN ARROW, HERE'S A BLOW-UP OF ONE OF THIS CRAZY ARTIST'S COMIC PAGES! I'LL STAKE MY REPUTATION AS AN EDITOR THAT NO LIVING ARCHER CAN PERFORM SUCH INCREDIBLE FEATS!

I'M NOT SO SURE! SPEEDY AND I WILL TRY OUT THE WIZARD ARCHER'S STUNTS IN *REAL LIFE!* THEN WE'LL SEE WHETHER THESE COMIC BOOK TRICKS ARE REALLY SO FANTASTIC!

THE WIZARD ARCHER

I'LL FIRE THIS TRICK ARROW THAT'LL INFLATE INTO THE SHAPE OF A FLYING HORSE!

NOW TO LASSO THE FLYING HORSE WITH A LARIAT ARROW!

AS THE WIZARD ARCHER MOUNTS HIS AERIAL STEED... NOW I CAN TAKE A SHORTCUT TO PURSUE THE PAY-ROLL BANDITS!

AH! THE WIND HAS DRIFTED MY FLYING HORSE INTO A PERFECT SPOT! HERE COMES THE GETAWAY CAR!

ONE DAY, IN STAR CITY, AS OLIVER QUEEN AND HIS YOUNG WARD, ROY HARPER, CHANGE INTO THEIR FAMED COSTUMES AS *GREEN ARROW* AND *SPEEDY...*

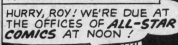

HURRY, ROY! WE'RE DUE AT THE OFFICES OF *ALL-STAR COMICS* AT NOON!

BOY! I'LL SURE ENJOY THIS VISIT!

IMAGINE MEETING THE WRITERS AND ARTISTS WHO WRITE AND DRAW MY FAVORITE COMIC HEROES!

I KNOW YOU'RE EXCITED, SPEEDY! BUT DON'T FOR-GET WE'VE BEEN INVITED TO *ALL-STAR COMICS* TO POSE FOR A POSTER TO COMBAT JUVENILE DELINQUENCY! LET'S GO!

MEANWHILE, ELSEWHERE IN THE CITY...

WISH ME LUCK, MOM! I'M SHOWING MY *WIZARD ARCHER* DRAWINGS TODAY TO THE EDITOR OF *ALL-STAR COMICS!* HE NEEDS A NEW FEATURE FOR ONE OF HIS MAGAZINES!

I'M SURE HE'LL LIKE *THE WIZARD ARCHER*, SON!

WELL, THE *GREEN ARROW* WAS MY INSPIRATION FOR *WIZARD ARCHER!* IF THE FEATURE'S ONE-TENTH AS TERRIFIC AS *GREEN ARROW*, IT'LL SELL!

YOU'VE ALWAYS WANTED TO BE A COMICS ARTIST, BILL! I'VE GOT A FEELING YOUR DREAM WILL COME TRUE *TODAY!*

AN HOUR LATER, AT THE OFFICE OF *ALL-STAR COMICS*...

THANKS, *GREEN ARROW!* WE'VE GOT YOUR PICTURE! IT'LL APPEAR ON THE BACK COVER OF NEXT MONTH'S ISSUES!

SPEAKING OF PICTURES, THAT MURAL IS THE GREATEST!

LET ME SHOW YOU AROUND, SPEEDY!

ALL-COM OFF!

HERE'S WHERE THE WRITERS TURN OUT THE SCRIPTS FOR THE COMIC BOOK STORIES!

GOSH, WHAT TERRIFIC IMAGINATIONS THEY MUST HAVE TO KEEP COMING UP WITH NEW IDEAS ALL THE TIME!

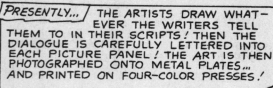

PRESENTLY...

THE ARTISTS DRAW WHATEVER THE WRITERS TELL THEM TO IN THEIR SCRIPTS! THEN THE DIALOGUE IS CAREFULLY LETTERED INTO EACH PICTURE PANEL! THE ART IS THEN PHOTOGRAPHED ONTO METAL PLATES... AND PRINTED ON FOUR-COLOR PRESSES!

GOSH, MR. SLOAN, AS THE EDITOR OF *ALL-STAR COMICS*, YOU MUST DO A TERRIFIC AMOUNT OF WORK!

BUT IT'S FUN, SPEEDY! I DISCUSS THE STORIES WITH WRITERS, REVIEW THE ART-WORK, AND LET OUR PRODUCTION MANAGER DO THE REST!

2

THAT'S IN ADDITION TO MY PET HOBBY... DAIRY FARMING! I'M PRESIDENT OF THE DAIRY LEAGUE CONVENTION THAT'S OPENING ITS EXHIBIT TOMORROW AT THE STAR CITY FAIR GROUNDS! I'M RESPONSIBLE FOR EVERY EXHIBIT, EVERY BUILDING...

PARDON, MR. SLOAN! THERE'S A YOUNG ARTIST, BILL NIXON, TO SEE YOU!

EDITOR MR SLOAN

OH, I ALMOST FORGOT! THIS ARTIST IS TRYING TO SELL ME A FEATURE WITH AN *ARCHER HERO!* I WISH YOU'D SEE IT, *GREEN ARROW!* I'D VALUE YOUR OPINION!

OKAY, SLOAN! LET'S LOOK AT IT!

SHORTLY AFTER...

IMAGINE MEETING *YOU* HERE, *GREEN ARROW!* WHY, YOUR FEATS WERE INSPIRATION FOR THE *WIZARD ARCHER!*

BAH! WHAT THE *GREEN ARROW* DOES IS *REAL!* BUT YOUR CHARACTER IS ABSOLUTELY *INCREDIBLE!* LOOK AT THE CRAZY, UNBELIEVABLE THINGS HE DOES!

M. SLOAN, EDITOR

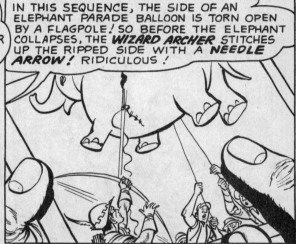

IN THIS SEQUENCE, THE SIDE OF AN ELEPHANT PARADE BALLOON IS TORN OPEN BY A FLAGPOLE! SO BEFORE THE ELEPHANT COLLAPSES, THE *WIZARD ARCHER* STITCHES UP THE RIPPED SIDE WITH A *NEEDLE ARROW!* RIDICULOUS!

HERE'S ANOTHER STUPID SCENE! THE *WIZARD ARCHER* CONFRONTS THREE CROOKS ACROSS A BILLIARD TABLE! OUR HERO HAS ONLY ONE ARROW LEFT! HOW CAN HE KNOCK OUT THREE VILLAINS WITH *ONE* SHAFT? HE STUDIES THE BILLIARD BALLS!

HE QUICKLY NOTES THE POSITION OF THE BALLS! HE FIRES AN ARROW! THE SHOT IS PERFECTLY ANGLED! ONE BALL HITS *THREE* BALLS, WHICH FLY OFF THE TABLE... KNOCKING OUT *THREE* CROOKS! INCREDIBLE! SIMPLY INCREDIBLE!

3

NOW COMES THE PAYOFF! SOME MINERS ARE TRAPPED IN A CAVE-IN! UNLESS FRESH AIR AND RESCUERS REACH THEM IMMEDIATELY, THEY'LL DIE! WHO PERFORMS THIS *IMPOSSIBLE TASK?* THE *WIZARD ARCHER*, NATURALLY!

M. SLOAN EDITOR

HE FIRES A *DRILL ARROW* RIGHT THROUGH THE GROUND... INTO THE MINE SHAFT! WHAT *NONSENSE!* OUR READERS WOULD LAUGH THEMSELVES SICK AT SUCH FANTASTIC ROT!

SORRY, SON! NO SALE! YOUR HERO, THE *WIZARD ARCHER*, IS TOO UNCONVINCING! DON'T YOU AGREE WITH ME, *GREEN ARROW?*

THE POOR KID'S SO DISAPPOINTED! I MUST HELP HIM!

I'M NOT SURE, SLOAN!

TELL YOU WHAT! WE'RE GOING OUT ON PATROL! IF SPEEDY AND I FIND WE CAN'T USE THESE THREE STUNTS IN *REAL* SITUATIONS, I'LL AGREE WITH YOU!

EXCELLENT! AND IF YOU *DO* DUPLICATE 'EM, I'LL ACCEPT *WIZARD ARCHER* FOR PUBLICATION!

LATER, AS THE *ARROW-CAR* CRUISES ALONG STAR CITY PARKWAY...

WE MUST MAKE THOSE COMIC BOOK ARROW TRICKS COME TRUE FOR THAT YOUNG ARTIST!

G.A.! LOOK! THERE'S OUR *FIRST* CHANCE TO PROVE HIM RIGHT!

THE SIDE OF THAT NAVY BLIMP HAS BURST OPEN! HELIUM IS GUSHING OUT! THE BLIMP WILL FALL RIGHT INTO PARKWAY TRAFFIC, UNLESS...

U.S. NAVY

YES, SPEEDY!--UNLESS WE USE *NEEDLE ARROWS* JUST AS THE *WIZARD ARCHER* DID!

INSTANTLY, AS THE TWINS OF THE TWANGING BOW FIRE TWO STEEL *NEEDLE ARROWS...*

JUST AS THE *WIZARD ARCHER* STITCHED UP THAT ELEPHANT BALLOON, WE'LL SEW UP THE BLIMP BEFORE ALL THE GAS LEAVES IT...AS IF WE WERE USING AN AERIAL SEWING MACHINE!

ZIPP!

ZIPPP!

SHORTLY AFTER, AT *ALL STAR COMICS...*

FLASH! THE *GREEN ARROW* JUST SAVED A FALLING BLIMP FROM DISASTER BY SEWING UP ITS TORN SIDE WITH *NEEDLE ARROWS!*

G-GOOD GRIEF! HE DUPLICATED THE *WIZARD ARCHER'S* STUNT IN REAL LIFE!

THE GOLDEN AVENGER

THAT AFTERNOON, AS THE *GREEN ARROW* PURSUES A FLEEING HOLDUP GANG...

T-THEY'VE PUNCTURED OUR TIRES! QUICK! RUN INTO THE PLANETARIUM! WE'LL HIDE THERE!

PLANETARIUM

MINUTES LATER...

HOLD UP, SPEEDY! THOSE THIEVES HAVE TAKEN REFUGE IN *"SPACE"*! THEY CAN SPRING A TRAP THAT'LL REALLY SEND US OUT OF THIS WORLD!

YES, G.A... BUT LOOK CLOSELY! DOESN'T THE LAYOUT OF THESE PLANETS *REMIND* YOU OF SOMETHING?

URANUS

VENUS

SATURN

MARS

EARTH

A WALK THROUGH SPACE

MERCURY

MOON

GASP! OF COURSE! THAT *BILLIARD TABLE* SCENE IN THE *WIZARD ARCHER!* IF WE CAN ONLY HIT THE *"EARTH"* BALL WITH ENOUGH FORCE AND GIVE IT THE PROPER *SPIN!*

LET'S DO IT!

EARTH

MARS

VENUS

5

SHORTLY...
OUR **BILLIARD STICK** ARROWS ARE HITTING THE "CUE BALL" HARD ENOUGH TO **DETACH** IT FROM ITS FIXED POSITION!

EARTH
CLICK!
CLICK!

NEXT...
POW!
YEEOWW!
YIIII!
CRACKK!
A PERFECT SHOT, SPEEDY! ONE BALL IS CAROMING INTO ANOTHER, SENDING 'EM FLYING IN **ALL** DIRECTIONS! WE'VE RACKED UP THE CROOKS!

LATER THAT DAY...
Y-YOU **WHAT?** SAY THAT AGAIN, **GREEN ARROW!**
WE USED THAT ARTIST'S BILLIARD BALL STUNT, SLOAN, TO CAPTURE FIVE GUNMEN! ONE MORE STUNT TO GO!

AMAZING AMAZON COMICS

BUT AS THE DAY WEARS ON INTO NIGHT...
SOUNDS LIKE OUR MISSILE ARROW, WHOSE RADAR DEVICE TRACKS DOWN A MOVING OBJECT WHEREVER IT GOES, JUST OVERTOOK THAT JEWEL THIEF!
YES, SPEEDY! BUT WE STILL HAVEN'T FOUND AN OCCASION TO DUPLICATE THE **WIZARD ARCHER'S** THIRD STUNT!
JEWELRY
YEEOW!
POW!

AT MIDNIGHT...
OUR STEEL-TIPPED **STEP-LADDER ARROWS** WILL LET US RESCUE THAT INJURED STORK THAT SETTLED DOWN ON TOP OF THE OBELISK AND CAN'T FLY AWAY! BUT IT LOOKS LIKE WE WON'T BE ABLE TO "RESCUE" THAT YOUNG ARTIST!

6

MAYBE HIS DRILL ARROW STUNT **WAS** A LITTLE TOO WILD TO BE ENCOUNTERED IN REAL LIFE! POOR KID! HIS **WIZARD ARCHER** FEATURE WILL NEVER BE PRINTED!

NEXT DAY, AT THE FAIR GROUNDS, AS **GREEN ARROW** MEETS EDITOR SLOAN, WHO IS ALSO THE PRESIDENT OF THE DAIRY LEAGUE...

WELL, SLOAN, YOUR EDITORIAL OPINION WAS CORRECT! WE COULDN'T DUPLICATE **WIZARD ARCHER'S** THIRD TRICK!

NATURALLY! IT WAS SHEER ACCIDENT THAT YOU DID THE FIRST **TWO** STUNTS!

MILK

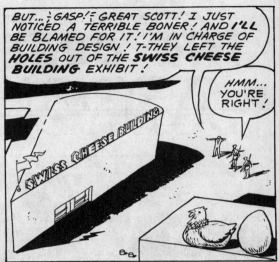

BUT...≟GASP!≟ GREAT SCOTT! I JUST NOTICED A TERRIBLE BONER! AND I'LL BE BLAMED FOR IT! I'M IN CHARGE OF BUILDING DESIGN! T-THEY LEFT THE **HOLES** OUT OF THE **SWISS CHEESE BUILDING** EXHIBIT!

HMM... YOU'RE RIGHT!

SWISS CHEESE BUILDING

SECONDS LATER...

I'M RUINED! THE CHEESE MANUFACTURERS WILL BECOME THE LAUGHINGSTOCK OF THE EXHIBIT! I...≟GASP!≟...**GREEN ARROW!** WHAT'RE YOU **DOING?**

RESCUING YOU FROM AN EMBARRASSING SITUATION, SLOAN!

TWANGG!

TWANG!

REMEMBER THE **WIZARD ARCHER'S DRILL ARROWS** THAT RESCUED THE ENTOMBED MINERS? WELL, WE'RE **DRILLING HOLES** IN YOUR SWISS CHEESE BUILDING TO RESCUE **YOU** FROM A COSTLY BONER!

RRRR

RRRR!

PRESENTLY...

THANK GOODNESS THE PUBLIC ISN'T ADMITTED TILL NOON! THE MAINTENANCE CREW WILL PUT THE FINISHING TOUCHES TO THOSE HOLES YOU DRILLED! HOW CAN I THANK YOU ENOUGH?

BY GIVING THE **WIZARD ARCHER** CREDIT FOR A VERY PRACTICAL STUNT!

⑦

THE NEXT DAY, AT **ALL-STAR COMICS**...

YES...BILL NIXON'S **WIZARD ARCHER** IS JUST WHAT WE NEED! LOGICAL AND NOT A BIT...ER... UNBELIEVABLE!

PSST...BILL! HEREAFTER, TUNE DOWN THAT IMAGINATION! WE MIGHT NOT **ALWAYS** MATCH YOUR CHARACTER'S TRICKS!

THAT DAY WILL NEVER COME FOR THE **TRUE** WIZARDS OF ARCHERY.

END

Up from the core of the earth they came, strange creatures that staggered a city! And behind this invasion lurked a fantastic secret that GREEN ARROW and SPEEDY hoped to reveal, when they drew their famed bowstrings against...

the MENACE of the MOLE MEN

AS HE DRAWS HIS GUN AND AIMS...

HUH.?...THEY'VE GOT WEAPONS, TOO-- BUT THEY'RE RETREATING!

SOME MOMENTS AFTER THE WEIRD CREATURES DEPART...

I TELL YOU THEY WERE TWO ALIENS--AND THEY CAME CRASHING UP THROUGH THE FLOOR! THEN THEY SCAMPERED AWAY WITHOUT A FIGHT! SEND A SQUAD CAR OVER...YOU'LL SEE!

MEANWHILE, ALONG A ONCE-QUIET STREET...

GREAT GUNS! WHAT'S HAPPENING?

CR-RACK!

REMORE CAR SERVICE

A CAVE-IN!... AND I-LOOK WHAT CAUSED IT!

NOT LONG AFTERWARDS, AT THE HOME OF WEALTHY OLIVER QUEEN AND HIS WARD, YOUNG ROY HARPER...

OH, OH...THE ARROW-SIGNAL FROM POLICE HEADQUARTERS, ROY! INTO YOUR COSTUME--HURRY!

RIGHT, OLIVER!

DONNING THE COLORFUL GARB OF GREEN ARROW AND SPEEDY, EMERALD ARCHERS OF JUSTICE, THEY RACE FORTH INTO THE NIGHT...

WHAT'S THE REPORT ON THE POLICE RADIO, G.A.?

SOMETHING ABOUT INVADING MOLE MEN! THEY'RE ATTACKING THE TOWERS BUILDING NOW!

AND AT THIS VERY MOMENT, IN THE DESERTED TOWERS BUILDING...

CR-RASH

SHORTLY, AS THE *CATAPULT* IN THEIR *ARROW-CAR* SENDS THE BATTLING BOWMEN UP TO A WINDOW OF THE BUILDING...

SPEEDY! LOOK!

THE *MOLE MEN!* IT WAS NO GAG!

INSTANTLY, THEY RELEASE A HAIL OF ARROWS...

WE'LL KEEP 'EM BUSY SO THEY CAN'T DRAW THEIR WEAPONS-- THEN SNARE THEM WITH SOME *ROPE ARROWS!*

NO, WAIT... THEY'RE BACKING UP! THEY DON'T *WANT* TO FIGHT!

ZING!

ZING!

INTO THEIR VEHICLE AND BACK THROUGH THE SHATTERED FLOOR GO THE MYSTERIOUS FIGURES...

WHAT DID THEY WANT? WHY DID THEY CRASH THROUGH HERE, TO THE THIRD FLOOR-- THEN LEAVE?

I DON'T KNOW... ACCORDING TO THE POLICE, THIS IS THEIR PATTERN WITH EACH RAID!

BY MORNING, THE STRANGE INVASION IS ON EVERYONE'S LIPS...

WHAT *IS* THE MYSTERY?... THAT'S WHY I'VE SUMMONED ALL OF YOU HERE!

DAILY BUGLE! WHAT'S THE MYSTERY OF THE MOLEMEN?

GREEN ARROW, THESE MEN ARE LEADING SCIENTISTS IN VARYING FIELDS! THEY'VE GOT AN ASTOUNDING THEORY ON THE SO-CALLED *MOLE MEN!*

YES... WE ASK YOU-- AS WE'VE ASKED THE POLICE-- NOT TO MOLEST THEM IN ANY WAY!

3

YOU MEAN--LET THEM *RUN FREELY* THROUGH THIS CITY?

WELL, CONSIDERING THAT THEY'RE *ALIEN CREATURES*-- WITHOUT ANY WISH TO *FIGHT* US --THEY DO WARRANT SPECIAL ATTENTION!

WE FEEL THAT THEY'VE BORED UP FROM THE *EARTH'S CORE,* INVESTIGATING *OUR* WORLD.! IF WE SHOW THEM WE'RE FRIENDLY, WE MAY SOON HAVE NEW, PEACE-LOVING NEIGHBORS!

I'M INCLINED TO AGREE, *GREEN ARROW!* AFTER ALL, THEY'VE SHOWN NO HOSTILITY, AND -- WAIT, THE PHONE...

R-RING!

THE *MOLE MEN* HAVE STRUCK AGAIN.! THIS TIME, THEY'VE POPPED UP RIGHT IN *CITY SQUARE!*

CHECK!... WE'RE ON OUR WAY!

SHORTLY AFTERWARDS, AT THE SITE OF THE LATEST INVASION...

PEACEFUL OR NOT, *SPEEDY,* THIS TIME THEY'RE HAVING COMPANY WHEN THEY FLEE BACK INTO THEIR HOLE!

SOON...

THEY GOT BACK INTO THE MACHINE-- AND LEFT!

YES... AND DID YOU NOTICE THE *WRECKED JEWELRY SHOP* UP ABOVE?

WHAT ARE YOU DRIVING AT, G.A.?

THE POLICE REPORT MENTIONED THAT AMONG THE MISSING ITEMS AT THE MUSEUM WAS A RARE, MILLION-DOLLAR PAINTING...

...AND AN ARMORED CAR IS STILL MISSING WHERE THE SECOND INVASION OCCURRED! THE SAME GOES FOR A SAFE -- IN THE TOWERS BUILDING!

GOLLY! DO YOU THINK...?

SUDDENLY...

THEY'RE COMING AT US WITH THEIR BORING MACHINE!

AND OUR ARROWS ARE BOUNCING RIGHT OFF THE THING!

USE THE PAINT ARROWS! WE'LL SMEAR THAT BUBBLE AND CUT OFF THEIR VISION...THAT SHOULD STOP THEM.

ZIP! ZIP!

BUT AS THE MOLE MEN EMERGE...

ACETYLENE GUNS! KEEP BACK, SPEEDY... THOSE THINGS CAN BURN THROUGH ROCK AND METAL!

BZZZZZT!

BZZZZT!

DEFTLY, THE ACE ARCHERS FIRE ANOTHER PAIR OF UNIQUE SHAFTS...

THAT'S IT... HIT THEIR PISTOL BARRELS WITH *RIVETING* ARROWS!

TWANG

ZING!

ZING!

TWANG-G

NOW-- THE *NET ARROWS*... AND THAT WINDS UP A GOOD "FISHING" TRIP!

AND UPON REMOVING THE HELMETS OF THE "ALIENS"...

WELL, WHAT DO YOU KNOW?... *ORDINARY CROOKS!*

NOT QUITE ORDINARY, *SPEEDY!* THEY WORKED A GREAT HOAX, HOPING THEY'D BE LEFT ALONE WHILE THEY PULLED ROBBERIES ALONG WITH THEIR "ACCIDENTAL" HOUSE-WRECKING!

IN A WAY, THOUGH, THEIR WISH WILL COME TRUE! NO ONE WILL BOTHER THEM FOR A LONG, LONG TIME-- AT *STATE PRISON!*

THE END

THE GREEN ARROW

THE GREEN ARROW

A LONG-SLUMBERING VOLCANO SUDDENLY ERUPTS, AND FROM IT EMERGES A TERRIFYING BEAST FROM ANOTHER ERA! BUT ON HAND ARE THOSE ARTFUL ARCHERS, *GREEN ARROW* AND *SPEEDY*, WHO ENCOUNTER ONE OF THEIR TRULY STRANGEST ADVENTURES WHEN THEY DRAW BOW AGAINST...

the CREATURE from the CRATER

IT'S VACATION TIME FOR OLIVER QUEEN AND YOUNG ROY HARPER, BUT NEITHER IS AWARE OF THE STRANGE EVENTS WHICH LIE LURKING AHEAD...

GOLLY, MR. RAINSEFORD! IT WAS SWELL OF YOU TO INVITE US DOWN HERE TO YOUR PLANTATION!

WELL, ROY, EACH YEAR I BRING ALONG A MEMBER FROM THE CLUB BACK HOME DOWN HERE! THIS YEAR WAS OLIVER'S TURN-- AND YOU'RE A SPECIAL GUEST!

MAYBE WHILE YOU TAKE US ON THIS TOUR WE CAN SEE THE VOLCANOES!

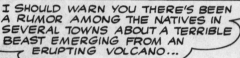

I SHOULD WARN YOU THERE'S BEEN A RUMOR AMONG THE NATIVES IN SEVERAL TOWNS ABOUT A TERRIBLE BEAST EMERGING FROM AN ERUPTING VOLCANO...

...BUT THAT'S SHEER NONSENSE! SO FAR AS I KNOW, THERE'S BEEN NO ERUPTION FOR SEVERAL CENTURIES...

VROOOM

WHAT IN THE WORLD WAS *THAT*?

I DON'T KNOW! LET ME TAKE A LOOK THROUGH MY BINOCULARS!

MY WORD! THE VOLCANO *IS* ERUPTING! AND THERE SEEMS TO BE SOMETHING STARTING TO COME OUT OF IT!

2

BEFORE THE PAIR CAN RECOVER...

WRR-AAA-GH!

WITH A FEW GIANT STRIDES, THE MONSTROUS MENACE FROM THE PAST DEPARTS, HEADING BACK TO THE VOLCANO, WHEN...

OKAY-- WE'RE OUT OF SIGHT NOW! GET THOSE ARROW-SHOOTING MEDDLERS INSIDE!

THUS, TO THEIR UTTER ASTONISHMENT, DO THE BATTLING BOWMEN LEARN THE INCREDIBLE SECRET OF THE "PREHISTORIC" MONSTER...

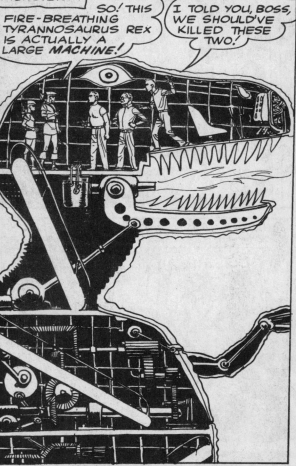

SO! THIS FIRE-BREATHING TYRANNOSAURUS REX IS ACTUALLY A LARGE MACHINE!

I TOLD YOU, BOSS, WE SHOULD'VE KILLED THESE TWO!

THERE WILL BE NO KILLING! MY IRON REX HASN'T CAUSED ONE DEATH, AND WE'LL MAINTAIN THAT RECORD!

SO -- YOU'RE THE INVENTOR OF THIS TRICKY APPARATUS!

"TRICKY APPARATUS" INDEED! THE IRON REX IS THE ENGINEERING MARVEL OF THE AGES! IT CAN SIMULATE EVERY MOVEMENT OF ITS ONCE-LIVING COUNTERPART, EVEN BREATHE FIRE!

BUT WHY DO YOU HAVE IT INVADE THE TOWNS...

5

...AND FRIGHTEN THE INHABITANTS? ALSO, HOW CAN IT SURVIVE INSIDE A LIVE VOLCANO?

I'LL TELL YOU, *GREEN ARROW!* WE MAKE 'EM LEAVE IN A HURRY, SO WE CAN COME DOWN AND LOOT THE PLACES!

I SEE! IN OTHER WORDS, THIS "MARVEL OF THE AGES" IS NOTHING MORE THAN ANOTHER GIMMICK FOR PULLING CRIMES?

DID YOU EVER SEE A GREATER "GIMMICK"? AS FOR SURVIVING WITHIN THE VOLCANO, I'LL SHOW YOU! WE ARE ENTERING THE CRATER NOW!

THERE! IS THE EXPLANATION CLEAR TO YOU NOW?

GOOD GRIEF! A PLANT DOWN HERE WHICH CREATES AN ARTIFICIAL ERUPTION!

OKAY-- THIS IS WHERE WE GET OUT! GET GOING-- DOWN THE STAIRS!

YOU'LL BE KEPT PRISONER! AFTER WE'RE FINISHED HERE, WE'LL DECIDE HOW TO DISPOSE OF YOU BEFORE GOING ON TO NEW AREAS TO CONQUER!

GIVE US YOUR BOWS AND ARROWS! I WARN YOU, WE DON'T WANT ANY FUNNY BUSINESS!

SURE--

6

BUT INSTEAD OF REMOVING HIS QUIVER, *GREEN ARROW* PLUCKS A *SMOKE SCREEN ARROW*, AND BREAKS ITS TIP...

YOW! I CAN'T SEE!

DOWN, *SPEEDY*-- AND FOLLOW ME! IT'S TIME FOR US TO MAKE *OUR* MOVE!

CRACK

HEAD UPSTAIRS FOR THE CONTROLS!

BULL'S-EYES IN BOTH BARRELS!

TWANG

TWANG

CRASH!

BAM

MOMENTS LATER, THE *IRON REX* AGAIN EMERGES FROM THE VOLCANO, BUT *THIS* TIME...

LOOK THERE! ISN'T THAT THE FAMED *GREEN ARROW* ATOP THE MONSTER'S HEAD?

THERE IS SOMETHING VERY PECULIAR ABOUT ALL THIS!

WE HAVE CAUGHT THE GANG, RECOVERED THE LOOT, AND SOLVED THE MYSTERY OF THE VOLCANO! BUT WHAT CAN WE DO WITH THIS MECHANICAL MONSTER?

PUT IT ON DISPLAY SOMEWHERE! KIDS WOULD SURE GET A KICK OUT OF IT!

7

LATER, AT RAINSEFORD'S PLANTATION...

GOOD HEAVENS--YOU TWO MISSED ALL THE EXCITEMENT! *GREEN ARROW* AND *SPEEDY* WERE HERE! THEY CAUGHT THE "BEAST" AND PROVED IT WAS A *MECHANICAL* THING!

MY WORD--THAT MUST'VE BEEN A SIGHT TO SEE!

The End

DAWN, AS BOLD CRIMINALS STRIKE AT A BANK VAULT...

HURRY UP! THE BLAST THAT OPENED THE VAULT SET OFF EVERY ALARM ON THE BLOCK!

RING-RING-G-G! CLANG! CLANG!

IT IS MINUTES LATER WHEN THE FAMED ARROWCAR SCREECHES TO A STOP OUTSIDE, ITS BUILT-IN CATAPULTS HURLING TWO FIGURES SKYWARD...

WITH THE POLICE BELOW, WE CAN AVERT ANY ROOF-TOP GETAWAY, SPEEDY!

BUT AS GREEN ARROW AND SPEEDY REACH THE ROOF...

G.A.--LOOK! THEY'RE ESCAPING IN BIG, SHATTERPROOF GLASS BUBBLES! OUR ARROWS CAN'T DO A THING AGAINST THEM!

HA, HA! WHAT A JOKE--TRYING TO BRING US DOWN WITH BOW AND ARROW!

THEY'RE GONE, G.A.-- WITH ALL THE LOOT!

THOSE BUBBLES WERE THE STRANGEST-LOOKING CRAFT I'VE EVER SEEN! WHERE IN DEUCE DID THEY COME FROM-- AND WHERE DID THEY GO?

LATER, WHEN THE ARCHERS RESUME THEIR EVERY-DAY IDENTITIES OF WEALTHY OLIVER QUEEN AND HIS YOUNG WARD, ROY HARPER ...

YOU MIGHT CALL THEM A NEW VERSION OF FLYING SAUCERS, OLIVER!

WHATEVER THEY ARE, ROY, WE MUST FIND OUT WHERE THEY CAME FROM-- AND WHO'S BEHIND THEM!

DAILY BUGLE
"BUBBLE BAND" ROBS BANK!
STRANGE CRAFT PUZZLES POLICE

2

IN THE DAYS THAT FOLLOW, MORE FANTASTIC BUBBLES APPEAR...

ALL RIGHT, GIVE ME THE BRIEFCASE OF DIAMONDS YOU'RE DELIVERING--THEN STAY WHERE YOU ARE TILL WE TAKE OFF!

EVEN A SPEEDING TRAIN IS NO MATCH FOR THE FANTASTIC SPHERES...

THAT WAS NEAT... THE BAGGAGE MAN DIDN'T KNOW WHAT HIT HIM!

AND EACH TIME GREEN ARROW AND SPEEDY INTERCEPT THE WEIRD BUBBLES...

OUR GRENADE ARROWS DON'T DO ANY GOOD EITHER, G.A.!

CRACK

BANG!

ZIP!

TWANG!

ONLY ONE THING TO DO NOW, SPEEDY... HIT THE TOPS OF THE BUBBLES WITH OUR PAINT ARROWS! WE'RE GOING TO TRY A WILD PLAN...

ZIP

ZIP!

TWANG!

AS THE ARROWS STRIKE WITH A THUD, RELEASING A QUICK-DRYING, BLACK PAINT...

WE'LL HITCH A RIDE NOW, WITHOUT THEM SEEING US! I'VE PUT THE ARROWPLANE ON AUTOMATIC CONTROL!... IT'LL SET ITSELF DOWN NEAR THE ARROW-CAVE!

3

BUT, AT THIS VERY MOMENT, IN A LABORATORY-HIDEOUT, MILES AWAY...

WELL, WELL... WE'VE GOT PASSENGERS THIS TIME... NONE OTHER THAN *GREEN ARROW* AND HIS IMPUDENT YOUNG PARTNER!

I'LL BRING THE SPHERES IN, EVER SO GENTLY, SO AS NOT TO SHAKE OFF MY "GUESTS!" I HAVE GREAT PLANS FOR THEM!

AS THE SPHERES APPROACH THE HIDEOUT, A GREAT ROOF SWINGS OPEN, AND...

G.A.! WE'RE CAUGHT!

YES--BY GREAT ROBOT HANDS!

NOW WE'RE BEING PLACED INTO ONE OF THE BUBBLES!

INDEED YOU ARE--FOR A VERY GOOD REASON!

BUT, BOSS... THEY'VE STILL GOT THEIR BOWS AND ARROWS!

YOU'VE SEEN HOW USELESS THEY ARE AGAINST US! *PSHAW!...* MERE ARROWS AGAINST THE ULTIMATE IN ELECTRONICS!

4

THIS GLASS IS THIN ENOUGH FOR YOU TO HEAR ME! LET ME EXPLAIN MY LITTLE OPERATION HERE--BEFORE I SEND YOU TO YOUR DOOM!

THE BUBBLES ARE PROPELLED BY THAT GREAT "BLOWER", WHICH CAN TRANSMIT A NEGATIVE OR POSITIVE CHARGE! TINY CELLS IN THE SPHERE'S WALL CONTAIN ONLY *NEGATIVE* CHARGES...

THUS, BY *TRANSMITTING* A NEGATIVE CHARGE, I CAUSE THE SPHERE TO *MOVE AWAY* FROM THE "BLOWER"... BY TRANSMITTING A *POSITIVE* CHARGE, I'M ABLE TO *BRING BACK* THE SPHERE!

BY VARYING THE CHARGES' *STRENGTH*, I CAN CAUSE THE SPHERE TO RISE, DESCEND, GO FAST OR SLOW--DEPENDING UPON WHAT I OBSERVE ON MY TV SCREEN!

RIGHT NOW, FOR EXAMPLE, I SEE HUGE ARTILLERY BEING SET UP TO GIVE US A RESOUNDING RECEPTION! ALAS, WEAPONS OF THAT SIZE *CAN* SHATTER MY SPHERES... SO I SHALL TAKE ADVANTAGE OF THIS OPPORTUNITY!

QUICKLY, THE ROBOT HANDS DEPOSIT THE SPHERE IN THE "BLOWER"-- AND MOMENTS LATER...

IT'S *YOU* THEY'LL BE FIRING AT, GREEN ARROW! HA, HA! HAVE A NICE TRIP!

BZZZT!

BZZZT!

WITH A ROAR, THE IMPRISONED ARCHERS GO HURTLING FORTH...

THAT WAITING ARTILLERY WILL BLAST US TO SMITHEREENS, UNLESS-- HOLD IT, SPEEDY!

WHROOSH

OUR CONFIDENT CAPTOR DOESN'T REALIZE WE CARRY AN ELECTRONICS ARROW! IT MAY BE THE KEY TO OUR RESCUE! HERE GOES...

ONCE RELEASED, THE CHARGED SHAFT STREAKS AROUND THE INNER WALLS OF THE SPHERE...

I DON'T UNDERSTAND, G.A.!

I'VE SET THE ELECTRONICS ARROW TO CHANGE THE SPHERE'S NEGATIVELY CHARGED CELLS TO POSITIVE! THAT SHOULD RETURN US TO THE TRANSMITTER!

SURE ENOUGH, WITHIN SECONDS...

IT WORKED!... AND WE GOT THE HATCH OPEN BY DISTURBING ITS FORCE FIELD WITH A MAGNETIC ARROW!

YES-- BUT WE'VE GOT TO PREVENT THEM FROM GETTING AT THE CONTROLS!

TWANG!

TWANG!

ZIP!

BOSS, THOSE ARROWS HAVE US PINNED! WE'RE STUCK!

RIGHT! CALL THE POLICE, SPEEDY!

AND SO, LATER, WHEN SQUAD CARS ROLL UP...

NICE WORK, GREEN ARROW... WE'LL START DISMANTLING THIS PLACE PRONTO!

HA, HA! COME TO THINK OF IT, THIS IS THE FIRST TIME I EVER HEARD OF SOMEONE BEING ARRESTED FOR SIMPLY-- BLOWING BUBBLES!

THE END

THE GREEN ARROW...

A MASTER OF COLOR, HE APPEARED FROM OUT OF THE NIGHT... AND SOON, THE WHOLE CITY WAS AGOG OVER THIS STRANGELY COSTUMED CONNIVER OF CRIMES! EVEN THE GALLANT GREEN ARROW SEEMED TO HAVE MET HIS MATCH IN...

the SINISTER SPECTRUM MAN

GREAT SCOTT! THAT RAY-- FROM THE *SPECTRUM MAN'S* HELMET... C-CAN'T SEE WHERE WE'RE GOING!

NIGHTTIME, AS A STRANGELY BEDECKED PLANE FLIES LOW OVER THE CITY...

ALIGHTING ATOP A RIVERSIDE WAREHOUSE, ITS TWO GREAT LIGHTS COME ON--LIKE THE GLARING EYES OF A GIANT BEAST...

WITHIN MINUTES...

FIRE! FIRE!

WHREEEEE!

RACING TO THE FIREFIGHTERS' AID COME *GREEN ARROW* AND *SPEEDY,* WHO MAKE A STARTLING DISCOVERY...

THIS IS NO FIRE, *SPEEDY!* IT'S AN *ILLUSION,* CAUSED BY THE RED LIGHTS OF AN AIRPLANE--ALONG WITH MISTS RISING FROM THE RIVER!

JUST THEN...

HELP! I'VE BEEN ROBBED!

THAT STRANGELY DRESSED MAN... HE MUST BE THE ROBBER!

CHECK CASHING SERVICE

AS THE FAMED ARCHERS WHIP ARROWS TO THEIR BOWS, THE WEIRD CRIMINAL TOUCHES A BUTTON ON HIS BELT...

G.A.--LOOK! HE SUDDENLY PRODUCED A GLOWING LIGHT, OUTLINING HIMSELF IN THE DARKNESS!

I DON'T UNDERSTAND... THAT ONLY MAKES HIM A BETTER TARGET FOR OUR *ROPE-ARROWS!*

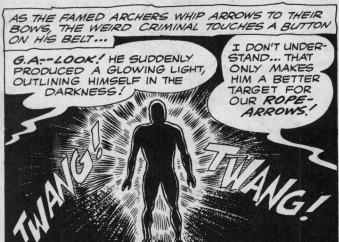

TWANG!   TWANG!

STRAIGHT TO THEIR MARK FLY THE UNIQUE SHAFTS, WHIRLING AROUND AND BINDING THE BRILLIANT FIGURE...

BUT WHEN *GREEN ARROW* AND *SPEEDY* DESCEND TO MAKE THE CAPTURE...

WHY... THIS IS NOTHING BUT AN *INFLATED DUMMY!*

HE PROBABLY HAD IT UNDER HIS CAPE, AND BLEW IT UP WHEN HE PRESSED HIS BELT BUTTON! IN THE SHARP GLOW, WE COULDN'T DETECT THE ACTION -- OR HIS ESCAPE!

2

MEANWHILE, IN THE HEART OF THE CITY...

OKAY, PETE...YOU TAKE THE BANK--I'LL TAKE THE JEWELRY STORE -- AND MIKE CAN HIT THE RARE STAMP COLLECTION!

EEEEK!

MUST BE THREE GUYS FROM A CIRCUS!

SHORTLY, IN THE BANK...

MAKE IT SNAPPY WITH THAT CASH!

AND IN THE JEWELRY STORE...

AND THE STAMP COLLECTOR'S...

BUT THAT'S MY RAREST COLLECTION-- WORTH THOUSANDS!

EXACTLY WHY I WANT IT!

BUT OUTSIDE EACH PLACE...

THEY'LL WALK RIGHT INTO OUR HANDS WHEN THEY COME OUT!

RIGHT...THEY WON'T GET FAR IN THOSE CLOTHES!

WHEN THE ACE ARCHERS ARRIVE...

THAT'S RIGHT, GREEN ARROW-- THREE OF 'EM ... AND WE'VE GOT 'EM TRAPPED!

HMM... SOUNDS JUST A LITTLE TOO EASY!

4

...AND WHAT BETTER WAY THAN WITH THE *ARROWCAR'S* CATAPULT!

BUT IN THAT VERY SAME MOMENT...

"HA, HA," VERY CLEVER, *GREEN ARROW*... BUT WATCH WHAT HAPPENS WHEN I TURN ON MY *COLOR DISTORTION RAY!*

WHAT--?

EEYOW! THE PLANE... THE SKY... ALL BROKEN UP INTO A CRAZY PATCHWORK!

QUICKLY, *SPEEDY...* THE *NET ARROWS!* FIRE THEM AT THE *SOUND* OF THE PLANE!

THE ARROWS, FIRED, RELEASE THEIR NETS, AND...

BLAST YOUR LUCK!

HAUL YOURSELF UP, *SPEEDY-* FAST!

6

YOU'VE HEARD OF MANY KINDS OF CLOCKS -- BUT DID YOU EVER HEAR OF A CLOCK THAT NOT ONLY RUNS BUT WALKS AS WELL? NO -- THIS IS NOT A RIDDLE BUT THE ELUSIVE SUBJECT OF THE SEARCH BY THOSE FAMED ARCHERS *GREEN ARROW* AND *SPEEDY,* WHO ATTEMPT TO STOP...

the CRIMES of the CLOCK KING

THE SANDS OF TIME ARE RUNNING OUT, *GREEN ARROW!*

AND WE'LL FALL INTO THE PIT BELOW!

HIGH UP IN THE PLUSH RITZ-ROYAL HOTEL, A GALA COSTUME BALL IS UNDER WAY...

...WHERE THE DISPLAY OF BIZARRE COSTUMES IS TAKEN FOR GRANTED...

HEY! LOOK AT THAT GUEST! I'LL BET HE ALWAYS KNOWS WHAT TIME IT IS!

JUST KEEP YOUR EYES PEELED FOR ANY ROBBERY ATTEMPT! THAT'S OUR JOB!

I KNOW WHICH WOMEN ARE WEARING THE MOST EXPENSIVE JEWELS -- BUT I CAN'T IDENTIFY THEM WITH THEIR MASKS ON! THEY'LL UNMASK AT MIDNIGHT, WHEN THIS CLOCK SOUNDS, BUT THAT'LL BE TOO LATE ...

... BECAUSE I AM TIMING MY ROOFTOP ESCAPE AT EXACTLY MIDNIGHT! I MUST GET THEM TO UNMASK *EARLIER* -- AND I THINK I KNOW HOW!

SUDDENLY, THE CLOCK TOLLS 12, THE SIGNAL TO UNMASK-- BUT THEN THE ROOM IS BLACKED OUT!...

WHAT HAPPENED? IT'S NOT MIDNIGHT YET!

EEEKK! MY DIAMOND TIARA! SOMEONE GRABBED IT!

AND SOMEONE TORE OFF MY NECKLACE! *HELP!*

UNOBSERVED IN THE CONFUSION, A LITHE FIGURE SPRINGS OUT A WINDOW...

LOOK -- OVER THERE AT THE WINDOW! IT'S THE MAN IN THE CLOCK COSTUME!

THAT WAS SIMPLE ENOUGH! I MOVED THE MINUTE HAND FORWARD!

MERE MOMENTS LATER, A FLAMING GREEN ARROW IS UNLEASHED FROM ATOP POLICE HEADQUARTERS AND LANCES THROUGH THE SKY...

WE'RE BEING PAGED BY THE LAW! INTO YOUR COSTUME, ROY!

AT THE CRIME SCENE, TWO FIGURES APPEAR-- *GREEN ARROW* AND *SPEEDY*, WHO CATAPULT FROM THEIR *ARROWCAR!*

THERE'S THE THIEF, *SPEEDY* -- UP THERE BY THE TOWER CLOCK!

GREEN ARROW, EH? I'LL FIX HIM!

2

THESE FIGURES EMERGE AT 12 SHARP! I CAN ESCAPE INSIDE THE TOWER, AND MAKE MY WAY BELOW--

AN INSTANT LATER, BOWS TWANG FURIOUSLY...

THAT'S IT, **SPEEDY!** USE YOUR ROPE ARROWS SO WE CAN TRUSS HIM UP!

ZIP! ZIP! ZIP!

BUT THEN...

BLAZES! THE CLOCK CRIMINAL FOOLED US-- BY DRAPING HIS COSTUME AROUND ONE OF THE FIGURES, THEN ESCAPING!

BONG!

WHAT'LL WE DO NOW?

GO HOME! THERE ISN'T A CLUE THAT WE CAN POSSIBLY FOLLOW!

IN THE DAYS THAT FOLLOW, THE **CLOCK KING** STRIKES AS IF ON SCHEDULE...

VOL II NO 5

CLOCK THUG ROBS TICK-TOCK CLUB

MINUTE-MAN SAVINGS BANK LOOTED

IDLE HOUR INN HELD UP

TELLER

3

THEN ONE NIGHT...

DID YOU NOTICE, EVERY ONE OF HIS JOBS HAS TO DO WITH *TIME*?

OR CLOCKS! WE'D BETTER STAY ON PATROL IN TOWN! MAYBE WE CAN FIGURE OUT HIS PATTERN!

LATER, IN AN EXCLUSIVE, DESERTED STREET...

HA! IF ANYBODY'S WONDERING WHAT THIS JOB HAS TO DO WITH *CLOCKS*... WELL, HE'LL SOON FIND OUT!

PROTECTED BY ALARM SIGNAL SYSTEM

AS THE DOOR IS FORCED OPEN, A CIRCUIT BREAKS...

THE BURGLAR ALARM! BUT I SHAN'T BE INTERRUPTED AFTER THE ARRANGEMENTS I MADE IN THE STORE NEXT DOOR!

R-R-RIN-NG!

ART ANTIQUES RARE WORKS OF ART

NO SOONER DOES HE SLIP INSIDE THAN...

R-RINNG! BONG! BONG! DING-DONG! DING!

FOR THE LUVVA MIKE -- ALL THOSE CLOCKS GOING OFF AT ONCE! THEY'LL WAKE UP THE WHOLE NEIGHBORHOOD!

ART ANTI RARE WOR AR

CLOCKS

BUT A FAMILIAR PAIR IS ALSO ATTRACTED BY THE LOUD RINGING...

IT'S FROM THE CLOCK STORE BELOW!

YES -- AND A LITTLE TOO OBVIOUS! OUR MAN IS AT WORK -- BUT NOT IN *THAT* PLACE

BONG DING-DONG DONG CLOCK WATCH BONG

LET'S TRY *THIS* ONE ... AND -- THERE HE GOES!

CONFOUND THOSE ARCHERS!

RARE PAINTINGS

4

THROUGH TWISTING, SHADOWY BACK STREETS THE BATTLING BOWMEN PURSUE THE FLEEING FIGURE...

ZIP!

ZIP!

...UNTIL HE DARTS INTO A DOORWAY...

COME IN, GENTLEMEN! WERE YOU LOOKING FOR ME?

NONE OTHER! THE POLICE WANT US TO BRING YOU IN!

ABRUPTLY...

I'M AFRAID YOU'LL HAVE TO CHANGE YOUR PLANS, GREEN ARROW!

G.A.! WE'RE FALLING--

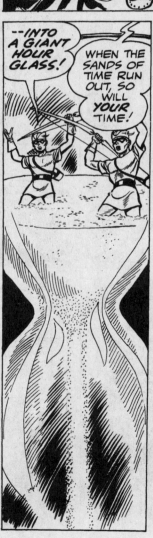

--INTO A GIANT HOUR GLASS!

WHEN THE SANDS OF TIME RUN OUT, SO WILL YOUR TIME!

THERE'S NO BOTTOM TO THE GLASS! THE SAND FALLS INTO A SPIKE-FILLED PIT!

AND THAT'S WHERE WE'LL FALL, UNLESS...

5

DESPERATELY, THE FAMED ARCHER FIRES A SHAFT AT THE CEILING...

NOT MUCH TIME LEFT, *SPEEDY!* MAYBE THE *SUCTION-CUP ARROW* CAN BAIL US OUT!

THE MOMENT THE ARROW HITS--AND HOLDS-- THE PAIR LEAPS...

HANG ON, LAD-- AND *CLIMB!*

THWUNK!

REACHING THE CEILING, *GREEN ARROW* CAUTIOUSLY PUSHES BACK THE TRAP DOOR, AND...

HUH? HOW DID YOU--?

NEVER MIND, MISTER! *YOUR* TIME'S RUN OUT!

ZING!

ZING!

IN A *SECOND,* WE'LL CALL THE POLICE... WHO'LL BE HERE WITHIN *MINUTES!* AND IN AN *HOUR,* YOU'LL BE IN JAIL!

LET'S JUST SAY, *G.A.,* HE'S GOING TO HAVE A LOT OF *TIME* ON HIS HANDS FROM NOW ON!

The End

THE GREEN ARROW

THE GREEN ARROW

IT TOOK ONLY A FRACTION OF A SECOND... BUT IN THAT BRIEF INSTANT, THE CRIME-FIGHTING CAREER OF *GREEN ARROW*, ACE ARCHER OF LAW AND ORDER, CAME CRUMBLING DOWN AROUND HIM! FOR HE WAS NOW AT THE MERCY OF...

THE SPY IN THE ARROW-CAVE

OKAY, *SPEEDY*... HEAD THE *ARROWCAR* OUT THROUGH OUR SECRET TUNNEL-EXIT!

HA, HA! IT'S NOT A SECRET ANY MORE, *GREEN ARROW*!

IN THE SECRET *ARROW-CAVE*, TWO FAMED CRIME-FIGHTERS PREPARE FOR THEIR NIGHTLY PATROL...

WE'LL HAVE TO BE ESPECIALLY WATCHFUL TONIGHT, *SPEEDY*!

I KNOW, *G.A.*-- FLINT MORGAN, THE ESCAPED EMBEZZLER, HAS BEEN SPOTTED SOMEWHERE IN THIS VICINITY!

THAT'S *ME* YOU'RE TALKING ABOUT, *GREEN ARROW* AND *SPEEDY*--OR SHOULD I SAY, *OLIVER QUEEN* AND *ROY HARPER*!

MORGAN! H-HOW DID *YOU* GET IN HERE?

WHILE RUNNING FROM THE LAW, I ACCIDENTALLY STUMBLED ACROSS YOUR SECRET TUNNEL-ENTRANCE! A ONE-IN-A-MILLION CHANCE, AND I'M THE LUCKY ONE! WHAT A HIDEOUT THIS'LL MAKE!

DON'T WORRY, THOUGH-- I'M NOT REALLY A BAD GUY! I WON'T REVEAL ANYTHING, AS LONG AS YOU LET ME STAY! SO GO AHEAD AND CHASE CROOKS... I WON'T GET IN YOUR WAY!

I--I GUESS WE'VE GOT NO CHOICE, *SPEEDY*-- AT LEAST FOR NOW! COME ON-- LET'S GO OUT ON PATROL!

AND AS THE ARCHERS START THEIR ROUNDS IN THE *ARROWCAR*...

WHEW! WHAT A SPOT MORGAN'S GOT US IN!

YES... THE MOMENT WE TURN HIM IN, OUR WHOLE CRIME-FIGHTING CAREER WILL BE OVER! THIS'LL TAKE SOME TALL THINKING, *SPEEDY*... WE'VE *GOT* TO COME UP WITH AN ANSWER!

MEANWHILE...

TALK ABOUT THE COMFORTS OF HOME! I CAN SPEND THE EVENING WATCHING *GREEN ARROW'S* NEWSREEL FILE!

I REMEMBER THAT CASE! THEY STOPPED THE *HELICOPTER ROBBERS* WITH A LOT OF *BALLOON ARROWS!*

AFTERWARD, AS FLINT MORGAN EXPLORES THE WONDERS OF THE *ARROW-CAVE*...

HMM... A MAP OF THE CITY--WITH *GREEN ARROW'S PATROL SCHEDULE* MARKED OFF ON IT! IT SHOWS WHERE HE AND *SPEEDY* PATROL EACH NIGHT, AND THE HOUR WHEN THEY'RE THERE!

2

WHAT AN OPPORTUNITY! I DON'T WANT TO HURT *GREEN ARROW*... I JUST WANT ENOUGH MONEY TO LIVE COMFORTABLY SOMEWHERE -- IN SECLUSION... AND THIS MAP IS MY TICKET!

THUS, A SHORT WHILE LATER, AFTER BORROWING SOME OF OLIVER QUEEN'S CLOTHING...

I KNOW JUST THE RIGHT GUYS TO APPROACH! THE BANCROFT GANG HAS HAD A LOT OF TROUBLE WITH *GREEN ARROW!* MIKE BANCROFT SHOULD BE A LIKELY CUSTOMER FOR THIS MAP!

SLIPPING BACK TO THE CITY, THE FUGITIVE APPEARS AT BANCROFT'S OFFICE, WHERE...

YOU MADE YOURSELF A NICE DEAL, MORGAN... $10,000 FOR A SIMPLE MAP OF THE CITY! ANYTHING ELSE TO SELL?

NO... THIS MONEY SUITS ME FINE!

BUT I'D PAY *TEN TIMES* THAT MUCH -- TO KNOW *GREEN ARROW'S SECRET IDENTITY!*

THAT'S **NOT** FOR SALE! LIKE I SAID -- I JUST WANT TO GO FAR AWAY... FROM PRISON AND FROM CRIME!

THE MOMENT MORGAN LEAVES...

OKAY, WE START OPERATIONS *RIGHT NOW!* TODAY'S WEDNESDAY -- AND *GREEN ARROW* WILL BE PATROLLING *HERE*, ON THE SOUTH SIDE...

CALL THE BOYS... TELL 'EM TO START A DECOY ACTION ON THE SOUTH SIDE, TO KEEP *GREEN ARROW* OCCUPIED, WHILE *WE* PULL A *REAL* CAPER -- ON THE *NORTH* SIDE!

3

WITHIN MINUTES, AS A BURGLAR ALARM RINGS SHRILLY IN A SOUTH SIDE SILVERWARE HOUSE...

THERE GO THE BURGLARS, *SPEEDY*... WE CAN STOP THEM EASILY, WITH *NET ARROWS!*

IT'S WORKING! WITH *GREEN ARROW* AFTER *US,* THE BOSS CAN HIT THAT NORTH SIDE JEWELRY STORE FOR A *BIG* HAUL!

TWIN *NET ARROWS* WHISTLE THROUGH THE AIR, AND...

HAW! WHAT DO YOU KNOW?... WE'RE *CAUGHT!*

IT'S WORTH IT, FOR WHAT THE BOSS'LL BE PAYING US!

ZIP   ZIP

SUDDENLY...

LOOK! THE *ARROW-SIGNAL* FROM POLICE HEAD-QUARTERS! *MORE* TROUBLE IS POPPING!

LET'S TURN THESE TWO OVER TO A PATROL-MAN--*FAST*--AND GET MOVING!

SOON, GUIDED BY POLICE RADIO, THE ARCHERS REACH THE NORTH SIDE, JUST AS...

WE MADE IT, BOSS! THEY'RE STUCK ON THE OTHER SIDE OF THE DRAW-BRIDGE!

THE TEN G'S WE GAVE FOR *GREEN ARROW'S* PATROL SCHEDULE HAS ALREADY PAID OFF!

BUT THE BATTLING BOWMEN AREN'T YET THWARTED--AS THE *ARROWCAR'S* CATAPULT SENDS THEM HURTLING SKYWARD...

BOING

THEY LAND ON THE RISING SEGMENT OF THE BRIDGE, PLANNING TO PUT THEIR BOWS TO WORK, BUT...

OW-W!... MY LEG!

GREAT GUNS! YOU'RE HURT!

4

DON'T BOTHER ABOUT ME NOW... GO ON AFTER THE CROOKS!

NOT TILL I GET YOU TO A HOSPITAL, LAD! THAT LEG IS BADLY INJURED!

AND SHORTLY AFTERWARD, AS FLINT MORGAN HEADS OUT OF THE CITY...

WHAT--? THE KID'S HURT! I DIDN'T WANT ANYTHING LIKE THIS TO HAPPEN! I ONLY WANTED MONEY... BUT ALL I CAUSED IS MORE HARDSHIP!

CONSCIENCE-STRICKEN, THE ESCAPED CONVICT ELBOWS HIS WAY THROUGH THE CROWD, AND...

WAIT! THIS IS MY FAULT! I SOLD SOME INFORMATION FROM THE ARROW-CAVE! BUT I'M TURNING OVER THE MONEY--AND SOME EVIDENCE... AND I'M GIVING MYSELF UP, TOO!

I'LL KEEP ALL YOUR SECRETS, GREEN ARROW! I'LL NEVER -- WAIT!... THERE'S BANCROFT AGAIN!

NOW'S OUR CHANCE TO GET GREEN ARROW! THE KID'S HURT-- AND WE'VE GOT THE DROP ON HIM!

BUT AS THE TOMMY GUN RATTLES...

NO!... I WON'T LET YOU -- I WON'T... OH-H-H...

BLAST IT! MORGAN RAN INTO THE LINE OF FIRE! LET'S GET OUT OF HERE!

BR-RAT-TAT-TAT

5

THE GREEN ARROW

the Amazing Miss ARROWETTE

GOLLY, G.A.-- THAT GIRL ARCHER STOPPED THOSE TWO CROOKS WITH A GIANT *HAIRPIN* ARROW!

AT THE *NATIONAL ARCHERY CONVENTION,* AN ANNUAL "QUEEN" IS CROWNED...

...AND IT IS THE JUDGES' DECISION THAT YOU BE CROWNED *MISS ARROWETTE,* BONNIE KING!

THANK YOU, SIR! THIS HAS BEEN MY LIFE-LONG AMBITION...

JUDGES

...IN MORE WAYS THAN ONE.' THIS TITLE ALLOWS ME TO FULFILL A DREAM I'VE HAD FOR YEARS!

BEWARE, *GREEN ARROW* AND *SPEEDY!*... FOR THIS YOUNG LADY'S "DREAM" INVOLVES *YOU!*

NEXT DAY, IN THE HOME OF WEALTHY OLIVER QUEEN AND HIS YOUNG WARD, ROY HARPER...

MUST BE A DAYLIGHT CRIME AFOOT, ROY... THAT *ARROW-SIGNAL* IS FOR *US!*

I'LL CONTACT POLICE HEADQUARTERS AFTER I'VE SLIPPED INTO MY COSTUME, OLIVER!

MINUTES LATER, THE SLEEK *ARROWCAR* STREAKS ALONG CITY STREETS, UNTIL IT STOPS SHARPLY, AND...

WE'RE IN TIME, *GREEN ARROW!* THE JEWEL THIEVES ARE JUST MAKING THEIR GETAWAY!

A FEW WELL-PLACED ARROWS SHOULD THWART THEIR PLANS, *SPEEDY!*

BUT SUDDENLY, FATE TAKES A HAND, AS...

EEYOW! A MANHOLE COVER -- AJAR! WHAT A TIME TO LOSE OUR FOOTING!

AND THAT GETAWAY CAR'S ALMOST ON TOP OF US...

YET EVEN AS THE CRIMINALS BEAR DOWN ON THE HELPLESS ARCHERS...

YIPES!... I -- I CAN'T SEE!

POOF

WHAT--? WE'VE GOT A *FRIEND* SOMEWHERE, *SPEEDY!* THAT STRANGE ARROW HAS OBSCURED THE DRIVER'S VISION... THEY'RE GOING TO CRASH!

AND AS THE GANG LEAPS TO SAFETY...

MY *POWDER-PUFF ARROW* KNOCKED OUT THEIR CAR ...NOW TO CAPTURE THEM WITH SOME *HAIRPIN ARROWS!*

CRASH

THUNK

GOLLY, *G.A.!* A *FEMALE ARCHER!*

LET'S GIVE HER A HAND, *SPEEDY!*

2

INSTANTLY, TWO MORE UNIQUE SHAFTS STREAK TO THEIR TARGETS...

OOF!

GOOD OLD *BOXING GLOVE ARROWS...* THEY NEVER FAIL! NOW, LET'S HAVE A TALK WITH THIS LADY ARCHER!

BUT... SH-SHE'S *GONE!* WHO WAS SHE, ANYWAY?

I DON'T KNOW, *SPEEDY...* AND ALTHOUGH I'M GRATEFUL TO THAT BOW-WOMAN, I HOPE WE'VE SEEN THE LAST OF HER! THIS ISN'T A GAME FOR GIRLS!

IT IS THE FOLLOWING DAY WHEN A BURGLAR ALARM SUMMONS THE FAMED ARCHERS INTO ACTION AGAIN...

IT'S *GREEN ARROW* AND HIS KID PAL!

SHOOT!... AND DON'T MISS!

IN THAT SAME INSTANT...

I'LL DISARM THEM, *GREEN ARROW!*

OUR MYSTERY ARCHER AGAIN!

BUT AS HER SHAFT STRIKES...

OH, DEAR! I SHOT MY *LOTION ARROW* BY MISTAKE!

G.A.!... CAN'T KEEP MY FEET!

SPLASH

... AND THEY'RE SLIDING DOWN THE CHUTE, RIGHT INTO THE CRIMINALS' TRUCK!

WHAT A BREAK! LOCK THE REAR DOORS, BOYS... WE'LL GET OUT OF HERE WITH *GREEN ARROW* AND *SPEEDY* AS OUR *PRISONERS!*

3

AND AS THE TRUCK ROARS OFF...

G-GOLLY... BECAUSE OF MY MISTAKE--MY INTERFERENCE--*GREEN ARROW* AND *SPEEDY* ARE IN CRIMINAL HANDS! ONLY ONE THING TO DO...

....MAKE THIS BLUNDER UP TO THEM! AND THE ONLY WAY I CAN DO THAT IS BY *RESCUING* THEM!

MY *HAIR TINT ARROW* WILL LEAVE A TRAIL OF DROPLETS THAT I CAN FOLLOW RIGHT TO THE CROOKS' HIDEOUT!

SPLAT

SHORTLY, AS THE TRUCK SCREECHES TO A STOP AT AN ABANDONED OIL REFINERY...

GIVE US YOUR GEAR-- THEN WE'LL ESCORT YOU TO OUR OWN PRIVATE JAIL...; HA, HA,;

LOOKS LIKE WE MADE A CLEAN GETAWAY... NOBODY'S ON OUR TAIL!

BUT UNSEEN BY THE CRIMINALS...

MY *MIRROR ARROW* IS SOARING OVER THE REFINERY NOW! AH ... I'VE SPOTTED SOMETHING!

THE CROOKS ARE IMPRISONING *GREEN ARROW* AND *SPEEDY* IN THAT HUGE, EMPTY OIL TANK! I'LL PAY A VISIT THERE TONIGHT...

4

LATE THAT NIGHT, A HUSHED VOICE REACHES THE EARS OF THE IMPRISONED ARCHERS...

GREEN ARROW... THIS IS MISS ARROWETTE! I'LL HAVE YOU OUT OF THERE IN NO TIME!

NO!... YOU'RE ENDANGERING YOURSELF! GET AWAY FROM HERE AND NOTIFY THE POLICE...

SUDDENLY...

EEK!.... I'VE LOST MY BALANCE!

OH, NO!... SHE'S FALLING!

BUT EVEN AS SHE TOPPLES, THE AGILE LASS SNAPS ANOTHER ARROW FROM HER SHEATH, AND...

W-WHY... SHE'S DESCENDING WITH A PARACHUTE OF SOME KIND!

CORRECTION, GREEN ARROW! THIS IS MY KERCHIEF ARROW... I KNEW IT WOULD COME IN HANDY SOME DAY!

UPON LANDING, HOWEVER...

I'M S-SO FRIGHTENED! I NEVER KNEW FIGHTING CRIMINALS COULD BE SO DANGEROUS!

I SURE WISH YOU'D THOUGHT OF THAT BEFORE, MISS ARROWETTE! NOW WE'VE GOT TO WORK OUT AN ESCAPE FOR THE THREE OF US!

HMM...

DON'T BE TOO ANGRY, SPEEDY... HER ARROWS MIGHT MAKE THAT POSSIBLE! CAN YOU EXPLAIN EACH ONE TO US, MISS?

OH, I'D BE GLAD TO, GREEN ARROW!

SOME MINUTES LATER, A BIZARRE SHAFT SOARS FROM THE TANK...

YOUR NEEDLE-AND-THREAD ARROW MAY DO THE TRICK, MISS! ARE YOU SURE IT'S STRONG ENOUGH TO SUPPORT OUR WEIGHT?

THE THREAD IS MADE OF THE STRONGEST WIRE I COULD BUY... AND THE STEEL NEEDLE WILL DIG ITSELF FIRMLY INTO THE GROUND!

5

THUS, ONE BY ONE, THE THREE CAPTIVES CLIMB FROM THEIR PRISON...

NOW TO GET YOU TO SAFETY, YOUNG LADY! *SPEEDY* AND I WILL DEAL WITH THESE CRIMINALS AFTER--

HOLD IT, *G.A.!* WE'VE GOT COMPANY!

OH!... THE CRIMINAL GANG! I'M AFRAID WE'RE FINISHED!

NOT YET, *MISS ARROWETTE!* THIS LOOKS LIKE A SPOT FOR ANOTHER OF YOUR SPECIAL SHAFTS!

POW

POW

WHICH ONE DID YOU FIRE, *G.A.*?

HER *HAIR NET ARROW!* LET'S HOPE IT WORKS THE WAY SHE DESCRIBED IT!

THE SOFT SHAFT STRIKES THE LEAD CRIMINAL, AND...

HUH--?

PERFECT!... IT BILLOWED OPEN INTO A HUGE HAIR NET! THEY'RE ALL TANGLED UP... LET'S GET 'EM!

AND SO, AFTER TAKING THE CROOKS IN TOW...

THANKS TO YOU, *MISS ARROWETTE,* WE CAME OUT OF THIS ALIVE! STILL, I CAN'T HELP THINKING...

I KNOW... CRIME-FIGHTING IS NOT FOR GIRLS LIKE ME!

WELL--YOU'RE RIGHT! FROM NOW ON, I'LL CONFINE MY BOWMANSHIP TO THE ARCHERY RANGE!

A WISE DECISION... AND YET, I CAN'T HELP WONDERING, *SPEEDY,* IF WE'VE REALLY SEEN THE *LAST* OF THIS GIRL ARCHER!

THE END

THE GREEN ARROW

FROM OUT OF THE COSMOS COMES A CRIMINAL WITH POWERS OF SUCH MAGNITUDE THAT THE BATTLING BOWMEN FIND THEMSELVES HELPLESS... UNTIL THEY ARE AIDED BY, AND AID, IN TURN...

GREEN ARROW'S ALIEN ALLY

W-WE C-CAN HARDLY M-MOVE! THE L-LIGHT IS CHANGING US TO ST-STATUES!

HURRY, *GREEN ARROW!* FIRE ME! YOU MUST FIRE ME-- IT IS YOUR ONLY HOPE!

AS TWO FIGURES--*GREEN ARROW* AND *SPEEDY*--KEEP VIGIL OVER A SLUMBERING CITY...

G.A.! LOOK DOWN THERE--AT THE JEWELRY STORE!

IT--IT'S ASTONISHING! THE DOORS HAVE BEEN *MELTED*, G.A.! AND THERE'S A MAN COMING OUT!

SSSSSS

IN AN INSTANT, THE BATTLING BOWMEN ARE SWINGING INTO ACTION ON THEIR ARROW-LINES...

HE'S SURE NOT IN A HURRY TO GET AWAY!

WE'LL FIND OUT IN A MOMENT WHAT HE'S GOT TO BE SO CONFIDENT ABOUT!

LANDING LITHELY, THE ARCHERS READY THEIR BOWS AS THE FIGURE TURNS...

??!!

GLANCING AT THE PAIR WHO ARE RASH ENOUGH TO STOP HIM, HE GESTURES AT AN OVERHANGING SIGN, AND...

LOOK OUT!

GREAT GUNS! DID YOU SEE WHAT HE DID TO MAKE IT COME CRASHING DOWN?

CR-RASH

INSTINCTIVELY, **GREEN ARROW** AND **SPEEDY** BRING THEIR FABULOUS BOWS INTO ACTION, FIRING A SCORE OF ARROWS...

WE'LL PIN HIS CLOTHES TO THE WALL TO HOLD HIM UNTIL THE POLICE ARRIVE!

ZIP ZIP

INDEED, NO SOONER IS THE JOB COMPLETED THAN...

G.A.! WHAT'S HAPPENING?

HE'S--HE'S *DISAPPEARING!*

2

MOMENTARILY STYMIED BY THE INCREDIBLE ESCAPE, THE FABLED BOWMEN MAKE THEIR WAY TO THE SECRET *ARROWCAVE*, WHERE A CHANGE OF COSTUME RETURNS THEM TO THEIR OTHER IDENTITIES -- OLIVER QUEEN AND YOUNG ROY HARPER...

HAVE YOU ANY EXPLANATION FOR WHAT HAPPENED, OLIVER?

NOT THE FAINTEST, ROY! I'M STUMPED!

YES, EARTHLINGS! *ANKOV'S* POWERS ARE FAR BEYOND YOUR COMPREHENSION!

SOMEBODY FOUND THE *ARROW-CAVE!*

HUH?

HERE I AM--SEE? I AM *VAN-JON*, FROM *THE PLANET OF TWO SUNS*, SEEKING ANKOV!

GREAT GUNS!

Y-YOU'RE FROM ANOTHER WORLD? HOW COME YOU SPEAK OUR LANGUAGE AND --

MORE THAN THAT, I KNOW YOUR SECRET IDENTITIES! THAT'S WHY I CAME HERE -- TO HELP YOU! YOU, IN TURN, CAN HELP ME!

I WAS ABLE TO DECIPHER YOUR LANGUAGE ON MY WAY HERE, WHEN I HEARD YOUR RADIO BROADCASTS! ALSO, I WAS ABLE TO OBSERVE MANY THINGS--ESPECIALLY YOUR FIGHT AGAINST CRIME!

THIS ANKOV YOU MENTIONED-- WHO IS *HE?*

3

ANKOV IS A NOTORIOUS CRIMINAL, FROM THE ASTEROID OF GIANTS! I AM A LAWMAN, AND I HAD CAPTURED HIM AND WAS TAKING HIM TO PRISON, WHEN HE ESCAPED!

BUT WITH ALL HIS POWERS, WHY SHOULD HE HAVE TO ROB A JEWELRY STORE?

FOR *DIAMONDS!* HE CAN MOLD THE SMALL GEMS INTO ONE GREAT STONE, AND GAIN EVEN GREATER POWERS BY REFLECTING STARLIGHT THROUGH IT!

BY THE WAY, HOW COULD A FELLOW AS TINY AS YOU CAPTURE SUCH A *LARGE* ONE?

I HAVE SOME POWERS OF MY OWN-- AS YOU SHALL SEE! BUT HURRY NOW, WE MUST GO AFTER ANKOV-- BEFORE IT IS TOO LATE!

OKAY! WE'LL CHANGE BACK TO OUR COSTUMES, AND BE RIGHT WITH YOU!

SHORTLY, AS THE *ARROWCAR* PURRS THROUGH DARK, SILENT STREETS...

STRAIGHT AHEAD, THEN TURN RIGHT AT THE NEXT CORNER! I'VE SENT OUT MENTAL WAVELENGTHS THAT HAVE LOCATED HIM AT HIS HIDEOUT! NOW, HERE'S MY PLAN...LISTEN CAREFULLY...

BUT AS THE VEHICLE CAREENS AROUND A CORNER...

HA, HA, HA, HA! WHERE ARE YOU, VAN-JON? I KNOW YOU COME FOR ME!

THAT *LIGHT!* IT'S *BLINDING!*

SCREEECH!

4

CAN'T SEE... CAN'T SEE!

DON'T WORRY ABOUT IT, *GREEN ARROW!* I ANTICIPATED SOME SUCH TRICK BY ANKOV! NOW, DO WHAT I TOLD YOU WHEN I OUTLINED MY PLAN...

THE *LIGHT...* IT'S "FREEZING" ME! I'M TURNING STIFF-- LIKE A STATUE!

YES, AND ALREADY YOUR FRIEND HAS BEEN "FROZEN"-- THE COMBINATION OF THE STARLIGHT AND A FORCE RAY SET UP BY ANKOV'S EYES! *HURRY! FIRE ME INTO THE AIR!*

THE BOWSTRING TWANGS... A SPLIT SECOND BEFORE THE ARCHER IS RENDERED IMMOBILE...

YOUR EARTHLING FRIENDS CAN NO LONGER HELP YOU, VAN-JON! WHERE ARE YOU? YOU'D BETTER TELL ME -- AND SURRENDER!

EH? WHAT IS CAUSING THE ARROW TO SPEED THIS WAY?

I AM, ANKOV!

KWHAMMA

VAN-JON! NO!

I HAD TO GET CLOSE TO YOU... BUT AVOID BEING STUNNED BY THE DIAMOND-LIGHT! AND NOW IT'S MY TURN TO TAKE THE OFFENSIVE!

ZIP!

ZIP!

ZIP!

5

THE GREEN ARROW

WHAT CHANCE HAS A SMALL BAND OF MEN, ARMED ONLY WITH BOWS AND ARROWS, AGAINST MODERN WEAPONS OF WAR? ONLY *GREEN ARROW* AND *SPEEDY,* ACE ARCHERS OF LAW AND ORDER, CAN ANSWER THIS QUESTION, AS THEY TRY TO TURN THE HAPLESS FORCE OF BOWMEN INTO...

THE MIGHTY ARROW ARMY

GOLLY-- OUR FRIENDS' ARROWS ARE BOUNCING OFF THAT TANK LIKE MATCHSTICKS, *GREEN ARROW!*

OUR *TRICK* ARROWS ARE THEIR ONLY HOPE, *SPEEDY*... BUT--WILL EVEN *THAT* BE ENOUGH?

HIGH OVER SOUTH AMERICA, A SLEEK PLANE CARRIES WEALTHY OLIVER QUEEN AND HIS YOUNG WARD, ROY HARPER, ON...

A 'ROUND-THE-WORLD TOUR! SOME VACATION, EH, OLIVER?

OH, OH...SOUNDS LIKE IT MAY BE INTERRUPTED, ROY!

...SPUT-SPUTTER

ENGINE TROUBLE! WHAT A TIME FOR THIS TO HAPPEN!

WE'LL LAND IN THAT CLEARING BELOW! BETTER CHANGE INTO OUR COSTUMES ... WE DON'T WANT TO GIVE AWAY OUR SECRET IDENTITIES, IF ANY-ONE'S DOWN THERE!

INDEED, SHORTLY AFTERWARD, AS THEY EMERGE IN THEIR COLORFUL GARB...

JEEPERS! A RECEPTION COMMITTEE!

AND HOSTILE!

ZOING!

BUT AS THEY MOVE TO FIGHT BACK...

WAIT! THESE TWO ARE NOT ENEMY SPIES... THEY ARE THE FAMOUS CRIME-FIGHTING AMERICANS -- GREEN ARROW AND SPEEDY!

WHEW! GLAD YOU RECOGNIZED US! NOW, WHAT'S THIS ALL ABOUT?

WAR, MY FRIEND! DICTATOR BRACATO IS ABOUT TO INVADE OUR PEACEFUL KINGDOM, WITH HIS SMALL BUT MODERN ARMY!

A DICTATOR, EH? BUT HOW DO YOU HOPE TO MATCH ARROWS AGAINST ARMOR AND RIFLES?

WE HAVE NO ALTERNATIVE... MODERN WEAPONS HAVE LONG SINCE BEEN BANNED BY OUR FRIENDLY PEOPLE!

THAT DOESN'T GIVE YOU MUCH HOPE! BUT MAYBE, IF SPEEDY AND I GAVE YOU AN ASSIST...

AT THAT MOMENT...

TO ARMS! ONE OF BRACATO'S PATROLS HAS CROSSED OUR BORDER!

THE INVASION'S STARTED! LET'S GET UP THERE AND LOOK THE SITUATION OVER!

BLAM POW POW

SHORTLY, ATOP A HILL...

GOLLY! THE DICTATOR'S RIFLEMEN HAVE THE ARCHERY SOLDIERS PINNED DOWN, G.A.!

OF COURSE...ARROWS ARE NO MATCH AGAINST BULLETS! BUT PERHAPS OUR CRIME-FIGHTING SHAFTS CAN STOP THE ATTACK!

2

NEXT INSTANT, AN AMAZING ARROW STREAKS OVER THE BATTLEFIELD...

WHY... THE ENEMY CANNOT SEE TO TAKE AIM! WHAT IS HAPPENING?

OUR *REFLECTOR ARROW* EJECTS MINUTE PARTICLES THAT REFLECT THE SUN'S RAYS AND FORM A BLANKET OF BRILLIANCE, SIR!

IT DISTORTS THEIR AIM!

RIGHT... AND NOW THAT WE'VE GOT THOSE TROOPS DAZZLED, LET'S SEE IF WE CAN'T MAKE THEM RETREAT!

AN AWESOME SOUND RENDS THE AIR, AS A FLIGHT OF SHAFTS FILLS THE SKY...

OUR *SUPERSONIC ARROWS!* THEY'RE DESIGNED TO TRAVEL AT ENORMOUS SPEED!

AND AS THEY BREAK THROUGH THE SONIC BARRIER, THE NOISE THEY MAKE SOUNDS VERY MUCH LIKE A SHELL EXPLODING!

BOOOM

KA-BOOM

AMAZING... BRACATO'S SOLDIERS WERE FOOLED INTO BELIEVING THEY WERE UNDER AN ARTILLERY BARRAGE! MY ARCHERS ARE CAPTURING THEM!

YES... BUT IF I KNOW DICTATORS, IT WILL TAKE MORE THAN THAT TO DAMPEN BRACATO'S DREAM OF CONQUEST! LET'S LOOK IN ON HIS ARMY...

AS THE *EMERALD ARCHER'S* UNIQUE *TV ARROW* SCANS THE ENEMY TERRITORY...

THE CAMERA'S PICKED UP A TANK MOVING THIS WAY, *G.A.*

WE'D BETTER TRY TO INTERCEPT IT... THAT IRON HULK COULD BLAST THE ARCHERY FORCE TO PIECES!

3

MINUTES LATER, ALONG THE TANK'S ROUTE...

BUT, *GREEN ARROW,* CAN EVEN YOUR SPECIAL SHAFTS STOP A TANK?

WE'LL SOON FIND OUT... WE'VE BEEN SPOTTED!

KA-BLAM

AS CANNON SHELLS ZERO IN ON THE TRIO...

THEY'LL HAVE US ON TARGET IN A SECOND... TAKE *COVER!*

OUR FIRST MOVE IS TO DRIVE THE CREW OUT OF THE TANK-- WITH THIS *SMOKE SCREEN ARROW!*

BOOM

BOOM

THE SMOKING SHAFT PLUMMETS THROUGH A TANK SLOT--AND MOMENTS LATER...

THERE GOES THE CREW... NOW LET'S FINISH OFF THE TANK!

I'LL ADMINISTER THE COUP DE GRACE...

JUST AS WE FIGURED... THE *FLARE ARROW* IGNITED THE TANK'S AMMO SUPPLY!

AND MY MEN ARE CAPTURING THE ESCAPING TANK CREW!

BOOM

YOU HAVE MADE US LOOK LIKE AN INVINCIBLE ARMY, *GREEN ARROW!* HOW CAN WE EVER THANK YOU?

BETTER SAVE THAT--UNTIL WE'RE *SURE* BRACATO IS BEATEN! AS I SAID BEFORE-- DICTATORS ARE A STUBBORN PACK!

INDEED, JUST ACROSS THE BORDER...

TWO AMERICAN ARCHERS MAKE FOOLS OF ONE OF MY PATROLS--DESTROY MY ASSAULT TANK! YOU ARE NOT OFFICERS... YOU ARE *IDIOTS!*

BUT, BRACATO-- THEY USED FANTASTIC ARROWS AGAINST US! WE WERE TAKEN BY SURPRISE!

4

VERY WELL -- THEN *I* SHALL LEAD THE NEXT ASSAULT ON LUANIA! I *DEFY* THESE BOLD YANKEE ARCHERS TO BEST BRACATO THIS TIME!

THAT NIGHT, FIERY FLASHES OF LIGHT DOT THE RIVER IN LUANIA, FOLLOWED BY...

BRACATO'S DESTROYER! HE SAILED DOWNRIVER TO LAY SIEGE TO OUR CAPITAL!

THIS WON'T BE EASY -- BUT WE'VE GOT TO SILENCE THOSE GUNS! C'MON... LET'S MAKE SOME FAST PLANS!

BOOM KA-BOOM

SOON, AN *ARROWLINE* ZOOMS OVER THE RIVER...

...AND WRAPS ITSELF AROUND THE DESTROYER'S GUARD RAIL!

THEN, SILENTLY, TWO FIGURES SLIP DOWN AN IMPROVISED BREECHES BUOY...

ONCE ABOARD, WE'VE GOT TO MAKE SURE THIS LINE ISN'T BROKEN!

SURE THING, *G.A.*!

BUT NO SOONER DO THE FAMED ARCHERS LAND AND RE-STRING THEIR BOWS THAN...

THE AMERICAN BOWMEN! *ALARM! ALARM!*

NOW WE'RE IN FOR IT! START DRAWING ARROWS, *SPEEDY!*

OUR *BOXING GLOVE ARROWS* TOOK CARE OF THE GUARDS... AND HERE COME OUR ARCHER FRIENDS DOWN THE *ARROWLINE!*

BUT IT'LL TAKE TIME FOR ALL OF THEM TO BOARD! MEANWHILE, WE'VE GOT PROBLEMS... *LOOK!*

5

THE GREEN ARROW

HE WAS THE STRANGEST FOE EVER TO CHALLENGE *GREEN ARROW* AND HIS YOUTHFUL PARTNER, *SPEEDY*... A CUNNING JUNGLE CREATURE, CAPABLE OF MATCHING THEM SHAFT FOR SHAFT! AND SOMEHOW, THE BATTLING BOWMEN HAD TO DEVISE A WAY TO BEAT...

The APE ARCHER

BONZO! STOP FIRING AT US! WE'RE YOUR OLD PALS-- *GREEN ARROW* AND *SPEEDY!* WE'VE COME TO HELP YOU!

ZIP!

ZIP!

TWANG- TWANG-TWANG!

AMONG THE SPECTATORS AT THE CIRCUS ONE NIGHT ARE OLIVER QUEEN AND YOUNG ROY HARPER...

LOOK AT BONZO, OLIVER! REMEMBER WHEN HE TURNED CRIMINAL?

I SURE DO, ROY-- HE GAVE US QUITE A TIME! I DON'T THINK EITHER OF US WILL FORGET *THAT* SCRAPE...

"THAT CROOKED TRAINER OF HIS, BART ROCKLAND, REALLY STARTLED US THE FIRST TIME WE MET..."

BONZO--THERE ARE THOSE TWO ARCHERS I WARNED YOU MAY SHOW UP...*GREEN ARROW* AND *SPEEDY!* SHOW 'EM YOUR STUFF!

G.A.--DO YOU SEE WHAT THAT CROOK HAS FOR AN ACCOMPLICE? AN APE-- APING *YOU!*

JEWELERS

"THEN..."

GREAT GUNS, *SPEEDY!* *WATCH IT!* HE FIRES ARROWS LIKE A MACHINE!

ZIP ZIP ZIP

THWUNK
THWUNK

"ONLY OUR CHLOROFORM ARROW COULD SUBDUE HIM..."

W-WHAT'S HAPPENIN' TO BONZO? HE'S GOIN' TO SLEEP!

S-S-S-S-SSSS-SS

ZIP

"AFTER CAPTURING ROCKLAND, WE GOT THE AMAZING STORY..."

...SO AFTER I TRAINED BONZO FOR THE CIRCUS, I TAUGHT HIM TO BE AN ARCHER IN CASE WE MET UP WITH YOU!

WELL, IT ALMOST WORKED! YOUR PET WAS AMAZING!

LISTEN TO THE APPLAUSE BONZO IS GETTING!

YES! HE SURE HAS BECOME A WONDERFUL PERFORMER SINCE HIS NEW TRAINER, LANCE, TOOK OVER AFTER ROCKLAND WAS JAILED! LANCE HAS DONE WONDERS FOR HIM!

TIGER! THERE'S A TIGER LOOSE!

RUN FOR YOUR LIVES!

I TURNED THE BEAST LOOSE TO ATTRACT EVERYONE'S ATTENTION WHILE I CARRY OUT MY PLAN!

ENTRANCE MAIN AR

2

IN AN INSTANT, OLIVER AND ROY DART INTO A DARK PASSAGEWAY, WHERE THEY SWITCH TO THEIR SECRET IDENTITIES OF *GREEN ARROW* AND *SPEEDY*... THEN, THEIR ARROW-LINES SEND THEM ARCHING THROUGH THE AIR TO COMBAT THE MENACING MARAUDER...

WE'VE GOT HIM CONCENTRATING ON US -- INSTEAD OF THE AUDIENCE!

BUT *NOW* WHAT'LL WE DO, G.A.? HE'S READY TO SPRING!

GRRRR

SHOOT YOUR NET ARROW, LAD!

GROW-W

EVEN AS THE SHADOW OF THE CATAPULTING BIG CAT FALLS OVER THEM, TWO BOWS TWANG SIMULTANEOUSLY...

TWANG

TWANG

IT'D BETTER WORK-- HE'S CLOSE... *TOO* CLOSE

THE NET ARROWS BLOSSOM OPEN JUST IN TIME, NETTING THE SNARLING, CLAWING FURY...

HE'S PRETTY MAD ABOUT THIS -- BUT HE ISN'T DANGEROUS ANY MORE! THE ANIMAL HANDLERS CAN TAKE OVER NOW!

GR-OOO-WW-R

3

SAY--WHERE'S BONZO? WHAT HAPPENED TO HIM?

I DUNNO! HE *WAS* HERE, AND SUDDENLY HE'S GONE! HE MUST'VE *ESCAPED!*

YES, OUT IN THE NIGHT, THE CLEVER SIMIAN PROWLS AMONG THE ROOFTOPS...

...UNTIL HE REACHES HIS DESTINATION! HE LEAPS FROM THE OPEN WINDOW AND...

EEK! THAT BEAST--! IT GRABBED MY DIAMOND NECKLACE!

MINE, TOO! AND MY BROOCH! *HELP!*

A HURRIED PHONE CALL TO THE POLICE SOON SENDS A GIANT ARROW FLAMING OVER THE CITY...

...SIGNALING THE BATTLING BOWMEN TO SPEED TO THE SCENE IN THEIR *ARROW-CAR...*

UP THERE, G.A..! IT'S *BONZO!*

WE'LL USE THE CATAPULT, *SPEEDY!* GET SET!

INSTANTLY, THE ARCHERS ARE HURLED SKYWARD...

BONZO! IT'S US-- YOUR OLD PALS! WE'VE COME TO HELP YOU DOWN!

4

HE'S OPENED FIRE ON US, *SPEEDY!* QUICKLY-- RELEASE YOUR PARACHUTE ARROW!

TWANG

TWANG

ZIP

IN A MOMENT, THEIR ASCENT IS ABRUPTLY HALTED, AS...

BONZO'S GOT THE BEST OF US! WE'LL HAVE TO MAKE A TEMPORARY RETREAT!

LANDING, THEY RETURN A BLISTERING COUNTERFIRE...

BANG-BANG

VOOM VOOM

OUR SMOKE AND FIRECRACKER ARROWS, *SPEEDY,* WILL CONFUSE BONZO--THEN WE'LL GO UP AND GET HIM!

BUT *A BALLOON ARROW,* FIRED BY THE BOLD BEAST, INFLATES AND CARRIES HIM OFF IN THE NIGHT WIND...

G.A.! LOOK!

SHORTLY AFTER BONZO FADES FROM VIEW...

I'M LANCE, BONZO'S TRAINER! DON'T HARM HIM, PLEASE! HE DOESN'T KNOW WHAT HE'S DOING!

DON'T WORRY-- THE CLOSEST WE'VE BEEN ABLE TO GET TO HIM IS HIS PAW-PRINTS!

NEXT NIGHT, IN THE CIRCUS' GLOOMY BASEMENT CORRIDORS...

HOLD IT-- RIGHT THERE! DON'T MAKE A MOVE!

5

JUSTICE LEAGUE of AMERICA

HEREBY ELECTS
GREEN ARROW

TO MEMBERSHIP FOR LIFE ~~ WITH ALL PRIVILEGES AND GRATUITIES INCLUDING THE WEARING OF THE SIGNAL DEVICE AND POSSESSION OF THE GOLDEN KEY WHICH PERMITS ENTRY INTO THE SECRET SANCTUARY, ITS LIBRARY AND SOUVENIR ROOMS. IT IS HEREBY FURTHER RESOLVED AND ACTED UPON, THAT ...

GREEN ARROW

SHALL RECEIVE A SPECIAL COMMENDATION FOR HIS EXPERT ASSISTANCE IN THE CASE WE HAVE ENTITLED ON OUR SCROLLS ...

Doom of the STAR DIAMOND!

WELCOME TO THE *JUSTICE LEAGUE*, GREEN ARROW!

SNAP!

The *Roll Call*

1. WONDER WOMAN
2. GREEN LANTERN
3. FLASH
4. J'ONN J'ONZZ
5. BATMAN
6. AQUAMAN
7. SNAPPER CARR
8. SUPERMAN

AND INCLUDING FOR THE FIRST TIME... GREEN ARROW

SEVERAL TRILLIONS OF MILES FROM EARTH, AN ALIEN SPACE-SHIP LURKS WITHIN THE SHELTER OF A BLACK, NEBULOUS CLOUD...

CARTHAN IS APPROACHING! READY STARBOARD DISINTOR-BEAMERS!

IN THE NEXT MOMENT, THE ONCOMING SPACESHIP HURTLES WITHIN FIRING RANGE...

INSIDE THE ATTACKING VESSEL CRIES OF UTTER AMAZEMENT RING OUT...

INCREDIBLE! CARTHAN'S SHIP'S BEEN DISINTEGRATED-- BUT CARTHAN HIMSELF IS UNHARMED! FIRE PORT BEAMERS AT HIM!

DISSOLUTION RAYS OF AWE-SOME POWER BATHE THE HELPLESSLY DRIFTING SPACE-MAN...

CARTHAN HAS BECOME INDE-STRUCTIBLE! THOSE RAYS WOULD TURN SOLID STEEL TO POWDER! BUT THEY DON'T HARM HIM! GET HIM INSIDE HERE WITH A SNATCH-BEAM!

SHORTLY, CARTHAN-- WARLORD OF THE SPACE-FLEETS OF THE PLANET DRYANNA-- STANDS A PRISONER OF HIS PLANETARY RULER...

XANDOR! I--I DON'T UNDER-STAND! WHY TRY TO DESTROY ME? I'VE JUST CONQUERED OUR ANCIENT ENEMIES THE SLYSSA FOR OUR PEOPLE!

EXACTLY! YOU'VE BE-COME A-- GREAT HERO!

MOST PEOPLE ON DRYANNA SECRETLY CALL ME A DICTATOR! YOUR FRIENDS--CERTAIN SCIENTISTS--HOPE TO OVERTHROW ME AND ESTABLISH A GOVERNMENT OF THE PEOPLE--AND WANT YOU TO LEAD THE REVOLT!

2

OBVIOUSLY I CANNOT LET YOU GO HOME TO *DRYANNA!* JUST AS OBVIOUSLY I CANNOT DESTROY YOU! ALL I CAN DO IS IMPRISON YOU! BUT BEFORE I DECIDE WHERE-- TELL ME! WHAT HAPPENED TO MAKE YOU INDESTRUCTIBLE?

IT OCCURRED WHILE I WAS ON A LONE SCOUTING MISSION AGAINST THE *SLYSSA...*

"*WHILE* I WAS INVESTIGATING A BARREN PLANET TO USE AS A POSSIBLE SUPPLY BASE AGAINST THE ALIENS, I WAS CAUGHT IN AN ELECTRO-MAGNETIC DISTURBANCE.."

I'VE NEVER ENCOUNTERED ANYTHING LIKE *THIS* BEFORE!

"*UPWARD* FROM THE HIGHLY MAGNETIC PLANET JETTED SPIRALS OF RAW ENERGY, TRAPPING ME BETWEEN THEM AND BATHING ME FOR AN HOUR IN UNIMAGINABLE RADIATIONS..."

WH-WHAT'S HAPPENING TO ME...?

AS *CARTHAN* CONCLUDES HIS TALE...

AS A RESULT OF THAT FREAK ACCIDENT, AN *AURA*--INVISIBLE IN NORMAL LIGHT--HAS SURROUNDED MY BODY... PROTECTING ME FROM DESTRUCTION! NOW TELL ME, WHAT DO YOU INTEND DOING WITH ME?

AFTER CONFERRING WITH HIS ADVISORS, *XANDOR* REACHES A DECISION...

I CAN'T IMPRISON YOU IN A JAIL OR ON A PLANET NEAR OUR HOME PLANET, FOR YOU COULD TELEPATH TO YOUR SCIENTIST FRIENDS FOR HELP! I HAVE ANOTHER PRISON UNDER CONSIDERATION, HOWEVER--A PLANET CALLED *EARTH!*

3

YOU'RE A VERY IMPORTANT MAN, SO WE'LL TREAT YOU WELL! I'M GOING TO GIVE YOU A FINE SPACESHIP--WHICH WILL BRING YOU ON AUTOMATIC CONTROLS TO EARTH! HOWEVER--

WE ARE *TELEPORTING* THREE GOLDEN BOX-MACHINES AHEAD OF YOU, DESIGNED TO KEEP YOU ON EARTH PERMANENTLY! SHOULD YOU DECIDE TO LEAVE EARTH, THESE MACHINES WILL CAUSE YOUR AURA TO BE COATED WITH *OPAQALUX*--BLINDING YOU FOREVER!

BLIND, YOU COULD NEVER FIND YOUR WAY BACK TO *DRYANNA*, OR LEAD A REVOLT AGAINST MY DICTATORSHIP! EARTH IS A DAWN ATOMIC ERA PLANET, WITHOUT SPACE-TRAVEL! SO YOU'LL GET NO HELP THERE!

IF YOU *REMOVE* THE METALLIC COVERINGS OF THE BOX-- MACHINES, YOU WILL BE ABLE TO ESCAPE FROM EARTH WITH YOUR EYESIGHT INTACT! *HOW-EVER*--THE INSTANT YOU RE-MOVE THE COVERINGS, YOU ACTIVATE THE MACHINES! ONCE TURNED ON...YOU'LL BE UNABLE TO TURN THE MACHINES OFF... AND THEY'LL DESTROY ALL HUMAN LIFE ON EARTH!

YOU'RE A HUMANITARIAN, *CARTHAN*! YOU WOULDN'T DELIBERATELY HARM THE BILLIONS OF PEOPLE ON EARTH-- EVEN TO SAVE YOURSELF! THAT'S WHY I'M SO CONFIDENT EARTH WILL BE YOUR HOME THE REMAINDER OF YOUR LIFE...

ON A WORLD TOO FAR AWAY FROM *DRYANNA* FOR *CARTHAN* TO TELEPATH FOR HELP, A SPACESHIP IS OUTFITTED FOR HIS JOURNEY TO EARTH..

THERE'S YOUR SPACESHIP, *CARTHAN*! IT WILL TAKE YOU AT SUPER-LIGHT SPEED TO EARTH WHETHER YOU WANT IT TO OR NOT! AFTER THAT-- YOU WILL BE YOUR OWN JAILOR! YOU WON'T DARE DOOM ANYONE ON EARTH-- SO YOU'LL LIVE OUT YOUR DAYS IN EXILE!

4

THROUGH HYPER-DIMENSIONAL GULFS, *CARTHAN* TRAVELS TOWARD EARTH AT MULTI—LIGHT SPEED...

AT LEAST I'LL BE ABLE TO STUDY THE CULTURE AND SOCIAL HABITS OF EARTH'S INHABITANTS WITH THE SUPER-INSTRUMENTS *XANDOR* GAVE ME!

AS HIS HIGHLY SENSITIVE COMPUTERS PICK UP REAMS OF INFORMATION...

WHY, THIS ISN'T AS BAD AS I FEARED! EARTH HAS A GROUP OF AMAZINGLY-ENDOWED HUMANS-- BANDED TOGETHER AND CALLING THEMSELVES THE *JUSTICE LEAGUE OF AMERICA*! ALL I NEED DO IS ASK THEIR HELP!

BUT AS HIS SPACESHIP, PROTECTED FROM DISCOVERY BY AN INVISIBILITY AND ANTI-RADAR BEAM, ENTERS EARTH'S ATMOSPHERE...

I *CAN'T* ASK FOR THE *JUSTICE LEAGUE'S* HELP! SOMETHING ABOUT MY *AURA* IS--*PREVENTING IT!* I WANT TO PLEAD FOR ASSISTANCE -- BUT I'M UNABLE TO DO SO!

WAIT! PERHAPS ALL ISN'T LOST! SUPPOSE I WERE TO PRETEND TO BE--*EVIL?* SUPPOSE I *DO* ACTIVATE THE MACHINES OF DESTRUCTION! SURELY THE *JUSTICE LEAGUE* WOULD FIND A WAY TO AVERT THE DOOMS! ONCE THAT WAS DONE, I COULD LEAVE EARTH WITH A CLEAR CONSCIENCE!

AT THIS MOMENT, IN THE HEADQUARTERS OF THE *JUSTICE LEAGUE OF AMERICA*, ITS MEMBERS ARE CONDUCTING A REGULARLY SCHEDULED MEETING--WITH *WONDER WOMAN* AS ROTATING CHAIRMAN...

WE'RE HERE TO DISCUSS THE ADMISSION OF NEW MEMBERS! DO YOU HAVE ANY SUGGESTIONS?

REMEMBER--ACCORDING TO OUR CONSTITUTION AND BY-LAWS--WE CAN ADMIT ONLY *ONE* NEW MEMBER AT A TIME!

HOW ABOUT *ADAM STRANGE?* HE'S ACHIEVED AN EXCELLENT RECORD!

YES, BUT *GREEN ARROW* HAS BEEN DOING FINE WORK FOR A LONG TIME!

HOW ABOUT THAT NEWCOMER IN *MIDWAY CITY?* HE'S KNOWN AS *HAWKMAN*...

AFTER AN HOUR OF FRIENDLY ARGUMENT...

MADAME CHAIRLADY, I MOVE THAT *GREEN ARROW* BE ELECTED UNANIMOUSLY!

I SECOND THE MOTION!

BUT BEFORE THE *AMAZON PRINCESS* CAN CALL FOR A VOTE...

ALL IN FAVOR OF *GREEN ARROW*... OHHH!

LIKE CRAZY, MAN! SPEAK-ING OF ARROWS.. WHERE'D *THAT* COME FROM?

IT APPEARED OUT OF THIN AIR!

IMPULSIVELY *BATMAN* LEANS FORWARD AND GRASPS THE SLENDER METAL SHAFT, AND AS HE DOES SO IT STARTS TO VIBRATE SOUND WAVES...

GREETINGS, *JUSTICE LEAGUE* MEMBERS! I HAVE NEWS FOR YOU! YOUR PROSPECTIVE NEW MEMBER-- *GREEN ARROW*--IS MY PRISONER!

WHO IN THE WORLD--?

HOW DID HE LEARN ABOUT US AND OUR SECRET MEETING PLACE?

MY NAME IS *CARTHAN*! I COME FROM A PLANET MANY LIGHT YEARS FROM EARTH, TO ADD YOUR PLANET TO THE OTHERS I HAVE CONQUERED!

EVEN AS I TELEPATH MY THOUGHTS TO YOU THROUGH THE COMMUNICATOR-ARROW I FASHIONED--THREE ENGINES OF DOOM ARE GOING INTO OPERATION AT THREE DISTANTLY SEPARATED LOCALITIES ON EARTH! YOUR OBVIOUS ASSIGNMENT IS TO SMASH THEM--IF YOU CAN!

6

"EVEN AT THIS INSTANT, MY ULTRA-WEAPONS ARE REMOVING THE METALLIC COVERINGS WHICH WILL SET THE DESTRUCTIVE ENGINES IN OPERATION.."

WHHIRRRRRRRRR!

"AS THESE ENGINES BEGIN FUNCTIONING, THEIR **ALPHA--OMICRON**-- AND **XI** RAYS WILL SEND A TRIO OF DISASTERS ACROSS THE EARTH! ONE MACHINE IS OUTSIDE **KEYSTONE CITY**--ANOTHER IN THE **PACIFIC OCEAN**, NORTHEAST OF AUSTRALIA-- THE THIRD AND LAST NEAR **ROME, ITALY**!"

WHEN THE VIBRATORY VOICE FADES AWAY...

BATMAN, THE TWO OF US MUST FIND THIS SPACE-VISITOR **CARTHAN**--AND FREE **GREEN ARROW**!

RIGHT! **J'ONN J'ONZZ**--I HAVEN'T BEEN TEAMED WITH YOU SINCE WE BATTLED **STARRO THE CONQUEROR**! LET'S HEAD FOR **KEYSTONE CITY**!

WE'LL HANDLE THE AUSTRALIAN SECTOR, **AQUAMAN**!

OKAY, **FLASH**!

THAT LEAVES ME TO TAKE CARE OF THE "ROMAN-MENACE"!

THE MEETING ROOM IS EMPTY, SAVE FOR THE WORRIED **SNAPPER CARR**! AS TIME SPEEDS BY, THE HONORARY JLA MEMBER GROWS MORE AND MORE RESTLESS, UNTIL...

I'M BLASTING OFF! I'M COMING UNGLUED FROM SUSPENSE! MAYBE BEATING FEET WILL CALM ME DOWN!

7

DOOM OF THE STAR DIAMOND CHAPTER 2

AS THE *AMAZON PRINCESS* AND THE *MARTIAN MAN-HUNTER* HURTLE TOWARD THE FIRST OF THE THREE DEADLY INSTRUMENTS OF DESTRUCTION PLACED ON EARTH, THEY DISCOVER THEIR WAY TO *KEYSTONE CITY* BARRED BY MONSTROUS INSECTS..

HAS YOUR *MARTIAN-VISION* SPOTTED THE *DOOM WEAPON* CARTHAN SPOKE OF, J'ONN J'ONZZ?

YES, *WONDER WOMAN!* IT'S HIDDEN NOT FAR FROM HERE-- BUT TO REACH IT, WE HAVE TO FIGHT A PATH THROUGH THOSE GIANT INSECTS!

EVEN AS HE SPEAKS, THE *MARTIAN SUPER-SLEUTH* REELS UNDER THE SAVAGE ATTACK OF A TITANIC WASP AND A BUMBLE BEE..

THEIR TREMENDOUS STINGERS CANNOT PENETRATE MY MARTIAN SKIN-- BUT THE REPEATED BLOWS ARE DAZING ME!

FIGHTING FURIOUSLY, HE BARRELS HIS WAY THROUGH THE SWARMING INSECTS...

CARTHAN'S WEAPON MUST BE SOME SORT OF GROWTH--CAUSING MACHINE! THE NEARER WE COME TO IT, THE LARGER ARE THE INSECTS--AND ANIMALS--AROUND IT! EVIDENTLY HUMANS ARE NOT AFFECTED!

8

THEN, AS LEAP-HOPPERS SPRING FORWARD...

I COULD BREAK FREE IF ONLY THE GLUEY SAP THESE LEAFHOPPERS ARE EXUDING WEREN'T COVERING MY EYES--AND STICKING MY LEGS TOGETHER...

UNABLE TO USE HIS JET-FLYING POWER, THE *JLA* MEMBER CRASHES TO THE GROUND JUST AS A HUGE FIREFLY LANDS BESIDE HIM...

A FIREFLY GIVES OFF *COLD LIGHT*--BUT WHEN IT'S *THIS* SIZE THE USUALLY NEGLIGIBLE AMOUNT OF HEAT IS ENOUGH TO SET FIRE TO DRY LEAVES!

THE FLAMES ARE SPREADING-- *WEAKENING ME!* IF THEY AREN'T STOPPED SOON, I'M DONE FOR!

FAR ABOVE THE HELPLESS MARTIAN, *WONDER WOMAN* FINDS THAT, SHE TOO, IS IN DIRE TROUBLE!

MERCIFUL MINERVA! THAT HUGE CAT THINKS MY *ROBOT-PLANE* IS A *BIRD...* AND IS TRYING TO CATCH IT!

NOW OTHER CATS ARE JOINING IT! I'VE GOT TO GET OUT-- SEND MY *ROBOT PLANE* INTO THE SKY--OUT OF DANGER--AND CONTINUE WITHOUT IT!

THEN--AS THE *AMAZING AMAZON* RIDES THE WIND CURRENTS...

A COOPER'S HAWK--SWOOPING DOWN ON ME!

GRIPPED IN THE CRUEL PINCERS OF THE POWERFUL TALONS, *WONDER WOMAN* STRUGGLES TO FREE HERSELF FROM THE CRUSHING PRESSURE...

THIS BIRD IS A PREDATOR--PREYING ON OTHER LIVING THINGS! IT'S HOLDING ME TOO TIGHTLY FOR ME TO BREAK LOOSE!

THEN, AS THEY PASS OVER THE FALLEN *MARTIAN MANHUNTER,* THE *AMAZON PRINCESS* YANKS LOOSE HER MAGIC LASSO...

MINERVA GUIDE MY AIM! THERE'S JUST ONE CHANCE FOR *J'ONN J'ONZZ* AND MYSELF! I MUST MAKE A PERFECT TOSS!

STRAIGHT AND TRUE FLIES THE GOLDEN ROPE! IT COILS AROUND THE ALMOST UNCONSCIOUS *J'ONN J'ONZZ* AND LIFTS HIM UPWARD...

WONDER WOMAN SAVED ME! I'M GETTING MY SUPER-STRENGTH BACK! NOW IT'S *MY* TURN TO GIVE *HER* A HELPING HAND!

THE *SUPER SLEUTH* INHALES WITH ALL HIS MIGHT! IN THE TERRIFIC DOWNDRAFT THUS CREATED THE *COOPER'S HAWK* FLUTTERS WEAKLY...

I DIDN'T INHALE HARD ENOUGH! I'LL HAVE TO TRY AGAIN!

10

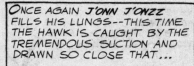

ONCE AGAIN *J'ONN J'ONZZ* FILLS HIS LUNGS--THIS TIME THE HAWK IS CAUGHT BY THE TREMENDOUS SUCTION AND DRAWN SO CLOSE THAT...

POW!

THE BLOW MADE THE HAWK OPEN ITS TALONS-- I'M FREE!

CARTHAN'S ENGINE IS HIDDEN IN THE DRY WELL JUST BELOW! START DOWN FOR IT, *WONDER WOMAN!*

AS THEY DIVE FOR THE WELL OPENING...

MORE INSECTS COMING AFTER US!

DIVE INTO THE WELL! I SEE A BIG BOULDER THAT WILL SEAL US SAFELY IN THE SHAFT!

AS *WONDER WOMAN* STUDIES THE STRANGE GROWTH MACHINE, *J'ONN J'ONZZ* SEALS THE ENTRANCE TO THE DRY WELL...

WHY, THERE DOESN'T SEEM TO BE ANY WAY TO SHUT THIS THING OFF! NO BUTTON OR LEVER OF ANY KIND!

I CAN HEAR ITS FAINT HUMMING WITH MY MARTIAN SUPER-EARS!

ONLY ONE THING TO DO-- *SMASH IT WITH OUR HANDS!*

TO THEIR UTTER AMAZEMENT, THE ALIEN ENGINE SEALS ITS TORN SIDES AND GAPING HOLES...

IT'S SOME QUEER KIND OF *SELF-RESTORING* METAL!

WHAT DO WE DO *NOW?* IF WE CAN'T SMASH IT-- AND IF THERE ISN'T ANY GADGET TO TURN IT OFF-- WE'RE BEATEN!

POP!

MEANWHILE, IN HIS INVISIBLE SPACESHIP, *CARTHAN* FOLLOWS THE ACTIVITIES OF THE *JLA* DUO ON HIS TELEVISION SCREEN...

THERE'S NO WAY FOR ME TO TURN OFF THE MACHINE--BUT THEY CAN DO IT, WITH THEIR SUPER-POWERS! IF ONLY THEY'D SEE THE *CLUE* IN THE WELL--TELLING THEM HOW TO GO ABOUT IT!

FRANTICALLY, *WONDER WOMAN* AND *J'ONN J'ONZZ* EXAMINE THE MYSTERY MACHINE UNTIL...

J'ONN J'ONZZ--LOOK! THIS IS THE METALLIC CLOTH COVER *CARTHAN* SPOKE ABOUT REMOVING! MAYBE IF WE REPLACE IT ON THE ENGINE IT WILL HELP US SOMEHOW!

AS THE COVER IS REPLACED...

SEE HERE! THE TOP OF THE COVER INDICATES IT SHOULD FIT SOME PART OF THE MACHINE--A MISSING PART--!IT MUST BE THE *OFF-ON HANDLE* WE'RE HUNTING! WHEN THE COVER WAS REMOVED, IT MIGHT HAVE CAUSED THE HANDLE TO DISAPPEAR INTO *ANOTHER DIMENSION*!

SECONDS LATER THE MAGIC LASSO IS TWIRLING SO FAST THAT IT SLIPS THROUGH THE DIMENSIONAL BARRIERS AND IS GUIDED BY *WONDER WOMAN'S* THOUGHTS INTO THE FOURTH-DIMENSION...

GOOD GIRL! YOU'VE CAUGHT IT! NOW GIVE IT A TUG TO THE LEFT--

NOTHING'S HAPPENING! I'LL TRY IT TO THE *RIGHT*!

AGAIN AND AGAIN THE *AMAZON* YANKS AT THE INVISIBLE LEVER UNTIL...

YOU DID IT! I CAN'T HEAR THE MACHINE HUMMING ANYMORE!

I PULLED THE HANDLE *DOWN* INTO THE MACHINE! THAT MUST HAVE TURNED IT OFF!

AND WHEN THE *JLA* DUO EMERGES FROM THE WELL...

IT WORKED ALL RIGHT! THE INSECTS ARE GROWING SMALLER AS THE EFFECTS WEAR OFF!

NOW THAT WE'VE DISPOSED OF THIS MENACE, LET'S START AFTER *CARTHAN*! I ONLY HOPE THE REST OF THE GROUP IS DOING AS WELL!

12

DOOM OF THE STAR DIAMOND — CHAPTER 3

RACING SO SWIFTLY THAT HIS FEET DO NOT BREAK THE SURFACE TENSION OF THE WATER--THUS ENABLING HIM TO STAY ABOVE THE SURFACE IN THE SAME MANNER THAT A FLAT STONE SKIPS ACROSS THE WATER--*FLASH* HURTLES TOWARD THE "DOWN UNDER" CONTINENT, PUSHING *AQUAMAN* BEFORE HIM...

IT WON'T BE LONG NOW, *AQUAMAN!* WE JUST WENT PAST THE *SOLOMON ISLANDS!* NOW WE'RE ENTERING THE *CORAL SEA!*

NEXT STOP-- AUSTRALIA!

AS THEY NEAR AUSTRALIA...

THE LAND IS SINKING INTO THE SEA!

THE MACHINE *CARTHAN* TOLD US ABOUT MUST BE CAUSING IT! DIVE DEEP, *AQUAMAN*-- SEE IF YOU CAN LOCATE THE MACHINE UNDER THE SURFACE!

THE ENTIRE CONTINENT WILL SOON BE UNDER THE SEA-- UNLESS I CAN DO SOME- THING ABOUT IT!

FASTER AND FASTER RACES THE **CRIMSON CAVALIER** AROUND THE ISLAND CONTINENT UNTIL HE BECOMES A MERE BLUR--

THE WATER FORMS A SOLID WALL--HELD UPRIGHT BY THE COMPRESSED AIR CREATED BY THE **FLASH'S** SPEED...

I DON'T KNOW HOW LONG I CAN MAINTAIN THIS FANTASTIC PACE! IT'S UP TO **AQUAMAN** TO LOCATE THAT MACHINE IN A HURRY!

BELOW THE COOL WATERS OF THE PACIFIC OCEAN, THE **RULER OF THE SEA** QUESTIONS THE DENIZENS OF THE DEEP...

YES, WE HAVE SEEN AN ODD ENGINE DOWN ON THE SEA BOTTOM!

BUT BE WARNED-- NOTHING ALIVE CAN GET NEAR IT!

WHEN **AQUAMAN** APPROACHES THE SUBMERGED METAL BOX, HE IS REPELLED VIOLENTLY BY THE STRANGE FORCES SURROUND-ING IT...

OOOF! WHATEVER REPELLING RAY THAT THING IS GIVING OFF HAS THE KICK OF A MISSOURI MULE! Hmmm-- IF I CAN'T GET NEAR IT-- MAYBE **FLASH** CAN!

SENDING OUT HIS MENTAL CALLS FOR HELP, THE **SEA SULTAN** IS SOON SURROUNDED WITH FINNY FRIENDS...

BUT FOR **FLASH** TO EXAMINE THE ENGINE, I'VE GOT TO GET IT UP TO THE SUR-FACE! MY FISH FRIENDS WILL HELP ME DO THAT!

SOON A GROUP OF OCTOPI IS BUSILY WEAVING A HUGE SEAWEED BASKET...

KEEP WORKING! YOUR BASKET MUST BE BIG ENOUGH AND STRONG ENOUGH BEFORE I CAN PUT MY PLAN INTO ACTION!

AT THE **GREAT BARRIER REEF**, SHORTLY AFTER, SWORDFISH ARE CUTTING AWAY AT CORAL...

I WANT A LONG, STRONG PIECE OF CORAL!

SINCE CORAL IS COMPOSED OF THE SKELETONS OF MARINE CREATURES, IT CAN BE SAFELY THRUST THROUGH THE REPELLING RAY OF THE DOOM MACHINE...

NOW--RAM THE CORAL INTO THE MACHINE--TILT IT ON EDGE-- AND **INTO THE SEAWEED BASKET!**

THEN, AS THE SHARKS SWIM UPWARD WITH THEIR BASKET-CATCH...

FLASH WILL BE RACING AROUND THE CONTINENT SO FAST--HE'LL OUT-RUN THE SOUND OF MY VOICE! HOW CAN I TELL HIM WE NEED HIS HELP? THAT I CAN'T GET NEAR THE MACHINE?

WHEN THE MACHINE IS SET UP ON SHORE...

MY ONLY HOPE IS TO HOLD UP SIGNS--JUST AS THEY DO AT AUTO-SPEED RACES--TO GIVE THE DRIVERS A MESSAGE!

I CAN'T GET [NE]AR THE [MA]CH!

AROUND AND AROUND THE CONTINENT RACES THE **SCARLET SPEEDSTER**, KEEPING THE WATER FROM INUNDATING THE LAND, UNTIL...

I CAN'T GET NEAR THE MACHINE!

NEED YOUR HELP TO TURN IT OFF!

AT THAT MOMENT IN **CARTHAN'S** SPACE-SHIP...

I HOPE **FLASH** REALIZES THAT HE HAS TO CAUSE COUNTER-VIBRATIONS TO NEUTRALIZE THE REPELLING VIBRATIONS OF THE BOX! IF HE CAN'T, THE BEAMS WILL EVENTUALLY PULL DOWN ALL THE LAND AREAS OF EARTH--TO THE BOTTOM OF THE SEA!

15

AS HE HURTLES AROUND THE *"DOWN UNDER"* CONTINENT, THE *FLASH'S* THOUGHTS KEEP LIGHTNING SPEED WITH HIS FLYING FEET AND-- THEN--

I'LL TAKE OUT A SPLIT-SECOND FROM MY CONTINENTAL "RUNAROUND" TO WORK UP VIBRATIONS TO COUNTER THE REPELLING VIBRATIONS!

BACK HE SUPER-SPEEDS AROUND AUSTRALIA, HOLDING AT BAY THE TONS OF SEA WATER...

DIDN'T WORK THE FIRST TIME! ON THE NEXT CHANCE I'LL TRY WHIRLING MY ARMS!

EACH SUCCEEDING EFFORT BY THE *SCARLET SPEEDSTER* MEETS WITH FAILURE UNTIL ...

YOU'VE HIT IT, *FLASH!* I CAN GET THROUGH! JUST KEEP ON SPINNING-- *UNTIL* I REACH THAT TURN-OFF HANDLE!

SECONDS LATER...

IT'S OFF!

GOOD! BUT UNTIL THE NORMAL BALANCE BETWEEN LAND AND SEA IS RESTORED-- I'D BETTER STAY ON THE JOB!

SHORTLY THEREAFTER...

MISSION ACCOMPLISHED, *FLASH!* EVERYTHING HAS BEEN RESTORED TO NORMAL!

NOW WE CAN GET AFTER *CARTHAN*-- TRUSTING THAT THE OTHERS HAVE DONE AS WELL AND WILL MEET US THERE!

16

DOOM OF THE STAR DIAMOND CHAPTER 4

AS THE **EMERALD WARRIOR** SWOOPS DOWN OUT OF THE SOFT ITALIAN SKY...

THIS IS A MOVIE SET, MADE FOR THE SCIENCE-FICTION PICTURE, "*GIANTS FROM GANYMEDE!*" THOSE GOLDEN GIANTS WERE PROPS USED IN THE MOVIE!

THOSE GIGANTIC GILDED STATUES HAVE **COME ALIVE!** SINCE THEY'RE **YELLOW** -- MY **POWER RING** CAN'T STOP THEM!

HURTLING ACROSS THE WAVES OF THE ATLANTIC -- OVER THE MIGHTY ALPS TO ITALY -- COMES **GREEN LANTERN!** AHEAD OF HIM LIES THE THIRD OF **CARTHAN'S** MYSTERIOUS DOOM MACHINES! BEFORE HIM ARE ALSO THE STRANGE AND DEADLY LIFE-FORMS THE MACHINE HAS CREATED!

THE MOVIE STUNT MEN MANIPULATED THEM BY WIRES FROM HELICOPTERS DURING THE FILMING OF THE PICTURE -- BUT THERE'S NO NEED FOR WIRES NOW! THEY'VE **COME ALIVE!**

17

MILITARY AND POLICE FORCES OF THE ITALIAN GOVERNMENT HAVE BEEN ATTACKING THE GILDED TITANS WITHOUT SUCCESS...

MY *POWER RING* IS JUST AS USELESS AGAINST THOSE "INVADERS" AS CANNON AND TANKS! SOME STRANGE FORCE FROM THE DOOM MACHINE HAS ENDOWED THEM WITH *LIFE!*

DODGING THE OUTSTRETCHED HANDS WHICH SEEK TO DESTROY HIM, THE *GREEN GLADIATOR* BEAMS HIS *POWER RING* AT A NEARBY FOREST...

ORDINARY METHODS WON'T OVERCOME THEM--SO I'LL TRY SOMETHING *EXTRA*ORDINARY!

FLAT GREEN BUZZSAWS FORM IN THE DEPTHS OF THE FOREST AND INSTANTLY BEGIN WORKING, SENDING HUGE SHOWERS OF SAWDUST INTO THE AIR...

BEHIND THEM, TREMENDOUS FANS ROTATE FASTER AND FASTER, SENDING THE SAWDUST TOWARD THE GILDED GIANTS WITH THE VELOCITY OF A HURRICANE...

I'VE GOT TO FORM A FINE LAYER OF SAWDUST OVER THE SURFACE OF THE GIANTS--THEN COAT MORE AND MORE LAYERS OVER THE FIRST ONE!

SOON THE "*GIANTS FROM GANYMEDE*" ARE THICKLY COATED WITH THE FINE PARTICLES OF WOOD SHAVINGS...

SINCE THE FIRST LAYER IS IN CONTACT WITH THE *YELLOW SURFACE* OF THE CREATURES, MY RING HAS NO EFFECT ON IT--BUT IT WILL WORK ON THE OTHER LAYERS COATING THEM!

18

UNDER THE RAYS OF THE **POWER RING** THE SAWDUST HARDENS INTO **PETRIFIED WOOD!** * ENCASED IN THICK SHELLS OF SOLID STONE, THE ALIEN TITANS CANNOT MOVE !

WELL, THAT TAKES CARE OF **THAT** TROUBLE !

Editor's Note : PETRIFIED WOOD IS CAUSED BY MINERALS REPLACING THE WOODY STRUCTURE OF TREES OR SAWDUST WITH QUARTZ !

TO HIS CHAGRIN, **GREEN LANTERN** LEARNS HIS TROUBLES ARE ONLY BEGINNING...

THE LIFE-FORCE THAT ANIMATED THE GIANTS IS WORKING ON OTHER SUBSTANCES NOW ! THOSE BUILDINGS-- STARTING TO MOVE ...

A PROBE BEAM FROM THE **POWER RING** STABS THIS WAY AND THAT TO LOCATE THE LIFE SOURCE ...

I MUST FIND THE MACHINE CREATING THOSE LIFE RAYS AND DESTROY IT ! ODD--MY RING CAN'T SEEM TO MAKE **CONTACT** WITH IT !

AND THEN...

THERE'S THE ANSWER ! THE LIFE RAY IS HIDDEN IN A **GOLDEN BOX** ! THE RING EMANATIONS CANNOT TOUCH IT-- BUT THEY'RE **GLOWING** TO ATTRACT MY ATTENTION !

GREEN LANTERN SURROUNDS THE GOLDEN BOX WITH A FORCE-FIELD BUT THE YELLOW LIFE RAYS EASILY PENETRATE IT...

IT'S NO USE ! THE RAYS THEMSELVES ARE GOLDEN ! BUT IT SHOULD BE SIMPLE ENOUGH FOR ME TO TURN THE MACHINE OFF WITH MY HANDS !

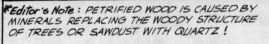

19

LUNGING FORWARD, HE REACHES FOR THE YELLOW HANDLE BUT AS HIS FINGERS WRAP AROUND IT...

OOOHH! THE LEVER IS *ELECTRIFIED!* I--I CAN'T MOVE!

FROM HIS *POWER RING* SHOOTS A GREEN BEAM...

MY FEET ARE GROUNDING ME! IF I COULD FREE THEM FROM CONTACT WITH THE EARTH, THE FLOW OF ELECTRICITY WON'T AFFECT ME TOO MUCH!

THE *POWER RING* SHOVEL DIGS AT THE GROUND UNDER *GREEN LANTERN'S* FEET UNTIL HE HANGS SUSPENDED IN THE AIR BESIDE THE BOX --

I'M FREE!...

WITH A WRENCH OF MIGHTY MUSCLES, HE YANKS DOWN ON THE LEVER...

I'VE SHUT IT OFF!

HIS MISSION ACCOMPLISHED, THE *JLA* MEMBER HEADS HOMEWARD...

I WONDER IF THE OTHERS MADE OUT AS WELL AS I DID? NO TIME TO FIND OUT, THOUGH! I'VE GOT TO LOCATE *CARTHAN* AND HIS SPACE-SHIP AND GIVE A HELPING HAND TO *SUPERMAN* AND *BATMAN!*

WHY--IT'S *GREEN ARROW!* I'LL HAVE YOU OUT OF THERE IN A JIFFY!

NO, *BATMAN-- WAIT!* DON'T TOUCH THOSE LIGHTS!

THOSE LIGHT BEAMS ARE ACTUALLY DEADLY RAYS! IF ONE OF THEM WERE TO TOUCH YOU, YOU WOULD DIE!

THESE RAYS WON'T BOTHER *ME, GREEN ARROW!* I'LL GET YOU OUT!

AS THE MAN OF STEEL EASILY PASSES THROUGH THE DEATH RAYS...

GOOD ENOUGH, *SUPERMAN--* BUT THAT DOESN'T SOLVE THE PROBLEM OF GETTING ME *OUT!*

I'VE ALREADY THOUGHT OF THAT! WATCH!

THE MIGHTY HANDS OF *SUPERMAN* RUN THROUGH THE SOLID LEAD FLOOR AS IF IT WERE MADE OF MELTING BUTTER...

THIS FLOOR IS OF SOLID LEAD TO PROTECT YOU FROM THE RAYS WHILE YOU'RE IN THE PRISON! HOWEVER, I CAN USE THAT SAME LEAD TO PROTECT YOU ANOTHER WAY!

SINCE THE RAYS CAN'T GET THROUGH LEAD I'LL MAKE A HUGE EGG IN WHICH TO CARRY YOU! THE RAYS WON'T TOUCH YOU AS I BRING YOU OUT!

MOMENTS LATER...

SUPERMAN! I'VE BEEN MIGHTY CURIOUS AS TO HOW *CARTHAN* CAPTURED *GREEN ARROW!*

GREAT WORK, SUPERMAN!

22

AS THE BUBBLE COALESCES, HARDENS-- THE FIVE JUSTICE LEAGUE MEMBERS FIND THEM- SELVES IMPRISONED INSIDE A GIGANTIC HOLLOW DIAMOND...!

THE JUSTICE LEAGUE CONSIDERS ME THEIR ENEMY! FOR MY OWN SAFETY I HAD TO IMPRISON THEM--UNTIL I CAN EXPLAIN!

THERE ARE TINY FLAMES ENCASED IN THE DIAMOND-- WEAKENING ME!

AND GOLDEN FLAKES-- WHICH MAKE MY POWER RING USELESS!

IF ONLY I KNEW WHERE TO STRIKE THE JEWEL AND SPLIT IT IN TWO--

THE STRESS POINT CAN ONLY BE DETERMINED FROM OUTSIDE THE GEM!

AT THIS MOMENT, SUPERMAN, BATMAN, AND GREEN ARROW DASH INTO THE ROOM...

OHH! THERE'S KRYPTONITE IN THAT DIAMOND--IT'S SAPPING ALL MY STRENGTH!

THERE'S THE ONE RESPONSIBLE FOR IT ALL--CARTHAN! I'LL HANDLE HIM!

SWIFTER EVEN THAN THE LUNGING BATMAN IS THE FLIGHT OF GREEN ARROW'S SLENDER SHAFT--WHICH BOUNCES HARMLESSLY OFF CARTHAN'S PROTECTIVE AURA AS.

NOTHING CAN HARM ME--CAN'T YOU UNDERSTAND?

I'VE GOT TO CONVINCE MYSELF!

KNOCKED OFF HIS FEET, CARTHAN FALLS BACKWARD AGAINST THE SPACESHIP MACHINERY...

WHY DIDN'T MY AURA PROTECT ME?

AS HE COLLAPSES TO THE FLOOR, THE SPACEMAN TOUCHES HIS HEAD...

OWW! MY HEAD-- IT HURTS! THEN--THAT MEANS MY PROTECTIVE AURA HAS SUD- DENLY FADED AWAY!

SOON CARTHAN IS POURING OUT HIS STORY TO AN ASTOUNDED BATMAN AND GREEN ARROW...

NOW AT LAST I CAN GO BACK TO MY OWN WORLD--BUT FIRST I HAVE TO RELEASE YOUR FRIENDS FROM THE DIAMOND! OH, *NO!* WHEN YOU THREW ME INTO THE MACHINERY, *BATMAN--I WRECKED IT!* I CAN'T RELEASE THEM!

WAIT! THERE'S A SLIM CHANCE! EVERY DIAMOND HAS A *STRESS POINT*...IF I KNEW WHERE THIS ONE IS, I COULD SPLIT THE DIAMOND WITH ONE OF MY DIAMOND-TIP ARROWS!

BEFORE CLEAVING A DIAMOND, DIAMOND-CUTTERS SOMETIMES SPEND MONTHS STUDYING IT TO DISCOVER THE *STRESS POINT!* WE DON'T HAVE THAT MUCH TIME! LUCKILY, I'VE MADE SUCH A STUDY ON MY HOME PLANET! AIM HERE, *GREEN ARROW!*

WITH SWEAT BEADING HIS FOREHEAD, THE *AMAZING ARCHER* DRAWS FAR BACK ON HIS GREAT BOW...

I MUST HIT THE DIAMOND *EXACTLY* AT THE POINT INDICATED-- WITH *EXACTLY* THE RIGHT FORCE-- OR I'LL DOOM *THE JUSTICE LEAGUE* TO A TERRIBLE FATE!

IN THE NEXT INSTANT THE BOWSTRING TWANGS AND...

GREAT SHOT, *GREEN ARROW!* YOU SAVED THEM!

25

FOR A FEW MOMENTS THERE IS WILD CONFUSION UNTIL *BATMAN* AND *SUPERMAN*, TOGETHER WITH *GREEN ARROW*, CONVINCE THE OTHER MEMBERS THAT THE MAN THEY THOUGHT WAS AN ENEMY IS REALLY A FRIEND...

MAYBE WE OUGHT TO GIVE YOU A HELPING HAND AGAINST *XANDOR!*

YOU'VE HELPED ME PLENTY! OVERCOMING *XANDOR* IS-- MY FIGHT!

LATER, AFTER *CARTHAN'S* SHIP HAS BEEN REPAIRED AND HAS DEPARTED FOR THE STARS, THE *JUSTICE LEAGUE* MEMBERS RETURN TO THEIR SECRET HIDE-OUT TO COMPLETE SOME UNFINISHED BUSINESS...

ALL IN FAVOR OF *GREEN ARROW* BEING ACCEPTED AS A NEW MEMBER SAY *AYE!*

AYE!

HEY, WAIT FOR ME-- I VOTE *AYE* TOO!

MAN, I'M COMING UNGLUED! CLUE ME IN ON WHAT HAPPENED AGAINST *CARTHAN!*

IF YOU'RE SAYING WHAT I *THINK* YOU'RE SAYING, *SNAPPER*- I WANT TO HEAR THEIR ADVENTURES, TOO! I WAS IN PRISON WHILE THEY WERE STOPPING THOSE DOOM MACHINES!

SNAP!

AND SO WHILE A POP-EYED *SNAPPER* "KEEPS THE MINUTES OF THE MEETING" EACH OF THE *JUSTICE LEAGUE* MEMBERS TELLS HIS STORY...

WHEN WE FOUND THE BOX HAD NO HANDLE...

SNAPPER, YOU HAVEN'T TAKEN DOWN A THING WE'VE SAID! GET WITH IT, MAN-- GET WITH IT!

SNAP!

SNAP!

*T*HE *JUSTICE LEAGUE OF AMERICA*, AUGMENTED BY *GREEN ARROW*, TAKES OFF ON ANOTHER SUPER-ADVENTURE IN THE NEXT ISSUE!

The End

26.

THE GREEN ARROW

THE GREEN ARROW

the CARTOON ARCHER

GOSH, *GREEN ARROW!* LOOK AT THAT! WHY IS THAT CARTOONIST MAKING US THE LAUGHINGSTOCK OF THE CITY?

DON'T WORRY, *SPEEDY!* I'M SURE EVERYBODY REALIZES THESE CARTOONS ARE NOT TRUE!

R CITY NEWS

THURSDA

VOL.10 NO. 20

GREEN ARROW GOOFS AGAIN!

OWW!

OHH!

FIRE MORE *BOOMERANG ARROWS,* BOYS! WE LOVE TO SEE YOU KNOCK YOURSELVES OUT!

Bill Rand

ONE NIGHT, AS *GREEN ARROW* AND *SPEEDY* TRY OUT THEIR NEW *ARROW-CAR...*

ATTENTION, ALL CARS! SEVERAL GUNMEN DISGUISED AS POLICE HAVE HELD UP THE SPORTS ARENA BOX-OFFICE! BE ON THE LOOKOUT FOR THEM!

MINUTES LATER...

THERE'S A SQUAD CAR, *G.A.!* LET'S ASK 'EM WHERE THEY LOOKED, SO WE DON'T COVER THE SAME ROUTE!

LOOK, BOYS! WE'VE RUN INTO *GREEN ARROW!* GUN HIM AS WE PASS HIS *ARROW-CAR!*

NEXT DAY, AT POLICE HEADQUARTERS...

SAY, *GREEN ARROW*, IT LOOKS LIKE RAME'S BROTHER IS TRYING TO MAKE A FOOL OF YOU!

WHAT ARE YOU TALKING ABOUT?

LOOK AT THIS CARTOON!

POLICE PRECINCT

POLICE

POLICE

I DON'T UNDERSTAND! WHY SHOULD HE KID US LIKE THAT?

GREEN ARROW AND SPEEDY TAKING OFF AT FIRST SIGN OF DANGER

SUGGESTION FOR GN ARROW! YOURSELF

BECAUSE HE MUST BE SORE AT YOU FOR JAILING HIS BROTHER! DID YOU NOTICE HIS SUGGESTIONS FOR "NEW" ARROWS?

YES! VERY CLEVER OF HIM!

SUGGESTION FOR *GREEN ARROW!* GET YOURSELF SOME NEW ARROWS!   Bill Rame

SPOTLIGHT ARROW...   TO FIND CROOKS IN THE DARK

OMELET ARROW...   TO BEAT UP THE "YEGGS"

SAP ARROW...   TO GUM UP A ROBBERY

OH, WELL! LET RAME ENJOY HIS JOKE! ALL THAT COUNTS IS THAT "SHORTY" RAME IS NOW BEHIND BARS! C'MON, *SPEEDY!* WE'VE WORK TO DO!

AND *THAT'S* NO JOKE!

TRASH

LATER THAT DAY, ELSEWHERE IN THE CITY...

G.A.-- THAT BUILDING SUDDENLY CAUGHT FIRE! LET'S SEE IF ANYONE IS TRAPPED INSIDE!

WAIT, *SPEEDY!* THAT FIRE STARTED SO FAST, IT'S MADE ME SUSPICIOUS! LET'S TAKE NO CHANCES! I'LL OPEN THE DOOR WITH THIS *BATTERING RAM ARROW!*

OKAY! I'LL GET OUR *FIRE EXTINGUISHER ARROWS!*

3

BUT AS THE MASSIVE MISSILE SMASHES THE DOOR...

:GASP: T-THE DOOR WAS BOOBY-TRAPPED! WE'D HAVE BEEN KILLED IF WE ENTERED!

AND THERE GO THE TWO WHO ARE RESPONSIBLE! LET'S PIN THEM DOWN!

KABOOMMM

THEY WON'T TELL US WHY THEY DID IT, BUT I'M SURE THEY TRIED TO GET EVEN WITH US FOR CAPTURING "SHORTY" RAME!

PWUTT!

PWUTT!

PWUTT!

NEXT MORNING, IN THEIR EVERYDAY GUISE OF OLIVER QUEEN AND HIS YOUNG WARD, ROY HARPER, THEY ARE GREETED WITH ANOTHER SURPRISE...

THE NERVE OF THAT BILL RAME! HE'S DRAWN ANOTHER CARTOON KIDDING US!

LET ME SEE IT!

STAR CITY NEWS

HMMM... HE'S SURE HOLDING A GRUDGE AGAINST US! HE KEEPS DISTORTING THE TRUTH TO MAKE IT SEEM AS IF WE'RE BUNGLERS!

BUT WHY DOES HE SUGGEST SPECIAL ARROWS, G.A.?

HOW TO PUT OUT A FIRE!

MONDA

SUGGESTION FOR GREEN ARROW! GET YOURSELF SOME

Bill Rame

I DON'T KNOW, SPEEDY! MAYBE IT'S HIS WAY OF MAKING FUN OF OUR TRICK ARROWS! BY MAKING SILLY SUGGESTIONS, HE RIDICULES OUR WEAPONS!

Bill Rame

SUGGESTION FOR GREEN ARROW! GET YOURSELF SOME NEW ARROWS!

SCEPTER ARROW...

OAR ARROW...

SKELETON ARROW...

TO CROWN GANGSTERS WITH

SO YOU CAN ROW AFTER SMUGGLERS

TO FRIGHTEN CROOKS

LATER THAT MORNING, AS THEIR PATROL TAKES THEM TO A SWANK SECTION...

HELP! POLICE! THIEVES!

THAT CAR PULLING AWAY FAST FROM THE CURB! LET'S OVERTAKE IT!

JEWEL

R-R-RRRR!

4

SHORTLY, AT BILL RAME'S STUDIO...

GASP! **GREEN ARROW!** **SPEEDY!** THANK GOODNESS YOU'VE COME! I THOUGHT Y-YOU'D NEVER ARRIVE!

BE RIGHT WITH YOU, RAME! FIRST WE'LL TAKE CARE OF YOUR GUESTS--

--WITH OUR **BOXING GLOVE ARROWS!**

THERE ARE MORE OF THEM! THEY HEARD THE COMMOTION AND ARE COMING OUT!

BUT AS **SPEEDY** FIRES ANOTHER SPECIAL ARROW...

Y!!!! W-WE'RE TRAPPED--

--LIKE THE POOR FISH YOU REALLY ARE! IT TOOK **GREEN ARROW** A LITTLE WHILE TO FIGURE OUT WHAT WAS GOING ON! BUT HE'LL EXPLAIN TO YOU!

TWANG!

WHEN I NOTICED THE FIRST LETTERS OF THE ARROWS BILL RAME SUGGESTED ALWAYS SPELLED **S.O.S.**, I REALIZED HE WAS TRYING TO SMUGGLE OUT A MESSAGE WITH EACH CARTOON! **SPEEDY** AND I WERE PREPARED FOR TROUBLE WHEN WE BARGED IN HERE!

SPOTLIGHT ARROW
OMELET ARROW
SAP ARROW

SCEPTER ARROW
OAR ARROW
SKELETON ARROW

SMOG ARROW
OIL ARROW
SLEEPING GAS

SALT ARROW
OPTICS ARROW
SHACKLE ARROW

NOBODY KNEW THAT AFTER YOU CAPTURED MY BROTHER, WHO RICHLY DESERVES PUNISHMENT, THE REST OF HIS GANG MADE MY PLACE THEIR HIDEOUT! THEY HELD ME HOSTAGE WHILE THEY ENJOYED MY MAKING FUN OF YOU!

BY REPEATEDLY ATTACKING YOU, I KNEW THAT SOONER OR LATER YOU'D CATCH ON AND COME HERE --NOT ONLY TO FREE ME BUT NAB THESE THUGS AS WELL!

WELL, RAME, YOU AND I HAVE THE LAST LAUGH! SPEEDY, CALL THE POLICE! TELL THEM WE HAVE A LOT OF COMPANY FOR "SHORTY"!

THE END

THE GREEN ARROW

She's back again!... That talented little lass who dreams of becoming a crime-fighting archer, just like her heroes, **GREEN ARROW** and **SPEEDY**! But now, she proves more than a mere headache to the battling bowmen, who have good reason to rue...

The RETURN of MISS ARROWETTE

SORRY, YOU TWO... I'M STOPPING YOUR EXHIBITION, WITH MY *NAIL FILE ARROW!*

OH, NO! *MISS ARROWETTE*-- INTERFERING AGAIN! AND THIS TIME, IT MAY COST THE *GREEN ARROW* AND *SPEEDY* THEIR LIVES!

SNIP!

SNIP!

Treasure Chest DONATIONS for CHARITY...

ON THE ROOF OF AN APARTMENT HOUSE, A PRETTY GIRL HALTS HER ARCHERY PRACTICE, TO GAZE LONGINGLY AT...

THE *ARROW-SIGNAL*... SUMMONING *GREEN ARROW* AND *SPEEDY* TO A CRIME SCENE! ;SIGH;; IF ONLY I HADN'T PROMISED TO KEEP OUT OF THEIR HAIR...

BUT, A SHORT WHILE LATER...

A *SECOND* SIGNAL! SOMETHING'S DELAYED MY ARCHER FRIENDS... AND THAT'S REASON ENOUGH TO TAKE MY *MISS ARROWETTE* COSTUME OUT OF MOTHBALLS!

SOON, AT *CITY AIRPORT*... *GREEN ARROW* AND *SPEEDY* FINALLY ANSWERED THE CALL-- BUT IT WILL TAKE MORE THAN THEIR *FIRECRACKER ARROWS* TO STOP THOSE ROBBERS!

POW

BOOM

BOOM

AND AS THE BLONDE BOW-GIRL SENDS A UNIQUE SHAFT STREAKING THROUGH THE AIR ...

HUH--? *BUBBLES*... MILLIONS OF 'EM! CAN'T SEE WHERE I'M GOIN'!

GRAB THEM, BOYS -- BEFORE MY *BUBBLE-BATH ARROW* LOSES ITS EFFECT!

AFTER THE HELPLESS CROOKS HAVE BEEN TAKEN IN TOW...

WELL, YOU'RE CERTAINLY FORTUNATE I CAME OUT OF RETIREMENT-- OR THE JAIL WOULD BE MINUS TWO THIEVES!

WE...UH... HARDLY EXPECTED TO FIND *YOU* HERE, *MISS ARROW-ETTE*, BUT THANKS FOR THE ASSIST, ANYWAY!

TELL ME, *GREEN ARROW*-- WHY DIDN'T YOU USE A MORE APPROPRIATE SHAFT, LIKE THE *BOXING GLOVE ARROW*, TO STOP THEM?

GUESS *SPEEDY* AND I WERE-- UH-- SLEEPING ON OUR FEET! WELL, SO LONG, *MISS ARROWETTE*... WE HAVE TO PRE-PARE FOR THE CHARITY FAIR LATER TONIGHT!

HMM... THEY MISSED THE FIRST *ARROW-SIGNAL* TONIGHT-- AND IT'S HARDLY LIKE THEM TO SELECT THE WRONG ARROW, UNLESS...

THAT'S IT!... *GREEN ARROW* AND *SPEEDY* MUST BE ILL-- AND TRYING TO KEEP THE FACT FROM THE UNDER-WORLD! I'VE GOT TO BE THEIR *GUARDIAN ARCHER*, UNTIL THEY'RE WELL AGAIN!

CONFOUND *MISS ARROWETTE*!... SHE BOTCHED UP OUR PLANS! I'VE GOT TO GET ON THE SHORT-WAVE RADIO, FAST!

POLICE CHIEF

WHY IS THE *POLICE CHIEF* SO ANGRY AT THE *GIRL ARCHER*?

2

THE ANSWER LIES IN HIS TRANSMISSION, ON A SECRET FREQUENCY, TO A CRIMINAL HIDEOUT SOMEWHERE IN THE CITY...

GREAT THUMPING BOWSTRINGS! WHAT A TIME FOR *MISS ARROWETTE* TO INTERVENE!

WHEN SHE FORCED US TO ARREST THOSE CROOKS ON THE SPOT, OUR PLANS WENT UP IN SMOKE, *GREEN ARROW!* I SAY, LET'S FORGET THE WHOLE THING...

NO, CHIEF! THIS IS THE ONLY WAY TO UNCOVER THE OUT-OF-TOWN HIDEOUT WHERE THE GANG IS STASHING ALL ITS LOOT!

WHEN WE GOT WIND OF THE UNDER-WORLD SCHEME TO REPLACE US WITH TWO IMPERSONATORS, WE *ALLOWED* OURSELVES TO BE CAPTURED, SO YOU COULD TAIL THE PHONIES TO THEIR SECRET LAIR!

WHERE DO THOSE BOGUS BOWMEN PLAN TO APPEAR NEXT, *GREEN ARROW?*

AT THE CHARITY FAIR... WHERE *WE* WERE SCHEDULED TO PERFORM TONIGHT!

HMM...ONCE THEY STEAL THE RECEIPTS, THEY'LL HEAD STRAIGHT FOR THEIR HIDEOUT, I GUESS!

YES... THEN YOU'LL BE ABLE TO MOVE IN ON THE WHOLE GANG, AND RECOVER ALL THE LOOT THEY'VE STOLEN IN THE PAST! GOOD LUCK, CHIEF!

BUT THAT EVENING, AS THE PHONY ARCHERS "PERFORM" AT THE WORLD CHARITY AFFAIR...

KEEP THE DONATIONS COMING, FOLKS! *SPEEDY* AND I WILL ENTER-TAIN YOU UNTIL THE TREASURE CHEST IS FILLED!

HMPH!... THEY HAVEN'T ATTEMPTED A TRICK SHOT ALL NIGHT! WHAT'S MORE, THERE'S SOMETHING STRANGE ABOUT-- *GOOD GRIEF!*

ABRUPTLY, A BIZARRE SHAFT KNIFES THROUGH THE AIR...

AN ARROW--SHAPED LIKE A HUGE NAIL FILE! IT'S CUT THE ARCHERS' BOWSTRINGS!

I'LL USE MORE THAN MY *NAIL FILE ARROW* ON THEM IF THEY DON'T TELL ME WHAT HAPPENED TO THE *REAL GREEN ARROW AND SPEEDY!*

3

GET A GRIP ON YOURSELF, *MISS ARROWETTE!* WHAT MAKES YOU SAY THESE TWO ARE IMPOSTORS?

SPEEDY'S HAIR...ANY WOMAN CAN SPOT A *DYEING JOB!*

WITH THAT, THE CRIMINALS WHIP OFF THEIR MASKS, AND...

DON'T TRY ANYTHING, CHIEF -- OR SOMETHING WILL HAPPEN TO THIS SMART GAL! GRAB THE MONEY CHEST, EDDIE!

OH, NO --! *MISS ARROWETTE'S* INTERFERED AGAIN ... AND THIS TIME, SHE'S PUT *HERSELF* ON THE SPOT!

TAIL US ... AND THIS GAL'S FIRED HER *LAST* ARROW! HEY, SISTER, STASH THAT PERFUME ... THIS IS NO TIME TO BE DOLLIN' UP!

SORRY...ONE OF MY-- ER -- NERVOUS HABITS!

AFTERWARD, WHEN THE STARTLING INCIDENT IS REPORTED TO THE CAPTIVE *GREEN ARROW* AND *SPEEDY*...

... AND THEN THEY ESCAPED, USING *MISS ARROWETTE* AS A SHIELD! AND DO YOU KNOW WHAT THE LITTLE FOOL WAS DOING?... DABBING HERSELF WITH *PERFUME!*

HMM ...THAT WAS AN ODD TIME TO PRIMP, CHIEF!

WELL, SHE'S SHOT OUR PLANS FULL OF HOLES! I'LL SPRING A RAID ON THEIR *CITY* HIDEOUT, AND FREE YOU TWO ...

NO TIME! *SPEEDY* AND I WILL BREAK OUT OF HERE AND JOIN YOU! THERE'S STILL A CHANCE OF FINDING THEIR OUT-OF-TOWN LAIR, CHIEF!

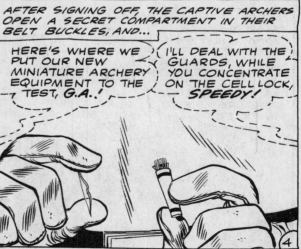

AFTER SIGNING OFF, THE CAPTIVE ARCHERS OPEN A SECRET COMPARTMENT IN THEIR BELT BUCKLES, AND...

HERE'S WHERE WE PUT OUR NEW MINIATURE ARCHERY EQUIPMENT TO THE TEST, G.A.!

I'LL DEAL WITH THE GUARDS, WHILE YOU CONCENTRATE ON THE CELL LOCK, *SPEEDY!*

4

THANKS, *MISS ARROWETTE*... THE SAME GOES FOR YOUR *HAIRPIN ARROW!*

ZUNK

AND SO, WITH THE GANG RENDERED HELPLESS...

I JUST CAN'T FIGURE IT... HOW'D YOU EVER FIND THE HIDEOUT, *GREEN ARROW?*

YOU CAN THANK *MISS ARROWETTE* FOR THAT! THAT PERFUME SHE USED EARLIER ...

"...IT WASN'T PERFUME AT ALL, AS WE DISCOVERED WHEN WE REACHED THE FAIR!"

SHE SPILLED SOME HERE AND --SNIFF-SNIFF-! IT SMELLS LIKE SOMETHING *BESIDES* PERFUME, *G.A.!*

YOU'RE RIGHT, *SPEEDY*... IT CONTAINS ONE OF THOSE ELECTRONICALLY TREATED CHEMICALS THAT WE CAN FOLLOW WITH OUR *BLOODHOUND ARROW!*

"IMMEDIATELY, WE SET THE CONTROLS ON THE LATEST ADDITIONS TO OUR ARROW ARSENAL ..."

THIS INTRICATE ELECTRONIC NETWORK RESPONDS TO THAT CHEMICAL ODOR -- EVEN FROM MILES AWAY -- JUST LIKE A *REAL* BLOODHOUND!

"WE THEN TOOK OFF IN OUR *ARROW-PLANE*, FIRED THE *BLOODHOUND ARROW*, AND KEPT TAILING IT..."

YESSIR, JUST LIKE A BLOODHOUND! IT'S LEADING US TO THE "PERFUME" THAT *MISS ARROWETTE* DABBED ON HERSELF!

AFTERWARD, WHEN THE POLICE ARRIVE, AND *MISS ARROWETTE* LEARNS OF HER ORIGINAL BLUNDER ...

OH, DEAR... THEN YOU ACTUALLY *WANTED* THE CRIMINALS TO ESCAPE -- SO THEY WOULD LEAD YOU TO THIS HIDEOUT! CAN YOU EVER FORGIVE ME?

WE'D FIND THAT DIFFICULT IF YOU HADN'T HELPED IN THEIR CAPTURE, *MISS ARROWETTE*... BUT IN THE FUTURE, *PLEASE* -- LEAVE THE CRIME-FIGHTING TO *US!*

The End

THE GREEN ARROW

OUT OF THE PAST COMES THE WONDROUS WEAPON TO LAUNCH A WILY THIEF ON THE ROAD TO FANTASTIC CRIMES! AND THE ONLY ONES WHO COULD CHALLENGE HIM ARE *GREEN ARROW* AND *SPEEDY*, WHO APPARENTLY FIND THEIR SUPERIOR AT LAST IN...

the MAN with the MAGIC BOW

IT SWEPT US UP, G.A.-- AND IS CARRYING US OFF INTO THE NIGHT!

HA, HA! WHAT CAN TWO SUCH PUNY ARCHERS DO AGAINST *ME*--WHEN I POSSESS THE *MAGIC BOW?*

MUNICIPAL MUSEUM

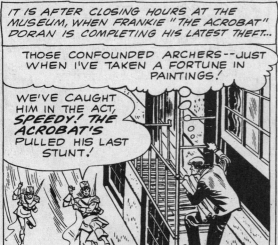

IT IS AFTER CLOSING HOURS AT THE MUSEUM, WHEN FRANKIE "THE ACROBAT" DORAN IS COMPLETING HIS LATEST THEFT...

THOSE CONFOUNDED ARCHERS--JUST WHEN I'VE TAKEN A FORTUNE IN PAINTINGS!

WE'VE CAUGHT HIM IN THE ACT, *SPEEDY!* THE ACROBAT'S PULLED HIS LAST STUNT!

DESPERATE TO ESCAPE, THE CUNNING CROOK HURTLES INSIDE AND RACES THROUGH A MAZE OF CORRIDORS UNTIL HE REACHES A DIMLY-LIGHTED WEAPONS ROOM...

M-MY CHEST-- SO M-MUCH PAIN! WH-WHO IS IT?

WHAT A BREAK-- HE CAN'T SEE ME! I DON'T KNOW WHAT'S WRONG WITH HIM-- BUT THAT'S JUST RIGHT FOR ME!

I'M *GREEN ARROW*, IN MY OTHER IDENTITY!

GREEN ARROW! WHAT A BIT OF GOOD FORTUNE! BUT WE MUST TALK QUICKLY! THE PAIN IS UNENDURABLE!

THIS ARRIVED ONLY TONIGHT! IT WAS FOUND BESIDE THE LAKE OF NEMI... *THE MAGIC BOW OF DIANA!* IT WAS LOST FOR CENTURIES... YOU MUST HAVE IT...READ LEGEND ON SCROLL... QUICKLY...

MAGIC BOW? LEGEND? SCROLL? WHAT'S THIS ALL ABOUT?

"...AND THE MAGIC BOW OF DIANA, WHOSE ARROWS WILL PERFORM AS ITS OWNER *DESIRES*, MUST INEVITABLY FIND ITS WAY TO ITS *TRUE* MASTER, THE GREATEST OF ARCHERS..."

WHAT KIND OF BALONEY IS *THIS?*

HEY! WHAT HAPPENED TO THE OLD GUY? HE CONKED OUT!

THIS WAY, *SPEEDY!* I SEE A LIGHT IN A ROOM UP AHEAD!

LEGEND OR NO LEGEND, THIS LOOKS LIKE A GOOD WEAPON! I'LL GIVE *GREEN ARROW* A TASTE OF HIS OWN MEDICINE!

AIMING IN THE DIRECTION OF *GREEN ARROW'S* VOICE, THE *ACROBAT* FIRES...

I DON'T KNOW IF I CAN HIT A BARN WITH THIS-- BUT I'D SURE LIKE TO SEE IT TAKE *GREEN ARROW* AND THAT BRAT RIGHT OUT OF HERE!

TWANGG!

2

ASTONISHINGLY, AS THE ARROW SPEEDS DOWN THE HALL, IT BEGINS TO GROW...

ZIP

...AND *GROW*...

ZIP

...AND *GROW*...

ZIP

...UNTIL IT SWEEPS BENEATH THE BATTLING BOWMEN, LIFTING THEM, AND CARRYING THEM INTO THE NIGHT...

G.A..! WHAT ON EARTH'S HAPPENING?

I--I DON'T KNOW! IT'S LIKE A NIGHTMARE!

ZIP

SPEEDY! FIRE YOUR *PARACHUTE* ARROW!

TWANG

TWANG

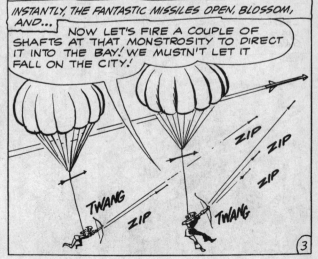

INSTANTLY, THE FANTASTIC MISSILES OPEN, BLOSSOM, AND...

NOW LET'S FIRE A COUPLE OF SHAFTS AT THAT MONSTROSITY TO DIRECT IT INTO THE BAY! WE MUSTN'T LET IT FALL ON THE CITY!

ZIP

ZIP

ZIP

TWANG

ZIP

TWANG

③

THEN, AFTER LANDING...

I DON'T KNOW WHAT HAPPENED IN THE MUSEUM -- BUT WE'RE GETTING BACK THERE -- *FAST* -- TO FIND OUT!

AND SHORTLY...

HE'S STILL ALIVE! CALL AN AMBULANCE AT ONCE!

I THINK THIS SCROLL EXPLAINS WHAT HAPPENED -- IF YOU CAN BELIEVE IN LEGENDS! WOWIE! THE MAGIC BOW OF DIANA!

LATER, AFTER THE STRICKEN CURATOR HAS BEEN TAKEN AWAY...

THIS MUST BE THE ANSWER, *SPEEDY*, FOR THAT GIANT ARROW! THE BOW IS TRULY MAGICAL!

WHAT A HEADACHE THAT'S GOING TO BE -- WITH THE BOW IN THE HANDS OF **THE ACROBAT!** OH, BOY!

SPEEDY HAS NEVER SAID A TRUER WORD, FOR ON THE FOLLOWING NIGHT...

I'D LIKE THIS ARROW TO GO STRAIGHT TO THE NATIONAL BANK, DRILL INTO THE VAULT, AND RETURN WITH SOME STACKS OF CASH!

TWAN-GG

WEAVING THROUGH THE SPIRES AND ROOFTOPS, THE ARROW SPEEDS UNERRINGLY TOWARDS ITS TARGET...

DRUGS

ZIP

...THE BANK VAULT...

...WHERE IT COMPLETES ITS MISSION!

BANK VAULT

ZIP

④

A MOMENT LATER... COME TO PAPA! GOOD--! NOW FOR A QUICK GETAWAY!

SHORTLY, AN ARROW FLASHING OVER THE CITY SIGNALS OLIVER QUEEN AND YOUNG ROY HARPER...

THAT'S US, ROY! LET'S SWITCH TO OUR COSTUMES AND FIND OUT ON THE POLICE RADIO WHAT'S UP!

MINUTES LATER, THE FAMED **ARROW-PLANE** ROARS OVER THE DARKENED CITY...

THE ARROW MUST'VE BEEN SEEN RETURNING TO ME, AND ALREADY **GREEN ARROW** IS ON MY TRAIL! BUT WHAT HAVE I TO WORRY ABOUT? I'M MORE THAN A MATCH FOR HIM NOW!

AS THE SLEEK CRAFT DIVES, A STRANGE DUEL BEGINS--THE UNIQUE ARROWS OF THE **GREEN ARROW** AGAINST DIANA'S!

HA, HA, HA! FIRST I'LL WARM UP WITH A COUPLE OF **ORDINARY ARROWS**... THEN...

SHOUTING A COMMAND, THE CUNNING CROOK FIRES ANOTHER ARROW, WHICH SPROUTS TWO GIANT "HANDS!"

NOW BRING THE **ARROW-PLANE** DOWN! I'LL HAVE THOSE PESKY ARCHERS AT MY MERCY! I'LL TEACH THEM NOT TO INTERFERE!

5

SUDDENLY...

HUH?

GOOD SHOT, *G.A.*! YOU GOT THE MAGIC BOW WITH YOUR *FISH-HOOK ARROW!*

ZIP

YOU TWO--? YOU WERE IN YOUR *ARROWPLANE,* FIRING DOWN AT ME! HOW DID YOU GET HERE?

WE WERE FOLLOWING YOU ALL THE TIME-- BY *BOAT*--WHILE DIRECTING THE *ARROW-PLANE* BY REMOTE CONTROL...

...WHICH ALSO RELEASED *ARROW-FIRING* DEVICES, WHICH WE HAD RIGGED!

6

AFTER A POLICE LAUNCH TAKES AWAY THEIR CAPTIVE...

SAY, THE BOW FINALLY WOUND UP WITH *YOU,* G.A.! MAYBE YOU'RE ITS TRUE MASTER, AS THE LEGEND SAYS!

PERHAPS, *SPEEDY!* BUT I THINK THE WORLD HAS SEEN THE LAST OF THE MAGIC BOW! I WEIGHTED IT DOWN, AND NOW OVER IT GOES, TO THE BOTTOM! IT'S MUCH TOO DANGEROUS TO HAVE AROUND!

THE END

THE GREEN ARROW

THEY WERE MEN FAMED FOR BRAVING UNTOLD DANGERS--AND NOW, THEY WERE ON THE MOST UNUSUAL--AND SINISTER-- VOYAGE OF THEIR LIVES! AND, NOT EXACTLY INVITED, BUT CERTAIN TO BE ON HAND, WERE GREEN ARROW AND SPEEDY, TO PLAY A STRANGE, PERILOUS ROLE IN...

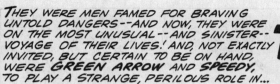

THE GREEN ARROW

the DEADLY TROPHY HUNT

I KNOW WE'VE GOT TO LEAD THIS RHINO ON A CHASE, GREEN ARROW-- BUT WHEN DO WE PLAY OUR TRUMP?

IN LESS THAN A MINUTE--IF WE'RE NOT TRAMPLED FIRST, SPEEDY!

BECAUSE OF AN UNUSUAL VOYAGE TO BE MADE BY FOUR MEN, NEWSMEN GATHER ON THE DECK OF A LUXURIOUS YACHT...

AS I UNDERSTAND IT, MR. BRIHILL, THIS IS YOUR YACHT...

YES, AND MY IDEA, TOO--THAT WE FOUR MEMBERS OF THE ADVENTURERS' CLUB TAKE A SPECIAL CRUISE!

EXCEPT FOR OLIVER QUEEN, HERE, WE'RE ALL FAMOUS FOR BRINGING BACK STRANGE TROPHIES FROM DANGEROUS CORNERS OF THE GLOBE!

YOU'VE CAPTURED SOME OF THE WORLD'S MOST *DANGEROUS GAME*, MR. BRIHILL, AND MR. ANDERSON HAS SET RECORDS, AS A SKIN DIVER, FOR SNARING DEADLY DENIZENS OF THE DEEP...

TRUE, AND TETLEY, HERE, HAS RECOVERED LOST *TREASURES* IN DEATH-DEFYING PLACES!

BUT YOU'VE ALL *RETIRED!* WHY THIS SUDDEN VOYAGE FOR *MORE* TROPHIES?

WE'VE BEEN STAMPED AS SELFISH MEN--WHO DID IT FOR PERSONAL GLORY AND REWARD! THIS TIME, OUR TROPHIES WILL GO TO *CHARITY!* THAT'S MY PLAN!

BY THE WAY, WHAT HAS *OLIVER QUEEN* DONE THAT'S FAMOUS?

FRANKLY, I'VE GOT THE REPUTATION OF AN *IDLE PLAYBOY!* THAT'S WHY I'M GOING ALONG, WITH MY YOUNG FRIEND, ROY HARPER... TO SEE WHAT *I* CAN DO FOR CHARITY!

TELL MR. BRIHILL WE'RE READY TO CAST OFF! ALL ASHORE, FOLKS!

THROUGH CHOPPY WATERS PLIES THE FABULOUS YACHT, TO WHERE, SOME DAYS LATER, ANDERSON--THE SKIN DIVER--ASKS TO DROP ANCHOR...

THE GIANT SQUID LURKS IN THESE WATERS! IF I CAN NET ONE, WE CAN TAKE IT BACK IN A SPECIAL TANK WE'VE GOT ABOARD! IT'LL BRING AN ENORMOUS PRICE!

MEANWHILE, IN THE CONCEALMENT OF THE SHIP'S HOLD...

TIME TO SWITCH TO *GREEN ARROW* AND *SPEEDY*, ROY! THAT'S WHY WE *REALLY* CAME ALONG... THESE MEN ARE NOT YOUNGSTERS ANYMORE... WE MAY HAVE TO *HELP* THEM, WITHOUT BEING SEEN, OF COURSE!

2

THE BATTLING BOWMEN DESCEND INTO THE DEPTHS NOT A MOMENT TOO SOON-- FOR DOWN BELOW, ANDERSON IS ALREADY ENSNARED IN A GREAT TENTACLE...

ANDERSON HASN'T SPOTTED US... BEFORE HE DOES, I'LL FIRE A *SMOKE ARROW!*

AS THE UNIQUE SHAFT STREAKS FROM THE SPECIAL BOW, EMITTING A GREAT, DARK CLOUD...

THE SMOKE LOOKS LIKE INK FROM THE SQUID! NOW TO PUT A COUPLE OF *ARROWLINES* INTO ACTION...

TWO MORE MISSILES, WITH STEEL LINES ATTACHED, SWIRL AROUND THE SQUID, ENSNARING IT HELPLESSLY...

AND BY THE TIME THE SMOKE STARTS TO CLEAR...

WE GOT ANDERSON'S NET ON THE THING-- AND CLEARED OUR LINES AWAY... HE'S IN CONTROL NOW... WE CAN SLIP BACK ABOARD!

SOME MINUTES LATER...

MY GREATEST CATCH-- AND MY LUCKIEST! IT SEEMED TOO EASY TO BE TRUE!

3

SO THE VOYAGE RESUMES--TILL FINALLY, THEY ANCHOR IN AN AFRICAN COVE...

I'M GOING AFTER THE *WHITE RHINO*--A RARE BEAST AND A KILLER! I'LL TRAP HIM--THEN YOU OTHERS CAN HELP ME TRANSPORT HIM TO THE SHIP!

BUT LATER, AFTER THE HUNTER AND HIS MEN HAVE PREPARED A PIT...

IT'S THE RHINO! WE DIDN'T FIND IT! *IT* FOUND *US!*

THERE ARE NO TREES CLOSE BY--AND WE DON'T WANT TO SHOOT THE BEAST! WHAT A DILEMMA!

UNSEEN BY THE OTHERS, HOWEVER, A SMALL ARROW ZIPS FROM THE BRUSH, EXPLODING HARMLESSLY--BUT ANNOYINGLY--AGAINST THE BEAST'S FLANK...

KBLAM

THE RHINO TURNS IN ANGER, AND CHARGES TOWARD TWO FIGURES THAT STIR IN THE BRUSH...

IT WORKED, *G.A.!* YOUR *FIRECRACKER ARROW* BROUGHT IT TOWARD *US!*

GOOD...NOW WE'LL HAVE IT CHASE US TOWARD THE PIT! THE OTHERS ARE FAR ENOUGH AWAY SO THAT WE CAN PULL OUR *BALLOON ARROW* STUNT!

MOMENTS LATER, BOTH FIRE SIMULTANEOUSLY, WHILE FALLING FLAT IN DEEP GRASS...

THIS HAD BETTER WORK, *SPEEDY*--OR WE'RE IN *REAL* TROUBLE!

ZIP! ZIP!

THE ARROWS LAND AT THE PIT'S EDGE, AND FROM EACH INFLATES A BALLOON OF CURIOUS DESIGN...

SSSS SSSS

④

SEEING THE LIFELIKE FIGURES, THE RHINO CHARGES...

GOOD ENOUGH, *SPEEDY!* NOW LET'S GET OUT OF HERE... THE OTHERS WILL RETURN SOON!

CRASH

SPLAT

SPLAT

TOWARD EVENING...

TALK ABOUT *LUCK*... THE RHINO CHASED US AWAY, THEN FELL INTO THE PIT!

LATER, ON DECK...

I TOLD YOU, BACK AT THE CLUB, THAT I KNEW THE LOCATION OF ONE OF THE RICHEST TREASURES! IT'S ODD, BRIHILL, BUT YOUR RHINO AND MY TREASURE TURN OUT TO BE IN THE SAME AREA...

SEE?...I CREPT OFF TO A RUINS NEAR HERE! THERE, AS AN OLD MAP SHOWS, I FOUND THESE PEARLS! OBSERVE -- THEY'RE THE SIZE OF A CONDOR'S EGGS!

INDEED THEY ARE, TETLEY-- AND I'LL TAKE THEM NOW!

BRIHILL! YOU CAN'T BE SERIOUS!

BUT I AM! THAT'S WHY I DRUMMED UP THIS "CHARITY" TROPHY HUNT... I WAS AFTER THAT TREASURE YOU KNEW ABOUT, TETLEY! CLEVER, EH?

BUT UNSEEN, OLIVER'S HAND STEALS TOWARD THE CAGE DOOR, AND...

THE RHINO IS LOOSE! LOOK OUT! RUN!

BRIHILL PLANS TO KILL US ANYWAY, SO THIS IS OUR ONLY CHANCE!

5

AS THE BEAST ESCAPES IN THE CONFUSION, A SWIFT CHANGE OF COSTUME IS FOLLOWED BY...

THAT'S IT, *SPEEDY*, THE *BOXING-GLOVE ARROWS*!

BIFF

POW

ANOTHER HAIL OF ARROWS PINS BRIHILL AND HIS MEN TO THE BULKHEAD--AND THEN...

UPON MY WORD--LOOK! THAT'S *GREEN ARROW!* HOW IN THE WORLD DID *HE* GET HERE?

WHEN A "FRIGHTENED" OLIVER QUEEN EMERGES FROM THE FOREST...

SAY!...I JUST CAME ACROSS *GREEN ARROW!* HE SAID HE SECRETLY FOLLOWED OUR SHIP IN HIS *ARROW-PLANE,* AS A PRECAUTION!

I SEE ROY WAS ABLE TO CHANGE IN TIME AND JOIN THEM! WE HAD TO MOVE FAST--BUT I THINK WE FOOLED 'EM!

6

WELL, THANKS TO *GREEN ARROW*-- WHEREVER HE IS NOW-- WE'VE STILL GOT TWO OF OUR TROPHIES FOR CHARITY, AND SOME CROOKS FOR THE POLICE! IT'S BEEN A GOOD HUNT!

AND I'M ONLY-- UH -- SORRY THAT *I* WAS SUCH A WASHOUT!

THE END

THE GREEN ARROW

HE WAS THE TOP ROOKIE ON THE POLICE FORCE--THE MOST PROMISING PATROL MAN EVER TO WALK A BEAT! BUT *GREEN ARROW* AND *SPEEDY*, EMERALD ARCHERS OF JUSTICE, KNEW HE HARBORED A TERRIBLE SECRET, WHICH MADE HIM...

THE GREEN ARROW

the COP who LOST his NERVE

LOOK, *GREEN ARROW!* FRED JENNINGS FROZE UP AGAIN-- AND LET THAT CROOK GET THE DROP ON HIM!

WE'LL HAVE TO RESCUE HIM AGAIN, *SPEEDY*, AND STOP THAT ROBBERY!

LATE ONE AFTERNOON, AFTER A THIEF HAS SLIPPED INTO A DIAMOND MERCHANT'S UNOCCUPIED HOME...

OH, OH... DIDN'T FIGURE AN ALARM WAS HOOKED UP TO THE SAFE! BETTER GRAB THE GEMS INSIDE AND SCRAM FAST!

CLANG CLANG CLANG CLANG

WITHIN MOMENTS, A SHIMMERING GREEN SHAFT GOES STREAKING FROM POLICE HEADQUARTERS OVER THE CITY...

LOOK, OLIVER... THE ARROW-SIGNAL!

THE POLICE ARE PAGING US... LET'S GO, ROY!

INSIDE, WEALTHY OLIVER QUEEN AND HIS WARD, ROY HARPER, UNDERGO A SWIFT CHANGE OF GARB...

WONDER WHERE *GREEN ARROW* AND *SPEEDY* ARE NEEDED THIS TIME?

WE'LL CHECK OUR POLICE RADIO ON OUR WAY OUT!

IN THEIR FAMED *ARROW-CAR*, THE CRIME-FIGHTING ARCHERS STREAK TO THE SCENE OF THE CRIME...

UP THERE, *G.A.* -- THAT PATROLMAN ALREADY HAS THE THIEF AT BAY!

GUESS WE WEREN'T NEEDED HERE AFTER ALL, *SPEEDY!*

BUT JUST THEN...

LOOK, G.A.! THE POLICEMAN SUDDENLY SEEMED TO FREEZE IN FRIGHT--AND DROPPED HIS GUN!

AND THE CROOK'S MAKING A BREAK! GET SET TO FLY, *SPEEDY!*

ME-OWW

A PRESS OF A BUTTON -- AND THE *ARROW-CAR'S CATAPULT* HURTLES THE PAIR TO THE ROOF...

BOINNG

THERE HE GOES, *SPEEDY*--TOWARD THAT FIRE-ESCAPE LADDER! YOU KNOW WHAT THIS CALLS FOR...

RIGHT... *BOOMERANG ARROWS*--WITH ATTACHED *ARROWLINES!*

TWANG

2

AS THE AMAZING SHAFTS STREAK TO THEIR TARGET...

A PACKAGE, NEATLY WRAPPED AND READY FOR DELIVERY TO THE POLICE!

YIPES!

ZIP

TH-THANKS, *GREEN ARROW*-- FOR RESCUING ME ON MY FIRST CASE!

WHY--YOU'RE FRED JENNINGS... YOU WERE TOP STUDENT AT THE POLICE ACADEMY! WHAT MADE YOU FREEZE UP THAT WAY, FRED?

IT--IT'S A LONG STORY! I *CAN'T* TELL YOU... AT LEAST--NOT TILL WE'RE ALONE!

AND SO, AFTER DELIVERING THEIR CAPTIVE TO THE CITY JAIL...

NOW, FRED, WOULD YOU CARE TO TALK ABOUT IT?

SIGH MIGHT AS WELL! UP ON THAT ROOF, A *CAT* SUDDENLY APPEARED! EVER SINCE I WAS A KID, I'VE HAD A DEATHLY *FEAR* OF CATS! YOU SEE, I'D BEEN IN AN AUTO ACCIDENT--NEARLY LOST MY LIFE...

... AND THE LAST SOUND I HEARD, BEFORE BLACKING OUT, WAS A CAT CRYING! SINCE THEN, I HAVEN'T BEEN ABLE TO SHAKE THIS FEAR!

WELL, WHAT D'YA KNOW! I GOT A FEW FRIENDS WHO'LL BE REAL GLAD TO HEAR THIS!

SHORTLY...

I TELL YA, ARTIE, I HEARD IT WITH MY OWN EARS! A COP WHO'S FRIGHTENED OF CATS... THINK OF WHAT WE CAN GET AWAY WITH ON *HIS* BEAT, HUH?

YEAH--I'M THINKING, WEASEL...IN FACT, I'M ALREADY MAKING PLANS!

MEANWHILE, IN THEIR SECRET *ARROW-CAVE*, THE ACE ARCHERS MAKE PLANS OF THEIR OWN...

GOLLY, *G.A.*-- I'D SURE HATE TO SEE FRED JENNINGS DROPPED FROM THE FORCE BECAUSE OF A *BOYHOOD FEAR!*

SAME HERE, *SPEEDY*... WE'VE GOT TO *HELP* HIM-- BY KEEPING AN EYE ON HIM, AND PREPARING A FEW *SPECIAL* ARROWS!

NEXT NIGHT, ON FRED'S BEAT...

THOSE MEN GETTING OUT IN FRONT OF THE *SPORTS MUSEUM*... WHAT WOULD *THEY* WANT THERE AFTER CLOSING TIME?

HOLD IT, ALL OF YOU! WHAT ARE YOU CARRYING IN THAT SACK?

WHY-- NOTHING, OFFICER...

...JUST *THIS!*

OH, NO--!

IT WORKED! GET HIS GUN, WEASEL, AND KEEP HIM COVERED!

A SHORT WHILE LATER, AS *GREEN ARROW* AND *SPEEDY* APPROACH ON THEIR USUAL PATROL...

LOOK, *G.A.* SOME MEN BREAKING INTO THE MUSEUM, AND ONE OF THEM'S GOT THE DROP ON FRED! I'LL TAKE CARE OF HIM WITH--

HOLD IT, *SPEEDY!* WE CAN HALT THIS THEFT EASILY ENOUGH... BUT LET'S SEE IF WE CAN *HELP FRED,* TOO!

NEXT INSTANT, AN EERIE SOUND FILLS THE AIR...

ME-EE-EOWW

HUH--? A CAT--?

4

...TO NET THE TRIO IN ONE FELL SWOOP!

YIPES!... I'M ALL TANGLED UP!

SO ARE OUR PLANS! WE'RE FINISHED!

A LITTLE LATER...

IT WAS AMAZING, *GREEN ARROW!* THAT CAT CAME DOWN FROM-- FROM *NOWHERE*... AND MAYBE SAVED MY LIFE!

IT DIDN'T EXACTLY COME FROM *NOWHERE*, FRED!

THE "CAT" WAS THIS *DUMMY*, ATTACHED TO A THIN SHAFT THAT YOU COULDN'T MAKE OUT--AND FIRED BY *ME!*

AND THE "MEOW" YOU HEARD WAS THE WIND WHISTLING THROUGH THIS SPECIALLY FLUTED ARROW--WHICH *I* FIRED!

WE FIGURED ONE WAY TO BREAK YOUR FEAR WAS TO HAVE YOUR LIFE *SAVED* BY A CAT!

WELL, IT WORKED FINE, *GREEN ARROW!* SEE?... I'M NOT EVEN FRIGHTENED BY THIS ONE THE CROOKS UNLEASHED AT ME! FROM NOW ON, KITTY, WE'RE *PALS!*

the End

THE DICTIONARY DEFINES "BOOBY" AS A STUPID FELLOW! IT ALSO DEFINES "TRAP" AS A SURPRISE SNARE TO CATCH AN INTENDED VICTIM! PUT THE TWO TOGETHER..."BOOBY TRAP"... AND YOU GET A SURPRISE TECHNIQUE TO BEAT A FOOL! HOW THE UNDERWORLD USED THIS METHOD TO GET RID OF *GREEN ARROW* IS THE AMAZING STORY OF...

THE GREEN ARROW

the BOOBY-TRAP BANDITS

THAT'S THE END OF *GREEN ARROW*... BOOBY-TRAPPED BY A BOMB HIDDEN UNDER THAT MANHOLE COVER!

BARRROOOMMMM

ONE NIGHT, AS A WAREHOUSE WATCHMAN MAKES HIS ROUNDS...

I KNEW I HEARD A DOG BARKING! WHO COULD'VE TIED HIS LEASH THERE? I'LL SET HIM FREE...

WOOF! WOOF! WOOF!

BUT AS HE UNTIES THE LEASH...

;GASP! TH-THE LEASH WAS ATTACHED TO A *GAS GRENADE!* ;GASP! BY REMOVING IT, I SET OFF AN EXPLOSION!

PWSSSSTTT

C-CAN'T KEEP MY EYES OPEN... SO SLEEPY... OHHHHHH...

OUR DOG-LEASH BOOBY TRAP WORKED... THE WATCHMAN'S OUT COLD! NOW WE CAN EASILY BREAK INTO THE BUILDING!

ZZZZ

LATER, WHEN THE POLICE, TOGETHER WITH *GREEN ARROW* AND *SPEEDY,* ARRIVE...

IMAGINE... I WAS ASLEEP WHILE THEY CRACKED THE WAREHOUSE SAFE!

THAT'S HOW THE *BOOBY-TRAP BANDITS* WORK! FIRST, THEY SPRING A TRAP TO KNOCK OUT ANY OPPOSITION... THEN COMES THE ROBBERY! BUT ONE DAY, *WE'LL* SURPRISE *THEM!*

NEXT MORNING, AS AN ARMORED CAR PROCEEDS DOWN A MIDTOWN STREET...

JIM! THAT GUY... HE'S FALLING RIGHT INTO OUR PATH!

I--I CAN'T AVOID HITTING HIM!

SCREECH

AS THE GUARDS LEAP OUT TO HELP THE "INJURED MAN"...

SAY... THIS ISN'T A HUMAN BEING... IT'S A *DUMMY!*

NO, FRIEND--*YOU'RE* THE DUMMY... FOR FALLING FOR OUR GAG! DROP YOUR HOLSTERS!

GASP! IT'S A *STICKUP!*

SHORTLY...

NOW WE'LL DRIVE THIS TANK TO OUR PRIVATE GARAGE, AND--

THERE COMES THE *ARROW-CAR!* THOSE BOW-AND-ARROW BOYS MUST'VE BEEN PATROLLING THE AREA!

THE ARMORED CAR IS SPURTING AHEAD... THEY MUST KNOW WE'RE AFTER THEM, *SPEEDY!*

DON'T WORRY, *G.A.*-- MY *COLORED DISC ARROWS* WILL STOP THEM!

TWANG

TWANG

AJAX TRUCKING

2

SEE?... AS THE DARK DISC COVERS THE GREEN LIGHT, THE *RED* DISC IS PASTED TO THE LOWER LIGHT, ALL THE VEHICLES AHEAD OF THE ARMORED CAR WILL OBEY THE RED SIGNAL AND STOP!

NICE WORK, *SPEEDY!* NOW FIRE SOME *GAS PELLET ARROWS* THROUGH THE SLITS OF THE ARMORED CAR!

RIGHT!

SHORTLY...

WE'LL OPEN THE DOOR WITH OUR *ACETYLENE TORCH ARROWS*, THEN WE'LL HAND-CUFF OUR TWO SLEEPING BEAUTIES!

THAT AFTERNOON, IN THE HIDEOUT OF THE *BOOBY-TRAP BANDITS*...

BLAST *GREEN ARROW!* I KNEW THAT SOONER OR LATER HE WOULD INTERFERE! WELL, OUR JOB NOW IS TO GET RID OF HIM!

SURE, BOSS! GOT ANY BRIGHT IDEAS?

HE HAS, INDEED... AND THAT EVENING, AT THE HOME OF WEALTHY OLIVER QUEEN AND HIS WARD, ROY HARPER...

LOOK, OLIVER... THE *ARROW-SIGNAL!* POLICE HEADQUARTERS WANTS US TO CONTACT THEM!

CHECK, ROY... LET'S GET INTO OUR COSTUMES!

AFTER SWITCHING TO THEIR FAMED SECRET IDENTITIES...

THE POLICE GOT A PHONE CALL TIP THAT THE *BOOBY-TRAP BANDITS* WILL RAID THE *FORT FANTASTIQUE NIGHT CLUB*... SO WE'LL INVESTIGATE!

3

AFTER PARKING THE **ARROW-CAR**...

HMM... THE STRETCH OF SAND LEADING TO THE DRAWBRIDGE LOOKS SUSPICIOUSLY SMOOTH! IT WON'T HURT TO BE CAUTIOUS ... AFTER ALL, WE DON'T KNOW **WHO** PHONED IN THAT TIP!

HEY! WHAT'S THE IDEA OF FIRING THAT SHOWER OF ARROWS?

JUST A HUNCH, **SPEEDY**... **HIT THE SAND!**

TWANG

TWANG

TWANG

G-GOLLY!... THE WHOLE AREA IS FULL OF **MINES!**

RIGHT, **SPEEDY**... THAT ANONYMOUS TIP WAS JUST A COME-ON, TO LURE US INTO AN AMBUSH!

BAM

BAMM

BAM

SHORTLY, AS THE BATTLING BOWMEN APPROACH THE "FORT"...

GREEN ARROW OUTWITTED US! WE INTENDED TO TOUCH OFF THOSE MINES BY ELECTRONIC SIGNALS, AS HE CROSSED THE SAND... NOW WE'LL HAVE TO TRY ANOTHER WAY!

LOOK OUT, **SPEEDY!**... THEY'RE DUMPING US INTO THE MOAT!

WE'LL BLAST 'EM AS THEY STRUGGLE IN THE WATER!

BUT BEFORE THE GUNMEN CAN FIRE...

I'LL MAKE 'EM JOIN US, WITH THIS **NET ARROW!**

WHASSS

④

ONE OF THE FISH IS GETTING AWAY, *SPEEDY!*

YIIII!

DON'T WORRY, *G.A.--* I'LL SHOW HIM WHERE TO GO!

I'LL FIRE THESE SMALL *NEON LIGHT ARROWS* AHEAD OF HIM!

TWANG

TWANG

IN THE DARK, HE WON'T SEE THAT THE NARROW WALK GOES AROUND THE CORNER OF THE *FORT!* HE'LL JUST FOLLOW THE "SIGNS"...

EEEOOOWW!

SOUNDS LIKE I GOT HIM, *G.A.!*

LATER, AT THE BANDIT HIDEOUT...

HERE IS OUR TV MOBILE-UNIT COVERING THE ARREST OF MEMBERS OF THE *BOOBY-TRAP GANG!* ACCORDING TO *GREEN ARROW,* THEIR LEADER STILL REMAINS AT LARGE!

BAH!... I'LL SEE TO IT THAT THEY REPORT THAT ARCHER'S DEATH TOMORROW!

NEXT MORNING, AS *GREEN ARROW* PATROLS THE SUBURBS...

TOO BAD *SPEEDY* MUST ATTEND SCHOOL TODAY, AS ROY HARPER... I COULD USE HIS SHARP EYE TO HELP ME SPOT THE *BOOBY TRAPPERS!* THEY'LL UNDOUBT-EDLY MAKE ANOTHER ATTEMPT ON OUR LIVES! HMMM ...

MEN AT WORK

5

THAT NIGHT, BACK IN THE SAME AREA...

IT WORKED! *GREEN ARROW* RESPONDED TO A PHONY S.O.S. ON THIS STREET... ONLY TO BE BLOWN UP BY A BOMB I SET OFF UNDER THAT MANHOLE COVER!

BARROOM

THAT'S THE END OF *GREEN ARROW!* I BOOBY-TRAPPED HIM--BUT GOOD!

WAIT, BOSS--! SOMETHING'S TURNING THE CORNER!

HANDCUFF ARROWS!

ZIP

GREEN ARROW AND *SPEEDY!* THEN ¿GASP! WHO ARE THOSE FIGURES IN THE BURNING CAR?

THAT'S ONLY A *FAKE ARROW-CAR,* OUTFITTED WITH DUMMIES, AND RADIO-CONTROLLED FROM *MY* CAR! YOUR MEN GOOFED THIS MORNING, WHEN THEY INSTALLED A MANHOLE COVER ALONG MY PATROL ROUTE! I KNEW THERE WAS *NO* MANHOLE SYSTEM IN THIS ENTIRE AREA...

...SO I DECIDED TO BECOME A HUMAN BOOBY TRAP...FOR *YOU!*

THE END

THE GREEN ARROW

AN ODD LOOK IN HIS EYES... A WEIRD NOISE IN HIS EARS... AND INEVITABLY **CATASTROPHE** FOLLOWED PROFESSOR MARLO'S WARNINGS! AND **GREEN ARROW**, EMERALD ARCHER OF JUSTICE, WAS SOON TO RUE THE DAY WHEN HE MET...

THE GREEN ARROW

the MAN who FORETOLD DISASTER

UP THERE!... THAT ESCAPED GORILLA-- CLIMBING THE FIRE-ESCAPES!

GREAT SCOTT!... HOW COULD YOU HAVE FORE-SEEN THIS? AND HOW ARE **WE** GOING TO CAPTURE THAT BEAST?

OVERHEARING A CONVERSATION, CURIOUS PATRONS PAUSE AT THE GATES OF AN OUTDOOR EXHIBITION...

BUT I TELL YOU, SOME TERRIBLE **CATASTROPHE** IS ABOUT TO HAPPEN HERE! PLEASE-- ALERT YOUR EMERGENCY CREW AND THE POLICE!

PLEASE, MISTER! CUT IT OUT ALREADY! YOU'VE BEEN RAVING THAT WAY FOR TEN MINUTES!

FLYING MACHINES PAST & PRESENT

BUT SHORTLY, INSIDE...

EEK!... THAT BALLOON TORE LOOSE FROM ITS MOORINGS!

AND MY TWO CHILDREN ARE TRAPPED IN IT!

A DISTRESS CALL GOES OUT--AND SOON, THE FAMED *ARROW-PLANE* COMES STREAKING TO THE RESCUE...

SET THE CONTROLS ON AUTOMATIC, *SPEEDY*, WHILE I DEFLATE THAT BALLOON!

RIGHT, *G.A.!*

SWIFT AS BULLETS, SHARP-TIPPED SHAFTS PIERCE THE GAS-LADEN CANVAS...

THUNK

HISSSSS

NOW TO GUARANTEE A SAFE DESCENT-- VIA OUR *PARACHUTE ARROWS!*

MOMENTS LATER, DOWN BELOW...

AMAZING!... *GREEN ARROW* AND *SPEEDY* BROUGHT THAT RUNAWAY BALLOON TO A PERFECT LANDING!

AFTERWARD, AT POLICE HEADQUARTERS, *THE EMERALD ARCHERS* HEAR A STRANGE STORY...

SO YOU *FORETOLD* THAT NEAR-DISASTER-- AND YOU CLAIM YOU CAN KEEP DOING IT, EH? HOW, PROFESSOR MARLO?

A BUZZING IN MY EARS... IT STARTS WHEN I APPROACH THE SITE OF ANY COMING CATASTROPHE-- AND CONTINUES UNTIL THE CATASTROPHE OCCURS!

2

YOU SEE, I DABBLE IN ANCIENT ALCHEMY! YESTERDAY, THERE WAS A SMALL EXPLOSION IN MY LABORATORY! I INHALED THE FUMES OF A NUMBER OF STRANGE CHEMICALS-- AND THIS WAS THE RESULT!

HMPH!... SOUNDS CRAZY TO ME! STILL, THE NEXT TIME YOU GET THAT BUZZING, CALL ME RIGHT AWAY...

...AND MEANWHILE, SOMMERS, PLEASE DON'T PRINT THIS STORY YET! MARLO DOESN'T NEED CURIOUS CROWDS FOLLOWING HIM!

SURE THING, CAPTAIN! I'LL HOLD OFF TILL HE'S... ER... NORMAL AGAIN!

SOME DAYS LATER, AS A CURIOUS CARAVAN WEAVES THROUGH THE CITY...

BUZZING IN MY EARS... ANOTHER CATASTROPHE IS ABOUT TO TAKE PLACE! I--I MUST INFORM THE POLICE!

THE CALL ALSO BRINGS **GREEN ARROW** AND **SPEEDY** TO THE SCENE, JUST AS...

UP THERE... THAT GORILLA ESCAPED FROM ITS CAGE!

SWIFTLY, THE ACE ARCHERS GO INTO ACTION...

A COUPLE OF **NET ARROWS** SHOULD TANGLE HIM UP SO HE CAN'T MOVE!

3

NOW, TWO HARMLESS *SLEEPING GAS ARROWS,* TO TAKE THE FIGHT OUT OF HIM!

SPLAT

SHORTLY, AS A CRANE LOWERS THE UNCONSCIOUS BEAST TO ITS CAGE...

IT HAPPENED THE SAME WAY... THAT BUZZING LASTED UNTIL THE MOMENT THE GORILLA ESCAPED!

INCREDIBLE... BUT THIS TIME, IT PROBABLY HELPED SAVE A LOT OF LIVES AND PROPERTY, PROFESSOR!

JUST THEN, HOWEVER...

GREEN ARROW! WHILE WE WERE ALL BUSY HERE, THREE CROOKS ROBBED THE *CITY BANK!* MADE A CLEAN GETAWAY, THEY DID!

WHAT? HOW COULD THEY HAVE KNOWN WE'D BE TIED UP ELSE-WHERE?

RETURNING TO THEIR SECRET *ARROW-CAVE,* THE PUZZLED ARCHERS RESUME THEIR EVERYDAY IDENTITIES OF WEALTHY OLIVER QUEEN AND HIS WARD, YOUNG ROY HARPER...

THAT ROBBERY *COULDN'T* BE A COINCIDENCE, OLIVER... IT WAS TOO WELL-TIMED!

RIGHT, ROY... SOMEONE-- MAYBE MARLO HIM-SELF--TIPPED OFF THE CROOKS! NEXT TIME HE GETS A PREMONITION, WE'LL HAVE TO USE A DIFFERENT STRATEGY!

IT IS THE FOLLOWING MORNING, WHEN THE *ARROW-SIGNAL* AGAIN SUMMONS THE BATTLING BOWMEN...

YOU KNOW THE PLAN, *SPEEDY...* WHILE YOU RIDE IN A SQUAD CAR, I'LL HANG BACK AT HEADQUARTERS IN THE *ARROW-CAR!*

RIGHT, G.A.! WE'LL STAY IN TOUCH BY BELT-RADIO!

MINUTES LATER, AFTER THE PAIR HAS SEPARATED...

OH, OH... THAT REPORTER, PETE SOMMERS -- A POSSIBLE SUSPECT! HE'S LEAVING HEADQUARTERS IN A BIG HURRY-- HEADING FOR HIS CAR! I'D BETTER KEEP ON HIS TAIL...

4

AND WHILE UNIQUE *FIRE-FOAM ARROWS* QUELL THE FLAMES...

... TWO ORDINARY SHAFTS WHIZZ BETWEEN TREES, AND...

WHAT--? THEY GOT US PINNED! WE'RE FINISHED!

THUD

THUNK

AND SO, AS THEY TAKE THE CROOKS IN TOW...

OFFICER, YOU'LL FIND A THIRD MAN OUT COLD, BACK AT THEIR HIDEOUT ON DENTON ROAD... BUT, *SPEEDY*, YOU HAVEN'T TOLD ME -- WHAT BROUGHT *YOU* HERE?

SIMPLE, *G.A.*! MARLO HAD ANOTHER PREMONITION OF DISASTER-- IT WAS THIS FIRE!

6

YES--BUT THIS TIME, THE BUZZING IN MY EARS STOPPED *LONG BEFORE* THE FIRE STARTED! MY POWER, I'M AFRAID, HAS WORN OFF FOREVER!

TOO BAD YOU CAN'T DUPLI-CATE THE FORMULA THAT PRODUCED IT, MARLO ... IT WAS A MIGHTY HANDY POWER, WHILE IT LASTED!

the END

THE GREEN ARROW

THE GREEN ARROW

WITH AMAZING PRECISION, THEY SWOOP DOWN ON PLACES WHERE NO SANE MAN COULD HOPE TO PULL OFF A SUCCESSFUL CRIME! FOR THEY ARE EXPERTS, GUIDED BY A MASTERMIND WHO BRINGS **GREEN ARROW**, EMERALD ARCHER OF JUSTICE, TO THE BRINK OF DISASTER, WHEN HE TACKLES...

THE CASE OF THE CRIME SPECIALISTS

BUT AS THE *ARROWPLANE* ALIGHTS ON ITS PONTOONS...

LOOK--G.A.!... HE'S DIVING OVERBOARD!

TOWARD THAT *PERISCOPE!* THERE'S A *SUBMARINE* WAITING FOR HIM, *SPEEDY!*

HE GOT AWAY FROM US, ALL RIGHT--

THIS IS THE SECOND SUCCESSFUL CRIME, IN THE PAST WEEK THAT STARTED OUT LOOKING IMPOSSIBLE! I'M BEGINNING TO SUSPECT IT WON'T BE THE *LAST* ONE!

INDEED, SOME HOURS LATER, IN A SECLUDED HIDEOUT...

EXCELLENT WORK, MY FRIEND! I WAS SURE A CRIMINAL WITH YOUR UNIQUE TALENT COULD PULL OFF THIS HEIST!

I SPENT YEARS BUILDING UP THIS FILE OF CRIME SPECIALISTS! THEIR SPECIAL SKILLS WILL ENABLE ME TO PULL JOBS THAT ORDINARY CROOKS WOULDN'T DREAM OF ATTEMPTING!

SKIN-DIVERS

ACROBATS

STUNT FLYERS

CHEMISTS

THE FOLLOWING WEEK, AS THE *ARROW-SIGNAL* AGAIN SUMMONS *GREEN ARROW* AND *SPEEDY* TO ACTION...

ACCORDING TO THE POLICE RADIO, G.A., A BURGLAR RAIDED THE NEW GOLD DISPLAY AT THE *MONARCH EXHIBITION BUILDING!* BUT THAT PLACE IS SO WELL GUARDED, WHO'D DARE TRY A ROBBERY THERE?

THE SAME MAN WHO PULLED THOSE LAST TWO "IMPOSSIBLE" JOBS, *SPEEDY!*

REACHING THE SCENE, THE *ARROW-CAR'S* CATAPULT SENDS THE ARCHERS HURTLING UP...

THERE'S ONLY ONE SAFE ESCAPE ROUTE FROM HERE -- AND THAT'S WHERE WE'RE HEADING!

2

SURE ENOUGH, AS THEY REACH THE TOP...

JUST AS I FIGURED... HE'S ESCAPING BY **HELICOPTER!**

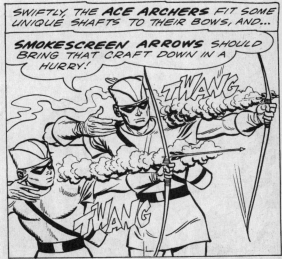

SWIFTLY, THE **ACE ARCHERS** FIT SOME UNIQUE SHAFTS TO THEIR BOWS, AND...

SMOKESCREEN ARROWS SHOULD BRING THAT CRAFT DOWN IN A HURRY!

TWANG

TWANG

NEXT MOMENT...

ALL THAT SMOKE... CAN'T SEE! GOTTA LAND BACK ON THE ROOF!

HISSSSS

SHORTLY...

SO MUCH FOR **THIS** "IMPOSSIBLE" CRIME! SOON AS WE FIND OUT WHO THEIR BOSS IS, WE CAN WRAP UP THE CASE!

BUT, AT POLICE HEADQUARTERS...

THE MASTERMIND BEHIND THESE CRIMES, WHOEVER HE IS, HAS THESE BOYS WELL-TRAINED, **GREEN ARROW!** WE CAN'T GET A WORD OUT OF THEM!

TOO BAD...THAT MEANS WE CAN EXPECT MORE OF THE SAME!

RETURNING HOME, **GREEN ARROW** AND **SPEEDY** RESUME THEIR EVERY-DAY IDENTITIES OF WEALTHY OLIVER QUEEN AND HIS WARD, ROY HARPER...

GOSH, OLIVER... AT THIS RATE, WE CAN NEVER BE SURE THAT WE'VE STOPPED THAT CRIME WAVE!

I KNOW, ROY... WE MUST FIGURE OUT SOME WAY TO TRACK DOWN THE MASTERMIND HIMSELF!

3

MEANWHILE...

SO **GREEN ARROW** OUT-SMARTED TWO OF MY EXPERTS, EH? WELL, HE'LL PAY FOR THAT! I CAN'T **WAIT** TILL HE TRIES TO STOP ME AGAIN!

THE CRIME CHIEF'S CHANCE COMES A FEW DAYS LATER, WHEN...

WHAT'S UP, VOLAR?

A NEWS ITEM...ABOUT A VALUABLE GOLD STATUE BEING DELIVERED TO THE CITY MUSEUM TONIGHT, UNDER THE PERSONAL SUPERVISION OF **GREEN ARROW**! WE'LL SEE NOW WHO'S THE CRAFTIER OF THE TWO OF US!

THAT NIGHT, A STRANGE CARAVAN WEAVES ITS WAY TOWARD THE CITY...

WITH **SPEEDY** TRAILING US IN THE **ARROW-CAR** AND **GREEN ARROW** KEEPING A LOOKOUT IN THE **ARROW-PLANE**, IT WOULD **REALLY** BE IMPOSSIBLE TO ROB US!

BUT SUDDENLY...

WOW! THAT CAR CAME OUT OF NOWHERE... AND KNOCKED THE ARMORED TRUCK OUT OF COMMISSION!

CRASH

¡COUGH-COUGH! **TEAR GAS GRENADES!** GOT TO ¡CHOKE! RETREAT!

POOF

POOF

A MOMENT LATER, THE **ARROW-PLANE** NOSES DOWNWARD, DISCHARGING A HAIL OF SHAFTS...

TAKE CARE OF THAT NUISANCE UP THERE...AND HURRY!

4

THE GREEN ARROW

NO TASK WAS TOO DIFFICULT FOR HIM-- NO RISK TOO GRAVE! FOR HE HAD AN URGENT MISSION TO ACCOMPLISH-- A MISSION WHICH MADE *GREEN ARROW* AND *SPEEDY*, BATTLING BOWMEN OF LAW AND ORDER, VERITABLE PARTNERS OF...

THE MAN WHO DEFIED DEATH

GOLLY, *G.A.*-- WHAT MADE FRED COME OUT ON A NIGHT LIKE THIS?

BEATS ME, *SPEEDY!* RIGHT NOW, OUR JOB IS TO *RESCUE* HIM!

PATROLLING THE CITY, ONE DAY, *GREEN ARROW* AND *SPEEDY*, EMERALD ARCHERS OF JUSTICE, STOP TO JOIN A WATCHING THRONG...

LOOK AT THAT MAN-- ABLE TO FLY WITH THAT PROPELLER MACHINE ON HIS BACK, *G.A.!*

AMAZING, *SPEEDY*... AMAZING!

SUDDENLY...

PUT, PUT, PUT

THE THING'S OUT OF COMMISSION! HE'S GOING TO *FALL!*

ONLY ONE WAY TO BREAK HIS FALL, **SPEEDY**... WITH **BALLOON ARROWS!** AIM AT THE MACHINE!

TWANG TWANG TWANG

WITH AMAZING ACCURACY, THE UNIQUE **BALLOON ARROWS** EXPAND AS THEY STREAK TO THEIR TARGET...

... AND BRING HIM TO A SAFE LANDING!

AFTERWARD, AS REPORTERS INTERVIEW THE DARING FLIER...

I'M GRATEFUL TO **GREEN ARROW** FOR SAVING MY LIFE, GENTLEMEN—BUT I PLAN TO TRY EVEN RISKIER STUNTS, IF NECESSARY! YOU SEE, THE INVENTOR OF THAT MACHINE PAID ME WELL TO TEST IT...

... AND I NEED LOTS **MORE** MONEY TO PAY FOR MY SON'S OPERATION! SO IF ANYONE HAS ANY RISKY PROPOSITIONS, HE CAN CALL ON ME—FRED JENKINS—FOR THE RIGHT PRICE, OF COURSE!

THE NEWSPAPER PUBLICITY GETS WIDE ATTENTION, AND, THE FOLLOWING NIGHT, SOME MILES OUT TO SEA ...

ONLY A FOOL WOULD BE OUT IN A SMALL CRAFT ON A NIGHT LIKE THIS, **SPEEDY!** NO WONDER HE HAD TO SEND OUT AN **S.O.S.!**

2

AS THE **ARROW-PLANE** REACHES ITS DESTINATION, **FLARE ARROWS** LIGHT UP THE SCENE...

GREAT SCOTT! AN ABANDONED FREIGHTER, WRECKED ON THOSE REEFS-- AND A SMALL CABIN CRUISER, **TRAPPED** NEARBY!

THERE'S A **MAN** ON THAT CRUISER!

OUR FIRST JOB IS TO FREE THAT CRUISER...

...WITH **TNT ARROWS!**

BOOM

BOOM

NOW, WITH A PAIR OF **ARROWLINES...**

... WE CAN TOW IT SAFELY TO SHORE!

3

BACK ON LAND, SOME TIME LATER...

FRED JENKINS--! WHAT IN THE WORLD WERE YOU DOING OUT THERE?

THE OWNERS OF THAT WRECKED FREIGHTER WILL PAY ME A FORTUNE FOR THE CARGO OF GEMS I SALVAGED FOR THEM! THANKS AGAIN FOR YOUR ASSIST, GREEN ARROW!

RETURNING HOME, THE ARCHERS RESUME THEIR EVERYDAY IDENTITIES OF WEALTHY OLIVER QUEEN AND HIS WARD, ROY HARPER...

GOSH, OLIVER... I FEEL SORRY FOR FRED JENKINS--BUT WE CAN'T BE AROUND TO PROTECT HIM ALL THE TIME!

I KNOW, ROY... LET'S JUST HOPE HE GETS THAT MONEY HE NEEDS AS FAST AS POSSIBLE!

FRED'S NEXT CHANCE COMES A FEW DAYS LATER, WHEN...

JENKINS TO DR. DAVIS...YOUR CLIFF CLIMBER WORKS FINE! LET ME KNOW WHEN YOU WANT ME TO TEST THE ARMS OF THIS CONTRAPTION!

SOON...

I'M NOW PARKED OUTSIDE THE CHANEY ESTATE...SHALL I START DOWN?

NO-- HERE ARE YOUR ORDERS...

AND SOME DISTANCE AWAY, IN THE HEAD-QUARTERS OF DR. DAVIS WHO HAD HIRED FRED...

CHANEY HAS AN ORIENTAL JADE STATUE ON THE GROUNDS, WORTH MILLIONS! YOU'LL USE THE VEHICLE'S ARMS TO STEAL IT!

WHAT?? I CAN'T DO THAT!

YES YOU CAN-- AND YOU WILL! I SENT YOU THIS WAY BECAUSE THE REGULAR ROAD UP THAT MOUNTAIN IS GUARDED AT THE BOTTOM! WHAT'S MORE, I SEALED YOU IN! IF YOU DON'T STEAL THAT STATUE, I NEED ONLY PRESS A BUTTON THAT WILL BLOW UP THE VEHICLE, WITH YOU IN IT! THINK IT OVER!

4

LONG, TENSE MINUTES ELAPSE-- TILL FINALLY...

I COULDN'T WAIT ANY LONGER! BESIDES, NOW THAT I KNOW MY INVENTION WORKS, I CAN BUILD ANOTHER AND OPERATE IT MYSELF!

BOOM

BUT JUST THEN...

YOU WON'T BUILD ANOTHER ONE WHERE YOU'RE GOING, DAVIS!

GREEN ARROW AND SPEEDY... AND FRED JENKINS!

A COUPLE OF BOXING GLOVE ARROWS BRINGS 'EM DOWN FOR THE COUNT OF TEN!

UGH!

AND A PLAIN, REGULAR SHAFT FINISHES THE BOSS HIMSELF!

THUNK

I--I DON'T UNDER-STAND... HOW DID FRED ESCAPE DEATH?

IT HAPPENED WHEN HE SPOTTED SPEEDY AND ME KEEPING TABS ON HIM FROM THE ARROW-PLANE...

"THINKING FAST, HE SENT US A QUICK MESSAGE..."

HE RIPPED UP THAT FLAGPOLE AND TURNED IT UPSIDE-DOWN! AN UPSIDE-DOWN FLAG IS THE STANDARD *DISTRESS* SIGNAL, *SPEEDY!* SOMETHING'S WRONG! WE'D BETTER GET HIM OUT OF THERE!

"A FEW *ACETYLENE TORCH ARROWS* DID THE TRICK, BURNING OFF THE DOME..."

HE'S FREE! NOW LET'S PICK HIM UP AND HEAR WHAT THIS IS ALL ABOUT!

THE REST YOU CAN FIGURE OUT YOURSELF! BY THE TIME YOU BLEW UP YOUR INVENTION, WE WERE ON OUR WAY OVER HERE!

AS FOR YOU, FRED, THE REWARD YOU'LL GET, FOR HELPING TO CAPTURE THIS GANG, SHOULD BE MORE THAN ENOUGH TO COVER YOUR SON'S OPERATION!

I DON'T MIND TELLING YOU-- THAT LAST JOB CURED ME OF RISKS FOREVER!

The End

THE GREEN ARROW

THE GANGSTERS WERE ABLE TO ANTICIPATE *GREEN ARROW'S* EVERY MOVE -- AND BLOCK HIM WITH A BETTER ONE OF THEIR OWN! FOR THERE SEEMED NO WAY OUT OF THIS TRAP, BECAUSE *THE BATTLING BOWMAN* HAD MADE HIMSELF A...

DUPE OF THE DECOY BANDITS

GREEN ARROW IS TURNING BACK! ¡HA, HA!¡ THE FOOL PLAYED RIGHT INTO OUR HANDS AGAIN!

BOOOM

ALONG A DESERTED STRETCH OF ROAD, ONE NIGHT, A UNIQUE EXPERIMENT REACHES ITS CLIMAX...

THE SEDAN'S COMING OFF THE BRIDGE! NOW -- RADIO THE PROBLEM!

HALT, GREEN ARROW! THE CROOKS HAVE JUST BLOWN UP THE BRIDGE! HOW WILL YOU STOP THEM NOW?

SCREECH

WE'LL HAVE TO STOP 'EM COLD, *SPEEDY*-- WITH *FIRECRACKER ARROWS!*

TWANG

TWANG

IT WORKED! WE DAZZLED THE DRIVER SO THOROUGHLY, HE WAS FORCED TO STOP!

CRACK CRACK CRACK CRACK CRACK

SCREEECH

AND AS THE FLEEING MEN TRY TO ESCAPE...

A COUPLE OF *BOLO ARROWS* BRINGS THEM DOWN! BY THE TIME THEY UNTANGLE THEMSELVES, THE POLICE WOULD HAVE THEM IN TOW!

EEYOW!

MINUTES LATER, THE OBSERVATION COPTER LANDS, AND...

THIS COMPLETES YOUR SERIES OF PSYCHOLOGICAL REACTION TESTS, *GREEN ARROW!* THANK YOU... YOUR COOPERATION HAS HELPED OUR STUDIES IMMENSELY!

WE'RE ALWAYS GLAD TO PARTICIPATE IN WORTH-WHILE EXPERIMENTS, PROFESSOR DAWES!

IT IS THE FOLLOWING DAY WHEN A *REAL* CRIME BRINGS *GREEN ARROW* AND *SPEEDY* IN AIRBORNE PURSUIT...

THE POLICE RADIO SAID THE CROOKS RAIDED THIS WATERFRONT WAREHOUSE, THEN MADE THEIR GETAWAY IN A *SUBMARINE!*

ONLY ONE WAY TO LOCATE THAT SUB, *SPEEDY*...

...WITH *DEPTH CHARGE ARROWS!*

2

THESE **SLEEP ARROWS** RELEASE A HARMLESS GAS THAT'LL PUT YOU ALL INTO A PLEASANT SLEEP FOR A FEW MINUTES!

OH-H-H...

SPLAT

WHY--THIS THING IS **EMPTY!** THEY DIDN'T STEAL ANYTHING!

AND THEIR STUNT WAS ANOTHER DUPLICATE OF ONE OF OUR PSYCHO-LOGICAL TESTS! WHAT WERE THEY **REALLY** UP TO?

THE ANSWER COMES A FEW MINUTES LATER, AS...

GREEN ARROW! WHILE THIS PAIR KEPT YOU BUSY HERE, ANOTHER ROBBERY WAS PULLED OFF-- SUCCESSFULLY--JUST A FEW BLOCKS AWAY!

AND DIDN'T WE KNOW IT! WE'LL GET PAID PLENTY FOR THIS, WHEN WE GET OUT OF JAIL!

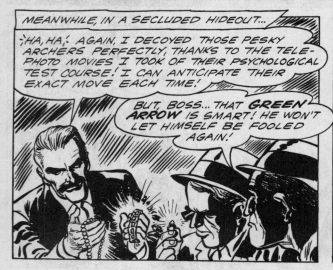

MEANWHILE, IN A SECLUDED HIDEOUT...

¡HA, HA! AGAIN, I DECOYED THOSE PESKY ARCHERS PERFECTLY, THANKS TO THE TELE-PHOTO MOVIES I TOOK OF THEIR PSYCHOLOGICAL TEST COURSE! I CAN ANTICIPATE THEIR EXACT MOVE EACH TIME!

BUT, BOSS...THAT **GREEN ARROW** IS SMART! HE WON'T LET HIMSELF BE FOOLED AGAIN!

WON'T HE? FOR MY NEXT JOB, I'VE PLANNED A LITTLE PSYCHOLOGY OF MY OWN AGAINST **GREEN ARROW!** LISTEN...

4

A WEEK LATER, FOLLOWING A BANK ROBBERY ALARM...

HERE COMES THE *ARROW-CAR*...I'LL START FLIPPING THAT *TNT!*

JUST AS WE FIGURED, *SPEEDY*... ANOTHER DUPLICATION -- ANOTHER DECOY MANEUVER!

BOOM

SCREECH

YOU CAN SLOW DOWN, ALFIE... JUST LIKE THE BOSS SAID, THERE WON'T BE ANY *FIRECRACKER ARROWS!*

YEAH... *GREEN ARROW* IS TURNING BACK TO CHECK WHERE A REAL CRIME WAS COMMITTED! ;HAW! THAT'S RICH!

INDEED, NOT LONG AFTER, BACK AT THE HIDE-OUT... ;HA, HA; MY PSYCHOLOGY WORKED LIKE A CHARM! *GREEN ARROW* OUT-SMARTED HIMSELF -- BECAUSE THIS TIME, THERE WAS *NO* DECOY!

YEAH...;HA, HA; HE MUST STILL BE OUT LOOKIN' FOR A DIFFERENT GETAWAY STUNT!

WRONG, GENTLEMEN! I'M RIGHT BEHIND YOU!

HUH? IT'S *THEM!*

OW-W-W!

I DON'T GET IT... YOU COULDN'T HAVE FOLLOWED MY MEN OVER THAT BLOWN-UP BRIDGE! AND THEY SAW YOU TURN BACK!

WE *DID* TURN BACK--BUT WE WEREN'T TAKING ANY CHANCES EITHER...

"... SO EVEN BEFORE THEY REACHED THE BRIDGE, WE MADE SURE WE WOULDN'T LOSE THEM PERMANENTLY!"

IN THE DARKNESS, THEY WON'T NOTICE ME FIRING *TRAIL ARROWS*... AND THE SOUND OF THEIR CAR ENGINE WILL DROWN OUT THE ARROWS' IMPACT!

"OUR *TRAIL ARROWS* ATTACHED THEMSELVES BY MEANS OF SUCTION CUPS-- AND FROM REAR SACS, THEY DRIPPED A MARKER DYE THAT WE FOLLOWED STRAIGHT TO THIS PLACE!"

THAT WAY, WE DIDN'T HAVE TO WORRY ABOUT WHETHER THEY REALLY WERE DECOYING US OR SIMPLY USING REVERSE PSYCHOLOGY! IT ALSO LED US TO THE REST OF YOUR LOOT!

BAH!... AND I THOUGHT THAT *I* HAD GOTTEN *YOU* TO OUTSMART YOURSELF!

The End

THE GREEN ARROW

THE GREEN ARROW

HE WAS ONE OF *GREEN ARROW'S* GREATEST FANS--AND HIS ONE BIG WISH WAS TO SEE THE EMERALD ARCHER IN ACTION! AND YET, WHEN HIS WISH CAME TRUE, WHO COULD BELIEVE THE STORY OF THE BOY WHO CLAIMED TO BE...

GREEN ARROW'S SECRET PARTNER

MY PLAN WORKED! *GREEN ARROW* AND *SPEEDY* ARE NETTING THOSE CROOKS!

AT A NEIGHBORHOOD *GREEN ARROW FAN CLUB,* A THRILLING MEETING GETS UNDER WAY...

AND NOW, TERRY BURNS WILL TELL US ABOUT A PERSONAL ADVENTURE HE CLAIMS HE HAD WITH *GREEN ARROW* YESTERDAY!

I STILL CAN'T BELIEVE IT HAPPENED TO ME...

"IT ALL STARTED WHILE I WAS RIDING HOME FROM SCHOOL..."

OH, OH--SOME TROUBLE UP AHEAD, AND... THE *ARROW-SIGNAL!* THE POLICE ARE CALLING *GREEN ARROW* AND *SPEEDY!*

WHEEEEE...

DRUGS

POLICE

"I REACHED *CITY BANK* JUST IN TIME TO SEE..."

CROOKS--WITH STRANGE GADGETS STRAPPED TO THEIR BACKS! WHAT ARE THEY FOR?

"AS THE POLICE BEGAN FIRING, I GOT MY ANSWER..."

ONE-MAN JET UNITS! THEY'RE *FLYING* AWAY!

"JUST THEN, WHO DO YOU THINK ARRIVED?"

GREEN ARROW AND *SPEEDY!*

SCREECH

"WHAT A SIGHT, AS *GREEN ARROW* RELEASED A *SMOKESCREEN* ARROW..."

TWANG

"...WHICH BLINDED ONE OF THE GANG, FORCING HIM DOWN!"

"WHILE *SPEEDY*, WITH *FIRECRACKER ARROWS*, DAZZLED ANOTHER ONE..."

YOW!... CAN'T SEE WHERE I'M GOING! GOTTA LAND!

CRACK!

CRACK!

CRACK

2

"FINALLY, WITH A PAIR OF *ARROWLINES*, TIED TO *BOOMERANG ARROWS*, THEY BROUGHT DOWN TWO MORE..."

GOLLY!... WAIT TILL THE CLUB HEARS ABOUT THIS!

SOME OF THEM GOT AWAY, *SPEEDY!* SOON AS WE TURN THIS BUNCH OVER TO THE POLICE, WE'LL SEARCH FOR THE OTHERS!

"I WAS ANXIOUS TO TELL YOU ALL ABOUT IT--BUT, ON THE WAY HOME, SOMETHING CAUGHT MY EYE..."

HMM... I'D BETTER SEE WHAT'S CAUSING THAT SMOKE!

"AS I TURNED INTO A CLEARING, I GOT MY ANSWER..."

THE REST OF THE GANG! THAT SMOKE CAME FROM YOUR JETS!

HUH? GET THAT KID!

INTO THE TRUCK, SONNY! WE CAN'T HAVE YOU BRINGIN' THE LAW HERE, TO PICK UP OUR TRACKS!

NO-- PLEASE...

3

NOW THAT WE'RE STUCK WITH A WITNESS, WHAT ARE WE SUPPOSED TO DO WITH HIM?

WE'LL LET THE BOSS DECIDE! HIS YACHT IS DOCKED AT PIER 4... HE WANTS US TO JOIN HIM THERE AT DAWN!

"WE DROVE TO THEIR TEMPORARY HIDEOUT-- AN ABANDONED CABIN ON THE OTHER SIDE OF THE WOODS..."

"HA, HA! GOOD THING THE BOSS PAYS US WELL! THE OTHER BOYS-- THE ONES WHO WERE CAPTURED-- WOULD RATHER SPEND TIME IN JAIL THAN REVEAL THIS PLACE OR THE YACHT!"

WHY NOT? THEY'LL BE WELL REWARDED WHEN THEY'RE RELEASED!

"AT THAT MOMENT...!"

HEY-- LOOK! THERE'S THE ARROW-PLANE... PROBABLY SEARCHING FOR US!

SO WHAT? THE TRUCK'S WELL HIDDEN IN THE WOODS... FROM UP THERE, THIS PLACE STILL LOOKS DESERTED!

"THEY WERE SO BUSY WATCHING THE PLANE, THEY FORGOT ALL ABOUT ME..."

GREEN ARROW-- RIGHT OVERHEAD! THIS IS MY LAST CHANCE... GOT TO ALERT HIM... BUT HOW-- HOW?

"SUDDENLY, I SAW MY OPPORTUNITY, AND..."

HEY!... THE KID'S FIDDLING AROUND WITH A JET UNIT!

HE'S TRYING TO SET IT OFF! STOP HIM!

"BUT THEY REACHED ME A SPLIT SECOND TOO LATE..."

CRASH

WHOOSH

4

"AND THAT'S WHEN THE PANIC SET IN..."

COME ON -- INTO THE TRUCK! BY THE TIME THAT PLANE LANDS, WE CAN BE FAR AWAY FROM HERE!

"EVEN I THOUGHT THEY'D ESCAPE AGAIN-- BUT ALL AT ONCE, THE *ARROW-PLANE* BANKED LOW, AND..."

TWANG!

NET ARROWS! YOU'RE TRAPPED -- FINISHED!

"AND AFTER THE ACE ARCHERS LANDED..."

THEY CAPTURED ME, *GREEN ARROW*... THEIR BOSS IS STILL AT LARGE... HE'LL BE WAITING FOR THEM AT--

WHOA, SON... ONE THING AT A TIME!

AND SO WITH THE INFORMATION I GAVE THEM, *GREEN ARROW* AND *SPEEDY* WERE ABLE TO PICK UP THE BOSS OF THE GANG, TOO!

WHAT ARE YOU HANDING US, TERRY?

5

HERE'S THE STORY OF THAT BANK HOLDUP! IT SAYS NOTHING ABOUT YOU, OR THAT CABIN-HIDEOUT, OR THE CAPTURE OF THE OTHERS! YOU TRYING TO MAKE YOURSELF OUT A HERO OR SOMETHING?

BUT-- BUT...

CITY COURIER ★★
GREEN ARROW CAPTURES BANK BANDITS
BUT 3 MANAGE TO ESCAPE

TERRY *IS* A HERO, BOYS! EVERY WORD HE SAID IS TRUE!

GREEN ARROW HIMSELF!

YOU SEE, IF WE HAD REVEALED OUR CAPTURE OF THE OTHER THREE GANGSTERS, THEIR BOSS MIGHT NOT HAVE WAITED FOR THEIR DAWN RENDEZVOUS...

SO WE ASKED TERRY TO KEEP THE STORY QUIET UNTIL *WE* WENT TO THAT YACHT ON PIER 4!

AS SOON AS WE MADE THE CAPTURE, *GREEN ARROW* TOLD TERRY AND THE NEWSPAPERS THEY COULD REVEAL THE *FULL* STORY!

Daily Bugle
YOUNG HERO HELPS CAPTURE REMAINING BANDITS

TERRY BURNS

THE END

THE GREEN ARROW

THE GREEN ARROW

BEHIND HIS DESK AT HEADQUARTERS HE SITS, DREAMING OF HIS PAST GLORIES AS A FIGHTING COP! SUDDENLY, HE SEES HIS CHANCE TO GET BACK INTO ACTION, LITTLE DREAMING THAT THE LIVES OF *GREEN ARROW* AND *SPEEDY* WILL REST IN THE HANDS OF...

the TOO-OLD HERO

;HA, HA; WHEN THE *ARROW-CAR* REACHES OUR LAND MINE, THERE'LL BE TWO DEAD ARCHERS!

GREEN ARROW IS RIDING TO HIS DOOM -- AND THERE'S NO WAY TO STOP HIM!

IN AN EXCLUSIVE PET SHOP, TWO CROOKS ATTEMPT A DARING DAYLIGHT HOLDUP, WHEN SUDDENLY...

HUH?

STAND WHERE YOU ARE! DROP YOUR GUN OR I START SHOOTING!

MY GUN? SURE-- FETCH IT, FIDO!

WOOF

YOW!

THAT'LL KEEP HIM BUSY AWHILE... C'MON!

BUT JUST THEN...

THOSE SACKS OF DOG MEAL SHOULD DO THE TRICK, *SPEEDY*-- WITHOUT HURTING ANYBODY!

IT'S *GREEN ARROW!*

TWANG

TWANG

AND AS THE SHARP-POINTED SHAFTS STRIKE THEIR TARGETS

HELP!

FUSSSHH

HA, HA, WE'LL HELP YOU, ALL RIGHT-- STRAIGHT TO THE POKEY!

AFTERWARD, AT POLICE HEADQUARTERS...

THAT WAS A FINE CAPTURE, *GREEN ARROW!* AS FOR PATROLMAN DONLEY, THIS WAS THE THIRD TIME IN THE PAST YEAR THAT YOU NEARLY LET YOUR MAN GET AWAY!

TH-THEY KIND OF TOOK ME BY SURPRISE, CHIEF...

FORGET IT... YOU'VE STILL GOT A FINE RECORD! IN FACT, WE'RE GOING TO PROMOTE YOU TO SERGEANT, GIVE YOU A DESK OF YOUR OWN, AND...

KICKING ME UPSTAIRS, EH? WELL *SIGH*, GUESS AN OLD COP LIKE ME SHOULD BE GRATEFUL FOR SUCH A "PROMOTION!"

2

AFTERWARD, AS THE ACE ARCHERS RESUME THEIR EVERYDAY IDENTITIES OF WEALTHY OLIVER QUEEN AND HIS WARD, ROY HARPER...

CHARLIE DONLEY TOOK HIS NEW DESK JOB REAL HARD, OLIVER!

NATURALLY, ROY... CHARLIE WANTED TO GO OUT LIKE A *HERO*-- NOT LIKE A COP WHO'S *LOST* HIS SKILL!

IT IS SOME DAYS LATER, WHEN *GREEN ARROW* AND *SPEEDY* AGAIN GO INTO ACTION...

THE ROBBERY WAS REPORTED IN THE *ROOF RESTAURANT*, ATOP THIS BUILDING! HOPE WE'RE NOT TOO LATE, *G.A.!*

GET SET, *SPEEDY*... I'M GOING TO RELEASE THE *CATAPULT!*

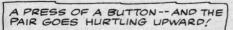

A PRESS OF A BUTTON -- AND THE PAIR GOES HURTLING UPWARD!

BOIN-NNG

THERE THEY ARE -- MAKING THEIR GETAWAY!

QUICK, *SPEEDY*... YOUR *BOLO ARROW!*

EEYOW!

AT LEAST WE'VE GOT TWO OF 'EM!

3

BY THE TIME THEY TAKE THEIR CAPTIVES IN TOW...

THE OTHER TWO ARE ESCAPING IN THEIR CAR, G.A.!

ONLY ONE WAY TO KEEP ON THEIR TAIL NOW... WITH A *RADAR ARROW!*

ITS SIGNAL CAN BE PICKED UP ON THE *ARROW-CAR'S* SPECIAL RADIO! IT'LL LEAD US TO WHEREVER THEY GO, AFTER WE TURN THEIR PALS OVER TO THE POLICEMAN ON THE BEAT!

2E 712

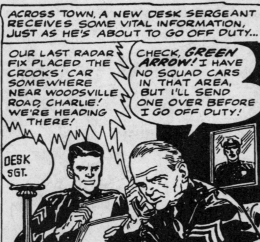

ACROSS TOWN, A NEW DESK SERGEANT RECEIVES SOME VITAL INFORMATION, JUST AS HE'S ABOUT TO GO OFF DUTY...

OUR LAST RADAR FIX PLACED THE CROOKS' CAR SOMEWHERE NEAR WOODSVILLE ROAD! WE'RE HEADING THERE!

CHECK, *GREEN ARROW!* I HAVE NO SQUAD CARS IN THAT AREA, BUT I'LL SEND ONE OVER BEFORE I GO OFF DUTY!

DESK SGT.

ALERT YOUR RELIEF, CHARLIE! SOON AS WE PINPOINT THE LOCATION, WE'LL RADIO IN!

SIGH WISH I WERE GOING WITH YOU, *GREEN ARROW!*

HMM... I KNOW A SHORT-CUT FROM HERE LEADING TO A RISE OVER *WOODSVILLE ROAD!* NO HARM IN RIDING OUT THERE AND SEEING WHAT *I* CAN SPOT!

SPEEDING OUT TO THE AREA, THE PLUCKY POLICEMAN ARRIVES IN TIME TO SEE...

TWO MEN -- DIGGING UP THE ROAD! WHAT ARE THEY UP TO?

4

LUCKY WE SPOTTED THAT *RADAR ARROW* WHEN WE STOPPED TO CHECK THE TIRES! IF *GREEN ARROW* COMES THIS WAY, THIS LAND MINE WILL *STOP HIM DEAD...;HA, HA,!*

YEAH... SOONER OR LATER, I KNEW WE'D PUT THAT LITTLE PIECE OF EMERGENCY EQUIPMENT TO GOOD USE!

THEY'RE SETTING A TRAP! GOT TO STOP THEM--

BUT AT THAT MOMENT...

OH, OH...THERE COMES THE *ARROW-CAR* NOW! HEAD FOR COVER--AND WE'LL WATCH THE BIG BOOM!

GREAT SCOTT!/... GOT TO WORK FAST!

GRABBING A NEARBY ROCK, CHARLIE TAKES QUICK CAREFUL AIM, AND...

MUST BE RIGHT ON TARGET-- WON'T GET ANOTHER CHANCE...

BOOOM

GREAT SHAFTS!/... WHERE'D **THAT** COME FROM?

GREEN ARROW! THE OTHER TWO CROOKS...THEY'RE HIDING IN THOSE WOODS!

WHAT--? NICE GOING, CHARLIE!

5

THE GREEN ARROW

A FLAMING GREEN ARROW BLAZES ACROSS NIGHT SKIES, SUMMONING TWO BOLD ARCHERS TO TAKE UP THEIR BATTLE WITH CRIME! BUT THIS TIME *GREEN ARROW* AND *SPEEDY* ARE CHALLENGED BY AN INVINCIBLE FOE AS THEY RUN HEADLONG INTO...

The IRON ARCHER!

WE WERE CLOSE ENOUGH TO THE SCENE TO CATCH THEM RED-HANDED, *G.A.!* THEY'RE STEALING VALUABLE FURS!

GREEN ARROW AND THE KID, EH? HERE'S WHERE WE TURN *IRON ARCHIE* LOOSE!

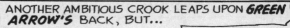

BOOMERANG ARROWS, ATTACHED TO ARROW LINES, CIRCLE THE ANKLES OF THE CROOKS, SPILLING THEM...

TWO DOWN!

ANOTHER AMBITIOUS CROOK LEAPS UPON *GREEN ARROW'S* BACK, BUT...

THREE DOWN!

THEN, FROM THE WAITING TRUCK, ROLLS A MIGHTY METAL THREAT, ITS GREAT BOW TWANGING LOUDLY...

I DON'T KNOW WHAT IN BLAZES THAT THING IS, *SPEEDY*, BUT *GET OUT OF THE WAY!*

TWANG! TWANG! TWANG! ZING! TWANG-G-G! ZING! CLANKETY-CLANK-CLANK!

AS THE FAMED ARCHER DODGES LEFT AND RIGHT, THE HAIL OF ARROWS TRACKS HIM RELENTLESSLY...

WHOO-EE! THAT IRON BABY IS SHEER *MURDER!*

ZING! ZING! ZING!

TAKING MOMENTARY COVER, THE BATTLING BOWMAN FIRES A ROPE-ATTACHED ARROW AT THE SKY...

MAYBE I CAN USE A LITTLE STRATEGY AGAINST THAT METAL MENACE!

ZING! ZING! ZING!

UP HIGH, THE ARROW INFLATES ITSELF, FORMING A GREAT *BALLOON*...AND THEN...

AN *AERIAL ATTACK* MIGHT KNOCK THAT BABY OUT!

TWANG!

BUT AN ARROW FROM *IRON ARCHIE* SHATTERS THE BALLOON, ENDING THE "STRATEGY"...AND THE FIGHT...

WHROOSH!

BLAST IT -- THEY'RE GETTING AWAY!

HA, HA! SO YOU WANTED TO SELL YOUR INVENTION TO AN ARCHERY RANGE FOR PEANUTS? INSTEAD OF TEACHING STUDENTS HOW TO BECOME BOWMEN, "PROF", *IRON ARCHIE* WILL MAKE A MILLION FOR US!

2

AND, TO BOOT, WE'VE GOT *GREEN ARROW* MYSTIFIED WITH A LITTLE SECRET HE'LL NEVER TUMBLE TO!

ON THE FOLLOWING NIGHT, IN THEIR EVERYDAY IDENTITIES AS WEALTHY OLIVER QUEEN AND YOUNG ROY HARPER, THE TWO ARCHERS PONDER A PUZZLING QUESTION...

DID YOU SEE THE WAY THAT THING'S ARROWS KEPT AFTER ME? I CAN'T UNDERSTAND HOW A *MECHANICAL* DEVICE CAN KEEP FIRING AT A MOVING MAN!

BUT THERE IS LITTLE TIME TO PONDER PUZZLES, BECAUSE...

I'LL BET THEY'RE AT IT AGAIN!

LET'S GO!

IN SHORT MOMENTS THEY AGAIN BECOME THE BATTLING BOWMEN, AND ARE SOON ROLLING FROM THE SECRET *ARROW-CAVE*...

WE'LL PICK UP THE DETAILS FROM HEADQUARTERS ON THE *ARROWCAR* RADIO, AND SHOOT RIGHT TO THE TROUBLE AREA!

WHICH MIGHT NOT DO US A LOT OF GOOD-- IF IT'S THAT GANG WITH THE METALLIC PET!

SOON...

WHAT A NERVE THEY'VE GOT-- ROBBING THE BIGGEST JEWELRY STORE IN TOWN... AND RIGHT ON THE MAIN STREET!

③

BUT IT HARDLY TAKES "NERVE"-- NOT WITH THE GREAT *IRON ARCHIE* ON HAND...

LOOK OUT! HERE IT COMES AGAIN!

TWANG! TWANG!

CLANKETY- CLANK CLANK!

AND ONCE MORE, AS THE BOWMEN LEAP FROM THE *ARROWCAR*...

BLAZES! HE'S GOT *ME* ON TARGET AGAIN!

ZING!

ZING!

I CAN KNOCK DOWN *SOME* OF HIS ARROWS-- BUT HE'S TOO FAST FOR ME TO STOP *ALL* OF THEM!

I'LL EXPLODE SOME OF HIS ARROWS WITH "ANTI-MISSILE" ARROWS, WHILE I GET BACK TO THE *ARROWCAR* FOR COVER, AND...

NOW I'VE HAD IT! HE'S CUT MY QUIVER LOOSE!

BR-RAM!

BR-RAM!

THEN...

WAIT A MINUTE! WHAT'S *THIS*?

ZING! ZING! ZING! ZING! ZING! ZING!

BEFORE THE FABLED BOWMAN CAN RECOVER HIMSELF...

THEY GOT AWAY AGAIN-- THANKS TO THAT IRON HEADACHE!

WAIT A MINUTE! SOMETHING REAL STRANGE JUST OCCURRED!

WHEN MY QUIVER FELL, I STOOD ONLY A FEW FEET FROM IT--AND NOT *ONE* ARROW WAS SHOT AT *ME*! EVERY ONE WAS FIRED AT THE *QUIVER*! I THINK I'VE GOT THE ANSWER TO THE *IRON ARCHER*!

BUT, AS THE CROOKS ATTEMPT FLIGHT...

ALL RIGHT, *SPEEDY!* WE'LL FINISH THE JOB *IRON ARCHIE* STARTED!

THEN, WITH THE GANG IN TOW...

IN THAT FIRST FIGHT YOUR MAN JUMPED ON MY BACK AND PLANTED THIS *ELECTRIC EYE* IN MY QUIVER! *IRON ARCHIE* WAS RIGGED TO FIRE AT THAT EYE... WHICH BROUGHT HIS ARROWS AT *ME!*

I TUMBLED TO THE GIMMICK WHEN HE KEPT FIRING AT MY FALLEN QUIVER! THEN I PUT THE EYE ON THE *GLIDER ARROW*... AND THIS BIG IRON BABY HELPED US ROUT YOU -- WHEN HE FIRED AT THE *GLIDER ARROW!* NICE JOB, IRON ARCHIE!

⑥ THE END

THE GREEN ARROW

ARCHERS -- BENEATH THE *SEA?* YES, IN THIS ISSUE, WHILE *AQUAMAN,* MONARCH OF THE MARINE-WORLD, TAKES TO LAND, IT'S OFF TO THE WATERY DEPTHS WITH *GREEN ARROW* AND *SPEEDY!* BUT THE BATTLING BOWMEN HAVE A FEW UNDERWATER STUNTS OF THEIR OWN AS THEY COME FACE TO FACE WITH THE SURPRISING ... AND DEADLY MENACE OF...

the *HUMAN SHARKS*

THE GREEN ARROW

STUNG WITH DEFEAT, THE BOWMEN EMERGE--DRIPPING WET...

WOW! THE WAY THEY ZIPPED OFF IN THOSE LITTLE UNDER-WATER MACHINES!

WELL, AT LEAST WE NOW KNOW HOW THE CROOKS WHO HAVE BEEN PULLING THESE RECENT JOBS SUDDENLY "VANISHED" WHEN WE WENT AFTER THEM!

YES--WE'RE DEALING WITH A CLEVER GANG OF UNDERWATER CROOKS! AND THOSE OUTFITS THEY WEAR MAKE THEM LOOK LIKE SHARKS!

BUT SHARKS CAN BE CAUGHT, SPEEDY! AND WE'VE GOT THE MEANS TO DO IT-- THANKS TO THE OVERTIME WE'VE BEEN PUTTING IN ON THE PROJECT IN THE ARROWCAVE!

LATER IN THE DAY, AS WEALTHY OLIVER QUEEN AND YOUNG ROY HARPER, THE TWO MAKE THEIR WAY TO THE SECRET ARROW-CAVE BENEATH THE QUEEN ESTATE ...

THIS IS THE BABY, ROY--THE ANSWER TO THE UNDERWATER GANG!

WELL, THE NEXT TIME THEY STRIKE, WE'LL FLOAT OUT THE ARROW-SUB AND PAY BACK A FEW SURPRISES!

BZZZT BZZT

BZZT! BZZT! BZZT!

THE GREEN ARROW LIGHT SIGNAL-- EMERGENCY CALL FROM HEAD-QUARTERS!

TUNE IN ON THE SHORT-WAVE AND SEE WHAT'S UP, WHILE I FINISH UP HERE!

THEN...

A FREIGHTER WAS JUST BOARDED IN THE HARBOR BY MEMBERS OF THE SO-CALLED SHARK GANG! THEY LOOTED THE SHIP, THEN VANISHED BENEATH THE WAVES!

2

ONCE AGAIN IN THEIR FAMILIAR COSTUMES, THE BOWMEN WORK SWIFTLY...

THE UNDERGROUND RIVER BENEATH HERE LEADS TO THE BAY! WHILE I LOWER THE *ARROW-SUB*, FETCH OUR SPECIAL *AQUA-BOWS* AND ARROWS!

RIGHT AWAY!

THOSE LITTLE UNDERWATER MACHINES THEY USED WERE VERY FAST! DO YOU THINK WE CAN...

...OUTRACE THEM? JUST WAIT AND SEE!

RELEASED FROM HER MOORINGS, THE *ARROW-SUB* SINKS DEEP INTO THE RIVER...

LIGHTS ON, *SPEEDY*!

A BRILLIANT YELLOW BEAM SUDDENLY STABS KNIFE-LIKE THROUGH THE MURKINESS...

NOW-- FULL STEAM AHEAD!

I SEE IT-- THE OUTLET INTO THE BAY!

HEAR HER? THIS BABY PURRS LIKE A CUDDLED KITTEN!

AND THERE, WHERE THE UNDERGROUND RIVER EMPTIES INTO THE LARGER BODY OF WATER, THE *ARROW-SUB* BURSTS FORTH FROM THE DARKNESS OF THE TUNNEL...

G.A.! OVER THERE! DO YOU SEE WHAT I SEE?

I SURE DO! LIGHTS OUT-- WE DON'T NEED THEM OUT HERE!

BUT SHORTLY THE *ARROW-SUB* IS SPOTTED-- BY SHARK GANG MEMBERS...

HEY! WHAT'S THAT THING COMIN' AT US?

DON'T KNOW--BUT YOU BETTER TELL THE BOYS UP ON DECK ABOUT IT!

BUT BEFORE THE **SHARK-MOBILES** CAN RETREAT...

NOW, **SPEEDY**--HURRY! THE **"JELLYFISH" ARROWS!**

ROGER!

LOOK OUT--SOMETHIN' CRAZY IS COMIN' AT US **AGAIN!**

ZING

ZING

THE **"JELLYFISH"** MISSILES FIND THEIR TARGETS... AND INSTANTLY COVER THEM WITH A GOOEY, STICKY SUBSTANCE...

YOW! MY SHARK-MOBILE IS TRAPPED IN **GLUE!** I CAN'T MOVE!

WHAT IN BLAZES ARE THE BOYS ON THE BOAT DOIN' TO HELP US? **DO** SOMETHIN', YOU GUYS UP THERE!

BUT THE CROOKS ON THE BOAT **ARE** DOING SOMETHING--THEY'VE GONE "FISHING" WITH TWO GREAT HOOKS SUSPENDED FROM CABLES... AND THEY'VE MADE A CATCH!

WE'VE GOT 'EM!

AND ABOVE...

ALL RIGHT--HAUL 'EM IN! WE'LL FIX THEIR WAGONS FOR GOOD!

YOU CAN RELAX DOWN THERE, BOYS--WE'VE GOT **GREEN ARROW** AN' THE KID! YEAH-- SURE ...WE'LL HAVE 'EM ABOARD IN A MINUTE!

BUT A "FISH" ON A LINE IS NOT NECESSARILY A "FISH" LANDED--AS THE FAMED ARCHER IS ABOUT TO PROVE...

WAIT'LL THEY TRY OUR **"ELECTRIC EEL" ARROWS,** EH? THEY SHOULD PROVIDE QUITE A SHOCKER! AIM AT THE CABLES, THEN UNHOOK YOURSELF!

ZIP

ZIP

5

YEEOW! I'VE BEEN STUNG BY SOMETHIN'!

AND BELOW...

JUST A MILD SHOCKER--BUT ENOUGH TO DO THE TRICK! WELL, THE PATROL BOATS WE SUMMONED FROM THE SUB SHOULD BE ALONG ANY MINUTE NOW TO FINISH THE "SHARK HUNT!"

AND WE CAN DELIVER THE CROOKS THAT ARE DOWN HERE!

A SHORT WHILE AFTERWARDS...

QUITE A BIG CATCH WE MADE, G.A.! ANY MORE DOWN BELOW?

NO, SIR! YOU MIGHT SAY THESE WATERS ARE SHARK-FREE AS OF NOW! AND -- THERE WERE NO BIG ONES THAT GOT AWAY!

WHEN THE LAW DEPARTS WITH THE DAY'S BIG HAUL, THE TWO TEMPORARY AQUA-ARCHERS WEND THEIR WAY HOME AGAIN, THROUGH THE TUNNEL THAT LEADS TO THE ARROWCAVE...

...AND THOUGH THEY ARE NOT TRULY MEN OF THE DEEP, EVEN AQUAMAN WOULD APPROVE OF THE JOB THEY DID THIS DAY!

THE END

THE GREEN ARROW

THE GREEN ARROW

"*THE BOY IS SERIOUSLY ILL,*" THEY TOLD THE BATTLING BOWMEN, "*AND PERHAPS YOU'RE* THE ONLY ONES WHO CAN MAKE HIM WELL AGAIN!*" THIS THEN BECOMES A **DOUBLE** CHALLENGE FOR **GREEN ARROW** AND **SPEEDY,** WHO MUST NOT ONLY PROVIDE A REMEDY FOR RAMPANT CRIME, BUT ALSO COME UP WITH...

A CURE FOR BILLY JONES

CAPTURED! WHAT A DISAPPOINTMENT FOR BILLY!

YES--INSTEAD OF CURING HIM -- WE'LL JUST MAKE HIM SICKER WITH OUR FAILURE!

LEE ELIAS

ON THE GRASSY BACK YARD OF HIS HOUSE, YOUNG BILLY JONES SITS, STARING MOODILY INTO... NOWHERE...

FRANKLY, I DON'T KNOW *WHAT* I CAN PRESCRIBE FOR BILLY! HE IS VERY ILL-- LISTLESS, DESPONDENT--IT'S AS IF HE DOESN'T *CARE* TO LIVE!

HE DOESN'T BOTHER WITH HIS PORTABLE TV, OR HIS BOOKS! AND HE STOPPED HIS FAVORITE SPORT--ARCHERY! HE DOESN'T EVEN READ ABOUT HIS HEROES ANY MORE-- *GREEN ARROW* AND *SPEEDY!*

HMMM... LET ME SEE THAT BOOK!

TALES OF GREEN ARROW AND SPEEDY

HIS HEROES, EH? I HAVE AN IDEA! I'LL NEED YOUR APPROVAL, OF COURSE-- BUT WE *MIGHT* HAVE HERE A *CURE* FOR BILLY!

ANYTHING YOU SAY, DOCTOR-- *ANYTHING!*

TALES OF GREEN ARROW AND SPEEDY

SOME TIME LATER, IN RESPONSE TO A SUMMONS, THE FAMED BOWMEN APPEAR AT HEADQUARTERS...

THIS MAY BE YOUR MOST IMPORTANT CASE, *GREEN ARROW!* SEE THAT LAD IN THERE?

YES, CHIEF-- WHAT'S WRONG WITH HIM?

THE BOY NEEDS SOME SORT OF REVITALIZATION, *GREEN ARROW!* HE IS NO LONGER INTERESTED IN LIFE! HE IS GRIPPED BY A MYSTERIOUS MALADY THAT DEFIES MEDICAL SCIENCE--BUT *YOU* MIGHT BE ABLE TO CURE HIM! YOU AND *SPEEDY* WERE HIS HEROES!

CHIEF OF POLICE

IF YOU CAN TAKE HIM WITH YOU ON A CASE, HE MIGHT SNAP OUT OF IT AND GET WELL AGAIN!

I SEE... WE CAN SET UP A *MOCK* CRIME, WITH HIRED ACTORS! IT WOULDN'T BE DANGEROUS--AND YET HE WOULD SEE US IN ACTION!

NO--NOTHING PHONEY! IF BILLY FOUND OUT, IT WOULD BREAK HIS HEART! IT'S *GOT* TO BE THE REAL THING!

BUT... THE *DANGER!*

HE WILL BE AS SAFE WITH YOU AND *SPEEDY* AS HE'D BE WITH A PLATOON OF SOLDIERS!

AND SO, TOWARD EVENING, YOUNG BILLY BECOMES A PRIZE PASSENGER IN THE FAMED *ARROWCAR*...

SEE, BILLY? WE'VE GOT A *BOW* FOR YOU--AND SOME *TRICK ARROWS!* EACH ARROW IS MARKED--SO YOU CAN'T MAKE A MISTAKE!

MMM...

SO FAR, THE BOY DOESN'T SHOW ANY INTEREST AT ALL!

2

THE TWO ARROWS THEN INFLATE--BECOMING HUGE BALLOONS, AND...

YOU'LL HAVE TO ADMIT, BILLY-- THIS IS A SENSATIONAL CATCH!

MMMM...

GOLLY, *G.A.!* WE DIDN'T IMPRESS HIM ONE BIT!

LOOKS AS IF WE'RE *FAILURES,* DOESN'T IT?

THEN SUDDENLY...

SPEEDY-- LOOK OUT!

;HA, HA, HA; THEY WALKED RIGHT INTO THE "MOUTH" OF OUR TRAP!

HE'S RIDDEN IN THE *ARROWCAR*-- ZIPPED UP FROM THE *CATAPULT*-- AND HAS SEEN US USE TRICK ARROWS TO CATCH CROOKS... AND HE ISN'T EVEN *INTERESTED!* WE MIGHT AS WELL RETURN HIM TO HEAD-QUARTERS WITH THESE CROOKS AND ADMIT DEFEAT!

THIS IS LIKE WINNING -- YET LOSING!

CLANG

THAT'S IT! WE'VE GOT 'EM! OKAY, TOM, LET'S MOVE IN!

SHORTLY...

KEEP YOUR GUNS HANDY-- DON'T GIVE THEM A CHANCE TO USE THEIR BOWS AND ARROWS!

OKAY, MIKE OPEN HER UP!

4

"HA, HA!"- WE WERE WATCHING YOU PROWL THE STREETS, *WAITING* FOR YOU! WHEN YOU "CAUGHT" PETIE AND JOE, YOU MERELY CAUGHT OUR *BAIT!* EASY--DON'T TRY ANYTHING!

CAPTURED! WE WERE SO BUSY WITH THE PROBLEM OF BILLY, WE WERE OFF-GUARD! NOW WE'VE TURNED OUT TO BE A COMPLETE BUST FOR HIM! WHAT A DISAPPOINTMENT!

WE'LL BE A DISAPPOINTMENT TO A *LOT* OF PEOPLE, ONCE THIS GANG UNMASKS US AND REVEALS OUR SECRET IDENTITIES!

BUT AMAZINGLY, FROM THE SHADOWS, COMES A *BOLO ARROW,* SNAPPING THE GUNS FROM THE HANDS OF THE CROOKS...

WELL! LOOK AT *THAT!*

WHAT'S HAPPENIN'?

ZIP

AND THEN COME OTHER ARROWS-- FROM THE BOW OF NONE OTHER THAN *BILLY JONES* -- THE *FIRECRACKER ARROW,* THE *BOXING GLOVE ARROW,* THE *ROPE-TRIP ARROW...*

TWANG

CRACK CRACK CRACK

ZIP

ZIP TWANG

TWANG

YEEOW-W-W! CLEAR OUTA HERE!

5

AT THIS MOMENT...

HO-- JUST A *KID!* I'LL TAKE THAT BOW AWAY FROM HIM LIKE TAKIN' CANDY FROM A BABY!

BUT BEFORE HE CAN MAKE HIS MOVE...

THE *COBWEB ARROWS* ARE HANDY TO CATCH A STRAY *"FLY,"* EH, *SPEEDY?* WRAP UP THESE HOODLUMS WHILE I TAKE CARE OF MIKE BELOW!

YOW! I'M TRAPPED!

THE NEXT MORNING... AT HEADQUARTERS...

...AND THEN I REALIZED THAT *GREEN ARROW* AND *SPEEDY* WERE IN TROUBLE BECAUSE THEY WERE WORRIED ABOUT *ME!* I HAD TO DO SOMETHING...

AND YOU DID *PLENTY,* BILLY-- YOU HELPED CAPTURE THE GANG! COME ON-- I'M TAKING YOU HOME...

FROM THAT DAY ON, THE BACK YARD OF THE JONES HOUSE BECOMES A BEEHIVE OF ACTIVITY...

...WITH THE HELP OF *GREEN ARROW* AND *SPEEDY,* WHO TRULY PROVIDED A QUIVER OF CURE FOR BILLY JONES!

THE END

QUICKLY SWITCHING TO HIS *SPEEDY* COSTUME, THE BOY ARCHER GOES TO THE *ARROW-CAVE* TROPHY ROOM BELOW, AND RETURNS WITH...

THE *G.A. DUMMY!* WE USED IT AS A DECOY IN THE CASE OF THE *"TV IMPERSONA-TIONS!"* MAYBE IT'LL WORK AGAIN!

I HOPE SO, LAD! I WISH I COULD GO WITH YOU--BUT I CAN'T EVEN *WALK!*

AND SOON, THE FAMED *ARROWCAR* WHEELS THROUGH SHADOWY STREETS...

I'LL KEEP THE DUMMY BEHIND THE WHEEL, WHERE *G.A.* USUALLY SITS-- BUT THE EMERGENCY DUAL CONTROL ENABLES ME TO DRIVE!...THE *HQ* REPORT SAID THE *BIRDMAN GANG* WAS STRIKING AGAIN, AT THE PLAZA JEWELRY STORE...

AT THE SCENE OF THE ROBBERY, A MAN ROCKETS TOWARD A WAITING HELICOPTER ON A NEARBY ROOF...

HA! MY JET-PROPULSION UNIT MAKES A GETA-WAY SIMPLE!

APEX JEWELRY

CLANG CLANG CLANG CLANG CLANG CLANG

BUT AS THE JET-PROPELLED *BIRDMAN* REACHES THE ROOF, A SECOND FIGURE HURTLES SKYWARD...

LET'S GET GOIN'!

EH? THE ARCHER KID! *GREEN ARROW* AND *SPEEDY* ALREADY ON MY TAIL!

APEX JEWELRY

2

A *"BOXING-GLOVE"* ARROW, CARRYING A HEAVYWEIGHT WALLOP, EASILY DISPATCHES THE JET-EQUIPPED *BIRDMAN*...

WHOMP

ZNNG

UNH...

AND AS THE GETAWAY COPTER RISES, A *"ROPE ARROW"* LOOPS OVER THE CRAFT'S WHIRLING ROTOR-BLADES...

THE KID HAS SNARLED OUR PROP! WE'VE GOTTA MAKE A FORCED LANDING!

CHUG-CHUG-CHUGGA

CHARM INC

THEN...

THE POLICE HAVE ARRIVED! I'D BETTER GET DOWN THERE-- BEFORE THEY SEE JUST *WHAT* IS SITTING IN THE *ARROWCAR!*

POLICE

IN A MOMENT, THE YOUTHFUL ARCHER DESCENDS WITH THE OTHER PRISONER...

HERE'S THE OTHER ONE, OFFICER! *G.A.* AND I WILL KEEP THIS JET-BOOSTER AS A SOUVENIR FOR OUR TROPHY ROOM!

SURE THING, *SPEEDY!* IT'S LITTLE ENOUGH REWARD FOR THE JOB YOU AND *GREEN ARROW* DID TONIGHT!

THEN, AS THE *ARROWCAR* DEPARTS...

GOT TO MAKE THIS LOOK GOOD... WAVE *"GOODBYE,"* MR. DUMMY...

SO LONG, FELLOWS! AND THANKS...

3

LATER THAT NIGHT...

THE DUMMY'S OKAY... SO LONG AS I KEEP IT IN THE *ARROWCAR*... BUT IT'S TOO WEIGHTY AND AWKWARD TO CARRY WITH ME WHEN I'M IN ACTION!

HMM... I THINK I HAVE AN IDEA...

NEXT DAY...

AND THERE WE HAVE IT-- A *GREEN ARROW BALLOON DUMMY!*

AND I CAN LIFT IT WITH ONE FINGER! *BOY!* THIS IS MORE LIKE IT!

YOU CAN KEEP IT ATTACHED TO YOU, AND CONTROL IT BY THIS *INVISIBLE STRING!* WHEN YOU PULL IT CLOSE TO YOU, AND RUN WITH IT, IT'LL APPEAR TO BE "RUNNING" WITH YOU!

WELL, WE'RE *REALLY* GOING TO FOOL 'EM NOW, AREN'T WE, MR. DUMMY?

THAT NIGHT, WHEN THE FIERY *GREEN ARROW* SIGNAL APPEARS AGAIN...

JUST RELAX, OLIVER-- MY NEW FRIEND AND I WILL HANDLE THIS! SEE YOU SOON...

THE LAD'S LUCK *CAN'T* HOLD OUT FOREVER! BLAST IT! IF ONLY I COULD DO SOMETHING TO MAKE *SURE* NOTHING HAPPENS TO HIM...

RACING FROM A DOWNTOWN BANK, THE REMAINING MEMBERS OF THE *BIRD GANG* MOUNT "AIR-CARS," AND...

BANK

PSHHH PSHHH PSHHH

④

BUT AT THAT MOMENT...

OKAY, PAL... HERE'S WHERE WE PUT ON OUR SHOW WITH SOME *BATTERING-RAM* ARROWS!

TWANG! TWANG! TWANG!

ZING

ZING

HEY! *GREEN ARROW* AN' THE KID AGAIN... THEY'RE UPSETTIN' US!

CLONK

CLONK

CLONK

REGAINING THEIR FEET AFTER FALLING, THE CROOKS THROW A HAIL OF LEAD AT THE DODGING "ARCHERS"...

LET 'EM HAVE IT, BOYS!

PING-

PING

POW

POW

TWANG! TWANG!

KA-REEE

POW!

A RICOCHETING SHOT FINDS ITS MARK, AND...

WHAT--? THEY GOT THE BALLOON DUMMY!

POP!

AND BEFORE THE SURPRISED ARCHER CAN RECOVER...

LOOK AT THAT! IT WASN'T *GREEN ARROW* AT ALL! IT WAS JUST A *BALLOON!* LET'S GET THE *KID!*

THE FAT'S IN THE FIRE NOW! I'M IN A TOUGH SPOT-- AND EVEN IF I GET OUT OF IT, EVERYBODY'LL KNOW SOMETHING'S WRONG WITH G.A.!

SUDDENLY...

THEY'VE GOT ME CORNERED-- HUH? WHAT'S THAT?

YOW!

PLUNK!

PLUNK!

PLUNK!

⑤

THEY ALL STARE IN ASTONISHMENT AT...

GREEN ARROW!??

NICE JOB, LAD-- DECOYING THEM WITH THE BALLOON, WHILE I GOT OVER HERE TO SURPRISE 'EM!

ULLPS!

I CAN'T SEE!

TWANG TWANG TWANG

POW!

SSSSS

AFTER A NEARBY SQUAD CAR RACES TO THE SCENE...

WELL... HOW?

I WAS WORRIED ABOUT YOU, SO I MADE MY WAY WITH CRUTCHES TO THE TROPHY ROOM AND GOT THE JET-BOOSTER YOU BROUGHT IN LAST NIGHT! THEN I TOOK THE ARROW-COPTER OUT, LEFT IT IN THE AIR ON AUTOMATIC CONTROLS, AND JETTED DOWN!

BUT HOW COULD YOU EVEN STAND HERE WITHOUT CRUTCHES... OH! I SEE!

AN OLD MOVIE TRICK...I STRAPPED MY BAD LEG UP LIKE THIS, AND WORE A PHONY LEG! BUT LET'S GET THE ARROW-COPTER! I'VE GOT TO GET BACK IN BED... DOCTOR'S ORDERS, YOU KNOW!

THE END

THE GREEN ARROW

IT'S GONE BACK INTO A TUNNEL OUT THERE! BLAZES-- THAT MACHINE HAS GOT TO HAVE A HUMAN BRAIN BEHIND IT, AND...

AND THAT MEANS SOME-ONE HAS DISCOVERED THE ARROW-CAVE! ALSO, WHO-EVER'S IN CONTROL OF IT COULD BE A GREAT MENACE!

CLICK! CLICK! WHIRR!

WITHIN MOMENTS, OLIVER AND ROY DON THEIR GREEN ARROW AND SPEEDY COSTUMES, AND IN THE ARROWCAR THEY TAKE TO THE TUNNEL...

THIS TUNNEL'S BEEN FRESHLY MADE! THE MACHINE PARTLY BORED THROUGH IT--AND PARTLY BURNED THROUGH IT! WE'LL TRY FOLLOWING ITS TRAIL!

AND DESPITE THE ERRATIC, TWISTING PATH OF THE MACHINE, THE FAMED ARCHERS FINALLY ARRIVE AT THE END OF THE TRAIL--A SPRAWLING UNDERGROUND LAB!

WHAT'S GOING ON DOWN HERE? HMM--THIS MAN SEEMS TO HAVE BEEN HURT!

GREEN ARROW--AND SPEEDY! I'M PROF. FRANKLIN--SOME CROOKS SURPRISED ME AND TOOK MY MASTERPIECE... VULCAN! THEY DROVE IT UP THE RAMP TO THE SURFACE...

"VULCAN"?

YES--AN ALMOST PERFECT MACHINE I INVENTED! SCIENTIFIC PARTIES CAN EXPLORE INNER EARTH IN VULCAN, TAKING CONTINUOUS PICTURES OF WHAT IT SEES THROUGH CAMERAS IN ITS EYES! I WAS EXPERIMENTING WITH A NEW REMOTE-CONTROL UNIT BUT VULCAN, I'M AFRAID RAN WILD...

YOU CAN SEE FROM THIS GRAPH THE ERRATIC ROUTE IT TOOK BEFORE IT REVERSED ITS PATH AND RETURNED HERE!

HMM..VULCAN BROKE INTO OUR ARROW-CAVE-- AND THAT GRAPH SHOWS ITS SECRET LOCATION!

2

HAVE NO FEAR, *GREEN ARROW!* I'LL DESTROY THE GRAPH, AND YOU CAN HAVE THE TUNNEL FILLED IN! YOUR SECRET WILL BE SAFE!

BUT *VULCAN* HAS A PHOTO-RECORD OF OUR SECRET IDENTITIES-- BECAUSE WE WEREN'T WEARING OUR CRIME-FIGHTING COSTUMES!

THERE'S NO TIME TO SPARE! WE'VE GOT TO GET THAT FILM! WE'RE GOING AFTER *VULCAN!* SEE YOU LATER...

THIS IS MY FAULT! I SHOULDN'T HAVE LET *VULCAN* GET OUT OF CONTROL...

MEANWHILE, *VULCAN'S* CAPTORS PLUNGE UNDERGROUND IN THEIR STOLEN MACHINE, AS THEY NEAR A CERTAIN DESTINATION...

HA! THIS IS LIKE PLOWING THROUGH SOFT SNOW! THIS BABY'S GONNA BE WORTH ITS WEIGHT IN *DIAMONDS*-- GET IT?

WHR-A-WHRA-WHR-A-WHR-R

YEAH-- THAT JEWELRY STORE'S GONNA BE IN FOR A BIG SHOCKEROO WHEN WE COME BUSTING UP THROUGH THE FLOOR OF ITS MAIN SHOWROOM! AND HANG ON-- WE'RE ALMOST THERE!

A FEW MOMENTS LATER--IN THE JEWELRY STORE...

GOOD GRIEF! WHAT'S HAPPENING-- AN *EARTHQUAKE?*

R-RUMBLE!

WHR-A-WHR-R-R

3

THEN... UH-- DIDN'T KNOW THERE WAS ANOTHER CROOK INSIDE! NOW IT'S MY FAULT THEY'RE ESCAPING!

BUT MAYBE THEY'RE *NOT*!

TWANG!

WHR-A-WHR-R-R

I FIRED A *SIGNAL* ARROW AT *VULCAN*! NOW WE CAN TRACK IT DOWN TO THEIR HIDEOUT WITH THE *ARROWCAR'S* SPECIAL RADIO!

CLICK

LATER, UNAWARE OF THE TELL-TALE SIGNALS FROM THE ARROW CLINGING TO THE MACHINE, THE CROOKS REACH THEIR HIDEOUT...

WAIT A MINUTE! *GREEN ARROW* MENTIONED SOMETHING ABOUT SOME *FILMS*! GET 'EM-- AN' WE'LL SEE WHY *HE* WAS SO ANXIOUS TO HAVE 'EM!

SURE-- RIGHT AWAY!

AT THAT INSTANT, TWO FIGURES APPEAR...

THOSE ARCHERS AGAIN...*ULLPS*!

BUT INSIDE MIGHTY *VULCAN*, THE THIRD CROOK TURNS THE MACHINE'S GREAT HEAD TOWARD THE ARCHERS, AND...

I'LL FIX 'EM--WITH A HOT-FOOT SPECIALTY!

WHROOSH!

AND FOR A TERRIFYING INSTANT, THE ARCHERS SEEM DOOMED!

LOOKS AS IF WE LOSE THIS TIME, LAD! THE OTHER CROOKS GOT TO SAFETY--INSIDE *VULCAN*-- AND THERE'S NO WAY OUT OF THIS FOR US!

WHROOSH

5

BUT, AMAZINGLY, THE GREAT *VULCAN* SUDDENLY SWINGS AROUND IN ANOTHER DIRECTION...

WHR-R

CLICK! CLICK!

HEY! WHAT'S HAPPENING? I CAN'T CONTROL THE THING!

THEN...

IT'S *GONE*, G.A.! WHY DID IT LEAVE-- AND WHERE DID IT GO?

HMM...I THINK WE'LL FIND ALL THE ANSWERS AT PROF. FRANKLIN'S LAB! AND WE'D BETTER GET THERE FAST!

SHORTLY AFTERWARDS, AT THE LAB...

OKAY, PROF-- NOW THAT WE'RE HERE -- START TALKING! WHAT'S IN THESE FILMS?

HERE'S YOUR ANSWER!

TWANG TWANG

AND AFTER TWO *ROPE* ARROWS MAKE SHORT WORK OF THE CROOKS...

FEELING THAT I WAS TO BLAME FOR ALL THIS, I KEPT WORKING ON MY REMOTE-CONTROL SYSTEM TILL I GOT IT IN SHAPE --AND BROUGHT *VULCAN* BACK!

THAT'S EXACTLY WHAT I FIGURED! AND YOU *NOT* ONLY SAVED OUR SECRET IDENTITIES-- YOU SAVED OUR LIVES!

THE END

THE GREEN ARROW

TWO SENSATIONAL DISAPPEARANCES DURING THE NIGHT EXPLODE INTO MORNING HEADLINES...

DAILY GRAPH

TWO SCIENTISTS MISSING!

RANKIN, INVENTOR OF ANTI-GRAVITY CRAFT, AND FORSYTHE, DISCOVERER OF Z-RAY, VANISH DURING NIGHT!

PROF. RANKIN   DR. FORSYTHE

ANOTHER CASE FOR OLIVER QUEEN AND YOUNG ROY HARPER--ACTUALLY FAMED ARCHERS GREEN ARROW AND SPEEDY...

FORSYTHE, APPARENTLY, DISAPPEARED ON HIS WAY HOME! RANKIN, POLICE SAY, WAS PICKED UP BY SOMEONE AT HIS FACTORY! THAT'S WHERE WE'LL START OUR INVESTIGATION!

LATER ON, AT THE SITE WHERE THE SECRET ANTI-GRAVITY RAY MODEL WAS PRODUCED...

PROF. RANKIN MUST'VE PUT UP A FIGHT, GREEN ARROW! THERE WERE SIGNS OF A FIERCE STRUGGLE!

HMM... AND WHAT HAVE WE HERE?

PINE NEEDLES--AND YET THERE'S NOT A PINE TREE WITHIN MILES OF HERE! THEY GROW ONLY IN THE MOUNTAINS TO THE WEST, AND COULD'VE DROPPED OUT OF A TROUSER CUFF DURING THE STRUGGLE HERE!

THAT NIGHT, THE FAMED ARROW-PLANE FLIES IN A CRISS-CROSS PATTERN OVER THE TOWERING MOUNTAINS...

IF RANKIN AND FORSYTHE WERE TAKEN BY CROOKS WHO HAVE A HIDEOUT IN THESE HILLS, THEY WON'T SUSPECT WE'RE AWARE THEY'RE HERE! WE MIGHT EVEN SPOT A LIGHT...

G.A.! DOWN THERE! I DO SEE A LIGHT!

SILENTLY MANEUVERING THE ARROW PLANE, GREEN ARROW SETS THE CRAFT ON AUTOMATIC PILOT, AND...

FORSYTHE AND RANKIN! THEY'RE BEING HELD IN THAT TOWER!

2

GREEN ARROW!

'SHHH--NOT A SOUND! WE'VE GOT TO GET YOU IN THE *ARROW PLANE* BEFORE THINGS START POPPING AROUND HERE! HURRY!

IN ANOTHER MOMENT...

HEY! LOOK UP THERE! *GREEN ARROW'S* GOT THE CAPTIVES! BLAST 'EM!

THEY'VE SPOTTED US!

BANG! BLAM!

NOW THAT FORSYTHE AND RANKIN ARE SAFE, *SPEEDY*-- WE CAN OPEN UP ON THOSE CROOKS!

TWANG! TWANG! TWANG!

ZIP! ZIP!

ZIP! ZIP!

BLANG!

HEAD FOR THE TUNNELS! THOSE ARROWS ARE COMING TOO FAST!

BANG!

BLAM! KPOW!

THEY'VE TAKEN COVER INSIDE THE MOUNTAIN! THE POLICE CAN ROUND THEM UP LATER ON! WE'LL TAKE YOU TWO HOME-- AND GET YOU A STRONGER SECURITY GUARD!

THE GANG WAS GOING TO FORCE US TO REVEAL PLANS FOR MY ANTI-GRAVITY CRAFT AND FORSYTHE'S Z-RAY MODEL! IT WAS A CLOSE CALL!

THE NEXT DAY, A SPECIAL ADMIRER OF THE BATTLING BOWMEN THRILLS TO THE ACCOUNT OF THE RESCUE-- BONNIE KING, ALSO KNOWN AS MISS ARROWETTE...

THE FOLKS AT THIS CHARITY BENEFIT ARE GETTING A NICE BONUS--*GREEN ARROW* AND *SPEEDY* PERSONALLY TELLING OF THEIR ADVENTURE! HOWEVER...

3

...I'M WORRIED! *GREEN ARROW* COULDN'T HAVE BEEN *EXAGGERATING* HIS STORY, AND YET--THERE'S SOMETHING WRONG! I HATE TO ASK HIM ABOUT IT... BUT I *MUST!*

THEN...

BONNIE! BONNIE KING! WELL, *MISS ARROWETTE*, IT'S BEEN SOME TIME SINCE WE LAST MET!

I KNOW! BUT FIRST, THERE'S SOMETHING I MUST CLEAR UP!

IF THE CROOKS FIRED SO MANY SHOTS AT YOUR *ARROW-PLANE*, AS YOU SO VIVIDLY DESCRIBED, WHY ISN'T THERE AT LEAST ONE BULLET HOLE PRESENT?

I DON'T KNOW! WE HAVEN'T YET HAD TIME TO LOOK!

A HURRIED INSPECTION OF THE CRAFT REVEALS...

SHE'S *RIGHT*, G.A.! NOT A SINGLE BULLET HOLE! THOSE CROOKS *COULDN'T* HAVE MISSED THE *ARROW-PLANE* WITH *EVERY* SHOT!

UNLESS...UNLESS THE SHOTS WERE *BLANKS!* AND THAT WOULD MEAN WE WERE VICTIMS OF A TERRIFIC HOAX!

RANKIN'S FACTORY IS CLOSEST TO US! I'VE GOT A FEW QUESTIONS TO ASK HIM! AND *YOU*, BONNIE--UH--*MISS ARROWETTE*--KEEP YOUR PRETTY POWDERED NOSE *OUT* OF THIS!

INDEED! IF *YOU'RE* GOING TO RANKIN'S PLACE, *MISS ARROWETTE* IS GOING TO VISIT FORSYTHE!

SOON...

A VERY CUNNING PLAN TO STEAL THE INVENTIONS! THOSE CROOKS CAPTURED THE TWO INVENTORS, AND DELIBERATELY LEFT A CLUE TO THEIR HIDEOUT!

THEN THEY FIRED BLANKS AT US WHILE WE RESCUED THEIR TWO PRISONERS... G.A.! UP AHEAD-- LOOK!

4

RANKIN! *PROF. RANKIN!* YOU'RE NOT SUPPOSED TO TAKE THE CRAFT OUT OF HERE!

THAT'S NOT RANKIN-- THE "PRISONERS" WE RESCUED WERE DOUBLES OF THE CAPTURED INVENTORS!

TOO LATE TO STOP THE PHONEY RANKIN, *SPEEDY!* TAKE OVER--I'M FIRING A *MAGNETIC ROPE-ARROW,* AND...

TWANG!

THE MAGNETIC ARROW STRIKES AND STICKS!

...I'M "HITCHING" A RIDE! I'LL CLIMP UP TO THE CRAFT, AND...

THE CRAFT RISES HIGHER-- HIGHER--AND AS THE FAMED BOWMAN MAKES HIS WAY UP THE ROPE...

HA, HA! *GREEN ARROW* DIDN'T KNOW I SPOTTED THE *ARROW-CAR* COMING UP--AND SAW HIM TRY TO GET A FREE RIDE! WELL, HE'S GOT IT--STRAIGHT BACK TO THE GROUND!

UH-UH... TROUBLE...

5

BUT MEANWHILE, BONNIE KING HAS SWITCHED TO THE ROLE OF *MISS ARROWETTE*, AND HAS ALREADY ARRIVED AT THE OTHER FACTORY...

THE TWO MEN IN THAT CAR ARE ACTING SUSPICIOUS... THEY KEEP LOOKING AROUND, AS IF EXPECTING SOMEONE...

SECONDS LATER...

NOW, I'LL... WAIT-- WHAT'S *THAT?*

NO, NO, MR. FORSYTHE! *MR. FORSYTHE!* COME *BACK,* SIR!

A STRANGE VIBRATING SOUND IS HEARD, FOLLOWED BY A HISSING NOISE -- AND THEN...

HA! THAT WAS SIMPLE! THE Z-RAY BURNED THROUGH THE WALL LIKE A TORCH THROUGH PAPER!

GOOD THING I CAME HERE! THAT MUST BE FORSYTHE'S *DOUBLE!* I MUST STOP HIM!

SWIFTLY, MISS ARROWETTE FIRES A *LOTION-ARROW* WITH DEADLY ACCURACY...

TWANG!

WHAT--?? I SLIPPED ON SOMETHING!

SPLASH!

THE PRETTY BOW-GIRL MOVES QUICKLY...

I'VE GOT TO GET THIS Z-RAY DEVICE AND GET OUT OF HERE FAST!

7

BUT... NO, YA DON'T, LADY-- JUST HAND THE LITTLE TOY OVER TO US!

GOOD WORK! PUT HER IN THE CAR! WE'LL TAKE HER WITH US!

AND IT IS NOT TOO LONG AFTERWARDS THAT THE BATTLING BOWMEN ARRIVE...

I COULDN'T STOP MR. FORSYTHE, *GREEN ARROW!* HE BURNED THIS HOLE THROUGH THE WALL--AND GOT AWAY WITH THE *Z-RAY!*

AND APPARENTLY *MISS ARROWETTE* WAS HERE, AND TRIED TO STOP HIM! SHE USED HER *LOTION ARROW...*

ALPHA-OMEGA CO.

SHE USED SOMETHING *ELSE* TOO! *MISS ARROWETTE* MUST'VE ANTICIPATED TROUBLE-- SHE AFFIXED HER *MASCARA ARROW* TO THE BACK BUMPER OF THE GETAWAY CAR-- AND IT'S LEFT A TRAIL FOR US!

WHAT ARE WE WAITING FOR?

LATER ON, AT THE GANG'S NEW HIDEOUT, NESTLED IN THE FOOTHILLS...

NOW THAT THE BOYS HAVE GONE OUT ON THE BIG JOB WITH THE Z-RAY AND THE ANTI-GRAVITY CRAFT, THE MASQUERADE IS OVER!

YOU FORGET ONE THING, MISTER! YOU'VE STILL GOT TO DEAL WITH *GREEN ARROW!*

DON'T REACH FOR STRAWS! *GREEN ARROW* HASN'T THE SLIGHTEST IDEA WHERE THIS NEW HIDEOUT IS LOCATED!

IN THAT CASE YOU'D BETTER EXPLAIN WHO THAT IS COMING THROUGH THE DOOR!

GREEN ARROW AND THE KID! *IMPOSSIBLE!* GUN 'EM DOWN, BOYS!

8

TWO GUNS ARE RAISED, AIMED--AND FIRED--BUT...

ZIP

ZIP

WHAM!
WHAM! CRRACK!

BOLO ARROWS BRING THE CROOKS DOWN AS THEY ATTEMPT TO FLEE...

HEY! I'M TRIPPED UP!

WHIRRA- WHIRRA- WHIRRA!

AND SOON...

WE OVERHEARD THE OTHER MEMBERS OF THE GANG PLAN THE BIG JOB! THEY'RE GOING TO ROB THE GEM COLLECTION IN THE *CLOUD TOWER!*

WE'LL GO AFTER THEM AT ONCE! BY THE WAY...UH -- THANKS FOR THE CLUE THAT LED US HERE! THE *MASCARA ARROW* DID THE TRICK!

DUSK SOON FALLS, AND AT THE *CLOUD TOWER...*

THE GEM COLLECTION ON THE TOP FLOOR IS GUARDED FROM THE ROOF AND FROM THE GROUND FLOOR! THEY CAN SEE ANYBODY APPROACH BY AIR-- OR ENTER THE BUILDING! BUT THEY CAN'T SEE US-- IN THIS ANTI-GRAVITY JOB, ASCENDING CLOSE TO THE BUILDING'S *SIDE!*

AS THE CRAFT IS KEPT HOVERING IN MID-AIR, TWO CROOKS ENTER A ROOM THROUGH A WINDOW-- AND THE Z-RAY DOES ITS WORK ON A VAULT...

THERE'S AN ESTIMATED *TEN MILLION* WORTH OF ROCKS IN THAT VAULT... AND IN ANOTHER MINUTE IT'S GONNA BE OURS!

BETTER NOT COUNT ON THAT!

ZZZZZ

9

A LONG CHASE NEARS ITS CLIMAX IN THE MOUNTAINS OF PERU, WHERE THOSE FAMED ARCHERS, *GREEN ARROW* AND *SPEEDY,* HAVE TRACKED AN ELUSIVE CRIMINAL...

MIGHTY MICRO HAS A FEW DAYS' START ON US, *SPEEDY*--AND IS SOMEWHERE UP AHEAD! WE DON'T KNOW WHICH TRAIL HE TOOK AT THIS FORK-- SO WE'LL HAVE TO SPLIT UP!

WE'LL SAVE TIME THAT WAY, AND WE CAN RENDEZVOUS BACK HERE LATER ON!

THAT WILY MAGICIAN-CROOK HAS ALWAYS KEPT A JUMP AHEAD OF US! LET'S HOPE THIS TIME WE NAIL HIM!

WELL, HIS MAGIC STUNTS PULLED A LOT OF CRIMES FOR HIM-- TILL WE SMASHED HIS GANG! WE CAN'T EASE UP UNTIL *MICRO* HIMSELF IS BEHIND BARS! I'LL TAKE THE RIGHT FORK-- YOU TAKE THE LEFT!

AS THE YOUNG ARCHER CAUTIOUSLY MAKES HIS WAY UP THE TRAIL, HIDDEN FIGURES WATCH HIM...

NOTHING SO FAR-- NOT A SIGN OF ANYBODY!

THEN...

YA-HA-A-A

ZING

ZING

GREAT GUNS! THEY LOOK LIKE ANCIENT *INCAS!* AND I'M NOT JUST "SEEING THINGS!" THOSE SPEARS ARE FOR *REAL!*

2

IN THE NEXT SPLIT SECOND, THE BATTLING BOY BOWMAN TWANGS OFF A VOLLEY OF ARROWS AT HIS ATTACKERS, AND...

THIS SHOULD STIR UP QUITE A STORM, AND GIVE 'EM SOMETHING TO THINK ABOUT.' THE *FIRECRACKER ARROW--* THE *RAIN* AND *LIGHTNING ARROWS!*

POW

TWANG TWANG TWANG

FLEE--*FLEE!* HIS ARROWS SHOW GREAT MAGIC! THE STRANGER MUST HAVE MIGHTY POWERS--LIKE THE *WIZARD!*

THAT DID THE TRICK, ALL RIGHT! BUT THEY MENTIONED A "*WIZARD!*" THAT *COULD* BE *MICRO!* I'D BETTER FOLLOW THEM! I MIGHT FIND SOMETHING TO REPORT BACK TO G.A.!

MEANWHILE, ON THE OTHER TRAIL, *GREEN ARROW* HIDES IN THE CLEFT OF A HILLSIDE AS HE HEARS VOICES...

SOMEONE'S COMING DOWN A PATHWAY FROM ABOVE! NO USE SHOWING MYSELF TILL I SEE WHO IT IS!

MICRO--WITH SOME...SOME INCA INDIANS THAT SEEM TO HAVE STEPPED OUT OF HISTORY!

ALL GOES WELL, MY FRIENDS! NO MORE SECRET MEETINGS WILL BE NECESSARY! I THINK WE ARE READY TO TAKE OVER THE TRIBE NOW!

3

THEN, AS THEY PASS, THE ARCHER EMERGES...

TOO MANY OF THEM FOR ME TO TRY TAKING MICRO! BESIDES, A COUPLE OF THOSE INDIANS MIGHT'VE BEEN HARMED!

MY BEST BET IS TO "INFILTRATE"-- FIND OUT WHAT MICRO IS UP TO --AND THEN EXPOSE HIM FOR WHAT HE IS! HMM-- HE MENTIONED A "SECRET MEETING"-- SO THERE MUST BE A MEETING PLACE UP HERE!

SOON...

A HUT! WITH LUCK, I MIGHT FIND JUST WHAT I NEED TO CARRY OUT MY "INFILTRATION" PLAN...

THEN, INSIDE THE HUT...

EXACTLY WHAT I NEED FOR A COSTUME! I CERTAINLY CAN'T GO PROWLING AROUND DOWN THERE IN MY OWN CLOTHES-- I'D BE SPOTTED AT ONCE! BUT I CAN POSE AS ANOTHER INCA!

FROM AN INNER LINING OF HIS QUIVER, THE FAMED ARCHER BRINGS FORTH A DISGUISE KIT...

I WON'T HAVE TIME TO MEET SPEEDY BACK AT THE RENDEZVOUS POINT AND EXPLAIN MY PLAN! I'LL JOIN HIM AFTER I FIND OUT WHAT MICRO IS UP TO...

SOON, WITH THE MAKE-UP COMPLETE, THE DISGUISED ARCHER STEPS FORTH...

I'LL HIDE MY OWN COSTUME, AND BOW AND ARROWS, TILL LATER! NOW, I'LL MAKE MY WAY INTO THE TOWN BELOW...

4

LATER ON, IN THE ANCIENT VILLAGE, A CROWD GATHERS IN THE SQUARE TO WATCH *MIGHTY MICRO* -- "THE WIZARD" -- PERFORM...

COME ONE -- COME ALL -- SEE THE MIGHTY WIZARD PERFORM GREAT FEATS OF MAGIC!

WHEN I'M FINISHED IMPRESSING THE NATIVES WITH MY "FANTASTIC" POWERS," THE CHIEF WON'T HAVE A CHANCE AGAINST ME...

I'LL WIN *TWO* VICTORIES! IF *GREEN ARROW* AND THE BOY APPEAR, THE NATIVES WILL SIDE WITH ME! AFTER THAT, I CAN GET WHAT I CAME FOR -- THE FABULOUS CEREMONIAL GEMS -- AND PULL A "VANISHING" ACT! WELL, MY AUDIENCE HAS GATHERED -- HERE GOES...

FIRST, THE WILY MAGICIAN PRODUCES AN UNLIT LAMP -- AND THEN, BY MERELY PASSING HIS HAND ACROSS IT...

A SIMPLE PARLOR TRICK -- A PELLET DROPPED INTO A SPECIAL CHEMICAL MIXTURE IN THE LAMP, AND...

WHROOOSH

AMAZING -- AMAZING! THE WIZARD CREATED A FLAME-CREATURE! HIS POWERS ARE GREAT, INDEED!

HOLD! WHAT HAVE WE HERE? WHO IS THIS STRANGER?

A *STRANGER* AMONG US! DOES ANYBODY KNOW HIM?

A *STRANGER*, EH? OR IS HE *REALLY* A STRANGER? *GREEN ARROW* WOULD BE JUST SMART ENOUGH TO DISGUISE HIMSELF IN ORDER TO CATCH ME BY SURPRISE...

5

THIS MAN MIGHT BE WEARING A *DISGUISE!* HE MIGHT BE ONE OF THE TWO EVIL MEN I SPOKE OF WHO FOLLOW ME--AND WISH TO DO ME HARM!

I'M GOING TO REMOVE YOUR DISGUISE--AND IF YOU *ARE* GREEN ARROW, I'LL KNOW YOUR REAL IDENTITY!

I HAVEN'T GOT A CHANCE! IN ANOTHER MOMENT HE'LL HAVE THE MAKE-UP OFF-- AND LOOK INTO THE FACE OF *OLIVER QUEEN*--GREEN ARROW'S OTHER IDENTITY!

BUT SOMEONE ELSE HAS STOLEN UNSEEN INTO THE VILLAGE...

MICRO CALLED THAT NATIVE "GREEN ARROW!" CAN IT BE--? CAN IT *POSSIBLY* BE?

AND IN THE SPLIT SECOND BEFORE *MICRO'S* HAND REACHES THE TELL-TALE MAKE-UP, A BOW TWANGS, AND...

TING

YA-HU-U-U! LOOK! A MAGIC ARROW!

THEN...

AH! A BOY STRANGER!

WE SAW HIM EARLIER TODAY! HE IS THE ONE WITH THE MAGIC BOW!

SPEEDY! GREEN ARROW'S KID ASSISTANT!

6

SENSING AN ABRUPT UPSET OF HIS SCHEMES, *MICRO* TRIES FOR A QUICK VICTORY...

YES! THEY ARE THE TWO WHO SEEK TO TAKE ME AND MY WONDROUS POWERS AWAY FROM YOU! I COMMAND YOU -- DESTROY THEM!

WAIT! THIS MAN YOU HOLD IS A GREAT WARRIOR! HE CAME HERE WITH ME TO *HELP* YOU...

HE LIES -- I'LL KILL HIM!

I'LL -- I'LL...

BLAM

INSTANTLY SPEARS ARE READIED BY MICRO'S FOLLOWERS...

FINISH THEM OFF!

HOLD! HOLD, I SAY!

AND AT THAT MOMENT, THE CHIEF OF THE INCAS MAKES HIS APPEARANCE...

LISTEN NOT TO THE WIZARD! I STILL COMMAND HERE!

TLAHUANA SPEAKS!

LISTEN TO OUR CHIEF!

THE TWO STRANGERS WILL GET A FAIR TRIAL! WE HAVE HEARD WHAT THE WIZARD HAD TO SAY -- NOW THE STRANGERS MAY HAVE *THEIR* SAY!

I CAN'T AFFORD A SHOWDOWN -- NOW! I'VE GOT TO GET THE CEREMONIAL JEWELS AND GET OUT OF HERE!

7

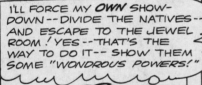

I'LL FORCE MY *OWN* SHOW-DOWN--DIVIDE THE NATIVES--AND ESCAPE TO THE JEWEL ROOM! YES--THAT'S THE WAY TO DO IT-- SHOW THEM SOME "WONDROUS POWERS!"

WAIT! WILL YOU LISTEN TO YOUR CHIEF--WHO IS OLD AND FEEBLE? OR WILL YOU LISTEN TO ME--*THE WIZARD?* I HAVE AMAZING POWERS THAT WILL DO MUCH FOR YOUR TRIBE--

THEN, DROPPING ANOTHER PELLET INTO THE LAMP'S CHEMICAL MIXTURE, *MICRO* SUMMONS UP A THICK DARK SMOKE CLOUD...

I CAN TURN DAYLIGHT INTO DARKNESS! I CAN PEFORM WONDERS YOU'VE NEVER SEEN! FOLLOW ME -- ALL YOU WHO WOULD SIDE WITH THE WIZARD!

YA-EEE! DAYLIGHT *HAS* TURNED TO DARKNESS!

FOLLOW THE WIZARD! FOLLOW THE WIZARD!

AND WHEN THE "DARKNESS" VANISHES...

THE WIZARD IS GONE--ALONG WITH HIS FORMER FOLLOWERS, AND SOME OTHERS WHO WERE IMPRESSED BY HIS POWERS!

BEFORE I MATCH "MAGIC" WITH HIM, I NEED MY BOW AND ARROWS! I WON'T BE GONE LONG!

SOON, WHEN THE ARCHER RETURNS...

THE WIZARD HAS GONE TO THE GREAT CHAMBER TO DON THE CEREMONIAL CLOTHES... TO BECOME CHIEF!

HE DOESN'T WANT TO BECOME CHIEF! I LEARNED A LITTLE ABOUT THE CERE-MONIAL CLOTHES WHEN I FIRST CAME TO YOUR VILLAGE IN DISGUISE! HE WANTS THE CEREMONIAL *JEWELS* THAT GO WITH THE CLOTHES!

8

THEN WHY DIDN'T HE SIMPLY *STEAL* THEM? WHY DID HE TRY TO WIN OVER MY PEOPLE?

STEALING THE JEWELS WOULDN'T HAVE BEEN SO EASILY DONE--WITHOUT THE AID OF YOUR PEOPLE! HE KNEW HE HAD TO WIN THEM OVER SO THEY COULD *HELP* HIM! HE'S FOOLED THEM--SO FAR! FOLLOW US!

AT THE ENTRANCE OF THE GREAT CEREMONIAL HALL...

STAND HERE-- WE DON'T WANT YOU HARMED! ALSO, WE DON'T WANT THOSE WHO FOLLOWED "THE WIZARD" HARMED! THIS IS A JOB WE MUST DO ALONE!

INSIDE...

YOU ARE TOO LATE, *GREEN ARROW!* ALREADY I HAVE BEEN MADE CHIEF-- AND CAN DO AS I PLEASE NOW! DON'T ARGUE WITH THESE SPEARS-- THEY CAN BE DEADLY!

WE HAVE ONLY ONE ANSWER FOR THEM-- AND FOR YOU!

THIS IS OUR ARGUMENT, SPEEDY! THE ROPE ARROWS...

TWANG TWANG

ZING

ZING

YA-EEEEE!

WHAT MAGIC!

ZING

9

THE GREEN ARROW and MANHUNTER from MARS

CHAPTER 1

A MYSTERIOUS CAPSULE HOVERING IN THE NIGHT SKY OVER **STAR CITY**... A STARTLING PRISON BREAK BY CONVICTS WHO COULD'VE ESCAPED AT ANY TIME... A DARING RAID ON A QUIET MUSEUM JUST TO STEAL A ROCK! WHAT TERRIBLE THREAT LIES BEHIND THESE BIZARRE EVENTS? WHAT INCREDIBLE DANGER BRINGS **GREEN ARROW AND THE MARTIAN MANHUNTER** TOGETHER TO JOIN FORCES WHEN EVERY ALARM CRIES..."**WANTED--THE CAPSULE MASTER!**"

NO, **MANHUNTER!** EVEN **YOUR** POWERS WON'T WORK! **YOU'LL** WIND UP A PRISONER, TOO!

THROUGH A BREAK IN THE CLOUDS OVER **STAR CITY,** A STRANGE CAPSULE DESCENDS AND HOVERS UNSEEN IN THE NIGHT SKY...

IT IS ONLY MINUTES LATER WHEN A GLITTERING GREEN SHAFT GOES STREAKING OVER THE ROOFTOPS...

LOOK, OLIVER... THE **ARROW-SIGNAL!**

IN THE MANSION OF WEALTHY OLIVER QUEEN AND HIS WARD, ROY HARPER, A SWIFT CHANGE OF GARB...

INTO YOUR COSTUME, ROY... THE POLICE NEED AN ASSIST-- FROM **GREEN ARROW** AND **SPEEDY!**

CHECK, OLIVER!

IN THEIR SLEEK **ARROW-CAR,** THE TWO FAMED CRIME-FIGHTERS RACE TO THE SCENE OF THE EMERGENCY--THE CITY PRISON...

THE ALARM...! MUST BE A JAILBREAK, G.A.!

WE'D BETTER GET IN THERE FAST, **SPEEDY!**

MOMENTS LATER, IN THE PRISON YARD...

THREE CONVICTS GONE FROM THEIR CELLS, **GREEN ARROW**--THEIR CELL DOOR BARS TWISTED LIKE TAFFY! BUT WE'RE SURE THEY HAVEN'T GOTTEN OVER THE WALLS YET!

AND THEY WON'T, WARDEN, ONCE WE ZERO IN WITH OUR ARROWS!

JUST THEN...

THERE THEY GO--ALONG THE WALL!

A VOLLEY OF **BOLAS ARROWS** OUGHT TO TANGLE UP THEIR FEET! LAUNCH, SPEEDY!

2

STRAIGHT TO THEIR MARKS FLY THE UNIQUE SHAFTS...

YIIII!

BUT, WITH ONE DEFT STROKE...

WHY..THEY BROKE THAT STRONG NYLON CORD LIKE IT WAS THREAD!

I'LL HAVE TO ORDER THE TOWER MACHINE GUN TO OPEN FIRE!

HOLD ON, WARDEN...THESE STUN ARROWS WILL HALT THEIR ESCAPE!

DESIGNED TO DELIVER A KAYO BLOW AT SLIGHTEST CONTACT, TWO STUN ARROWS BUZZ FROM THE ACE ARCHERS' BOWS, AND...

TWO DOWN -- ONE TO GO!

BUT EVEN BEFORE THEY CAN RELOAD...

G.A.! I--I'M SEEING THINGS! THOSE TWO GOT UP AGAIN! NOBODY'S EVER RECOVERED SO QUICKLY FROM A STUN ARROW!

I--I CAN'T BELIEVE IT! THEY LEAPED 40 FEET TO THE STREET-- AND VANISHED!

PRETTY UNUSUAL CONVICTS! LET'S SEE THOSE CELLS, WARDEN!

3

WHAT'S MORE, ALL THEY TOOK WAS A *METEOR* THAT WAS FOUND IN THE CITY PARK, TWO YEARS AGO! JUST A CHUNK OF ROCK... WHAT COULD THEY WANT WITH IT?

LOOK AT THIS, MR. CURATOR, AND MAYBE YOU'LL HAVE YOUR ANSWER!

THE IMAGE OF SOME GADGET INSIDE THE METEOR, ON THE CASE CARDBOARD!... LIKE A PHOTOGRAPHIC IMPRINT! THAT METEOR MUST'VE BEEN *RADIO-ACTIVE*, TO MAKE SUCH A "PHOTO"!

BUT... BUT HOW COULD ANYTHING BE *INSIDE* THE METEOR?

I'M NOT SURE--BUT THIS PROVES IT WAS THERE, AND THOSE STRANGE THIEVES KNEW IT, TOO! THAT'S WHAT THEY WERE *REALLY* AFTER! COME ON, *SPEEDY!*

RETURNING TO THEIR HEADQUARTERS, *GREEN ARROW* REACHES A GRIM DECISION...

SPEEDY, WE NEED HELP ON THIS CASE--AND THE BEST MAN TO GIVE IT TO US IS...THE *MARTIAN MANHUNTER!*

THE *MANHUNTER?* I...I DON'T GET IT, *G.A.!*

UNLESS I'M BADLY MISTAKEN, ONLY ONE THING EXPLAINS THESE HAPPENINGS! THE ESCAPED CONVICTS' STRANGE POWERS -- THE DESCRIPTION OF THE MUSEUM THIEVES... THE DISGUISE WE FOUND IN THE CELL...

...THEY ALL ADD UP TO OUR QUARRY BEING *ALIENS* -- MAYBE, EVEN, *MARTIANS!* WHO BETTER TO HELP US CORRAL THEM THAN THE *MARTIAN MANHUNTER!*

ABRUPTLY...

I CAN'T HAVE MEDDLERS SPOILING MY PLANS!

YIIII! SOMETHING'S GOT THE BOAT!

TOSSED INTO THE SWIFT CURRENT, ONE OF THE HARBOR PATROLMEN MANAGES TO FREE HIS FLARE PISTOL AND...

THE FLAMING *ARROW-SIGNAL* IS QUICKLY SPOTTED BY THREE PAIRS OF EYES, ON THEIR FIRST JOINT PATROL...

TROUBLE BELOW, MAN-HUNTER!

MY *MARTIAN VISION'S* LOCATED IT, NEAR THE WATERFRONT! FOLLOW ME!

DOWNWARD ZOOMS THE *MARTIAN MANHUNTER*, FOLLOWED BY THE FAMED *ARROW-PLANE*...

YOU WERE RIGHT, *GREEN ARROW*-- THERE *IS* A MARTIAN GANG LOOSE IN *STAR CITY!* I'LL HANDLE THEM--!

GO, MAN-HUNTER!

LIKE A THUNDERBOLT, *J'ONN J'ONZZ* PLUNGES INTO THE THUGS FROM HIS HOME PLANET...

LOOK AT THE *MANHUNTER* IN ACTION, *G.A.!* THAT WAS A GREAT IDEA, CALLING HIM IN!

7

BUT THEN...

YOU FORGET-- *OUR* POWERS ARE AS GREAT AS *YOURS* HERE, *MANHUNTER!* HOLD HIM... I'LL MAKE VULKOR A GIFT OF OUR FELLOW MARTIAN!

OOF!

THE *MANHUNTER'S* IN TROUBLE, *SPEEDY!* HANG ON... WE'RE GOING DOWN TO GIVE HIM AN ASSIST!

BUT SUDDENLY...

G.A.! SOMETHING'S GOT US... A GIANT METAL ARM!

AND THERE'S THE JOKER WHO'S CONTROLLING IT!

AS POWERFUL BOWSTRINGS TWANG...

SMOKE ARROWS SHOULD BLIND HIM LONG ENOUGH FOR US TO FREE THE PLANE FROM HIS GRASP!

BUT AS THE CHEMICALLY-LOADED SHAFTS LET OFF THEIR TIMED BURSTS OF THICK SMOKE...

A JET OF AIR FROM THAT NOZZLE IS BLOWING THE SMOKESCREEN INTO NOTHING-NESS!

THE TAIL OF THE *ARROW-PLANE*... IT'S BEING RIPPED OFF!

8

... PARA-ARROW! IT WILL LET THE PLANE DOWN LIGHT AS A FEATHER! BUT WE'RE OUT OF THE FIGHT!

WHILE DOWN BELOW...

THERE!... OUR MARTIAN PLAYMATE CAN'T TAKE A STRONG DOSE OF EQUAL MARTIAN MUSCLE!

WE'RE GOING INTO A DEATH DIVE!

OUR ONE CHANCE IS A KING-SIZED...

THE MARTIAN MANHUNTER BEATEN TO HIS KNEES!... GREEN ARROW AND SPEEDY THWARTED! WHO IS THIS VULKOR? FOR WHAT EVIL AMBITION IS HE USING HIS GREAT POWERS? THE ANSWER--IN CHAPTER TWO, ON THE NEXT PAGE FOLLOWING!

THE GREEN ARROW

CHAPTER 2

THE DESERTED DOCK IN **STAR CITY** IS THE SCENE OF A STARTLING EVENT... THE TRIPLE DEFEAT OF THE MIGHTY **MANHUNTER FROM MARS** AND THE **EMERALD ARCHERS!**

HOW... HOW COULD IT HAPPEN? VULKOR AND HIS MARTIAN THUGS **BEAT** US, G.A.!

REMEMBER, WE'RE NOT FIGHTING ORDINARY CRIMINALS, **SPEEDY!** YOU ALL RIGHT, **MANHUNTER?**

YES... I OVERLOOKED THE FACT THAT I WAS MEETING POWERS EQUAL TO MY OWN, FOR THE FIRST TIME SINCE I'VE BEEN LIVING ON EARTH! WE'VE GOT TO WORK OUT A PLAN OF ACTION, **GREEN ARROW!**

... BUT FIRST, I'LL HAVE TO PAY A VISIT TO MARS AND FIND OUT MORE ABOUT THIS VULKOR!

GOOD... MEANWHILE, **SPEEDY** AND I WILL RETURN TO THE **ARROW-CAVE** AND START DESIGNING SPECIAL SHAFTS TO COMBAT HIM! WE'LL ALL RENDEZVOUS THERE AS SOON AS YOU RETURN!

SWIFTLY, THE **MARTIAN MANHUNTER** RETURNS TO HIS SECRET MOUNTAIN CAVE, WHERE...

THE ROBOT BRAIN, WITH WHICH THE LATE PROFESSOR ERDEL ACCIDENTALLY BROUGHT ME TO EARTH... I MUST USE IT ONCE MORE, TO RETURN TO MARS!

COMPLEX CIRCUITS HUM, TELEPORTING THE ALIEN SLEUTH ACROSS THE VOID OF SPACE, ON AN INCREDIBLE JOURNEY TO HIS HOME PLANET...

10

AND SOON, AT A MEETING OF THE *ALL-MARTIAN COUNCIL...*

YOU ASK OF VULKOR, *J'ONN J'ONZZ?* HE IS THE ARCH-CRIMINAL RECENTLY ESCAPED FROM LONG IMPRISONMENT!

HE IS NOW ON EARTH, MY EXILE HOME! TELL ME-- WHAT CRIME WAS HE IMPRISONED FOR HERE?

OUR SCIENTISTS DEVELOPED A SUPER-WEAPON TO DEFEND OUR PLANET...VULKOR STOLE THE ONLY SUCCESSFUL WORKING MODEL, WHICH WAS DESTROYED WHEN HE WAS CAPTURED IN A FURIOUS FIGHT!

WE HAVE NEVER BEEN ABLE TO DUPLICATE THE WEAPON-- IT IS GONE FOR-EVER! WITHOUT IT, VULKOR IS NO REAL THREAT TO US!

PERHAPS...BUT ON EARTH, HE AND HIS HENCHMEN ARE A MENACE, WITH THEIR MARTIAN POWERS AND VULKOR'S ATOM-POWERED CAPSULE!

SINCE WE HAVE NO OFFICIAL RELATIONS WITH EARTH, WE CAN-NOT SEND OUR SPACE POLICE TO HELP YOU!

I KNOW... IT IS UP TO *ME* TO BRING VULKOR TO JUSTICE, WITH THE HELP OF MY EARTH ALLY-- *GREEN ARROW!*

AT THIS VERY MOMENT, BACK ON EARTH, IN THE PATROLLING *ARROW-CAR...*

NO SIGN OF VULKOR AND HIS BUNCH *ANYWHERE!*

ALL THE SAME, I WISH THE *MANHUNTER* WERE BACK! OUR NEW ARROWS LOOK GREAT-- BUT HOW WILL THEY *WORK* AGAINST THAT ALIEN MENACE?

11

JUST THEN...

ALL UNITS! ALIEN GANG ATTACKING HIGHMOUNT RADAR STATION!

VULKOR'S STRUCK AGAIN! HERE WE GO!

WITHIN MINUTES, THE POWERFUL VEHICLE STREAKS INTO THE COUNTRYSIDE, WHERE...

IT'S THEM, ALL RIGHT! THEY'RE WRECKING THE RADAR STATION... BUT WHY?

NO TIME FOR QUESTIONS, SPEEDY! READY ONE OF OUR NEW SHAFTS!

AS TWIN BOWSTRINGS TWANG...

IT'S THOSE TWO FOOLS, USING ANTIQUE EARTH WEAPONS AGAINST US! THEY CAN'T HARM US!

BUT AS THE HARMLESS-LOOKING SHAFTS SPEED PAST THE CRIMINAL ALIENS...

OW-W!

Y¡¡¡!

WHAT IS HAPPENING TO THE POWERFUL MARTIAN THUGS THAT THEY CRUMPLE LIKE RAG DOLLS UNDER THE ATTACK?

YES, SPEEDY, THE ULTRA-HIGH FREQUENCY SOUND WAVES THEY EMIT ARE HITTING THE ACUTE MARTIAN EARS LIKE SHOCK WAVES! KEEP FIRING!

THESE SONIC ARROWS WORK REAL SLICK, G.A.!

12

THEIR EARTH EARS TOTALLY "DEAF" TO THE HIGH-PITCHED SOUNDS, **GREEN ARROW** AND **SPEEDY** PREPARE TO ROUND UP THEIR ALIEN FOES, WHEN...

G.A.! SOME KIND OF FORCE FROM ABOVE STOPPING OUR ARROWS! THE SONIC WAVES ARE BLOCKED!

IT COULD ONLY BE...

...VULKOR! HEAD FOR THE ARROW-CAR!

AS THE ACE ARCHERS BEAT A SWIFT, STRATEGIC RETREAT...

AT GREEN ARROW'S COMMAND, HIS YOUNG ALLY HITS THE BRAKE PEDAL AND TWISTS THE WHEEL, AS...

VULKOR'S RIGHT ON OUR TRAIL, G.A.! WE CAN'T OUT-RACE THAT SKY BUS OF HIS!

I DON'T INTEND TO! WHEN I GIVE THE SIGNAL, BRAKE HARD, AND SPIN THE ARROW-CAR COMPLETELY AROUND!

VULKOR DIDN'T EVEN KNOW ABOUT OUR CATAPULT SEATS! AND HERE'S SOMETHING ELSE HE DOESN'T KNOW ABOUT...

WHEN THAT ACID ARROW HITS, IT'LL EJECT AN ACID SPRAY THAT WILL MELT THE CAPSULE AND GROUND THAT MARTIAN CRIMINAL FOR GOOD!

13

BUT...

THE CAPSULE!.. IT'S SURROUNDED BY AN *INVISIBLE FORCE FIELD!* THE ACID'S SPLATTERING HARMLESSLY AGAINST IT!

AN IN THE SAME INSTANT...

OH, NO! G.A.'S CAUGHT BY VULKOR'S CAPSULE ARM!

BUT THE PLIGHT OF THE MASTER ARCHER IS SEEN BY POWERFUL EYES FROM ABOVE...

I RETURNED NOT A MINUTE TOO SOON...

LET'S SEE IF THAT CAPSULE CAN RESIST A HIGH-SPEED DRILL!

BUT AS *J'ONN J'ONZZ*, WHIRLING WITH INCREDIBLE SPEED, WHIZZES TOWARD THE FLEEING CAPSULE...

OOF!

IT'S THE *MANHUNTER!* HE DIDN'T KNOW THAT FORCE FIELD WAS THERE, EITHER!

14

THE **MANHUNTER'S** TUMBLED LIKE A ROCK INTO THAT LAKE!.. AND THERE GOES VULKOR, WITH **GREEN ARROW** A PRISONER!

UNABLE TO HELP HIS SENIOR PARTNER, **SPEEDY** RACES TO THE LAKESIDE...

MANHUNTER-- ARE YOU OKAY?

YES...THAT FORCE FIELD AROUND VULKOR'S CAPSULE STUNNED ME, BUT THE WATER REVIVED ME!

THE PAIR QUICKLY RETURNS TO THE RADAR SITE, WHERE THE POLICE HAVE ARRIVED...

GREEN ARROW A PRISONER OF THIS ALIEN CRIMINAL?... UNBELIEVABLE! HIS GANG TOOK OFF WHEN HE DID! BUT WHAT WERE THEY AFTER? THEY'VE **WRECKED** THIS NEW RADAR SET-UP!

YOU SAY THIS IS A **NEW** STATION, CHIEF?

YES, **MANHUNTER!** BUILT LESS THAN A YEAR AGO TO AID PLANES COMING INTO **STAR CITY AIRPORT!**

SHORTLY, BACK AT THE **ARROW-CAVE...**

WHAT'S OUR NEXT MOVE, **MANHUNTER?** WE'VE GOT TO RESCUE **G.A.** SOMEHOW!

I KNOW, **SPEEDY!** BUT IT MAY HELP US IF WE CAN FIGURE OUT JUST WHAT VULKOR'S UP TO! WHERE'S THE "PHOTO IMAGE" THAT RADIOACTIVE METEOR MADE?

AS THE SPACE SLEUTH STUDIES THE STRANGE, ACCIDENTAL PICTURE...

WHAT DO YOU SEE, **MANHUNTER?**

I'M REMEMBERING, **SPEEDY...** REMEMBERING THE QUICK GLIMPSE I GOT, WITH MY **MARTIAN VISION**, OF THE **INSIDE** OF THAT METAL CHEST VULKOR'S GANG SALVAGED FROM THE RIVER!

15

AS THE *MANHUNTER* GAZES AT THE "PHOTO IMAGE," HIS MEMORY SUPERIMPOSES OVER IT WHAT HE SAW INSIDE THE METAL CHEST...

GREAT STARS!

WHAT... WHAT IS IT, *MANHUNTER?*

THE SUPER-WEAPON THE MARTIAN LEADERS SAID VULKOR STOLE... IT *WASN'T* DESTROYED! VULKOR IS *SALVAGING* IT--PUTTING IT BACK TOGETHER PIECE BY PIECE!

WHEN HIS HENCHMEN TOOK REFUGE ON EARTH, YEARS AGO, THEY MUST'VE DISMANTLED THE WEAPON AND HID EACH PIECE IN A DIFFERENT PLACE! THE FIRST PIECE WAS INSIDE THAT METEOR... THE SECOND IN THAT CHEST IN THE RIVER!

AND... AND THE *THIRD* PART?

MY HUNCH IS IT WAS BURIED ON *HIGHMOUNT HILL,* BEFORE THAT RADAR STATION WAS BUILT! THAT'S WHY VULKOR'S GANG WAS TEARING THE PLACE APART--SEARCHING FOR IT!

DON'T WORRY ABOUT *ME,* YOU TWO! I'LL BE WITH YOU EVERY INCH OF THE WAY!

G.A.! YOU'RE *FREE!* BUT... *HOW?*

IF VULKOR PUTS THAT WEAPON TOGETHER AND USES IT--EITHER HERE OR ON MARS--IT COULD BE CATASTROPHIC! THAT RADAR STATION... IT'S THE KEY TO EVERYTHING!

BUT WHAT ABOUT *GREEN ARROW?* CAN WE RISK TACKLING VULKOR WHILE *G.A.'S* IN HIS HANDS?

16

*"VULKOR'S CAPSULE WAS PASSING OVER **STAR CITY'S** OUTSKIRTS, WHEN A SUDDEN DIP CAUSED IT TO GRAZE A HIGH TENSION WIRE..."*

*"I SAW MY CHANCE, AND I LEAPED FROM THE CAPSULE! A **PARACHUTE ARROW** GOT ME SAFELY DOWN..."*

I GUESS VULKOR WAS TOO BUSY, NAVIGATING OUT OF TROUBLE, TO PURSUE ME!

GREAT TO HAVE YOU BACK, **GREEN ARROW,** BECAUSE YOUR BOW IS GOING TO PUT AN END TO VULKOR'S CRIMINAL CAREER! HERE'S MY PLAN...

LATE THAT NIGHT--ON **HIGH-MOUNT HILL**...

MY CAPSULE COULD BE DETECTED BY THIS RADAR ANTENNA! IT'S SAFER IF I SEARCH FOR THE WEAPON'S FINAL PART ON FOOT!

WITH A DETECTION DEVICE, THE MARTIAN CRIMINAL PROBES THE GROUND AROUND THE LARGE GRID ANTENNA...

THERE'S SOMETHING BURIED HERE! THE WEAPON'S THIRD PART... THE PIECE THAT WILL ENABLE ME TO BECOME MASTER OF MARS!

CLICK! CLICK!

CLICK!

17

CHAPTER 3

AGAIN AND AGAIN, THE MASTER MARKSMAN FIRES AN *INCENDIARY ARROW* AT HIS MARTIAN FRIEND...

THE FLAMES... THEY'RE WEAKENING ME!

¡HA, HA,¡ THEY JOINED FORCES TO COMBAT ME-- BUT I'VE TURNED ONE AGAINST THE OTHER!

MEANWHILE, WHERE *SPEEDY* HAS BEEN WATCHING FROM CONCEAL-MENT...

GA'S TURNED AGAINST *MANHUNTER!* GOT TO DOUSE THOSE FLAMES WITH A *FOAM ARROW* BEFORE THE *MANHUNTER* IS OVER-COME!

THE SHAFT WHIZZES UNERRINGLY TO ITS MARK--AND FROM ITS HEAD, QUICK-REACTING CHEMICALS SPREAD A BLANKET OF FOAM...

THEN, AS THE YOUNG ARCHER REACHES THE HILLTOP...

MAN-HUNTER! ARE YOU ALL RIGHT?

YES, *SPEEDY,* THANKS TO YOUR QUICK ACTION! VULKOR ESCAPED IN THE CONFUSION WITH THE WEAPON'S LAST PART... ALL BECAUSE *GREEN ARROW* TURNED ON ME!

WHAT CAME OVER YOU? WHAT MADE YOU TURN THOSE INCENDIARY SHAFTS AGAINST *ME?*

WHAT? I DON'T *REMEMBER* DOING IT! I SEEMED TO HAVE *BLACKED OUT* FOR A TIME!

19

DISMAYED BY *GREEN ARROW'S* STARTLING BEHAVIOR -- DEFEATED AGAIN BY THEIR ARCHFOE, IT IS A GRIM TRIO THAT RETURNS TO THE *ARROW-CAVE*...

I STILL CAN'T FIGURE MY BLACKING OUT AND TRYING TO HARM THE *MANHUNTER!* IT'S AS IF -- AS IF I WERE DOING VULKOR'S DIRTY WORK FOR HIM!

LISTEN! THE SPECIAL SHORT-WAVE'S PICKING UP SOMETHING!

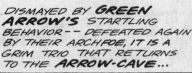

BZZZT CRACKLE

... *BZZT*... MY PLAN TO TURN *GREEN ARROW* AGAINST THE *MAN-HUNTER* WORKED! NOW WE WILL TEST THE WEAPON HERE ON EARTH, BEFORE USING IT TO CONQUER MARS!

VULKOR! BUT HOW...?

IN ALL THAT EXCITEMENT, I FORGOT TO TELL YOU... I FIRED A TINY *RADIO TRANSMITTER ARROW* THAT HOOKED INTO VULKOR'S CLOTHING AS HE FLED THE RADAR SITE!

GREAT WORK, *SPEEDY!* NOW WE'VE GOT A SECRET PIPELINE TO THE ORDERS HE'S GIVING HIS GANG! SHHH!

THE TEST ON THE *NEPTUNE* WILL PROVE THE WEAPON'S POWER -- AND THE *MANHUNTER* WON'T BOTHER US, SINCE I CAN EASILY TURN *GREEN ARROW* AGAINST HIM AGAIN... *BZZZZT... CRACKLE...*

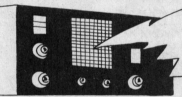

HEAR THAT? VULKOR IS EXERTING SOME CONTROL OVER YOU, *GREEN ARROW!* BUT *HOW?* AND WHAT'S THIS *"NEPTUNE"* HE'S GOING TO TEST THE WEAPON ON?

ISN'T IT THAT NEW ATOMIC-POWERED LINER, DUE TO MAKE ITS FIRST CRUISE TOMORROW?

RIGHT! VULKOR MUST BE PLANNING TO BLAST THE *NEPTUNE* AT SEA! WE'VE GOT TO STOP HIM -- NO MATTER WHAT!

YOU TWO WILL HAVE TO FIGHT HIM... *WITHOUT ME!* THIS MYSTERIOUS CONTROL VULKOR HAS OVER ME... I KNOW NOW HE REALLY LET ME ESCAPE SO HE COULD USE ME AGAINST YOU!

20

NEXT DAY, AS THE NEW NUCLEAR SUPER-SHIP PLOWS THE WAVES, AMONG THE JOYOUS PASSENGERS ARE THREE GRIM FACES THAT HIDE SECRET IDENTITIES...

I HOPE OUR PLAN WORKS, MANHUNT-- I MEAN, JOHN JONES! IF WE FAIL, THIS SHIP AND EVERYONE ON IT MAY BE DOOMED!

THEY'D BE DOOMED ANYWAY, OLIVER, IF WE DIDN'T TRY TO DEFEAT VULKOR THIS ONE LAST TIME!

DESPITE OUR PLAN, I'M STILL WORRIED THAT I'LL ONLY BE A THREAT TO YOU BOTH....!

OLIVER! JONES! LOOK UP THERE... IT'S -- VULKOR!

A SWIFT, SECRET CHANGE OF GARB -- AND MOMENTS LATER...

KAAACHUNNG

HANG ON!... WE'VE GOT TO GET AIRBORNE FAST!

WHILE UP ABOVE...

SO! THOSE LAWMEN ARE FOOLISHLY RISING TO CHALLENGE ME AGAIN! I'LL QUICKLY TURN GREEN ARROW AGAINST THE MAN-HUNTER, AND STOP THEIR ATTACK!

AS THE ARROW-PLANE STREAKS PAST THE CAPSULE...

GREEN ARROW EJECTING, AND READY TO FIRE A SHAFT AT US! I'LL STOP THAT...

21

EH--? I'VE PRESSED THE CONTROL THAT MAKES HIM MY PUPPET, BUT-- IT'S *NOT WORKING!* HE'S FIRING AT THE *CAPSULE!*

AS A SHOWER OF METAL FOIL, FROM THE ARROW, FORMS A GLITTERING CLOUD AROUND THE CAPSULE...

THOSE MAGNETIZED PARTICLES ARE SUPPOSED TO ABSORB THE FORCE FIELD AROUND VULKOR'S VEHICLE! I'LL SOON FIND OUT IF THEY WORKED!

SWEEP AWAY THESE PARTICLES... VULKOR'S ORDERS! FORGET *GREEN ARROW--* HE CAN'T HURT US!

TIME TO SHOW THOSE THUGS HOW WRONG THEY ARE, WITH A BLAST OF *MARTIAN SUPER-BREATH!*

GREEN ARROW WITH ALIEN SUPER-POWERS? HOW IS THIS POSSIBLE?

THE FORCE FIELD AND MY MEN, BLOWN AWAY BY *GREEN ARROW?* I--I DON'T UNDER- STAND!

22

IT CLICKED, *G.A.!* LETTING THE *MANHUNTER,* DISGUISED AS YOU, ATTACK WHILE WE STAYED OUT OF RANGE, CANCELLED VULKOR'S MYSTERY CONTROL OVER YOU!

CHECK, *SPEEDY!* NOW IT'S TIME TO LEND THE *MANHUNTER* A HAND!

YET EVEN WHILE THE *ARROW-PLANE* DARTS IN TOWARD THE SCENE...

STAY BACK!...OR I WILL USE THE WEAPON TO DESTROY THE SHIP BELOW!

EVEN WITH MY POWERS, I CAN'T MAKE A MOVE!

BUT UNNOTICED BY THE CAPSULE OCCUPANTS...

WHIZZZ

THE THIN, STRONG WIRE, ATTACHED TO THE *SUCTION CUP ARROW,* IS YANKED TAUT, AND...

FULL THROTTLE, *SPEEDY!*

CHECK!

AND LIKE A GIANT GAME OF CRACK-THE-WHIP IN THE SKY...

SWISHH

23

BUT NEXT MOMENT...

THE WIRE'S SNAPPED!.. AND HERE COMES VULKOR'S GANG!

QUICKLY, *SPEEDY!*... SOME *FIREWORKS ARROWS!*

INSTANTLY, A HAIL OF UNIQUE SHAFTS ENGULFS THE ONCOMING SPACE CRIMINALS...

JUST LIKE THE FOURTH OF JULY!

IT'S NO HOLIDAY FOR THOSE MARTIANS, *SPEEDY!* FIRE IN *ANY* FORM IS THEIR NEMESIS!

MEANWHILE, INSIDE THE TUMBLING CAPSULE...

I'LL DESTROY THAT SHIP... THE EARTH... MARS--!

YOU'LL *NEVER* FIRE THAT WEAPON, VULKOR!

DOWN INTO THE UNCARING GRIP OF THE SEA TUMBLES THE CAPSULE...

24

THE GREEN ARROW

THE GREEN ARROW

COME TO THE CIRCUS! MEET *FUNNY-ARROW*, THE ARCHER-CLOWN--THE BUMBLING BOWMAN WITH A THOUSAND LAUGHS! BUT BE WARY-- BEHIND THE PEALS OF LAUGHTER LURKS A MYSTIFYING CRIME PLOT THAT LAUNCHES *GREEN ARROW* AND *SPEEDY* ON A TWISTING TRAIL OF SURPRISES, AS THEY SEEK TO UNMASK...

The SECRET FACE of FUNNY-ARROW!

PRESENTING THE SENSATION OF THE CIRCUS CIRCUIT... *FUNNY-ARROW* AND HIS *HILARIOUS* SPOOF OF *GREEN ARROW!*

HA, HA, HA, HA, HA, HA! HE'S A *RIOT*, ISN'T HE?

CLAP- CLAP- CLAP- CLAP!

ARROW WHEEL BARROW

AMONG THOSE AT THE CIRCUS THAT HAS OPENED IN *STAR CITY* ARE OLIVER QUEEN AND YOUNG ROY HARPER, ACTUALLY *GREEN ARROW* AND *SPEEDY*...

I'VE READ ADVANCE NOTICES ABOUT THIS LAMPOON OF *GREEN ARROW*--IT'S SUPPOSED TO BE GOOD!

WELL, THE COSTUME IS SURE A DILLY!

THE CROWD IS SILENT NOW, AS...

THERE--YONDER I SEE A GREAT FIRE AT THE DOCKS! THIS CALLS FOR MY *SQUIRT-GUN ARROW!*

THAT'S A TAKE-OFF ON G.A.'S *RAIN-ARROW!*

FROM A SMALL DOOR IN HIS EXAGGERATED QUIVER, THE ARCHER-CLOWN DRAWS A FOOLISHLY SMALL -- AND ODD-SHAPED -- ARROW...

AWAY-- TRUSTY *SQUIRT-GUN ARROW!* OFF YOU GO TO THE DOCKS, TO PUT OUT THE FIRE!

TWANG!

BUT LIKE A ZIG-ZAGGING BAT, THE MISSILE FLUTTERS OFF-COURSE, AND...

YOW-W-W!

KA-SPLASH!

HA, HA, HA, HA!

CLAP! CLAP! CLAP!

FOLLOWING THE DRENCHING "MISHAP," THE BUMBLING BOWMAN HAS THE CROWD IN STITCHES WITH A COLORFUL DISPLAY OF SCREWBALL ARROWS... UNTIL...

HA, HA, HA! HO, HO! I TELL YOU, HE'S A RIOT!

NOW... MY *GREATEST* ARROW... IN WHICH I COMBINE THE *ROPE-BOOMERANG* AND *BALLOON ARROWS*... WATCH ME NAIL THAT CROOK!

HE FIRES AT THE CARDBOARD TARGET...

TWANG-G!

BUT...

ZING!

HA, HA, HA, HA!

ZIP!

ZIP!

ZIP!

2

IN ANOTHER SPLIT-SECOND, THE ARCHER-CLOWN IS A HELPLESS PRISONER--

HE MISSED THE TARGET BY A COUNTRY MILE-- HA, HA, HA!

AND HE TRAPPED HIMSELF-- WITH HIS OWN ARROW! NOW HE'S BEING HOISTED AWAY! HO, HO, HO!

THAT'S SOME FINALE, OLIVER! IT'S THE FUNNIEST LAMPOON I'VE EVER SEEN!

I AGREE!

HO, HO, HO! CLAP- CLAP- CLAP- HA, HA, HA!

SHORTLY AFTER THE ROARING LAUGHTER SUBSIDES...

I WANT TO SAY IN ALL SERIOUSNESS, THAT WHEN I ENTERTAIN YOU FOLKS WITH MY SPOOF ON GREEN ARROW, IT'S ONLY BECAUSE I'VE LONG ADMIRED HIM AND SPEEDY AS GREAT CRIME-FIGHTERS!

WHEREVER THEY ARE, I HOPE THEY'LL FORGIVE MY LITTLE ACT! NOW--LET'S HAVE A CHEER FOR GREEN ARROW AND SPEEDY!

HURRAY!

YAHOO!

CLAP- CLAP!

NEXT DAY, STILL CHUCKLING OVER THE BUNGLING BOWMAN'S PERFORMANCE, THE BATTLING BOWMEN PATROL STAR CITY IN THE ARROWCAR, WHEN ...

MY MESSENGER-PIGEON ARROW-- USED FOR SENDING "SECRET MESSAGES" TO THE POLICE!

ZING!

FLAP-FLAP-FLAP!

FUNNY-ARROW AGAIN! LOOKS LIKE A PUBLICITY STUNT FOR HIS ACT!

3

THE FLAPPING MISSILE SEEMS TO HEAD OUT OVER THE CITY--BUT THEN WHEELS, AND...

HA, HA! THAT'S NOT A *MESSENGER* PIGEON-- IT'S A *HOMING* PIGEON!

SPLAT!

AND DID *HE* LAY AN *EGG!* HO HO, HO!

SUDDENLY, OVER THE PEALS OF LAUGHTER, A FRANTIC CRY...

HELP--HELP! I'VE BEEN ROBBED!

THAT CRY CAME FROM THE *JEWELER'S!* COME ON, *SPEEDY*-- THIS IS NO LAUGHING MATTER!

SWIFTLY ELBOWING THEIR WAY THROUGH THE CROWD, THE ARCHERS SPOT A FLEEING FIGURE...

OUT OF THE WAY--PLEASE GET BACK!

THERE HE GOES, *GREEN ARROW!* I WAS WATCHING THE ACT WHEN HE SURPRISED ME--AND STOLE MY JEWELRY!

JEWELRY M

AT THAT MOMENT...

I'LL LEND HIM A HAND IN STOPPING THE CULPRIT! OBSERVE MY AMAZING *JACK-IN-THE-BOX* ARROW!

TWANG!

BLAZES! *FUNNY-ARROW* HAS *REALLY* BUNGLED THINGS THIS TIME! THE CROOK'S GETTING AWAY!

4

THEN... GOLLY, *GREEN ARROW!* I'M SORRY THAT MY ACT GUMMED UP EVERYTHING!

WELL, WE MIGHT PICK UP THAT CROOK'S TRAIL AGAIN... BUT YOU CAN SEE HOW *DANGEROUS* THESE STREET EXHIBITIONS CAN BE!

THE CROOK TOOK ADVANTAGE OF THE BIG CROWD, AND YOUR ACT! HE MADE HIS HAUL WHILE ALL ATTENTION WAS ON *YOU* -- SO YOU'LL HAVE TO BE MORE CAREFUL ABOUT WHERE YOU PERFORM!

YOU'RE RIGHT! I *WILL* BE MORE CAREFUL!

BUT THEN, NOT MORE THAN AN HOUR LATER...

...AND FOR THE *SECOND* TIME TODAY, A CRIME WAS COMMITTED WHILE THE SENSATIONAL *FUNNY-ARROW* WAS GIVING A FREE PERFORMANCE! THIS TIME, A SUPER-MARKET AT...

JOT DOWN THAT ADDRESS! WE'RE HEADING OVER THERE!

AND IN JUST A FEW MOMENTS, AT THE SUPER-MARKET...

THAT CLOWN WAS OUT FRONT, AND AS WE ALL WATCHED THE ACT THIS MAN CAME IN, AND TOOK $9,000 WE HAD READY FOR DEPOSIT!

VERY COINCIDENTAL -- FIRST THE *GEMS,* NOW *THIS* -- WHILE *FUNNY-ARROW* WAS PERFORMING! WE'RE GOING OVER TO THE CIRCUS!

SHORTLY...

YES, I'M DANTON -- OWNER OF THE CIRCUS! BUT I KNOW *NOTHING* OF PUBLICITY STUNTS PERFORMED IN THE STREETS BY *FUNNY-ARROW!* LET'S TALK TO HALLEY, MY PRESS AGENT...

INSIDE...

HALLEY -- WHAT ABOUT THESE *FUNNY-ARROW* ANTICS AROUND TOWN! DID *YOU* PLAN THIS AS PART OF YOUR PUBLICITY CAMPAIGN?

NO! IT'S THE FIRST I'VE HEARD OF IT! WE'D BETTER HAVE A CHAT WITH *FUNNY-ARROW!*

THE *GREATEST* COLLECTION OF *ACTS* IN THE WO...

BUT UPON ENTERING THE ARCHER-CLOWN'S QUARTERS...

WOW! LOOKS LIKE OUR **ELEPHANTS** STAMPEDED THROUGH HERE!

I WAS OUT IN A CLEARING IN THE WOODS-- REHEARSING SOME NEW STUNTS--AND RETURNED TO FIND **THIS** MESS!

I HEARD ABOUT THE ROBBERIES OVER THE RADIO--BUT IT WASN'T ME! IT WAS THE PERSON WHO RANSACKED MY QUARTERS-- STOLE A SPARE COSTUME, AND SOME OF MY EQUIPMENT!

AN **IMPOSTOR**, EH?

YES--AND HE'LL DESTROY MY ACT--AND **ME**--UNLESS HE'S STOPPED!

WE'LL ALERT THE WHOLE CITY--AT ONCE!

BUT THE NEXT DAY, THE IMPOSTOR CLOWN STRIKES AGAIN, AT A FACTORY WHERE A PAYROLL IS BEING READIED...

HA, HA! THIS **VACUUM ARROW**, DRAWING IN ALL THAT MONEY, WILL MAKE **LOTS** OF LAUGHS -- FOR **ME**!

ZIP!

SHWOOOOO!

GREAT GUNS! KICK THE EMERGENCY ALARM!

AS THE ALARM CLANGS HARSHLY...

THAT ALARM CAN BE HEARD IN THE NEXT STATE-- I'D BETTER CLEAR OUT OF HERE! OH-OH... THERE COMES **GREEN ARROW**!

CLANG- CLANG- CLANG!

I SPOTTED HIM ON THE ROOF! THE CATAPULT SHOULD GET US UP THERE IN TIME!

BUT THE WILY ARCHER-CLOWN HAS BROUGHT A **BALLOON ARROW** INTO PLAY...

HA, HA! I DON'T MIND LAUGHING AT MY OWN JOKE-- WHEN IT'S ON **GREEN ARROW!**

HE OUT-SMARTED US-- FLOATED ACROSS TO THAT ROOF!

THE ARCHERS START TO TAKE UP SWIFT PURSUIT, BUT...

HOLD IT! LOOK WHAT OUR "IMPOSTOR" DROPPED!

A **HANDKERCHIEF!**-- THE SAME KIND **FUNNY-ARROW** HAD WHEN WE SPOKE TO HIM BACK AT THE CIRCUS!

YES! AND IT'S HARDLY LIKELY THAT AN IMPOSTOR WOULD HAVE STOLEN A **PERSONAL** ITEM LIKE THIS, THAT WASN'T USED IN THE ACT!

BOY! DID **FUNNY-ARROW** MAKE CHUMPS OUT OF US! LET'S GET BACK TO THAT CIRCUS!

SHORTLY, AT THE CIRCUS...

YOU LEFT SOMETHING AT THE FACTORY, **FUNNY-ARROW!** GET DRESSED! WE'RE GOING DOWN TO HEADQUARTERS!

LOOK! THERE'S THE LOOT! WE'VE GOT HIM COLD!

BUT SUDDENLY, FROM INSIDE HIS BELT, THE ARCHER-CLOWN WHIPS OUT AN ARROW...

AS AN OLD ARROW MAN YOURSELF, YOU SHOULD BE ON **GUARD** AGAINST TRICKS LIKE THIS!

CAN'T SEE-- COUGH COUGH --CAN'T SEE!

BY THE TIME THE BATTLING BOWMEN RECOVER...

FUNNY-ARROW JUMPED IN A CAR AND WENT HIGH-BALLING DOWN THAT BACK ROAD--FASTER THAN ANY **ARROW** I EVER SAW HIM SHOOT!

THANKS! LET'S GO, **SPEEDY!**

IN A FEW SHORT MOMENTS, THE SLEEK **ARROW-CAR** ROARS OVER THE TWISTING BACK ROAD...

THERE ARE NO TURNOFFS ON THIS ROAD FOR MANY MILES -- AND UNLESS **FUNNY-ARROW** HAS A SOUPED-UP CAR, WE'LL CATCH UP TO HIM!

AND THE WAY WE **TRUSTED** THAT GUY! WOW!

BUT THE TRAIL SUDDENLY ENDS...

THERE HE IS -- AT THAT OLD BARN! HE CHANGED TO MAKE A GETAWAY!

GREEN ARROW'S BOW INSTANTLY SNAPS -- TWANGS...

COULDN'T TAKE ANY CHANCES ON HIM PULLING ANOTHER TRICK! THE **BOXING-GLOVE ARROW** SHOULD FIX HIM!

THWACK!

THEN...

LOOK AT THIS!

SOME OF THE FACTORY PAYROLL HE STOLE! KEEP AN EYE ON HIM WHILE I LOOK INSIDE THE BARN!

TWO COSTUMES HERE -- HIS, AND THE "STOLEN" ONE! THE MISSING EQUIPMENT, TOO! THIS BARN HAS SERVED AS HIS **HIDEOUT!** HMMM...

AND AS THE ARCHERS HEAD BACK TO THE CIRCUS WITH THEIR CATCH...

I GUESS WE'VE GOT THIS CASE WRAPPED UP, EH?

YES -- **REALLY** WRAPPED UP!

LATER, WHEN AN UNFUNNY **FUNNY-ARROW** IS RETURNED TO THE CIRCUS...

AND **YOU** WERE SPOUTING AFTER EVERY ACT HOW YOU ADMIRED **GREEN ARROW**! HA! A FALSE COVER-UP FOR CRIMES YOU BLAMED ON AN IMPOSTOR!

I TELL YOU, I'VE BEEN FRAMED!

YOU'RE NOT TALKING THE WAY YOU DID WHEN YOU ESCAPED!

"ESCAPED"? I DIDN'T ESCAPE! SOMEBODY SNEAKED UP ON ME HERE-- KNOCKED ME OUT! THEN... I HAD JUST COME TO WHEN **GREEN ARROW** SAW ME, AND BELTED ME WITH HIS **BOXING GLOVE ARROW**!

LIKELY STORY! I SUPPOSE YOU WALKED IN YOUR SLEEP TO THAT BARN! REALLY, YOU'RE GETTING FUNNIER BY THE MINUTE!

I DON'T FIND IT SO FUNNY! I FIND IT VERY PAT-- **TOO** PAT!

THE CRIMES AT THE START POINTED TO **FUNNY-ARROW**! WHEN WE THOUGHT IT WAS AN IMPOSTOR-- THE CRIMINAL DROPPED **FUNNY-ARROW'S** HANDKERCHIEF... POINTING AGAIN TO **FUNNY-ARROW**!

THE LOOT IN THE DRESSING ROOM-- **PAT**! THE TRICK ARROW FOR ESCAPE-- THE "GETAWAY" OVER THE BACK ROAD--**PAT**! THE CAR STOPPED AT THE BARN--**FUNNY-ARROW** WANDERING OUT WITH SOME LOOT ON HIM-- THE EQUIPMENT IN THE BARN... ALL **PAT**!

TOO PAT, HALLEY! AND NOBODY **MENTIONED** THAT WE FOUND **FUNNY-ARROW** AT A BARN! HOW DID YOU HAPPEN TO KNOW IT? YOU ASKED HIM IF HE WALKED IN HIS SLEEP... TO A **BARN**!

9

YOU POSED AS FUNNY-ARROW! YOU PULLED OFF THOSE CRIMES, WITH A COHORT! YOU KNOCKED OUT FUNNY-ARROW AND TOOK HIM TO THE BARN, THEN DROPPED A CLUE THAT LED US BACK TO THE CIRCUS...

... WHERE YOU PULLED YOUR PREPARED SMOKE-ARROW TRICK, AND FLED IN THE CAR--ONLY TO HAVE US TRAIL YOU!

I'M GETTING OUT OF HERE!

YOU RAN THROUGH THE WOODS, JOINED YOUR WAITING COHORT-- AND CAME BACK HERE! FUNNY-ARROW CAME TO, AND WALKED RIGHT INTO OUR HANDS, AS YOU FIGURED...

SCRAM, MIKE! HE'S WISE TO US... ULLLPS!

10

AND WITH HALLEY IN TOW...

AND I SEE YOU GOT HIS HENCHMAN, FUNNY-ARROW! HE'S ONE OF HALLEY'S STAFF, ISN'T HE?

YEP! BUT THE FUNNIEST PART OF IT IS THAT I DIDN'T THINK I COULD ACTUALLY HIT A TARGET WITH THIS GIMMICK ARROW! YOU KNOW, G.A.--MAYBE I'M LEARNING REAL ARCHERY-- FROM YOU!

THE END.

THE GREEN ARROW

THE GREEN ARROW

ON TRAVELS THE **ARROW-CAR**, THROUGH AN EERIE FOG, CARRYING **GREEN ARROW** AND **SPEEDY** INTO AN ALIEN WORLD WHERE DANGER AND DARK MYSTERY AWAIT THEM -- WHERE THEY MUST SOMEHOW STOP A FANTASTIC WAR AND SEARCH FOR A WAY OUT OF...

the LAND of NO RETURN

DUSK... AND HOMEWARD BOUND *GREEN ARROW* AND *SPEEDY* SUDDENLY COME UPON A STRANGE PHENOMENON...

GOLLY, *G.A.,* FOG AHEAD! BUT WHAT'S THAT STRANGE LIGHT IN IT?

THERE'S SOMETHING COMING THROUGH THAT LIGHT! SOMETHING *BIG!*

WHA-- WHAT IS IT?

SOMETHING NOT OF OUR WORLD, THAT'S FOR SURE!

SUDDENLY THE CREATURE TURNS, EAGERLY RUSHING AT A HIGH TENSION TOWER, AND...

ELECTRICITY DOESN'T EVEN HARM IT!

IT LOOKS TO ME LIKE IT'S *ABSORBING* ELECTRICITY! NOTICE HOW MUCH FATTER THE CREATURE IS GETTING! I WONDER...?

TO TEST HIS THEORY, THE MASTER BOWMAN SPEEDS AN ARROW FORWARD AND...

YES-- THE CREATURE'S ACTUALLY ABSORBING THE ENERGY OF THE *EXPLOSIVE ARROW! SPEEDY,* THAT BEAST *FEEDS ON ENERGY!*

BLAMM!

UNEXPECTEDLY, THE CREATURE TURNS, HEADING BACK INTO THE EERIE FOG...

WE'VE GOT TO FOLLOW THAT CREATURE--DESTROY IT BEFORE IT BECOMES A MENACE!

INTO THE FOG SPEEDS THE *ARROW-CAR,* CARRYING *GREEN ARROW* AND *SPEEDY* INTO FANTASTIC ADVENTURE!

2

THROUGH THE FOG RIDES THE **ARROW-CAR**-- TO EMERGE UPON A FANTASTIC SIGHT!

HUH? **DAYLIGHT?** AND LOOK WHAT'S AHEAD! LAND VEHICLES--OF ALL TYPES--AND FROM ALL AGES!

AN ANCIENT ROMAN CHARIOT--A 17TH CENTURY FRENCH CARRIAGE-- A WESTERN COVERED WAGON-- A MEDIEVAL OXCART...

WHAT'S IT ALL MEAN?

I'VE NEVER SEEN PLANTS AND TREES LIKE THESE BEFORE! **WHERE ARE WE?**

I--I DON'T KNOW YET--BUT MAYBE WE'LL FIND OUT BY TRAILING THE CREATURE! COME ON!

NO USE! IT'S DISAPPEARED SOMEWHERE IN THAT HUGE CAVERN! DO WE GO AFTER IT?

TOO RISKY! WE CAN'T TELL WHAT KIND OF DANGER WE'D BE GETTING INTO!

SUDDENLY...

SEIZE THEM!

3

THAT MONSTER IS CERTAIN TO CRUSH ONE OF US!

NOT IF IT CAN'T SEE US! QUICK, *SPEEDY*-- THE *SMOKE-SCREEN ARROW!*

TWIN SHAFTS HISS FORWARD-- THEN ARROWHEADS EXPLODE AND...

THAT SMOKE WILL BLOCK THE GNORL'S VISION LONG ENOUGH FOR EVERYONE TO GET AWAY!

QUICK--THIS WAY! YOU AND THE BOY MUST NOT GO WITH THE *HILL PEOPLE* OR YOU WILL EVENTUALLY FALL UNDER THE INSIDIOUS INFLUENCE OF THEIR *METEOR!* I'LL EXPLAIN LATER!

LATER...

THAT FREAKISH FOG--IT'S REALLY A "DOORWAY" TO THIS DIMENSION...

YES--A "DOOR-WAY" THAT HAS APPEARED THROUGH THE CENTURIES IN MANY PLACES! THAT'S WHY VEHICLES OF DIFFERENT LANDS AND TIMES HAVE COME INTO THIS WORLD!

THOUGH EACH NEWCOMER HAD KEPT HIS OWN WAYS AND CUSTOMS, EVERYONE WAS HAPPY! THE UNIQUE ATMOSPHERE HERE KEPT PEOPLE FROM AGING! THERE WAS PLENTY OF FOOD! EXCEPT FOR THE GNORL, THERE WAS CONTENTMENT AND PEACE-- UNTIL THE *METEOR* FELL FROM THE SKY!

AS IT FELL, IT BURST INTO TWO PIECES! ONE PIECE LANDED AMONG THOSE PEOPLE WHO'D SETTLED ON THE HILL-- THE OTHER PIECE FELL AMONG THE VALLEY FOLK! YES -- TWO METEORS -- TWO HARBINGERS OF EVIL!

5

"EACH METEOR EMITTED AN EERIE ENERGY--AND PEOPLE WITHIN ITS AURA FELT EXHILARATED..."

IT GIVES US WARMTH--LIKE THE SUN!

IT MAKES ME FEEL SO GOOD-- SO HAPPY...

"BUT AS DAYS WENT BY, THE AURAS INSIDIOUSLY, GRADUALLY, MADE EVERYONE GREEDY!"

WHY SHOULD THE VALLEY PEOPLE HAVE A METEOR, TOO? WE MUST TAKE THEIR METEOR-- THEN THE TWO METEORS WILL MAKE US TWICE AS HAPPY!

BEFORE COMING TO THIS DIMENSION, I WAS A GOLD PROSPECTOR! I SAW GOLD MAKE MEN GREEDY, JUST AS THE METEORS DID! LUCKILY, BEING A HERMIT, I LIVED BEYOND THE VILLAGES AND METEORS, SO I WAS NOT AFFECTED...

BUT EVERYONE ELSE WAS, AND NOW THEY'RE AT WAR! HOW AWFUL!

LOOK! THE FOG IS STILL THERE! GO THROUGH IT! IT MAY BE YEARS BEFORE YOU HAVE ANOTHER CHANCE!

SPEEDY, IF WE LEAVE, WE LEAVE THE PEOPLE HERE MEN- ACED BY THE GNORL--AND BY WAR!

WE CAN'T RUN OUT ON THEM!

RIGHT! WE'VE GOT TO STAY-- THINK OF A PLAN THAT WILL DESTROY THE GNORL--AND ALSO BRING PEACE TO THE PEOPLE!

LATER AS SENTRIES GUARD THE METEOR OF THE HILL PEOPLE, SUDDENLY...

THE BOXING GLOVE ARROWS HAVE KAYOED THE GUARDS! LET'S MOVE, SPEEDY!

SWIFTLY A BUCKBOARD ROLLS FROM HIDING, THE METEOR IS HURLED ABOARD, AND THEN...

WE'VE GOT ONE METEOR! NOW FOR THE ONE IN THE VALLEY!

6

BELOW, THE VALLEY PEOPLE'S METEOR RESTS BEHIND A FENCE, WHEN...

OKAY, *SPEEDY*-- GET SET! I'M ABOUT TO CUT THE TRACES!

INSTANTS LATER, TWO FIGURES LEAP ASTRIDE THE FREED HORSES, AS THE BUCKBOARD ROLLS ON LIKE A JUGGERNAUT!

CRA-AA-SH!

NEARBY, HIDDEN BY BRUSH, THE HERMIT SUDDENLY DRIVES INTO THE OPEN, AND...

OKAY--EVERYTHING'S RIGHT ON SCHEDULE SO FAR!

RIGHT! NOW--LET'S TRANSFER BOTH METEORS INTO THE COACH!

THE TRANSFER IS COMPLETED SWIFTLY, AND THEN...

YAHOO!

AFTER THEM!

LATER, SOME DISTANCE AWAY FROM THE GNORL'S LAIR, THE METEORS ARE REMOVED AND LASHED TO A GREAT SHAFT OF WOOD!

THIS YOUNG TREE MAKES A GOOD ARROW!

7

THEN THE GREAT SHAFT IS NOTCHED TO A CROSS-BOW FIT FOR THE HANDS OF A TITAN!

OKAY -- NOW TO BAIT THE *GNORL!*

GREEN ARROW'S BOW TWANGS, AND...

THE ENERGY OF THE *FIRECRACKER ARROW* SHOULD BRING THAT HUNGRY BEAST OUT OF ITS LAIR!

TWANG-G

GET SET! HERE IT COMES!

LOOK WHO ELSE IS COMING! THE HILL AND VALLEY PEOPLE! WE'RE CAUGHT BETWEEN TWO FIRES!

8

SUDDENLY, LIKE A NOVA, THE GNORL BECOMES A SWELLING BURST OF ENERGY...

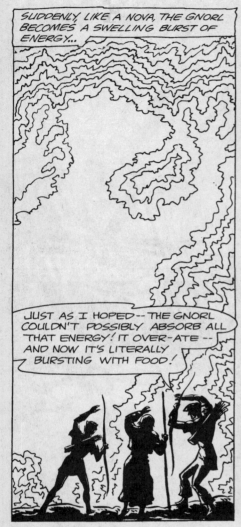

JUST AS I HOPED -- THE GNORL COULDN'T POSSIBLY ABSORB ALL THAT ENERGY! IT OVER-ATE -- AND NOW IT'S LITERALLY BURSTING WITH FOOD!

INSTANTS LATER, ALL THAT IS LEFT OF THE METEOR AND THE GNORL IS A MOUND OF LIFELESS DUST-- AND AS THIS HAPPENS...

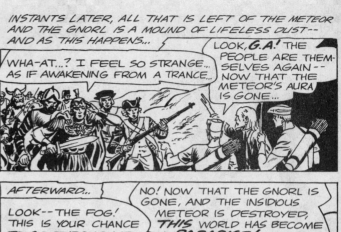

WHA-AT...? I FEEL SO STRANGE.. AS IF AWAKENING FROM A TRANCE...

LOOK, *G.A.!* THE PEOPLE ARE THEM-SELVES AGAIN -- NOW THAT THE METEOR'S AURA IS GONE...

AFTERWARD...

LOOK -- THE FOG! THIS IS YOUR CHANCE TO GO THROUGH IT-- ESCAPE INTO OUR WORLD!...

NO! NOW THAT THE GNORL IS GONE, AND THE INSIDIOUS METEOR IS DESTROYED, *THIS* WORLD HAS BECOME A *PARADISE!* WE ARE CONTENT TO REMAIN!

AND SO, THE ARCHERS RIDE ONWARD -- HOMEWARD BOUND -- THEIR INCREDIBLE ADVENTURE OVER AT LAST!

IT WAS OUR GREAT FORTUNE THAT YOU CAME HERE, *GREEN ARROW!* FAREWELL!

THE END

THE GREEN ARROW

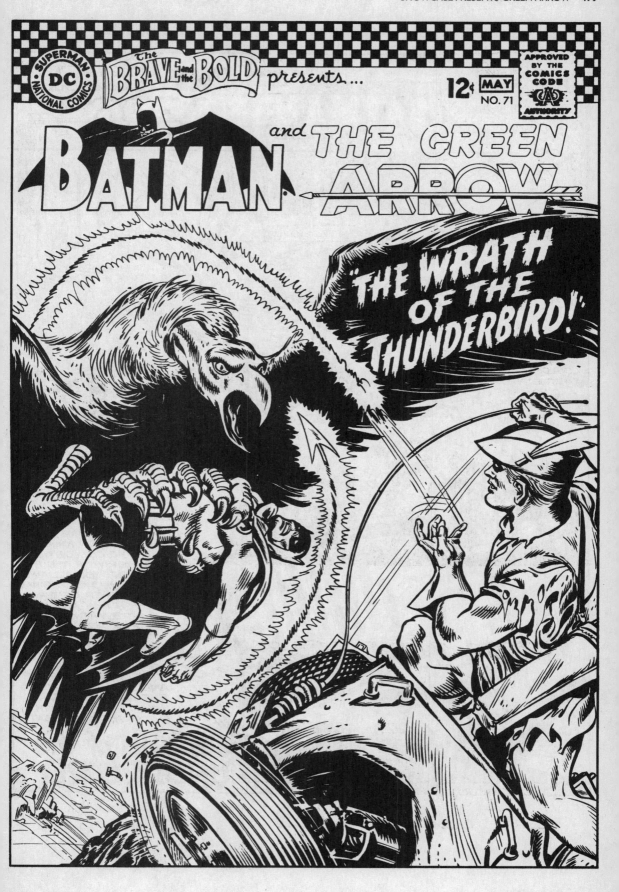

ON AN ELEVATED HIGHWAY ALONG THE RIVERFRONT IN GOTHAM CITY, TWO TRUCK DRIVERS ARE HAVING A DISPUTE OVER THE RIGHT OF WAY...

LOOK OUT, BUSTER! I'M COMIN' THROUGH!

CRASH!

YEEOWWW!

MOMENTS LATER, AS THE POLICE AND A FAMED, FAMILIAR FIGURE ARRIVE...

BATMAN! GOOD THING YOU ARRIVED! THE EMERGENCY TRUCK WON'T BE HERE FOR A FEW MINUTES--AND THAT POOR FELLA'S IN TROUBLE!

HELP ME! I CAN'T SWIM!

INSTANTLY, THE FAMED BAT-ROPE SNAKES OUT, AND...

THERE! WISH THEY WERE ALL AS EASY AS THIS!

SHORTLY...

T-THANKS, BATMAN...

FORGET IT, FRIEND! OFFICERS!... ARREST THAT OTHER DRIVER! HE CAUSED THE WHOLE THING...HE'S ONE OF TOM TALLWOLF'S THUGS!

SOMEBODY MENTION MY NAME?

2

NOT LONG AFTER, AS THE *CAPED CRUSADER* KEEPS AN APPOINTMENT...

ALMOST FORGOT THIS CALL... WONDER WHAT MY OLD FRIEND WANTS?

WHITEBIRD ENTERPRISES

SOON, IN A PENTHOUSE OFFICE...

HUH?

THUNK

HOLY PEACEPIPES, *JOHN WHITEBIRD!*... ARE YOU GOING ON THE WARPATH AGAIN?

HA! HA! NO, *BATMAN*... AND I'M AFRAID WITH THAT KIND OF SHOOTING, I'D BE A "DEAD MAN" PRETTY QUICKLY! I'M JUST *PRACTICING!*

PRETTY EXPENSIVE PRACTICE! WHAT'D THAT MASTERPIECE COST... TWENTY THOUSAND?

THIRTY THOUSAND... BUT REMEMBER, I'M A "HEAP BIG TYCOON"! MONEY I HAVE PLENTY OF... BUT MONEY'S A WHITE MAN'S ACHIEVEMENT! IT'S THE OLD *INDIAN* SKILLS I NEED NOW...

I DON'T DIG, JOHN!

OLD *STANDING BEAR*, CHIEF OF MY PEOPLE, THE *KIJOWAS*, HAS JUST DIED, OUT ON THE RESERVATION! A NEW CHIEF MUST BE CHOSEN SOON! THERE ARE TWO CONTENDERS FOR CHIEF... MYSELF--AND *TOM TALLWOLF!*

OH, NO!

BUT OH, *YES!* WHICHEVER OF US WINS THE CONTEST BECOMES CHIEF! AND THE EVENTS IN THE CONTEST ARE ALL THE OLD INDIAN SKILLS-- RIDING --WRESTLING-- JAVELIN TOSSING-- BOWMANSHIP... ALL EVENTS THAT I'M NO GOOD AT!

4

AS THE DAYS PASS, UNDER *BATMAN'S* EXPERT COACHING, THE MAN WHO WOULD BE CHIEF OF ALL THE KIJOWAS LEARNS QUICKLY...

BRAVO!

FROM A "SOFT" INDIAN, HE TURNS INTO A TOUGH, FEARLESS BRAVE...

BULL'S-EYE!

NOT BAD, JOHN...

NOT BAD-- AT ALL... *OOOF!*

BUT ONE MORNING, AS "CLASS" RESUMES...

ARCHERY? UH-OH... I'M AFRAID THAT'S NOT MY STYLE, JOHN!

BUT BOWMAN-SHIP'S THE MOST IMPORTANT PART OF THE CONTEST! THE SILVER-PLATED BOW OF THE KIJOWA CHIEFS IS THE SYMBOL OF THEIR POWER... AND TOM TALLWOLF IS A TERRIFIC ARCHER-THE BEST IN THE TRIBE!

HMM-- ONLY ONE THING TO DO... GET YOU ANOTHER TEACHER--A *SPECIALIST* IN ARCHERY! AND I THINK I KNOW JUST THE GUY--IF HE'S AVAILABLE! KEEP TRAINING, JOHN-- I'LL BE BACK AS SOON AS I CAN!

TWO DAYS LATER...

BATMAN! YOU'RE BACK-- WITH... WITH *GREEN ARROW?*

HELLO, JOHN WHITEBIRD-- HOW'S MY PUPIL?

7

THE GREEN ARROW

The WRATH of the THUNDERBIRD PART 2

How! YOU ARE HEREBY INVITED TO *GOTHAM STADIUM,* AS GUEST OF THE KIJOWA TRIBE, TO WITNESS THE CONTEST FOR CHIEF OF ALL THE KIJOWAS! PRICE: 12¢...THOSE PEOPLE IN THE STANDS PAID MORE-- LOTS MORE!

SCORE
TALLWOLF
WHITEBIRD

LET THE CONTEST, WHICH SHALL CHOOSE THE MOST WORTHY TO FOLLOW IN THE STEPS OF THE GREAT ANCESTORS, NOW BEGIN!

THOSE TWO CHIEF WANNABES ARE ABOUT TO START THE BALL GAME, PROMOTER!

EXCELLENT, CHECKS! IS EVERYTHING FIXED SO TALL- WOLF WILL TRIUMPH?

YOU BET, PROMOTER!

THE FIRST EVENT TO CHOOSE THE CHIEFTAIN OF ALL THE KIJOWAS IS *BROAD JUMPING...*

HERE I GO, INDIAN BROTHER... MAKE "HEAP BIG JUMP"! HA! HA!

10

BUT AS THE TWO BOWMEN FIRE AGAIN AND AGAIN...

SSWISH

THUNG THUNG

SSWISH

THUNG THUNG

SWISH

SWISH

TALLWOLF WINS! HE'S THE NEW CHIEF OF THE KIJOWAS!

WHAT SHOOTING! WHITEBIRD DIDN'T THREAD THE LOOP ONCE!

I--I'M SORRY, BATMAN... I LET YOU AND GREEN ARROW-- AND MY PEOPLE DOWN!

EASY JOHN... YOU DID YOUR BEST! I DON'T GET IT! IN TRAINING, I HAD JOHN HITTING THAT LOOP WITH HIS EYES CLOSED! SOMETHING'S FISHY HERE!

SO TOM TALLWOLF IS NOW CHIEF OF ALL THE KIJOWAS-- BUT THAT NIGHT...

NO DOUBT ABOUT IT-- THE GOAL POSTS WERE GIMMICKED WITH ELECTRO-MAGNETS CONCEALED IN THE UPRIGHTS AND REMOTELY POWERED!

YES, AND EACH TIME JOHN FIRED AN ARROW, THE JUICE WAS TURNED ON! THERE WAS JUST ENOUGH MAGNETITE ORE IN THOSE STONE ARROWHEADS TO DEFLECT THEM! THAT'S WHY YOU LOST, JOHN-- IT WAS FIXED!

BZZZZ...

I KNOW TALLWOLF IS UNSCRUPULOUS, BUT I NEVER THOUGHT HE'D STOOP TO THIS-- DISHONORING THE CHIEFTAINSHIP OF OUR PEOPLE!

HE MUST'VE HAD HELP... AND... WAIT! A LIGHT BURNING LATE IN THE OFFICE OF J. JAY JAYE, THE BIG PROMOTER! HAVING THE CONTEST HERE WAS HIS IDEA! G.A., CAN WE EAVESDROP ON THAT CUBICLE?

TWANG

SURE, BATMAN!

14

NOW, THE SMALL, TRANSISTORIZED RADIO TRANSMITTER ON THE ARROW IS PICKING UP THE CONVERSATION INSIDE THAT CUBICLE!

ALL RIGHT, TALLWOLF, SO I MADE YOU CHIEF! NOW I WANT A FAVOR IN RETURN... WHEN WE GO TO THE KIJOWA RESERVATION TO-MORROW, I WANT YOU TO GIVE ME THE SECRET OF-- THE THUNDERBIRD!

WHAT? THE THUNDERBIRD?

YES--AS CHIEF, YOU ALONE WILL HAVE THE POWER TO CALL FORTH THE THUNDERBIRD--AND I WANT IT!

YOU'RE CRAZY, PROMOTER! I'LL NEVER GIVE YOU THAT! IT'S A SACRED TRUST... NO CHIEF OF THE KIJOWAS WOULD EVER BETRAY IT!

SACRED TRUST? YOU ALREADY VIOLATED THAT, REDSKIN, BY CHEATING TO BECOME CHIEF! YOU'LL DO IT--OR I'LL EXPOSE THE WHOLE SCHEME TO YOUR PEOPLE... TO THE WORLD!

WHY, YOU ROTTEN PALEFACE... I WAS A FOOL TO TRUST YOU! NOW, I'VE GOT NO CHOICE!

THE THUNDERBIRD?... JUST WHAT IS IT, JOHN?

HEAP BIG TROUBLE, BATMAN! THE RESER-VATION... WE'VE GOT TO BE THERE WHEN TALLWOLF IS INSTALLED AS CHIEF!

SO THE FOLLOWING DAY, IN THE FAR WEST, ON THE KIJOWA'S VAST RESERVATION...

HERE-- THIS IS THE SPOT... I HOPE WE'RE IN TIME!

I HEAR VOICES... BELOW...

ALL RIGHT, TALLWOLF-- NOW YOU'RE CHIEF! CALL FORTH THE THUNDERBIRD!

15

THE GREEN ARROW

BATMAN? HE--HE'S NOT MOVING...

BUT A MOMENT LATER... *G.A.....* HELLO, OLD *BUDDY!* WHAT A *RIDE!* I WAS HALF-CONSCIOUS WHEN SOMETHING *FLASHED!*

THAT WAS MY *HOT-LINE ARROW!* I BANKED ON THE BIRD'S *POUCHY* SKIN, UNDER ITS TALON, ACTING AS A KIND OF INSULATOR FOR YOUR BODY!

GOOD THINKING! I KNEW YOU WERE A HOTSHOT BOWMAN, *G.A.*, BUT I DIDN'T KNOW YOU WERE ALSO AN EXPERT IN *ORNITHOLOGY!*

HERE COMES JOHN WHITEBIRD...

BATMAN! YOU'RE OKAY... WHAT A *RELIEF!* BUT *TALLWOLF* AND THE *PROMOTER* ARE ESCAPING -- AND WE'RE MAROONED, MILES FROM THE *RESERVATION!*

BUT JUST THEN... *WHUP-WHUP-WHUP*

LOOK! A *RANGER COPTER...* AND *TALLWOLF* IS INSIDE!

AFTER THE WHIRLYBIRD SETTLES ON THE MESA...

RELAX, *BROTHER!* I TURNED MY-SELF AND THE *BIG PROMOTER* IN TO THE AUTHORITIES -- AND THEY LET ME COME OUT HERE TO SEARCH FOR YOU! I WAS WRONG... *WRONG* ABOUT EVERYTHING! I MADE A BIG MISTAKE TRUSTING THE *PROMOTER...* YOU CAN'T TRUST PALE-*FACES!*

CORRECTION, INDIAN *BUDDY!* HERE'S *TWO* PALEFACES *ANYBODY* CAN TRUST... *GREEN ARROW* AND *BATMAN!*

I STAND *CORRECTED,* WHITEBIRD -- BUT *GOOD!*

23

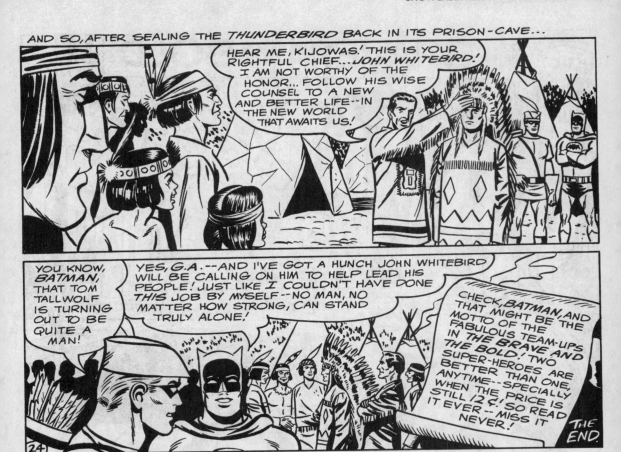

THE GREEN ARROW

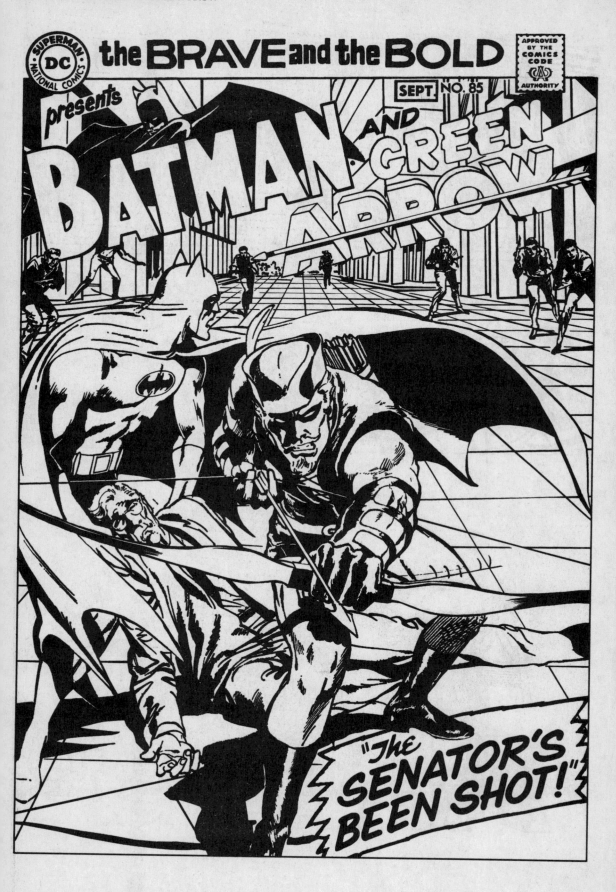

IF I AM ELECTED... I VOW TO SWEEP OUR STATE AND NATION FREE OF CRIME! TO THIS END, I DEDICATE THE REMAINING YEARS OF MY LIFE...

PAUL'S VOICE... WHAT KIND OF VERMIN WOULD SHOOT A MAN LIKE THAT...

AFTER **BATMAN** RESUMES HIS CIVILIAN IDENTITY, HE GOES DIRECTLY TO GOTHAM STATE HOSPITAL WHERE PAUL CATHCART LIES IN A COMA, BALANCED BETWEEN LIFE AND DEATH. FOR HOURS, BRUCE WAYNE AND THE SENATOR'S SON, EDMOND, STAND SILENT VIGIL AT HIS BEDSIDE...THEN...

MR. WAYNE... IT'S THE **GOVERNOR** CALLING, SIR...

YOU CAN USE THE PHONE IN THE CHIEF RESIDENT'S OFFICE!

...THANK YOU, NURSE! I'LL BE BACK SHORTLY, EDMOND.

BRUCE, THIS IS A TERRIBLE TRAGEDY, BUT PAUL WAS READY FOR IT. I HAVE HIS RESIGNATION FROM THE SENATE HERE, TO GO INTO EFFECT IF HE WERE UNABLE TO VOTE ON HIS ANTI-CRIME BILL. THE DOCTORS SAY HE WON'T BE ABLE TO VOTE. I'M APPOINTING A NEW SENATOR!

PAUL ALWAYS THOUGHT OF HIS VOTERS FIRST. OF COURSE EDMOND WOULD BE **MY** FIRST CHOICE TO REPLACE HIM!

BUT NOT **MINE**, BRUCE. HIS PSYCHIATRIC PRACTICE KEEPS HIM TOO BUSY. EDMOND IS NOT FULLY AWARE OF THE IMPORTANCE OF HIS FATHER'S BILL.

I'M ASKING YOU TO FINISH THE SENATOR'S TERM, BRUCE!

YOU... YOU MUST BE JOKING!

3

YOU WERE RIGHT, BRUCE! THIS WORKOUT IS RELAXING ME. I'VE BEEN UP TIGHT SINCE DAD WAS SHOT. I'M ALSO GLAD THE GOVERNOR APPOINTED YOU TO REPLACE DAD! THAT TAKES ANOTHER LOAD OFF MY MIND!

I'VE BEEN MEANING TO TALK TO YOU ABOUT THAT, EDMOND. I'M AFRAID I CAN'T TAKE THE APPOINTMENT!

CAN'T?? YOU'RE THE ONLY ONE WHO *CAN!* MY FATHER'S WHOLE CAREER HAS LED UP TO THAT ANTI-CRIME BILL...

YOU MUST!

YOU HAVE THE *TIME*... THE *MONEY*... THE *KNOWLEDGE!* MY FATHER LIES IN A COMA... CLOSE TO *DEATH!* *BATMAN* HAS SWORN NOT TO REST UNTIL HE HAS RUN DAD'S ATTACKERS TO THE GROUND...

AND *YOU* WON'T EVEN STAND UP AND BE COUNTED!

IT'S NOT BECAUSE I DON'T WANT TO, EDMOND...

BUT I GAVE AN OATH TO DO ANOTHER JOB-- *BATMAN'S* JOB, BECAUSE *I AM*... *BATMAN!*

7

THE GREEN ARROW

BRUCE... YOU... *BATMAN?* THIS ISN'T A DODGE...? NO, IT WOULDN'T BE! BUT WHY...

WHY TELL *YOU?* TWO REASONS! ONE: YOU *DESERVE* TO KNOW! TWO: I NEED YOUR HELP... YOUR ADVICE!

AS A PSYCHIATRIST, YOU'LL NEVER REVEAL MY SECRET, SO IT'S SAFE WITH YOU. BUT MY PROBLEM REMAINS.

OF COURSE, WHICH CAREER IS MORE IMPORTANT, SENATOR WAYNE'S OR BATMAN'S? ONE RULES OUT THE OTHER!

IT'S GOT ME TIED IN MENTAL KNOTS!

WITH WHICH CAREER CAN I DO THE MOST GOOD? THE CRIME BILL IS MOST IMPORTANT BUT...

...WHAT EFFECT WILL THE DISAPPEARANCE OF *BATMAN* HAVE ON CRIME AND CRIMINALS IN GOTHAM...?

AND ON *BRUCE WAYNE,* WHOSE PERSONALITY IS MOST IMPORTANT IN THIS CASE!

IT'S A DECISION ONLY *YOU* CAN MAKE, BRUCE! ALL I CAN DO IS GUIDE YOU!

YOU'VE ALREADY BEGUN TO CLEAR THE COBWEBS AWAY, ED! THANKS! WILL I SEE YOU AT THE HOSPITAL LATER?

8

THE FOLLOWING MORNING...

...I HEREBY ACCEPT THE OFFICE OF UNITED STATES SENATOR... TO SERVE THE PEOPLE OF THIS STATE...

MINOTAUR MADE UP MY MIND FOR ME! THE ANTI-CRIME BILL MUST PASS!

MEANWHILE, ABOVE A SMALL VOLCANIC ISLAND IN THE MEDITERRANEAN...

ED'S ABDUCTION DECIDED ONE THING... I'VE GOT TO PLAY GREEN ARROW FOR A WHILE ANYWAY!

SOON...

MINOTAUR'S PRIVATE YACHT... ENTERING A HIDDEN GROTTO!

TO TRAIL THE SLEEK CRAFT WITHIN THE GROTTO'S MAZE, GREEN ARROW FIRES AN ELECTRONIC TRACKER SHAFT...

THUNK

BEEP
BEEP BEEP
BEEP

12

SO, DR. CATHCART, YOU WILL TELL ME NOTHING OF BRUCE WAYNE AND OLIVER QUEEN?

I'M ONLY THEIR FRIEND, MINOTAUR... I KNOW NOTHING ABOUT THEIR POLITICS OR BUSINESS!

I THINK YOU *LIE!* I HAVE HAD MUCH MORE IMPORTANT MEN THAN YOU KILLED FOR DEFYING ME!

I'LL BET HE HAS... AND MY FATHER MAY YET BE ADDED TO HIS LIST!

A TRACKING DEVICE...SIR! WE FOUND IT AT THE REAR OF THE SHIP.

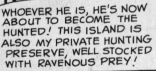

SO WE ARE BEING HUNTED! VERY INTERESTING BUT FOOLISH!

WHOEVER HE IS, HE'S NOW ABOUT TO BECOME THE HUNTED! THIS ISLAND IS ALSO MY PRIVATE HUNTING PRESERVE, WELL STOCKED WITH RAVENOUS PREY!

KRUNCH

BLAST! THE SIGNAL'S GONE DEAD! I'M LOST!

13

UH-OH! MINOTAUR'S LET LOOSE SOME PETS!

BLAZES! SOUNDS LIKE ... THEY'RE COMING...FROM ...EVERYWHERE!

14

GREEN ARROW--!

BATMAN? HOW... HOW IN BLAZES DID YOU FIND ME?

I TRACKED YOUR *JUSTICE LEAGUE* LOCATOR TRANSMITTER! I HOMED IN ON YOU LIKE A PIGEON!

BAD NEWS! BECAUSE I'M A LOST PIGEON, AND NOW THAT MAKES TWO OF US!

I THINK I CAN GET US OUT OF HERE!

BATMAN DOESN'T KNOW THAT AS OLIVER QUEEN MY BID ON "NEW ISLAND" MUST BE SUBMITTED WITHIN 48 HOURS OR MINOTAUR WINS THE CONTRACT BY FORFEIT!

CAN'T TELL HIM I'VE GOT TO BE IN WASHINGTON BY TOMORROW AFTERNOON TO VOTE ON THE ANTI-CRIME BILL!

WHY'D YOU BRING THE BAT DOWN?

IT'S GOING TO LEAD US OUT OF HERE!

SOON...

BATS HAVE THEIR OWN "RADAR" FOR NAVI-GATING ANY MAZE--OR NIGHT FLYING--AND WITH ONE OF OUR LOCATOR TRANSMITTERS SENDING SIGNALS BACK TO US...IT'LL LEAD US TO DAYLIGHT...AND MINOTAUR!

16

MEANTIME, AHEAD, WHERE THE MAZE EXITS...

BUT SINCE BRUCE WAYNE HAS BECOME SENATOR WAYNE AND OLIVER QUEEN STILL PERSISTS IN COMPETING WITH "ARGONAUT UNLIMITED," I HAVE ORDERED MY AMERICAN AGENTS TO DESTROY THEM BOTH!

IF HE ONLY KNEW HE'D ALSO BE DESTROYING *BATMAN* AND *GREEN ARROW!* HE MUSTN'T SUCCEED! BUT WHERE *IS GREEN ARROW?*

WHOEVER FOLLOWED US IN THE LABYRINTH, DR. CATHCART, IS NO MORE! MY BEASTS TOOK CARE OF ANY PURSUER!

GREEN ARROW... AND BATMAN!

JUST THOUGHT WE'D DROP IN!

HALT AND SURRENDER... OR THIS MAN DIES...INSTANTLY!

DON'T LISTEN... TAKE HIM!

17

18

KILL YOU--? I AM SIMPLY A BUSINESSMAN! IF YOU'RE JOKING --

I'M NOT, ESPECIALLY WHEN I SAY YOU ARE UNDER ARREST TO BE RETURNED TO AMERICA FOR YOUR CRIMES!

YOU CANNOT ARREST ME HERE IN A FOREIGN COUNTRY!

OH, EXCUSE ME! DIDN'T I MENTION... THIS IS LEGALLY AMERICAN SOIL...? HADN'T YOU NOTICED THIS IS THE *AMERICAN EMBASSY?*

MR. AMBASSADOR! THIS IS AN OUTRAGE!

INDEED, I QUITE AGREE. BUT I'M SURE MR. QUEEN AND THESE FEDERAL MARSHALS WILL BE GLAD TO RETURN YOU HOME WHERE YOU'LL BE CHEERFULLY WELCOMED!

YOU WILL NEVER GET ME TO AN AIRPORT, MR. QUEEN! MY MEN WAIT OUTSIDE!

AH-- BUT I HAVE THE ANSWER FOR *THAT!*

MOMENTS LATER, FROM THE EMBASSY ROOF...

20

AND AS THE CRUCIAL VOTE NEARS A CLOSE

I VOTE... NO!

THE BILL'S GOING TO LOSE... UNLESS WAYNE GETS HERE....!

SENATOR BRUCE WAYNE! YOUR VOTE, PLEASE! SENATOR WAYNE?

TOO LATE! NO SIGN OF HIM... THE BILL'S LOST!

FOR THE LAST TIME... IS SENATOR WAYNE VOTING ON THIS BILL?

HOPE NOBODY COMES IN--THERE'S NO SENATOR BATMAN IN CONGRESS --YET!

SENATOR WAYNE PRESENT AND VOTING, SIR! AND I VOTE... YES!

BRAVO! WE'VE WON!

22